Mangrove Ecosystems: A Global Biogeographic Perspective

Victor H. Rivera-Monroy • Shing Yip Lee
Erik Kristensen • Robert R. Twilley
Editors

Mangrove Ecosystems: A Global Biogeographic Perspective

Structure, Function, and Services

Springer

Editors
Victor H. Rivera-Monroy
Department of Oceanography
and Coastal Sciences
College of the Coast and Environment
Louisiana State University
Baton Rouge, LA, USA

Erik Kristensen
Department of Biology
University of Southern Denmark
Odense, Denmark

Shing Yip Lee
School of Environment
Griffith University
Southport, QLD, Australia

Robert R. Twilley
Department of Oceanography
and Coastal Sciences
College of the Coast and Environment
Louisiana State University
Baton Rouge, LA, USA

ISBN 978-3-319-62204-0 ISBN 978-3-319-62206-4 (eBook)
https://doi.org/10.1007/978-3-319-62206-4

Library of Congress Control Number: 2017955833

Printed on acid-free paper

This Springer imprint is published by Springer Nature
The registered company is Springer International Publishing AG
The registered company address is: Gewerbestrasse 11, 6330 Cham, Switzerland

Foreword

An international symposium on the biology and management of mangrove ecosystems (Walsh et al. 1975) took place at the East-West Center in Honolulu Hawaii between October 8 and 11, 1974. Mangrove experts from all over the world and in different stages of their professional careers were present at this symposium. As I listened to the comprehensive presentation on mangrove biogeography of V.J. Chapman, I had no idea of how important this meeting would turn out to be from the historical perspective of the study of mangrove wetlands. Chapman's presentation was based on his soon-to-be published encyclopedic work on mangrove vegetation (Chapman 1976), which culminated decades of research on mangroves from a natural history perspective. The proceedings of the meeting also included a memorial for William Macnae, the South African zoologist who passed away in 1975 and was known for his comprehensive research in the fauna and flora of the Indo-West Pacific mangrove forests (Macnae 1968). At the time of the Hawaii meeting, Sam Snedaker and I had completed a review that highlighted the application of ecosystem science to mangrove ecology (Lugo and Snedaker 1974). Cintrón et al. (1978) applied this systems perspective to mangrove zonation in arid environments and anticipated the importance of hurricanes to long-term processes in mangrove forests. Later, I tried to capture the ecosystem-level and ecophysiological challenges of mangrove zonation in relation to their successional status (Lugo 1980). Also present in Hawaii was B. J. Thom, who since the 1960s had been interpreting mangrove ecology in relation to geomorphological settings (Thom 1975). His work explicitly related mangrove ecosystem function to regional fluvial and geomorphological processes. The focus on mangrove research after Hawaii was clearly expanding to regional landscapes and long-term processes. The dissertations of W. Odum (1971) and E. Heald (1971) at the University of Miami had the same effect of expanding mangrove research to ecosystems and communities as close or as far as the detritus from mangroves could be traced.

Today, almost 50 years after the Hawaii meeting, mangrove research activity, the technology available for conducting mangrove research and social interest in mangrove environments has exploded. Ernesto Medina, Cathleen McGinley, and I recently reviewed some of the ecosystem-level and ecophysiological advances in mangrove research as well as some of the policy measures that best apply to mangrove ecosystems under Anthropocene conditions (Lugo et al. 2014, see also reviews in Lugo 2002 and Lugo and Medina 2014). Mangroves were in the past a scientific curiosity for their capacity to grow in seawater, but today, they are at the center of the global conservation discussion. This global attention is not due to any discovery unknown in the 1970s, or to any new functional characteristics of mangroves. What has changed is public perception of mangroves coupled to the advent of the Anthropocene, which places mangrove forests at the interface between built infrastructure, raising sea levels, and human needs.

Mangrove ecosystem research has evolved significantly since the Hawaii meeting, and there is so much new information available, hence the need for a new synthesis of the many studies that are dispersed in the scientific literature. Recent books about this ecosystem focus on its global area and distribution (Spalding et al. 2010), energetics (Alongi 2009), silviculture (FAO 1994, Saenger 2002), and the ecology of regional mangroves (Yañez-Arancibia and A. L. Lara-Domínguez 1999, Clough 1982). A comprehensive global synthesis is lacking, one that places mangroves in the context of the Anthropocene that new research tools allow us to assess. Such a synthesis would represent another step in the progression of mangrove research from natural history, to ecosystem level, to a landscape context, to ecophysiological detail, and now the global and biogeochemical levels. The publication of this book might represent that historic moment when mangrove research takes a turn toward greater insight and comprehension by exploring new scales of complexity (both biotic and abiotic). Only time will tell. The title *Mangrove Ecosystems: A Global Biogeographic Perspective* certainly fits the bill; it cranks up the global focus.

After the Introduction, Chap. 2 by N.C. Duke is titled *Revisiting Mangrove Floristics and Biogeography*. This chapter is one of those works that instantly become a classic of the mangrove literature due to their in-depth, rich, and authoritative content. The chapter is organized around ten generalized factors that mostly influence the biogeography of mangroves. Each mangrove taxon gets individual attention, and its evolutionary history is displayed, as are maps of the distribution of all the mangrove tree species in the world. In Chap. 3, *Biodiversity of Mangroves*, by Lee et al., we learn that the total species richness supported by mangrove ecosystems is two orders of magnitude greater than the number of mangrove tree species. In Chap. 2, it was reported that in the mangrove hotspot of the Indo-West Pacific, 54 mangrove tree species correspond to 500 coral and 5000 fish species. I was amused by the statement in Chap. 3 that research in mangroves is hindered by a large number of dangerous or disturbing wildlife that can bite and kill; they were referring to biting insects, crocodiles, tigers, and so on, which can make mangrove research an action adventure when combined with tidal bores, muddy terrain, and dense prop roots! But of greater concern to scientists is that the majority of entries in the group-by-group biodiversity tables in this book chapter are "ND," or no data.

Chapter 4, *Spatial Ecology of Mangrove Forests: A Remote Sensing Perspective*, by Lucas et al. reviews examples of remote sensing applications to mangrove forests worldwide. Authors advocate for the development of mangrove-dedicated remote sensing approaches and present superb images of mangrove landscapes.

Chapter 5, *Productivity and Carbon Dynamics in Mangroves*, by Twilley et al. is a comprehensive global review of carbon fluxes and storages in mangrove environments. The review is authoritative and summarizes a large data set. I was surprised to find that other book chapters make independent estimates of carbon fluxes rather than using those in Chap. 5. Chapter 6, *Biogeochemical Cycles: Global Approaches and Perspectives*, by Kristensen et al. focuses mostly on Australia and North America, where these kinds of data are collected. It also provides a superb level of detail on the sediments, a mangrove compartment that is usually treated as a black box in most mangrove studies. My favorite image of this review is the three-dimensional view of mangroves, which includes the atmosphere, lithosphere, and biosphere. I expect that this approach to mangroves will be instrumental to the future understanding of these ecosystems. Such an approach will require attention to ecosystem interfaces, especially with sediments, an interface between the hydrosphere and lithosphere. Interface work will in turn require studies at smaller molecular and microbial scales. These smaller scales are as challenging as the global scale and together form the basis of future mangrove research and understanding.

Chapter 7, *Climate Change*, by Jennerjahn et al. includes all expected anthropogenic effects on mangrove environments, but excludes the formation of novel mangrove forests as a result of global dispersal of mangrove species. The authors expect a reduction of mangrove services as a result of climate change and identify gaps in ecophysiological understanding relative to conditions in the Anthropocene. Chapter 8, *Mangroves and People: Local Ecosystem Services in a Changing Climate*, by Huxham et al. explains how mangrove carbon stored in the wood of an untouched forest is a desirable future for the global community, while for the local communities, the desirable future is burning that wood to satisfy their energy and cooking needs. This is the old dilemma between preservation and human needs, one that was debated when the conservation focus was on moist and dry forests and their use for fuelwood by needy people. This chapter is important for mangrove conservation because it underscores the usually neglected social-ecological issues, and it is also independent of other book chapters in relation to anthropogenic effects and future scenarios of climate change.

The social-ecological focus of Chap. 9 is stronger than in Chap. 8. In Chap. 9, *Anthropogenic Drivers of Mangrove Loss: Geographic Patterns and Implications for Livelihoods*, Chowdhury et al. use regional case studies to illustrate mangrove-dependent subsistence and poverty traps and relate conservation problems to large-scale use of mangroves by such industries as the global shrimp trade. Chapters. 8, 9, and 11, when dealing with problems of mangrove uses, do not address management solutions that have been documented for mangroves as possible mitigation avenues (below). It appears that the gap between academic study and active management remains open in mangroves.

In Chap. 10, *Mangrove Forest Restoration and Rehabilitation*, López-Portillo et al. review the experience in 90 sites around the world where mangrove restorations were attempted. My colleague Jack Ewel once said that restoration is the ultimate test for ecological understanding, and judging by the lack of success with mangrove restorations, our understanding of mangrove ecology must be limited.

Alternatively, restoration projects might be ignoring what we know about mangroves, which is why a significant portion of Chap. 10 addresses critical ecological theory and operational processes required for assuring successful mangrove restoration projects. To the recommendations in this chapter, I would add the need to eliminate normative thinking and terminology from this literature (i.e., "damage," "impact," "deteriorated," "better," "improved," "integrity," "alien," "exotic," etc.), which introduces bias to the evaluation of ecological conditions and ignores directional change and adaptability to prevailing environmental conditions.

Chapter 11, *Mangrove Macroecology*, by Rivera-Monroy et al. promotes macroecology as the approach to use to answer large-scale questions in the future. Ideally, macroecology will encompass all aspects of traditional ecological research: ecology, biogeography, paleontology, landscape ecology, and macroevolution. The fact that only two studies on macroecology of mangroves are available suggests that the future is wide open for this approach. Further research will determine the desirability and effectiveness of this approach.

This book was written at a time when the effects and consequences of the Anthropocene on mangrove ecosystems remain uncertain. The authors of this book are generally pessimistic about the future of mangrove forests, probably because they mostly focus on the areas where mangroves are in retreat. The knowledge that mangrove forests are expanding their territory (mentioned briefly in the book) does not alleviate the pessimism; it increases as authors also worry about the losing ecosystems, i.e., salt marshes or some other coastal community. The book focus is academic (except for Chap. 10) and the integration of the science recorded here with the management of mangrove stands, which has been partially captured by the FAO (1994) and Saenger (2002), is still open for synthesis.

A mangrove paradox is the apparent simplicity of the mangrove forest implicit in the single tree species monoculture zones nicely arrayed over the landscape, when in fact mangrove forests are very complex systems when viewed in three dimensions and temporal succession along endless gradients operating from the microscale of redox potentials in sediments to global latitudinal scales delimited by frequency of frost and strength of wave action on the appropriate substrates. As this book demonstrates, there are still many hurdles and unanswered questions before we can comfortably say that we understand mangrove ecosystems, and the leap into the global aspects of mangrove functioning further stretches the limits of our imagination. This book, however, points the way, much like how the Hawaii meeting led us into ecosystem level research. One of the lessons from the Hawaii meeting is that once the scientific engine is pointed and cranked, there is no turning back, nor limits to the insights to be gained.

USDA Forest Service International Institute of Tropical Forestry Ariel E. Lugo,
Río Piedras, PR, USA

Literature Cited

Alongi DM (2009) The energetics of mangrove forests. Springer, New York

Chapman VJ (1976) Mangrove vegetation. J. Cramer, Leutershausen

Cintrón G, Lugo AE, Pool DJ, Morris G (1978) Mangroves of arid environments in Puerto Rico and adjacent islands. Biotropica 10:110–121

Clough BF (ed) (1982) Mangrove ecosystems in Australia: structure, function and management. Australian National University Press, Canberra

FAO (1994) Mangrove forest management guidelines. Food and Agriculture Organization of the United Nations. FAO, Rome

Heald EJ (1971) The production of organic detritus in a south Florida estuary. Miami Sea Grant Technical Bulletin 6:1–110

Lugo AE (1980) Mangrove ecosystems: successional or steady state? Biotropica 12(supplement 2):65–72

Lugo AE (2002) Conserving Latin American and Caribbean mangroves: issues and challenges. Madera y Bosques 1:5–25

Lugo AE, Medina E (2014) Mangrove forests. In: Encyclopedia of natural resources: land. Taylor and Francis, New York, pp. 343–352. https://doi.org/10.1081/E-ENRL-120047500

Lugo AE, Medina E, McGinley K (2014) Issues and challenges of mangrove conservation in the Anthropocene. Madera y Bosques 20:11–38

Lugo AE, Snedaker SC (1974) The ecology of mangroves. Annu Rev Ecol Syst 5:39–64

Macnae W (1968) A general account of the fauna and flora of mangrove swamps and forests in the Indo-West-Pacific region. Adv Mar Biol 6:73–270

Odum WE (1971) Pathways of energy flow in a south Florida estuary. In: Sea grant technical bulletin 7. University of Miami, Miami

Saenger P (2002) Mangrove ecology, silviculture and conservation. Kluwer Academic Publishers, Dordrecht

Spalding M, Kainuma M, Collins L (2010) World atlas of mangroves. Earthscan, London

Thom BG (1975) Mangrove ecology from a geomorphic viewpoint. In: Walsh G, Snedaker S, Teas H (eds) Proceedings of the international symposium on biology and management of mangroves. University of Florida, Institute of Food and Agricultural Sciences, Gainesville, pp. 469–481

Walsh G, Snedaker S, Teas H (eds) (1975) Proceedings of international symposium on biology and management of mangroves. Institute of Food and Agricultural Sciences, University of Florida, Gainesville/Honolulu

Yañez-Arancibia A, Lara-Domínguez AL (eds) (1999) Ecosistemas de manglar en América tropical. Instituto de Ecología, A.C., Xalapa/Veracruz

Reviewers

Elizabeth Ashton
Jake Brenner
Benjamin M. Brown
Joseph R. Burger
Kyle Cavanaugh
Pat Dale
Ronald D. Delaune
Gustavo Duque Estrada
Aaron M. Ellison
Temilola Fatoyinbo
Colin Field
Dan Friess
Lucy Gillis
Gerald Alexander Islebe
Eric Keys
Ken Krauss
Ronald Loughland
Catherine Lovelock
Ariel Lugo
Ernesto Medina
Beth Middleton
Essam Yassin Mohammed
Christophe Proisy
Kerrylee Rogers
Peter Saenger
Christian J. Sanders
Martin Skov
Pierre Taillardat
Ludwig Triest

Acknowledgments

We are enormously grateful to the reviewers who agreed to contribute to this book with their time and suggestions; their comments and ideas improve the quality and scope of the book

We also thank Springer and the different Publishing Editors who help us to advance the preparation and production of this book. Particularly Janet Slobodien who accepted our initial idea about the need of this book among many other priorities, and for her help and patience to navigate the different steps to complete this project. We are also thankful to Elaina Mercatoris and Andrea Sandell for their editorial assistance and expertise.

- Victor H. Rivera-Monroy acknowledges the partial support in the preparation of the book by the following agencies: Florida Coastal Everglades Long-Term Ecological Research program through the U.S. National Science Foundation (NSF) grants DEB-9910514, DBI-0620409, DEB-1237517, NASA-JPL (LSU Subcontract# 1452878) project *"Vulnerability Assessment of Mangrove Forest Regions of the Americas"*, the U.S. Department of the Interior (DOI) South Central-Climate Science Center (SC-CSC) (Cooperative Agreement #G12 AC00002), and the NSF- Dynamics of Coupled Natural and Human Systems (CNH) Program *"Poverty Traps and Mangrove Ecosystem Services in Coastal Tanzania"* (grant #CNH-1518471). Special thanks to Barbara Hasek for providing invaluable editorial assistance and encouragement throughout this project.
- Shing Yip Lee thanks his students and collaborators for inspirations and discussions that have contributed to the work presented in this book.
- Erik Kristensen was supported by grants from the Danish Council for Independent Research (contract # 272-08-0577, 09-071369 and 12-127012) and from the Danish Council for Strategic Research (contract # 09-063190 and 12-132701).

Any opinions, findings, and conclusions or recommendations expressed in this book are those of the authors and do not necessarily reflect the views of the NSF, DOI/SC-CSC, NASA, JPL, the Danish Council for Independent Research or the Danish Council for Strategic Research.

Contents

Chapter 1
Introduction

Victor H. Rivera-Monroy, Shing Yip Lee, Erik Kristensen, and Robert R. Twilley

1.1 Relevance: A Short Story

Nine years ago, Bouillon et al. (2008) published a review paper where they analyzed the most recent data and information to improve previous estimates of global carbon budgets for mangrove wetlands proposed in the early 1990s and 2000s (Twilley et al. 1992; Jennerjahn and Ittekkot 2002; Duarte et al. 2005). Their objective was underscored by the increasing interest in assessing the ecological role of coastal wetlands as either carbon sinks (i.e., "blue carbon") and/or sources in the context of climate change, one of the most critical environmental issues of our time (Hopkinson et al. 2012). Bouillon et al. (2008) identified a major "missing" carbon flux when reconciling global mangrove primary productivity with major carbon sinks that included organic carbon export, sediment burial, and mineralization (Fig. 1.1). Interestingly, this "missing" flux represented >50% of the carbon fixed by mangrove wetlands and was equivalent to 30–40% of the estimated global riverine organic carbon input into the coastal zone (Bouillon et al. 2008) (Fig. 1.1). Based on these findings, the authors proposed several mechanisms that potentially could

V.H. Rivera-Monroy (✉) • R.R. Twilley
Department of Oceanography and Coastal Sciences, College of the Coast and Environment, Louisiana State University, Baton Rouge, LA 70803, USA
e-mail: vhrivera@lsu.edu

S.Y. Lee
Simon F S Li Marine Science Laboratory, School of Life Sciences, and Earth System Science Programme, The Chinese University of Hong Kong, Shatin, Hong Kong SAR, China

E. Kristensen
Department of Biology, University of Southern Denmark, Campusvej 55, 5230 Odense M, Denmark

V.H. Rivera-Monroy et al. (eds.), *Mangrove Ecosystems: A Global Biogeographic Perspective*, https://doi.org/10.1007/978-3-319-62206-4_1

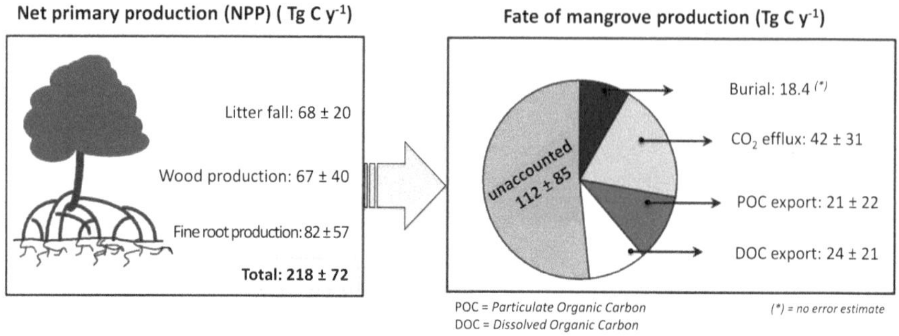

Fig. 1.1 Comparative analysis of published estimates of the fate of Net Primary Production (NPP) in mangrove wetlands showing the "missing" or unaccounted carbon flux (112 ± 85 Tg C y^{-1}) (Modified from Bouillon et al. 2008)

explain this discrepancy and "missing" component in the global budget, including net export of dissolved carbon into adjacent estuaries and coastal oceans. The quantification and understanding of these proposed mechanisms launched a number of research efforts in several coastal regions in the following years (e. g., Miyajima et al. 2009; Alongi et al. 2012; Breithaupt et al. 2012; Maher et al. 2013; Mukherjee et al. 2013; Alongi and Mukhopadhyay 2015; Stewart et al. 2015; Sippo et al. 2016; Ho et al. 2017).

In addition to the scientific relevance of the proposed research questions identified by Bouillon et al. (2008), their study also represented a valuable collaborative work among researchers from different countries and institutions from the new and old world. This collaboration was initially conceived in discussions during an international conference on mangrove ecosystems in Brisbane, Australia in 2006 (Mangrove Macrobenthos and Management meeting—MMM2). Thus, the issues and hypotheses discussed in the MMM2 meeting provided the template for the preparation of a proposal as response to a call by the Australian Research Council (ARC) in 2008. Recognizing the significance of comparative mangrove ecological studies at the global scale, a proposal to the ARC was submitted ("Mangrove Biodiversity and Ecosystem Services—A Global Assessment") by a number of authors who are now contributing to this book.

Although the proposal was not funded, the momentum in preparing the proposal helped to further identify knowledge gaps within carbon cycling in mangrove dominated-ecosystems and the need to advance a research agenda in different fronts, particularly in mangrove biogeochemistry and biodiversity assessment and conservation. This effort resulted not only in the preparation of the paper mentioned earlier, but also the consolidation of previous collaborations among researchers. One lesson learned from this interaction was the tremendous value of inter- and transdisciplinary studies to tackle mangrove ecological research questions at spatial scales relevant for the conservation and management of mangrove wetlands. Yet, another outcome, was the recognition of a major problem hindering the advance of mangrove socio-ecological research in the last two decades: the lack of funding, not only for global cross-comparative studies, but also at the country level, especially in subtropical and tropical countries with coastal regions having proportionally most of the total mangrove area (range: 83,495–137,760 km^2) (i.e., Asia: 42%; Africa: 20%, North and Central America: 15%; Oceania:12%: South America 11%) (Fig. 1.2; Giri et al. 2011; Hamilton and Casey 2016).

To our knowledge, no global, landscape level mangrove wetland research initiatives are presently funded by governments. Thus, most of the recent work at this scale, to address some of the most pressing problems in mangrove degradation and area loss, is merely the result of the collective effort of researchers coordinated by nongovernment organizations (NGOs) (e.g., International Union for the Conservation of Nature's (IUCN) Global Species Programme/Mangrove Specialist Group (MSG); Friess et al. 2016) and presented at scientific meetings (e.g., Mangrove Macrobenthos and Management meetings; such as the MMM3 (2012) in Sri Lanka and IUCN MSG symposia in Southeast Asia, and more recently in North America (MMM4, 2016, St Augustine, Florida USA)). Yet, most of these organizations and venues are basically considered a powerful platform for exchange of actions and research directions given the lack of own funding sources. Indeed, significant efforts are needed to orchestrate mechanisms to finance and support long-term studies in strategic regions around the world to warrant the continuity of studies at different spatial scales and geographical regions. This approach is needed not only to address local issues and develop an ecological theory on mangrove ecosystems, but also to develop applicable strategies at the regional and continental scale. Potentially, the outcome of such initiatives could be modeled at the continental level, for example by the US Long Term Ecological Research (LTER) Network (Kratz et al. 2003) and National Ecological Observatory Network (NEON) program in the USA (Keller et al. 2008). However, it is evident that these large-scale initiatives are expensive and require considerable organization efforts and time before they can be implemented (Hampton et al. 2013; Utz et al. 2013).

It is paradoxical that countries readily recognizing the social and economic value of mangrove wetlands, particularly those with a significant mangrove area (Fig. 1.2.), still suffer mangrove wetland degradation and loss; along with other closely interconnected coastal systems (i.e., seagrasses, coral reefs, marshes). One explanation to this current paradox is highlighted by the low percentage (~7%) of mangrove area currently classified as protected reserves (Giri et al. 2011; Hamilton and Casey 2016). Since the publication of the first global economic assessment of the value of mangrove wetlands (including tidal wetlands; US$ 1648 × 10^9) (Costanza et al. 1997), a number of studies have attempted to refine and update this figure in a variety of management contexts and methods (Costanza et al. 2014; Barbier 2016) to emphasize the economic importance of mangroves in the global economy. The attempts to refine and improve the significance of this economic value include not only the most visible and well-recognized mangrove ecosystem services around the world, e.g., fisheries and wood harvest (Twilley et al. 1998; Ronnback 1999), but also other services such as carbon sequestration, storm protection, and maintenance of water quality (Lee et al. 2014). Because these mangrove ecosystem services are well recognized, it is common to read in many mangrove papers published since the

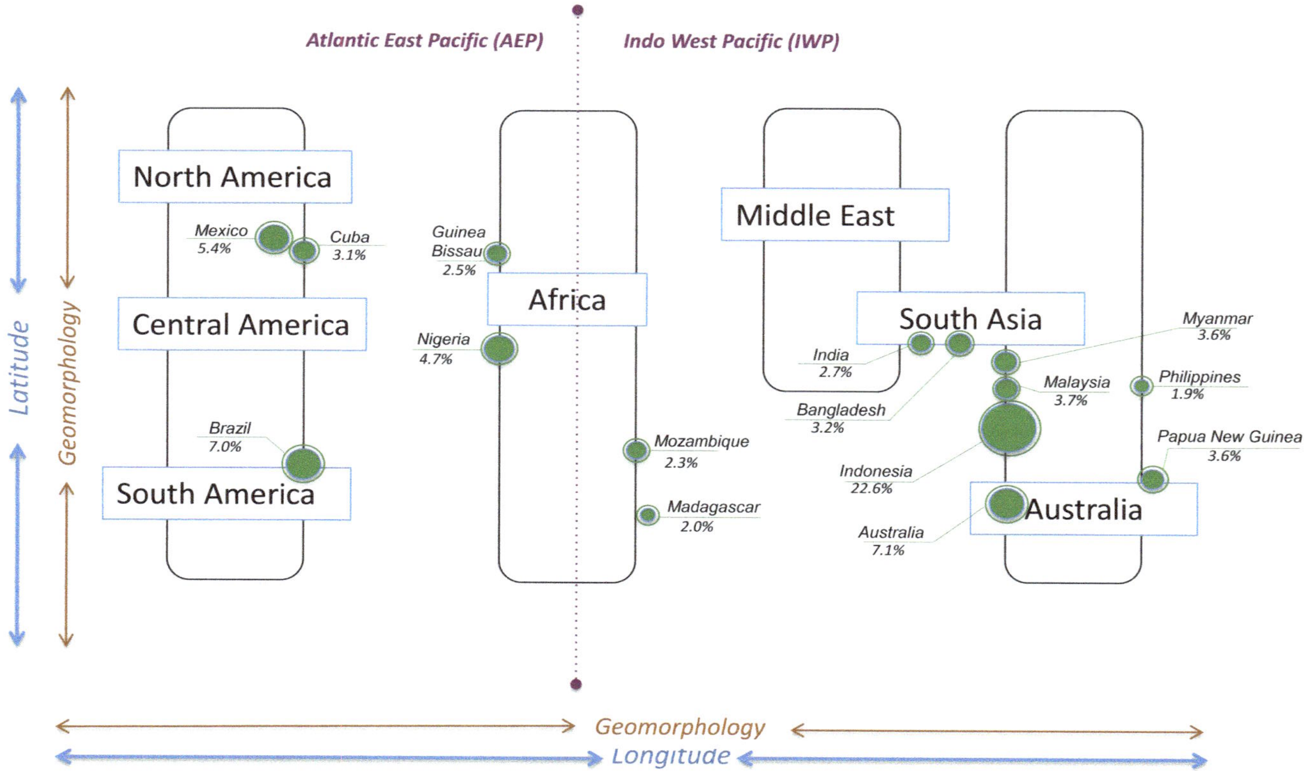

Fig. 1.2 The biogeographic (AEP, IWP) and regional (geomorphology) dimensions underpinning the comparative analysis of ecological and socioeconomic mangrove processes in this book. The circles represent the 15 countries contributing 75.4% of the total global mangrove area in 2000 (Data from Giri et al. 2011)

1980s, a long list of such services to strength and underscore the paper's importance and contribution. Yet, despite this qualitative listing, we still lack a comparative and comprehensive quantitative global assessment of the economic value of these ecosystems in the context of local and regional economies, especially in developing countries (e.g., McNally et al. 2011; Barbier 2016).

One of the issues to consider in the advancement of the global and regional economic valuation of mangrove ecosystem services, as well as the applicability of current proposed values, is the significant difference in ecosystem structure and function among various types of mangrove wetlands (sensu Lugo and Snedaker 1978) that thrive in diverse eco-geomorphic settings from subtropical to tropical latitudes (Fig. 1.2.). Actually, these biophysical attributes of mangrove ecosystems significantly influence, not only the quality of each ecosystem service, but also their quantity and availability in the long term (Ewel et al. 1998). Recent findings suggest that mangrove ecosystem threats and functioning, and thus their capacity for ecosystem services, differ between biogeographic regions and socioeconomic settings (Lovelock and McAllister 2013; Lee et al. 2014; Lovelock et al. 2014). Moreover, even down to the local level, differences in threats and drivers necessitate management considerations for specific coastal ecosystems and associated ecotypes (Rivera-Monroy et al. 2004; Jerath et al. 2016).

Another issue in the valuation of ecosystem services is the diverse socioeconomic context within which each country is embedded and how this affects decisions about what are the best strategies in managing its coastal resources, including mangrove wetlands. This is a major problem especially when considering the diversity of local drivers controlling mangrove deforestation in distinct coastal regions such as in East Africa, Central America, or Australia (Hamilton and Casey 2016). Certainly, there are numerous studies documenting the relative role of urban, agricultural, aquacultural, and infrastructure development in current mangrove loss rates, and their degree of impact on these rates (Walters et al. 2008). Yet, from an economic perspective, countries have opted for these development priorities, knowingly or unknowingly, of the major ecological (i.e., land use/change, total loss, pollution) and socially detrimental implications (e.g., poverty and economic inequality) (Bonifaz and Parks 1993; Rivera-Monroy et al. 2006). These negative outcomes are trade-offs between economic development and ecosystem conservation represented by the assessment of direct use (e.g., timber) and existence values (e.g., preserving biodiversity). The selected values include major challenges since, for example, maintenance of the present status is difficult to value. Furthermore, direct use values may not be priced adequately in current markets since in most cases, they often benefit social sectors beyond those who provide and pay for them (Frank and Schlenker 2016). Indeed, the lack of economic incentives to encourage mangrove wetland preservation (e.g., establishment of national parks or marine reserves) and the lack of accepted operational metrics to value carbon storage in wetlands are good overall examples of these daunting challenges (e.g., Jerath et al. 2016); even when carbon sequestration has become one of the more notable ecosystem services over the last decade or two in the context of climate change (Alongi 2011; Donato et al. 2011).

One of the major issues we currently need to address in mangrove ecological research is the causes (qualitative and quantitative) that drive current mismanagement and degradation of mangrove wetlands; one of the most productive and valuable coastal ecosystems in the world. This is a complex task, and we need to consider the complex interactions between social and ecological systems, including an evaluation of *"state of the art"* mangrove ecosystem science. We contend that adaptive science-based management (Berkes et al. 2000; Armitage et al. 2009; Granek et al. 2010) should be a major basis for protection, conservation, restoration, and management of mangrove wetlands in this century, especially under the uncertainty of future climate change scenarios. Thus, it is paramount to pause and evaluate what we know and need to know to improve our understanding of how mangrove wetlands function, and how this functionality and societal needs can be translated into sound management plans under various socioeconomic settings across the globe. We consider this book such a pause—an exercise in the analysis of our current knowledge of mangrove-dominated ecosystems that aims to provide a new research agenda for this century and that explicitly addresses current mangrove area loss risks and vulnerabilities.

1.2 Approach: Integration and Ecosystem Services

Just as in the case of the missing carbon sink mentioned earlier, we hope this book contributes to the consolidation of current and future interdisciplinary and transdisciplinary initiatives among researchers and countries with major stakes in mangrove conservation. A number of recent books on mangrove ecosystems are devoted to selected aspects of mangrove ecosystems. For example, the updated version of the *World Mangroves Atlas* (Spalding et al. 2011) provides a detailed analysis of global mangrove spatial distribution and regional forest extent. Similarly, *The Energetics of Mangrove Forests* (Alongi 2009) has a strong focus on material flow within and between different compartments of the mangrove ecosystem. *Coastal Wetlands: An Integrated Ecosystem Approach* by Perillo et al. (2009) deals with biophysical aspects of all coastal wetland types, with some coverage of the biogeographic or socioeconomic perspectives of mangrove ecosystems. Twilley and Day (2013) present a general overview of the ecology of mangroves in the second edition of *Estuarine Ecology* (Day et al. 2013).

This book complements these contributions and advances other research priorities aiming to (1) provide a scholarly and authoritative analysis of mangrove ecological processes, covering data at the local, biogeographic, and global scales with an emphasis on regions and countries holding the largest mangrove resources; (2) integrate ecological and socioeconomic perspectives on mangrove function and management using a system level hierarchical analysis framework; and (3) explore the nexus between mangrove ecology and the capacity for ecosystem services, with an emphasis on thresholds, multiple stressors, and local conditions that determine this capacity. The book is organized in eleven chapters, each addressing various aspects of mangrove ecology central to the delivery of ecosystem services. We aimed for a comparative approach, thus the book was prepared with the collaboration of a team of authors with research experience in five regions: the Neotropics, Africa, Middle East, Southeast Asia, and Australia (Fig. 1.2.). These regions encompass the major biogeographic (Atlantic East Pacific: AEP; Indo West Pacific: IWP) and socio-economic settings of mangrove distribution. Another major objective was to compile a comprehensive reference for managers and researchers dealing with the multifaceted and complex issues concerning local, regional, and global management of mangrove resources.

Chapter 2 by Duke (2017) sets the stage to help understand the current and past mangrove wetland distribution with a wide-ranging analysis of mangrove biodiversity patterns and evolution based on ancestral biogeography and existing floristics. This chapter discusses why mangrove plants manage to occur where they do by analyzing the key factors limiting their distribution. Duke also analyzes how each of those factors has changed during the evolution of the 80 species, within 18 family lineages and hybrids currently recognized. One of the major contributions of this chapter is the explanation of how mangrove distributional patterns are closely related to each genotype under a historical perspective. The author concludes by offering a novel hypothesis where geophysical occurrences over the last 100 million years are considered the major force in how mangrove species have dispersed, diversified, and evolved following common phylogenetic pathways.

Although plant species diversity is low in mangrove wetlands, as analyzed by Duke (2017, Chap. 2), when compared to other ecosystems (e.g., coral reefs, tropical rain forests), Lee et al. (2017, Chap. 3) addresses the high diversity of other organisms including decomposers, detritivores, and consumers that support key mangrove ecosystem services. The authors underscore the broad levels of key functional and structural biodiversity components of mangrove ecosystems in relation to major species assemblages such as macrobenthic invertebrates and finfish that contribute to key ecological processes. Lee et al. also perform a biodiversity comparison of selected assemblages associated with the two main biogeographic regions (i.e., IWP and AEP). Interestingly, the authors suggest higher species richness of finfish in the AEP systems when compared to the IWP region, even when considering latitudinal differences. This pattern seems to be the case in other biodiversity components as well, although further data and information is needed. The authors conclude that if this difference between regions is confirmed when more data is available, it may have implications for species assemblage function and, therefore, the ecosystem services they can provide. Additionally, this chapter contributes to the elucidation of the relationship between diversity and ecosystem function. It underscores how conservation and management of mangrove biodiversity require efforts at various levels of sociopolitical organization and the need for developing and implementing legal and economic instruments.

Lucas et al. (2017, Chap. 4) discuss the state-of-the-art tools needed to understand and evaluate mangrove spatial distribution and the consequences of historical and future natural and anthropogenic impacts in mangrove wetlands. The authors examine how the range of remote sensing data and its calibration (ground, airborne, and space borne instrumentation) has been used to describe the multiple dimensions of mangrove forests by focusing primarily on spatial scales, temporal frequencies, spectral responses, and three-dimensional state. They also explain how remote sensing data have been used to describe the structural complexity of mangrove environments, especially their connectivity with other habitats across a range of scales. Finally, the authors discuss strategies on how to use remote sensing data in long-term mangrove management and conservation programs. The benefits of this application in mangrove characterization, mapping, and monitoring programs is highlighted in specific study cases using instruments such as radar, lidar, and optical sensors from a wide range of locations, including in Australia, Southeast Asia, and Central America.

One of the ecosystem functional properties attracting increasing interest in research initiatives is the large-scale spatial assessment of net primary productivity (NPP). As mentioned earlier, remote sensing tools have since the 1970s been critical to determine not only the global mangrove wetland distribution (Giri et al. 2011; Kuenzer et al. 2011; Hamilton and Casey 2016), but also to estimate above mangrove biomass (e.g., Simard et al. 2006; Simard et al. 2008; Montesano et al. 2013). Since mangrove NPP is closely associated with structural variables such as biomass, the assessment of NPP is now a major research priority, particularly in the context of carbon cycling (i.e., blue carbon) and climate change (carbon uptake). Twilley et al. (2017, Chap. 5) evaluate mangrove NPP and carbon dynamics as related to the potential to sequester atmospheric carbon in above- and below-ground biomass and in the soil. The authors assess both NPP and carbon across different coastal environmental settings and emphasize global patterns of these ecosystem processes by comparing the AEP and IWP biogeographic regions. They also point out that the relative contribution of below-ground allocation into soil carbon storage and wood production to total NPP in mangrove wetlands have significant implications for the net carbon exchange balance. Twilley et al. provide examples of net carbon exchange to help determine the relative role of mangrove ecosystems in the global carbon budget and addressing the need for a better understanding of biomass allocation in these mangrove forests. This contribution advances our current understanding of the carbon cycling (Bouillon et al. 2008; Rivera-Monroy et al. 2013; Alongi 2014) and underscores the lack of comprehensive data in different geomorphic settings to determine how the fate of carbon export may influence net carbon exchange in the coastal zone. One major recommendation from this work is the need to obtain more information on how natural (i.e., tropical cyclones) and human disturbances (e.g., deforestation) controlling wetland recovery trajectories, may influence carbon flux in the coastal zone. This is a key component for determining the spatiotemporal role of mangrove wetlands as carbon sinks and sources. Indeed, these driver regimes are known as major sources of uncertainty in identifying the magnitude of carbon exchange between mangrove wetlands and both the atmosphere and adjacent estuarine/coastal waters (Alongi 2014).

Analogous to the approach by Twilley et al. (2017, Chap. 5), Kristensen et al. (2017, Chap. 6) also reviews the current understanding of the carbon cycling, but they also discuss the critical role of other elements (e.g., nitrogen, phosphorus, sulfur, iron, manganese) controlling NPP in mangrove ecosystems. In this chapter, the authors identify potential sources of variation in biogeochemical processes across different locations and analyze current advances in evaluating transformations of carbon and other elements in the context of mangrove conservation and management priorities. Their main objective was to identify knowledge gaps and research priorities across biogeographic regions and latitudes. One significant contribution of this review is the assessment of ecosystem services provided by mangrove wetlands through their biogeochemical functions, including: climate change mitigation, flood regulation, and water purification. The authors underline significant differences in mangrove functionality among regions that are difficult to explain from the current data availability. However, it is evident that this variation within and among mangrove forests depends on the hydrological regime, type of mangrove ecotype and local geomorphology. In fact, the spatial heterogeneity of redox processes caused by burrows and roots in mangrove sediments (i.e., oxygen pumping) is much more complex and variable in intertidal mangrove environments than in adjacent coastal and oceanic settings. The waterlogged and anoxic mangrove sediments promote slow decomposition, and the authors make the case that this allows significant carbon sequestration and long-term organic carbon accumulation in the sediments. Hence, mangrove wetlands have a strong climate change mitigation function that needs to be considered in coastal management plans. Therefore, the authors emphasize the need to translate current knowledge about the complexity of mangrove biogeochemistry (i.e., supporting ecosystem services) into robust and applicable performance measures in management programs including mangrove restoration and rehabilitation programs.

Indeed, climate change is one of the critical environmental issues of our time, and mangrove ecosystems are considered major players in ameliorating excess carbon in the atmosphere (Chaps. 5 and 6). However, mangrove responses can follow different trajectories depending on their location and environmental signature. Jennerjahn et al. (2017, Chap. 7) evaluate the response of mangrove ecosystems to possible outcomes of climate change, with regard to a set of categories including (1) distribution, diversity, and community composition; (2) physiology of flora and fauna; (3) water budget; (4) productivity and remineralization; (5) carbon storage in biomass and sediments; and (6) the filter function for elements beneficial or harmful to life. Based on this assessment, the authors identify regions most vulnerable to climate change. The four most important factors determining the response of mangrove ecosystems to climate change are sea level rise, an increase in frequency and/or intensity of storms, increases in temperature, and aridity. Jennerjahn et al. explain that although these changes may be beneficial for some mangrove forests at latitudinal distribution limits, they threaten forest structure and functions related to ecosystem services in most cases. The authors discuss the interaction of climate change with human interventions and how ecosystem services can be impacted. Based on this information, adaptation and management strategies are proposed. They also list a set of knowledge gaps that include, among others, the lack of information on the physiological response of mangrove plants and animals, especially on the response to interacting multistressors, and the need to increase public and decision makers awareness about the value of mangrove ecosystem services that contribute to decreasing the risk in mangrove loss as related to climate change.

Chapters 8 and 9 discuss socioeconomic interactions in the context of the whole socio-ecological system. One of the major issues in evaluating system interactions is the negative feedback between the availability of mangrove ecosystem services and the range of anthropogenic drivers promoting mangroves loss. Huxham et al. (2017, Chap. 8) reviews the multiple relationships among a variety of ecosystem services (e.g., provision of fuel, timber, fodder, crustacean, finfish, and shoreline protection services) with global patterns in biodiversity and poverty. The authors correlate higher floral and faunal diversity with a greater range of species exploited for fuel, timber, crustaceans, and coastal protection in the IWP region, compared with the AEP region. One finding from this analysis is that although poverty is a strong predictor for reliance on some local services (e.g., fuel wood), it is not related, for example, to finfish harvest or use. The association indicates that local people may be "liberated" from reliance on some services by increased income, but use other ecosystem services to generate this income. As underscored by other chapters, the vulnerability of these services to climate change depends on local geomorphological, biological, and social factors. In fact, forests with good supplies of sediment and fresh water, and fauna with relatively simple life-cycles will probably be more resilient to those threats. Huxham et al. point out that greater wealth (or investment) may permit people to shift from fishing natural populations to aquaculture and to show flexibility in the face of changing or reduced service provision. The authors conclude that economic development may increase local resilience to environmental change, but does not imply a reduction in the value, economic or ecological, of mangrove forests. It might, in fact, result in a shift in importance, often from provisioning towards regulating services and from less preferred to higher valued products.

Chowdhury et al. 2017 (Chap. 9) further elaborate the findings of Huxham et al. (2017, Chap. 8) by stressing that the human dimensions of mangrove-dominated ecosystems are vital to understand how drivers of mangrove losses interact at local levels. In this chapter, the authors review case studies of mangrove ecosystems to compare the fundamental drivers of regional mangrove losses. They present a systematic, synoptic review of case studies involving mangrove ecosystems from Africa, Asia, and Latin America to compare the fundamental drivers of mangrove losses at a global scale. The authors identify agriculture and aquaculture as major proximate sources of mangrove losses worldwide. Then, they focus their analysis on two significant drivers of mangrove losses: (1) mangrove-dependent subsistence economies and related poverty traps, and (2) the global shrimp trade. In this regional context, specific drivers are examined in Southeast Asia/China and Ecuador, which represent geographic regions that have experienced rapid mangrove losses in the last few decades. Extractive activities such as harvesting of timber and non-timber resources from mangroves are also linked to serious degradation of local mangrove resources, as is the significant increase in infrastructure development. Given the hierarchical level and degree of impact by anthropogenic drivers, the authors recommend the use of a coupled socio-ecological system approach to understand and quantify the bidirectional linkages between mangrove ecological dynamics and the constellation of anthropogenic drivers of mangrove change.

As a result of the significant net loss of mangrove wetlands and associated ecosystem services at the global scale, as described by Huxham et al. (2017, Chap. 8) and Chowdhury et al. (2017, Chap. 9), major initiatives and regional programs have been developed and implemented to restore and rehabilitate mangrove wetlands. Consequently, millions of dollars have been allocated in attempts to recuperate these valuable wetlands. However, the success in restoring structural and functional attributes of mangrove ecosystems has been mixed. Given the strategic importance of these management programs, Lopez-Portillo et al. (2017, Chap. 10) analyze current best practices and recommendations used in the implementation of mangrove rehabilitation and restoration (R/R) projects in the AEP and the IWP biogeographic regions in the last 20 years. The authors' approach is the analysis and classification of the sources of damage/impact, including their origin, as natural (siltation, erosion, the direct and indirect effect of tropical storms or tsunamis) or anthropogenic (pollution, land use policies, overharvesting, aquaculture, altered hydrology and hydroperiod) and their spatial extent. The authors find that the causes of damage were a complex mixture associated with erosion, hydrological impairment, deforestation, siltation, and land conversion for aquaculture and other land uses. Based on this analysis, Lopez-Portillo et al. conclude that a number of projects were implemented without an underlying science-based approach and were often ill prepared and unsuccessful. They underscore that a critical step is to develop a decision tree that operates as a guide to optimize the use of available funding in the development, implementation, and monitoring of R/R protocols. These protocols (e.g., Ecological Mangrove Rehabilitation) should be guided by a set of clear objectives, goals, and deadlines as part of a robust research agenda based on sound ecological theory and reliable monitoring practices, including the participation of local communities. Another recommendation by the authors is that the results of each R/R project, whether successful or not, should be published since any documentation could be a valuable source of data and information for future development of mangrove R/R practices and methods within the community of restoration ecology science. The chapter ends with a call for the continental level implementation of guidelines to advance international initiatives aimed at protecting and conserving mangrove ecosystems.

The final chapter (Rivera-Monroy et al. 2017, Chap. 11) addresses two key objectives of the book—first, an analysis to integrate ecological and socio-economic perspectives on mangrove function and management using a system-level hierarchical analysis framework; second, the exploration of the nexus between mangrove ecology and the capacity of mangrove ecosystems to sustain long-term ecosystem services. Here, Rivera-Monroy et al. propose that the discipline of macroecology can be used to embrace advancement and continue developing mangrove ecological theory regarding complex structural and functional patterns and to assess human impacts on mangrove ecosystems. The authors discuss the prospective utility of macroecology-based studies that could answer process-based ecological questions and help expand long-term ecological studies at regional and continental scales. They explain that macroecology uses statistical analyses to investigate large-scale universal patterns in the distribution, abundance, diversity, and organization of species and ecosystems, including the scaling of ecological processes and structural and functional relationships. Thus, transdisciplinary macroecology explores the boundaries where ecology, biogeography, paleontology, landscape ecology, and macroevolution come together. According to the authors' analysis, macroecology provides an explicit mechanistic ecological understanding of issues that deal with the distribution, abundance, energetics, and interaction networks of individuals and species across multiple spatial and temporal scales. Rivera-Monroy et al. use several examples to illustrate the utility of this framework, including the analysis of continental distribution of aboveground net primary productivity and carbon storage, and the variation in mangrove forest ecosystem structure and function in relation to macroclimatic drivers (e.g., temperature and rainfall regimes) and climate change. The chapter also includes a description of current trends in mangrove modeling approaches and their potential utility to test hypotheses about mangrove structural and functional properties. The authors emphasize that given the gap in relevant experimental work at the regional scale, mangrove restoration and rehabilitation projects can be considered macroecological studies that advance the critical selection and conservation of ecosystem services. The authors finally indicate that in the "epoch" of the Anthropocene, characterized by an unprecedented mangrove degradation and loss, macroecology can advance and provide information to maintain mangrove goods and services to society in the long term.

We foresee the contribution of the eleven chapters included in this book as a significant step forward in both closing the knowledge gap about mangrove structural and functional properties, and the development of an integrated research agenda for the implementation of global long-term socio-ecological studies in mangrove-dominated ecosystems. Overall, all the contributors reiterate the critical ecological, social, and economic importance of mangrove wetlands to society. This work promotes a strategic and operational global strategy to further advance the conservation of one of the most productive ecosystems in the world for future generations.

Acknowledgements VHRM participation in the preparation of this chapter was partially supported by the Florida Coastal Everglades Long-Term Ecological Research program through U.S. National Science Foundation (NSF) grants (DEB-9910514, DBI-0620409, DEB-1237517), NASA-JPL (LSU Subcontract# 1452878) project "Vulnerability Assessment of Mangrove Forest Regions of the Americas", the Department of the Interior – South Central Climate Science Center, Cooperative Agreement #G12 AC00002, and the NSF Dynamics of Coupled Natural and Human Systems program (grant#: CNH-1518471). We thank Dr. John Day for helpful comments on the manuscript.

References

Alongi DM (2009) The energetics of mangrove forests. Springer, Breinigsville

Alongi DM (2011) Carbon payments for mangrove conservation: ecosystem constraints and uncertainties of sequestration potential. Environ Sci Pol 14:462–470

Alongi DM (2014) Carbon cycling and storage in mangrove forests. Annu Rev Mar Sci 6(6):195–219

Alongi DM, Mukhopadhyay SK (2015) Contribution of mangroves to coastal carbon cycling in low latitude seas. Agric For Meteorol 213:266–272

Alongi DM, de Carvalho NA, Amaral AL, da Costa A, Trott L, Tirendi F (2012) Uncoupled surface and below-ground soil respiration in mangroves: implications for estimates of dissolved inorganic carbon export. Biogeochemistry 109:151–162

Armitage DR, Plummer R, Berkes F, Arthur RI, Charles AT, Davidson-Hunt IJ, Diduck AP, Doubleday NC, Johnson DS, Marschke M, McConney P, Pinkerton EW, Wollenberg EK (2009) Adaptive co-management for social-ecological complexity. Front Ecol Environ 7:95–102

Barbier EB (2016) The protective service of mangrove ecosystems: a review of valuation methods. Mar Pollut Bull 109:676–681

Berkes F, Colding J, Folke C (2000) Rediscovery of traditional ecological knowledge as adaptive management. Ecol Appl 10:1251–1262

Bonifaz M, Parks PJ (1993) Nonsustainable use of renewable resources – mangrove deforestation and mariculture in Ecuador. Am J Agric Econ 75:1299–1299

Bouillon S, Borges AV, Castaneda-Moya E, Diele K, Dittmar T, Duke NC, Kristensen E, Lee SY, Marchand C, Middelburg JJ, Rivera-Monroy VH, Smith TJ, Twilley RR (2008) Mangrove production and carbon sinks: a revision of global budget estimates. Global Biogeochem Cycles 22:GB003052

Breithaupt JL, Smoak JM, Smith TJ, Sanders CJ, Hoare A (2012) Organic carbon burial rates in mangrove sediments: strengthening the global budget. Glob Biogeochem Cycles 26:3

Chowdhury RR, Uchida E, Chen L, Osorio V, Yoder L (2017) Chapter 9. Anthropogenic drivers of mangrove loss: geographic patterns and implications for livelihoods. In: Rivera-Monroy VH, Lee S, Kristensen YE, Twilley RR (eds) Mangrove ecosystems: a global biogeographic perspective structure, function and ecosystem services. Springer, New York, pp XX–XX

Costanza R, dArge R, deGroot R, Farber S, Grasso M, Hannon B, Limburg K, Naeem S, Oneill RV, Paruelo J, Raskin RG, Sutton P, vandenBelt M (1997) The value of the world's ecosystem services and natural capital. Nature 387:253–260

Costanza R, de Groot R, Sutton P, van der Ploeg S, Anderson SJ, Kubiszewski I, Farber S, Turner RK (2014) Changes in the global value of ecosystem services. Global Environmental Change-Human and Policy Dimensions 26:152–158

Day JW, Crump BC, Kemp WM, Yañez-Arancibia A (2013) Estuarine ecology, Second edn. John Wiley & Sons, Hoboken

Donato DC, Kauffman JB, Murdiyarso D, Kurnianto S, Stidham M, Kanninen M (2011) Mangroves among the most carbon-rich forests in the tropics. Nat Geosci 4:293–297

Duarte CM, Middelburg JJ, Caraco N (2005) Major role of marine vegetation on the oceanic carbon cycle. Biogeosciences 2:1–8

Duke NC (2017) Chapter 2. Mangrove floristics and biogeography revisited: further deductions from biodiversity hot spots, ancestral discontinuities and common evolutionary processes. In: Mangrove ecosystems: a global biogeographic perspective structure, function and ecosystem services. Springer, New York, pp XX–XX

Ewel KC, Twilley RR, Ong JE (1998) Different kinds of mangrove forests provide different goods and services. Glob Ecol Biogeogr 7:83–94

Frank EG, Schlenker W (2016) Balancing economic and ecological goals what are the trade-offs between economic development and ecosystem conservation? Science 353:651–652

Friess DA, Lee SY, Primavera JH (2016) Turning the tide on mangrove loss. Mar Pollut Bull 109:673–675

Giri C, Ochieng E, Tieszen LL, Zhu Z, Singh A, Loveland T, Masek J, Duke N (2011) Status and distribution of mangrove forests of the world using earth observation satellite data. Glob Ecol Biogeogr 20:154–159

Granek EF, Polasky S, Kappel CV, Reed DJ, Stoms DM, Koch EW, Kennedy CJ, Cramer LA, Hacker SD, Barbier EB, Aswani S, Ruckelshaus M, Perillo GME, Silliman BR, Muthiga N, Bael D, Wolanski E (2010) Ecosystem services as a common language for coastal ecosystem-based management. Conserv Biol 24:207–216

Hamilton SE, Casey D (2016) Creation of a high spatio-temporal resolution global database of continuous mangrove forest cover for the 21st century (CGMFC-21). Glob Ecol Biogeogr 25:729–738

Hampton SE, Strasser CA, Tewksbury JJ, Gram WK, Budden AE, Batcheller AL, Duke CS, Porter JH (2013) Big data and the future of ecology. Front Ecol Environ 11:156–162

Ho DT, Ferron S, Engel VC, Anderson WT, Swart PK, Price RM, Barbero L (2017) Dissolved carbon biogeochemistry and export in mangrove-dominated rivers of the Florida Everglades. Biogeosciences 14:2543–2559

Hopkinson CS, Cai WJ, Hu XP (2012) Carbon sequestration in wetland dominated coastal systems – a global sink of rapidly diminishing magnitude. Curr Opin Environ Sustain 4:186–194

Huxham M, Dencer-Brown A, Diele K, Kathiresan K, Nagelkerken I, Wanjiru C (2017) Chapter 8. Mangroves and people: local ecosystem services in a changing climate. In: Rivera-Monroy VH, Lee S, Kristiansen YE, Twilley RR (eds) Mangrove ecosystems: a global biogeographic perspective structure, function and ecosystem services. Springer, New York, pp XX–XX

Jennerjahn TC, Ittekkot V (2002) Relevance of mangroves for the production and deposition of organic matter along tropical continental margins. Naturwissenschaften 89:23–30

Jennerjahn TC, Gilman E, Krauss KW, Lacerda LD, Nordhaus I, Wolanski E (2017) Chapter 7: mangrove ecosystems and climate change. In: Rivera-Monroy VH, Lee SY, Kristensen E, Twilley RR (eds) Mangrove ecosystems: a global biogeographic perspective structure, function and ecosystem services. Springer, New York

Jerath M, Bhat M, Rivera-Monroy VH, Castaneda-Moya E, Simard M, Twilley RR (2016) The role of economic, policy, and ecological factors in estimating the value of carbon stocks in Everglades mangrove forests, South Florida, USA. Environ Sci Pol 66:160–169

Keller M, Schimel DS, Hargrove WW, Hoffman FM (2008) A continental strategy for the national ecological observatory network. Front Ecol Environ 6:282–284

Kratz TK, Deegan LA, Harmon ME, Lauenroth WK (2003) Ecological variability in space and time: insights gained from the US LTER program. Bioscience 53:57–67

Kristensen E, Connolly RM, Otero XL, Marchand M, Ferreira TO, Rivera-Monroy VH (2017) Chapter 6. Biogeochemical cycles: global approaches and perspectives. In: Rivera-Monroy VH, Lee S, Kristensen YE, Twilley RR (eds) Mangrove ecosystems: a global biogeographic perspective structure, function and ecosystem services. Springer, New York, pp XX–XX

Kuenzer C, Bluemel A, Gebhardt S, Vo Quoc T, Dech S (2011) Remote sensing of mangrove ecosystems: a review. Remote Sens 3:878–928

Lee SY, Primavera JH, Dahdouh-Guebas F, McKee K, Bosire JO, Cannicci S, Diele K, Fromard F, Koedam N, Marchand C, Mendelssohn I, Mukherjee N, Record S (2014) Ecological role and services of tropical mangrove ecosystems: a reassessment. Glob Ecol Biogeogr 23:726–743

Lee SY, Jones EBG, Diele K, Castellanos-Galindo GA, Nordhaus I (2017) Chapter 3. Biodiversity. In: Rivera-Monroy VH, Lee S, Kristensen YE, Twilley RR (eds) Mangrove ecosystems: a global biogeographic perspective structure, function and ecosystem services. Springer, New York, pp XX–XX

Lopez-Portillo J, Lewis RR, Saenger P, Rovai AS, Koedam N, Dahdouh-Guebas F, Agraz-Hernández C, Rivera-Monroy VH (2017) Chapter 10. Mangrove ecosystems: a global biogeographic perspective structure, function and ecosystem services. In: Rivera-Monroy VH, Lee S, Kristensen YE, Twilley RR (eds) Mangrove ecosystems: a global biogeographic perspective structure, function and ecosystem services. Springer, New York, pp XX–XX

Lovelock CE, McAllister RRJ (2013) 'Blue carbon' projects for the collective good. Carbon Manage 4:477–479

Lovelock CE, Adame MF, Bennion V, Hayes M, O'Mara J, Reef R, Santini NS (2014) Contemporary rates of carbon sequestration through vertical accretion of sediments in mangrove forests and saltmarshes of south East Queensland, Australia. Estuar Coasts 37:763–771

Lucas R, Vázquez Lule A, Teresa Rodríguez M, Kamal M, Thomas N, Asbridge E, Kuenzer C (2017) Chapter 4. Spatial ecology of mangrove forests: a remote sensing perspective. In: Rivera-Monroy VH, Lee S, Kristensen YE, Twilley RR (eds) Mangrove ecosystems: a global

biogeographic perspective structure, function and ecosystem services. Springer, New York, pp XX–XX
Lugo AE, Snedaker SC (1978) The ecology of mangroves. Annu Rev Ecol Syst 5:39–64
Maher DT, Santos IR, Golsby-Smith L, Gleeson J, Eyre BD (2013) Groundwater-derived dissolved inorganic and organic carbon exports from a mangrove tidal creek: the missing mangrove carbon sink? Limnol Oceanogr 58:475–488
McNally CG, Uchida E, Gold AJ (2011) The effect of a protected area on the tradeoffs between short-run and long-run benefits from mangrove ecosystems. Proc Natl Acad Sci U S A 108:13945–13950
Miyajima T, Tsuboi Y, Tanaka Y, Koike I (2009) Export of inorganic carbon from two southeast Asian mangrove forests to adjacent estuaries as estimated by the stable isotope composition of dissolved inorganic carbon. J Geophys Res Biogeosci 114:G1
Montesano PM, Cook BD, Sun G, Simard M, Nelson RF, Ranson KJ, Zhang Z, Luthcke S (2013) Achieving accuracy requirements for forest biomass mapping: a spaceborne data fusion method for estimating forest biomass and LiDAR sampling error. Remote Sens Environ 130:153–170
Mukherjee J, Ray S, Ghosh PB (2013) A system dynamic modeling of carbon cycle from mangrove litter to the adjacent Hooghly estuary, India. Ecol Model 252:185–195
Perillo GME, Wolanski E, Cahoon DR, Brinson MM (2009) Coastal wetlands: an integrated ecosystem approach. Elsevier, Amsterdam
Rivera-Monroy VH, Twilley RR, Bone D, Childers DL, Coronado-Molina C, Feller IC, Herrera-Silveira J, Jaffe R, Mancera E, Rejmankova E, Salisbury JE, Weil E (2004) A conceptual framework to develop long-term ecological research and management objectives in the wider Caribbean region. Bioscience 54:843–856
Rivera-Monroy VH, Twilley RR, Mancera E, Alcantara-Eguren A, Castañeda-Moya E, Casas-Monroy O, Reyes F, Restrepo J, Perdomo L, Campos E, Cotes G, Villoria E (2006) Adventures and misfortunes in Macondo: rehabilitation of the Cienaga Grande de Santa Martaa lagoon complex, Colombia. Ecotropicos 19:72–93
Rivera-Monroy VH, Castañeda-Moya E, Barr JG, Engel V, Fuentes JD, Troxler TG, Twilley RR, Bouillon S, Smith TJ, O'Halloran TL (2013) Current methods to evaluate net primary production and carbon budgets in mangrove forests. In: DeLaune RD, Reddy KR, Megonigal P, Richardson C (eds) Methods in biogeochemistry of wetlands, Soil Science Society of America Book Series, pp 243–288
Rivera-Monroy VH, Osland MJ, Day JW, Ray S, Rovai AS, Day RH, Mukherjee J (2017) Chapter 11. Advancing mangrove macroecology. In: Rivera-Monroy VH, Lee SY, Kristiansen E, Twiley RR (eds) Mangrove ecosystems: a global biogeographic perspective structure, function and ecosystem services. Springer, New York, pp XX–XX
Ronnback P (1999) The ecological basis for economic value of seafood production supported by mangrove ecosystems. Ecol Econ 29:235–252
Simard M, Zhang KQ, Rivera-Monroy VH, Ross MS, Ruiz PL, Castaneda-Moya E, Twilley RR, Rodriguez E (2006) Mapping height and biomass of mangrove forests in Everglades National Park with SRTM elevation data. Photogramm Eng Remote Sens 72:299–311
Simard M, Rivera-Monroy VH, Mancera-Pineda JE, Castaneda-Moya E, Twilley RR (2008) A systematic method for 3D mapping of mangrove forests based on shuttle radar topography mission elevation data, ICEsat/GLAS waveforms and field data: application to Cienaga Grande de Santa Marta, Colombia. Remote Sens Environ 112:2131–2144
Sippo JZ, Maher DT, Tait DR, Holloway C, Santos IR (2016) Are mangroves drivers or buffers of coastal acidification? Insights from alkalinity and dissolved inorganic carbon export estimates across a latitudinal transect. Glob Biogeochem Cycles 30:753–766
Spalding EA, Kainuma M, Collins L (2011) World atlas of mangroves. ITTO, Kuala Lumpur
Stewart BT, Santos IR, Tait DR, Macklin PA, Maher DT (2015) Submarine groundwater discharge and associated fluxes of alkalinity and dissolved carbon into Moreton Bay (Australia) estimated via radium isotopes. Mar Chem 174:1–12

Twilley RR, Day JW (2013) Chapter 7. Mangrove wetlands. In: Day JW, Crump BC, Kemp WM, Yanez-Arancibia A (eds) Estuarine ecology. Wiley-Blackwell, Hoboken, pp 165–202

Twilley RR, Chen RH, Hargis T (1992) Carbon sinks in mangroves and their implications to carbon budget of tropical ecosystems. Water Air Soil Pollut 64:265–288

Twilley RR, Gottfried RR, Rivera-Monroy VH, Zhang W, Montaño M, Bodero A (1998) An approach and preliminary model of integrating ecological and economic constraints of environmental quality in the Guayas River estuary, Ecuador. Environ Sci Pol 1:271–288

Twilley RR, Castaneda-Moya E, Rivera-Monroy VH, Rovai AS (2017) Chapter 5. Productivity and carbon dynamics in mangrove wetlands. In: Rivera-Monroy VH, Lee S, Kristensen YE, Twlley RR (eds) Mangrove ecosystems: a global biogeographic perspective structure, function and ecosystem services. Springer, New York, pp XX–XX

Utz RM, Fitzgerald MR, Goodman KJ, Parker SM, Powell H, Roehm CL (2013) The National Ecological Observatory Network: an observatory poised to expand spatiotemporal scales of inquiry in aquatic and fisheries science. Fisheries 38:26–35

Walters BB, Ronnback P, Kovacs JM, Crona B, Hussain SA, Badola R, Primavera JH, Barbier E, Dahdouh-Guebas F (2008) Ethnobiology, socio-economics and management of mangrove forests: a review. Aquat Bot 89:220–236

Chapter 2
Mangrove Floristics and Biogeography Revisited: Further Deductions from Biodiversity Hot Spots, Ancestral Discontinuities, and Common Evolutionary Processes

Norman C. Duke

2.1 Introduction

Mangroves are dominant coastal shoreline habitats of tropical and subtropical regions of the world (Tomlinson 2016; Spalding et al. 2010). They comprise chiefly flowering trees and shrubs uniquely adapted to tidal conditions and a special combination of factors that influence coastal and estuarine shorelines, like seawater, periodic inundation and exposure, waves and wind, strong currents and runoff, and fine sediments (Duke et al. 1998). In such settings, mangroves take the form of distinctly vegetated and often densely structured habitats of verdant closed canopies that clad coastal margins and estuaries of the tropics and subtropics.

In tropical waters, mangrove stands are often positioned between two of the world's iconic ecosystems—coral reefs and tropical rainforests. These are each, biota-structured ecosystems of tropical shorelines, intimately linked by integrated roles in coastal and estuarine ecosystem processes. However, such links are threatened by human development and habitats like mangroves are in serious decline worldwide (Duke et al. 2007). These habitats are also further dependent on mobile biota uniquely adapted to the habitats' unusual and often dramatic physicochemical gradients. Developed over millennia, these linked and mutual relationships have become vital to the survival of each species (e.g., Mumby et al. 2004).

Mangroves are recognized further for their unique morphological and physiological adaptations for coping with salt, exposed conditions, saturated soils, and regular tidal inundation along this unique estuarine and coastal marine habitat. Some of these distinctive attributes include: exposed breathing roots above ground; extra stem support structures; salt-excreting leaves; low water potentials and high intracellular salt concentrations for maintaining favorable water relations in saline environments; and

N.C. Duke (✉)
James Cook University TropWATER Centre, Townsville, QLD, Australia
e-mail: norman.duke@jcu.edu.au

V.H. Rivera-Monroy et al. (eds.), *Mangrove Ecosystems: A Global Biogeographic Perspective*, https://doi.org/10.1007/978-3-319-62206-4_2

their unusual viviparous, water-dispersed propagules. Such a unique combination of attributes has enabled mangroves to dominate tropical shorelines worldwide. In this chapter, the floristics and biogeography of these amazing plants and habitat has been re-evaluated; this includes taking a fresh look at some of the key factors responsible for their unusual diversity and distribution worldwide.

2.2 Factors Influencing Mangrove Distributions

The genetic composition of todays' mangrove flora, while clearly subject to present-day climatic and geographical conditions (Duke 1995; Duke et al. 1998; Triest 2008), appears to be largely relict. Species are where they are today because of past events and circumstances. As such, todays' distributional patterns cannot always be explained by current-day deterministic factors alone. Moreover, each mangrove plant type is influenced by a combination of attributes including: individual physiologies, ecology, dispersal ability, propagule buoyancy and longevity, geological circumstances, evolutionary rates, and the genesis of each taxon.

The key criteria are summarized in the following ten generalized factors (ordered in 3 groupings) as the ones mostly influencing the biogeography and evolution of mangroves. These criteria are based on the observations of a number of authors, including Tomlinson (2016), Duke et al. (1998), and Triest (2008), and modified further here.

2.2.1 *Floristics and Biogeography*

1. *Diversity of Plant Types Makeup Mangrove Habitat.* The species numbers of 80 taxa are not necessarily high, but mangrove taxa are present in a broad cross-section of plant family lineages (Tomlinson 2016). They chiefly comprise a small but diverse selection of shrub and tree species from 18 plant families with around 69 species and 11 hybrid intermediates (Table 2.1). There are 32 genera represented with all but one being a flowering plant (Table 2.2).
2. *Variable Ranges for Different Species.* The distributional ranges of mangrove species vary considerably (Spalding et al. 2010; Duke 2011, 2013, 2014a). While some extant mangrove taxa occur worldwide, others have more local distributional ranges. The overall high concentration of mangrove species in the Indo-Malesian area today defines the chief diversity hot spot for mangroves, comparable with other shallow water, tropical marine habitats, like seagrass and reefal coral species (Hoeksema 2007). And likewise, a further secondary hot spot for mangroves occurs in the Caribbean Central American area. A general feature of mangrove taxa is their high incidence of hybrid species (as mostly infertile, intermediate forms), along with sibling species (species that are unusually similar in morphology and genetic makeup).

Table 2.1 Mangrove species of the world: Families and genera shaded are exclusively mangrove; Genera marked with an asterisk have been classified comprising their own family, named, Diospyraceae, Barringtoniaceae, Aegicerataceae, Nypaceae, Aegialitidaceae, and Pellicieraceae, respectively; Species underlined refer those of the Atlantic East Pacific region. Species in bold occur naturally in both regions; Reported hybrid taxa are shown with an "X" before the species name

Families with Mangroves	Family Relatives of Mangrove Taxa	Family Genera	Mangrove Genera	Non-Mangrove	Mangrove Spp.	Species of Mangroves
Acanthaceae	Black-eyed Susan, Shrimp Plants	250-300	*Acanthus*	30	2	*Acanthus ebracteatus* *Acanthus ilicifolius*
(ex Verbenaceae; or Avicenniaceae)	Grey Mangroves	1	***Avicennia****	0	8	*Avicennia alba* *Avicennia bicolor* *Avicennia germinans* *Avicennia integra* *Avicennia marina* *Avicennia officinalis* *Avicennia rumphiana* *Avicennia schaueriana*
Arecaceae	Palms	200	*Nypa**	0	1	*Nypa fruticans*
Bignoniaceae	Trumpet, Tulip Tree, Jacarandas	120	*Dolichandrone*	9	1	*Dolichandrone spathacea*
			Tabebuia	245	1	*Tabebuia palustris*
Fabaceae (or Bombaceae)	Baobab, Balsa, Kapok, Durian	31	*Camptostemon*	0	2	*Camptostemon philippinense* *Camptostemon schultzii*
	Cassia, Tamarind, Legumes	150	*Cynometra*	70	1	*Cynometra iripa*
(or Caesalpiniaceae)			*Mora*	19	1	*Mora oleifera*
			Muellera	3	1	*Muellera moniliformis*
Combretaceae	Combretum, Quiqualis	20	*Lumnitzera*	0	3	*Lumnitzera littorea* *Lumnitzera racemosa* *Lumnitzera* X *rosea*
			Laguncularia	0	1	*Laguncularia racemosa*
			Conocarpus	0	1	*Conocarpus erectus*
Ebenaceae	Ebony, Persimmons	3	*Diospyros**	400	1	*Diospyros littorea*
Euphorbiaceae	Castor Oil, Spurges	300	*Excoecaria*	35-40	1	*Excoecaria agallocha*
Lecythidaceae	Brazil Nuts	15	*Barringtonia**	40	1	*Barringtonia racemosa*
Lythraceae	Crepe Myrtle, Henna, Cuphea	25	*Crenea*	30	1	*Crenea patentinervis*
			Pemphis	1	1	*Pemphis acidula*
(or Sonneratiaceae)	*Duabanga*	2	*Sonneratia**	0	9	*Sonneratia alba* *Sonneratia apetala* *Sonneratia caseolaris* *Sonneratia griffithi* *Sonneratia* X *gulngai* *Sonneratia* X *hainanensis* *Sonneratia lanceolata* *Sonneratia ovata* *Sonneratia* X *urama*
Malvaceae	Silvery leaf trees, Hibiscus Cocoa, Kola, Bottle Trees	70	*Brownlowia*	2	1	*Brownlowia tersa*
			Heritiera	29	2	*Heritiera fomes* *Heritiera littoralis*
(or Sterculiaceae)			*Pavonia*	1	2	*Pavonia paludicola* *Pavonia rhizophorae*
Meliaceae	Mahogany, Rosewood	50	*Xylocarpus*	1	2	*Xylocarpus granatum* *Xylocarpus moluccensis*
Myrsinaceae	Turnip-wood, Mutton-wood	35	*Aegiceras**	0	2	*Aegiceras corniculatum* *Aegiceras floridum*
Myrtaceae	Eucalyptus, Bottlebrush, Guavas	80-150	*Osbornia*	0	1	*Osbornia octodonta*

(continued)

Table 2.1 (continued)

Pellicieraceae	Tea, Camellia, Franklinia	1	*Pelliciera**	0	1	*Pelliciera rhizophorae*
Plumbaginaceae	Sea Lavender, Thrifts	10	*Aegialitis**	0	2	*Aegialitis annulata* *Aegialitis rotundifolia*
Pteridaceae	Ferns	35	***Acrostichum***	**0**	**3**	***Acrostichum aureum*** *Acrostichum speciosum* *Acrostichum danaeifolium*
Rhizophoraceae	Crossostylis, Cassipourea	16	*Bruguiera*	0	7	*Bruguiera cylindrica* *Bruguiera exaristata* *Bruguiera gymnorhiza* *Bruguiera hainesii* *Bruguiera parviflora* *Bruguiera* X *rhynchopetala* *Bruguiera sexangula*
			Ceriops	0	5	*Ceriops australis* *Ceriops decandra* *Ceriops pseudodecandra* *Ceriops tagal* *Ceriops zippeliana*
			Kandelia	0	2	*Kandelia candel* *Kandelia obovata*
			Rhizophora	0	12	*Rhizophora* X *annamalayana* *Rhizophora apiculata* *Rhizophora* X *brevistyla* *Rhizophora* X *harrisonii* *Rhizophora* X *lamarckii* *Rhizophora mangle* *Rhizophora mucronata* *Rhizophora racemosa* ***Rhizophora samoensis*** *Rhizophora* X *selala* *Rhizophora stylosa* *Rhizophora* X *tomlinsonii*
Rubiaceae	Coffee, Gardinia, Quinine	500	*Scyphiphora*	0	1	*Scyphiphora hydrophylacea*

Table 2.2 Distribution of family, genera, and species in the two subregion hot spots of the world as the Indo West Pacific (IWP) and the Atlantic East Pacific (AEP); Summarized counts from Table 2.1

	IWP	AEP	Worldwide
Families	17	9	18
Genera	24	11	32
Species + hybrids	63	19	80
Hybrids	9	2	11
Species – Hybrids	54	17	69
Subspecific taxa	4	1	5

3. *Phylogenetic Relationships within Key Mangrove Genera.* It is important to have detailed information about the relationships amongst species of polyspecific mangrove genera (like *Avicennia, Bruguiera, Rhizophora, Sonneratia*). This is essential for identifying and understanding the genetic lineages and relationships amongst closely related taxa (Duke 1992, 1995, 2010; Duke et al. 2002; Duke

and Ge 2011; Lo et al. 2014). Such knowledge is further useful when ranking species by ancestral age, and in constructing phylogenies. There is also the idea, pursued in this treatment, of possible common patterns existing between genera where each entity may have evolved, diversified, and speciated in response to common overwhelming geophysical circumstances and drivers.

2.2.2 Extant Influencing Factors

4. *Dispersal by Sea.* Mangroves have mostly, *water-buoyant propagules* dispersed by water currents across seas and between estuaries, driven by wind, waves, tides, and ocean circulation (Rabinowitz 1978; Lo et al. 2014). Over half the mangrove species have unusually, well-developed viviparous propagules instead of seeds. By contrast, vivipary is rare in other plant ecotypes. But, one unanswered question remains: how does vivipary influence dispersal? It is clear that long distance dispersal (compare with Nathan et al. 2008) in mangroves is fundamentally important and well utilized, but there appear to be finite limits to both seed and viviparous propagule viability and longevity in transit (e.g., Drexler 2001). This is most evident however in the absence of native mangrove species in key locations, like the central Pacific Hawaiian Islands, prior to their introduction by people (Allen 1998).
5. *Topographic Elevation Range.* Mangroves are naturally restricted to a very *narrow elevation range* between mean sea level and the highest water levels of spring tides (Duke 1992; Duke et al. 1998). At lower limits of the range, they appear constrained by tidal inundation frequencies mostly less than 50% of the time, dependent on possible limitations during submergence with gas exchange of exposed roots. At higher intertidal margins, species appear constrained by desiccation as well as competition for light from supratidal upland specialists. To cope with conditions in the intertidal zone, mangrove plants have well-recognized specialized abilities to deal with salt and saturated airless soils, by their distinctive exposed air breathing root surfaces and lenticels.
6. *Ecological and Climate Conditions.* Mangroves are *constrained worldwide by extant climatic conditions*. Generally, mangroves are confined to sheltered tropical coastlines where mean monthly seawater temperatures drop only to around 20 °C (Duke et al. 1998). Based on observed different latitudinal limits (e.g., Duke 2006), there appear to be species–specific differences in temperature tolerances. However, a recent assessment has challenged this view, reporting no differences between widely distributed genera, *Avicennia* and *Rhizophora* (Quisthoudt et al. 2012). This view is however at odds with present understandings, and it may be limited by not accounting for observed variations in temperature tolerances amongst and between species; as an example, a notable cold-tolerant variant is known for *Avicennia marina*, var. *australasica* in south eastern Australia, and there are a number of other examples (Duke 1992, 1995, 2006). These observations emphasize the great importance of knowing the

ecophysiological tolerances of each genotype, especially those at distributional endpoints. Furthermore, mangrove taxa are often restricted further to areas of higher rainfall, as well as estuaries of larger riverine catchments (Duke et al. 1998). Mangroves can be more abundant and diverse in areas of unconsolidated sediments, like the mouths of larger estuaries, while also growing on hard substrates of rock and coral.

2.2.3 Dispersal Pathways and Evolutionary Processes

7. *Disjunctions of Distribution.* There are several major genetic *discontinuities amongst present-day mangrove distributions* worldwide (Duke 1995; Triest 2008; Takayama et al., 2013). Such disjunctions are marked by extant genetic anomalies where no current dispersal barriers exist. Disjunctions may be characterized by: the presence or absence of taxa, or the presence of sibling taxa on either side of an anomaly. It is also likely that relationships amongst taxa may be proportional to the magnitude and age of the disjunction, as well as to the specific characters of each plant involved—especially in their dispersal ability, propagule buoyancy, and longevity; also cs. Nathan et al. (2008). For these reasons, an ancient primary disjunction is recognized separating the Atlantic East Pacific (AEP) and Indo West Pacific (IWP), with distinct species, common genera, and separate families; while "younger" disjunctions, like that between SE Asia and Australia, is marked by sibling species, separate intra-specific forms, with notable common genera and families.
8. *Distributional Fluctuations.* Where climate conditions and sea levels change, the *distributional extent of individual mangrove species both expand and contract* (Duke 1995; Duke et al. 1998). This indicates that while mangroves are able to disperse and colonize unoccupied shorelines, they also at times have become extinct along other shorelines.
9. *Appreciable Geological Age.* Mangroves are known to exist from ancient *fossil records of macrofossils and pollen* dating back to the mid Cretaceous era, possibly up to around 100 million years ago (Churchill 1973; Plaziat et al. 2001). This is the case in varying degrees for extant common genera today. Where recognizable fossil taxa match existing mangrove plants, this reveals a genetically conservative situation where the presence of persistent genotypes as common species is indicative of relatively slow rates of genetic mutation and diversification over tens of millions of years. As such, species numbers today are indicative of the realised speciation rates for each lineage.
10. *Continental Drift Theory and Mangrove Evolution.* The dispersal and speciation of key mangrove angiosperm lineages took place at the same time as the *breakup of the massive supercontinent of Gondwana (*Duke et al. 2002; Triest 2008). Over the last 55 million years at least, large continental fragments have moved around the globe ferrying plants and animals with them. This explains

why plants with both low and high dispersal abilities may have comparable distributional ranges. Over time, oceans and seaways that once existed have closed, while land barriers have both emerged and disappeared. Understanding the occurrence of these barriers and their effectiveness, helps explain how water-borne mangrove plants might have dispersed to their current day distributional endpoints. Furthermore, in the Indo West Pacific (IWP), former Gondwanan southern land mass fragments, including Africa, India and Australia, have migrated dramatically northward to separately join with Asia. These massive movements must surely have seriously influenced the diversity and distribution of mangroves and other species throughout the region.

The relative importance and influence of each criterion explained here (1–10) is entirely species-dependent. And, for this treatment, criteria 1, 2, and 3 rely on best available *taxonomic discrimination* along with the most up to date records of worldwide *distributions*. Criterion 4 concerns dispersal by propagules, which can be overrated for long-distance dispersal if data on propagule buoyancy and longevity for individual taxa are unavailable. And, while vivipary in mangroves may enhance *long-distance dispersal*, its key function and value lies in its enhanced *success* for essential local *re-establishment, rejuvenation, and habitat turnover*. In all cases, dispersal ability must always be considered finite and limited. Criteria 4, 5, and 6, describe the predominant and major *physiological constraints* that particularly influence within-site species distributions, based chiefly on temperature, moisture, and tides. Criteria 7, 8, 9, and 10 all relate to biological *evolutionary processes in response to historical changes in geological and climatic conditions*, where these influence changes to genetic makeup with mutations and occasional speciation.

This treatment considers all these relevant, prominent features of each species entity by taking a closer look at individual occurrence records. Summary information is presented for each mangrove genus, listing all species and hybrids (Appendix 1). Detailed descriptions that accompany these observations include diagnostic characters along with current distribution maps of all mangrove taxa, as displayed in the e-book field guide, ***World Mangrove iD*** (Duke 2013, 2014a). Other extant influencing factors have been explained further and summarized by Duke et al. (1998).

2.3 Floristics and Distribution

Mangrove plants collectively form an ecological type rather than a genetic entity with a singular genetic makeup for coping with intertidal life. Amongst the 32 genera with mangrove species, each entity has separately refined its own obligate specialized attributes for the tidal wetland habitat. These features are manifest in entities with often distinctive anatomy, physiology, and functionality. Each genus has varying numbers of representatives for each of the 18 plant families (see Table 2.1; Duke 2011, 2014a). And, all but one are flowering plants; the odd one is

Acrostichum, the fern family Pteridaceae. Their collective occupation of the upper tidal zone is considered a relatively recent geological occurrence, mostly post Cretaceous. Otherwise, mangrove habitat is comparable to rainforests, with all families present in mangroves being present also in tropical rainforests.

Of these 32 plant genera, 17 are exclusive to mangrove habitat, including: *Acrostichum, Aegialitis, Aegiceras, Avicennia, Bruguiera, Camptostemon, Ceriops, Conocarpus, Kandelia, Laguncularia, Lumnitzera, Nypa, Osbornia, Pelliciera, Rhizophora, Scyphiphora,* and *Sonneratia*. The other 15 genera are less specialized, with both mangrove and non-mangrove entities present, including, *Acanthus, Barringtonia, Brownlowia, Crenea, Cynometra, Diospyros, Dolichandrone, Excoecaria, Heritiera, Mora, Muellera, Pavonia, Pemphis, Tabebuia,* and *Xylocarpus*.

As an example, *Pemphis* has a widespread IWP species, and a sole upland species located inland as an isolated population on the island of Madagascar (Tomlinson 2016). While others, like *Brownlowia, Diospyros*, and *Tabebuia*, have single mangrove species and a greater number of upland species (Duke 1992). In addition, unusually widespread and distinct hybrid intermediates are reported in five genera including, *Bruguiera, Lumnitzera, Sonneratia, Rhizophora*, and *Avicennia* (Huang et al. 2014; note that the recently reported *Avicennia* hybrid lacks morphological description). Overall, the total number of mangrove species in each genus is relatively low, being one or two. For relatively larger genera, the number of mangrove species plus hybrids worldwide does not exceed 12.

These relatively low levels of diversity are believed to be the result of the harsh and saline growth conditions present in intertidal habitats. The extreme conditions favor high levels of optimized efficiency for the survival and evolution of mangrove inhabitants (cs. Provine 2004). Such broadly defined factors support the underlying tenet in this treatment, which states that the resilience and survival of each species is largely dependent on its individual distribution, dispersal capabilities, levels of taxonomic divergence, and phylogeny. In this way, evolving local and regional environmental/ecological factors are expected to have a key role in defining the genetic characteristics that ultimately define all mangrove taxa known today.

All mangrove species in the world today are considered in this review, along with their current distribution maps, showing records of living, introduced, and fossil records (Duke 2013, 2014a; see Appendix 1). These maps provide the latest information available on the distribution of each entity. But, despite such outcomes, the description and characterization of mangrove plant types remains incomplete. Hence, there is an ongoing need to revise and update each entity, especially since occurrences are likely to change (Duke et al. 2007). There might also be new undescribed taxa. For example, with *Pelliciera*, while currently described as one species (Tomlinson 2016), it appears this taxon has subspecific forms within its limited Central American range (Castillo-Cardenas et al. 2005; plus personal observations). There are also notable taxonomic discrepancies and gaps in established larger genera, like *Rhizophora* in northern South America and West Africa (Ceron-Souza et al. 2010; Triest 2008).

2.4 Dispersal Pathways and Evolutionary Processes

Prior assessment of mangrove global biogeography (including, Duke 1992, 2011, 2014b; Tomlinson 2016; Triest 2008) shows a general consensus for two hot spots disproportionately spanning sub-regions of the Indo West Pacific (IWP) and Atlantic East Pacific (AEP). These are shown in Fig. 2.1 along with all distribution information summarized from the 70 species maps (excluding hybrids). While the IWP hot spot, also referred to as the Indo Australian Archipelago biodiversity hot spot (Cowman and Bellwood 2013), is about three times more diverse, there are significant species concentrations in the AEP spanning the Caribbean American Isthmus. Both hot spots form gradients in species richness driven by estuarine diversity along respective tropical shorelines. In this way, estuarine systems influenced by individual catchment runoff (Duke et al. 1998), act as diversity nodes to form species enclaves and gradients connecting intertidal habitats. At its center, the IWP hot spot hosts 54 mangrove species, along with around 500 species of coral and 5000 species of fish (Hoeksema 2007). By contrast, the AEP hot spot has 17 mangrove species and proportionately fewer associated shallow water marine species.

The genesis of mangrove hot spots seems almost certainly due to past changes in continental drift acting in concurrence with speciation and dispersal events driving vicariance and re-association (cs. Duke 1995; Duke et al. 2002). To have such a profound influence, these massive changes in global geography have coincided with

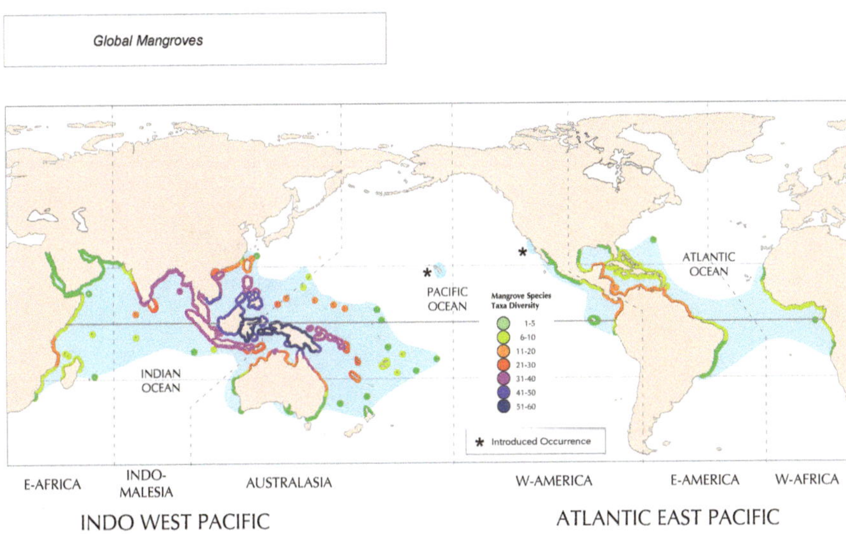

Fig. 2.1 The current world map of the distribution of mangroves (*blue shading*) showing diversity as numbers of specific taxa (species and nominal hybrids). Also identified are two outlying areas where mangrove taxa have been introduced in recent decades. Overall, there is notable separation in eastern and western global bioregions with two disproportionate diversity hot spots: the dominant Indo Australian Archipelago area north of Australia; and a lesser one, the Caribbean American Isthmus area north-west of South America

the evolution of early angiosperms to derive the dominant mangrove entities known today. Over time, the global displacement of continental fragments has changed dramatically to form today's distribution maps with biota occurring for the most part constrained according to current day circumstances. But, there are notable unexplained patterns sparking debate and speculation about their relevance to the evolution of mangroves.

The approach in this article has been to further review current evidence, identifying each vital clue to understanding how each mangrove entity might have evolved. The evidence (also see Fig. 2.1) includes: (1) areas of species occurrence and richness; (2) hot spots associated with clines in species diversity; and (3) unexplained discontinuities across hot spot gradients in the absence of apparent current-day influences on gene flow.

2.5 Dispersal Barriers

Extant mangrove distributions are collectively defined by a number of tangible barriers blocking water-borne gene dispersal (Duke et al. 1998; Duke et al. 2002; Triest 2008), including: (a) cold water (<20 °C in summer) mostly in higher latitudes; (b) land mass, mostly longitudinally applied; and (c) broad water expanses, also mostly longitudinal. The current locations of these three barrier types are displayed in Fig. 2.2. The effectiveness of land barriers and adverse cold temperatures are each

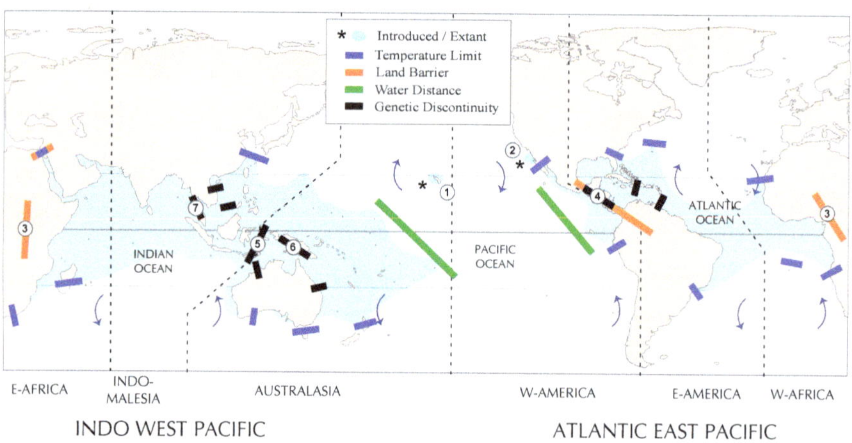

Fig. 2.2 Dispersal barriers of temperature, land, and water compared with genetic discontinuities restricting gene flow around coastal margins of the Atlantic East Pacific and the Indo West Pacific regions (Adapted from Triest 2008): The extant distributional range of mangroves with areas of outlying introduced populations are shown as blue shaded areas; The circumstances surrounding the four barriers (1–4) and three discontinuites (5–7) marked are discussed in the text

relatively self-evident. Mangroves are restricted to mostly tropical latitudes except in deviations created by respective oceanic circulation patterns affecting ocean shoreline occurrences (arrows show major oceanic current gyres).

Ocean water distances can be barriers to buoyant propagule dispersal where widespread mangrove taxa have viviparous, living propagules with distinctly limited survival times at sea (e.g., see Steinke 1986; Drexler 2001). While it is evident that seed-distributing mangrove types are less widespread (no families in common between IWP and AEP regions), it is also curious why species with greatly different survival abilities have comparable global ranges (both IWP and AEP), like *Rhizophora* and *Avicennia*. Their viviparous propagules can last up to 150 days, or less than 15 days, respectively. In the absence of more information on dispersal abilities for each mangrove entity, it suffices to say that water distance can be a barrier if propagules have limited capacity to stay afloat and remain viable. In support of this contention, there are other important clues displayed in two examples of species introductions, where modern day ranges have been artificially extended to isolated previously unoccupied mangrove-suitable habitat. These are evidence of active natural barriers.

In one case (Fig. 2.2, site 1), *Rhizophora mangle* was artificially established during the 1920s in the northern Pacific Hawaiian islands, accompanied by a second introduced species *Bruguiera sexangula* (Allen 1998). Their establishment and subsequent spread among these islands clearly demonstrated the suitability of previously unoccupied habitat. Hence, the only thing preventing their introduction to these islands before this was the ocean water surrounding them—for at least around 5000 years. This is tangible evidence of an existing water distance barrier for mangroves crossing the Pacific Ocean. Only two mangrove species, *Acrostichum aureum* and *Rhizophora samoensis*, and three genera (*Avicennia, Rhizophora, Acrostichum*) occur naturally on both sides of the barrier. And, there are significantly greater lower taxonomic level differences within respective subregions (see Table 2.2). The oddity in this, has been the westward dispersal of *R. samoensis*, which appears only possible naturally if facilitated by an ancient chain of volcanic islands (Schlanger et al. 1981; Schlanger and Premoli-Silva 1981). And, this suggestion has some credibility where there are corresponding distributional records of a small number of shallow water reef fishes across the southwestern Pacific (Cowman and Bellwood 2013). In the Atlantic however, water distances appear much less effective as a barrier. While this ocean has large ocean water distances, there appears to be qualified gene flow between eastern and western shorelines (Takayama et al. 2013). In support of this contention, all seven native species on the African west coast occur also in the Americas (Table 2.3). And, while four genera on the American east coast are not recorded on the African west coast, five common genera occur naturally on either side of the ocean. However, the depauperate diversity of mangrove assemblages in West Africa remains a mystery. It is proposed that this low diversity might reflect the relative stability of this area over the last 40–50 my.

In a second example (Fig. 2.2, site 2), a mangrove outlier population on the North American west coast was established with the introduction of *Avicennia marina* to

Table 2.3 Tentative ranking of mangrove genera (for the relevant species present in mangroves) by age (oldest to youngest, 1–32) based on earliest fossil evidence coupled with distributional extent across global regions and subregions (IWP = Indo West Pacific; AEP = Atlantic East Pacific; E Africa = East Africa; Indo Mal = Indo Malesia; Austral Asia = southern Asia to Australia; W Amer = West America; E Amer = East America; W Africa = West Africa), shown as numbers of species excluding hybrid intermediates. Dated (mya) fossil evidence were described by Plaziat et al. (2001), others (*) were deduced from Appendix 2

				IWP	IWP	IWP	AEP	AEP	AEP
	Mangrove Genera	Fossil Age (mya)	Spp	E Africa	Indo Mal[1]	Austral Asia[1,2]	W Amer[2,3]	E Amer[3,4]	W Africa[4]
1	*Nypa*	55	1		1	1	Introd	Introd	Introd
2	*Acrostichum*	55	3	2	2	2	2	2	2
3	*Rhizophora*	50	6	1	3	4	2	2	2
4	*Avicennia*	50	8	1	4	4	2	2	1
5	*Pelliciera*	50	1				1	1	
6	*Sonneratia*	50	6	1	6	4			
7	*Bruguiera*	50	6	1	5	6		Introd	
8	*Ceriops*	50	5	1	3	3			
9	*Heritiera*	40	2	1	2	1			
10	*Laguncularia*	25–40*	1		Introd		1	1	1
11	*Conocarpus*	25–40*	1			Introd	1	1	1
12	*Lumnitzera*	40*	2	1	2	2		Introd	
13	*Xylocarpus*	40*	2	1	2	2			
14	*Pemphis*	40*	1	1	1	1			
15	*Aegiceras*	40	2		2	1			
16	*Pavonia*	25–40*	2				2	1	
17	*Aegialitis*	25–40*	2		1	1			
18	*Acanthus*	25–40*	2		2	2			
19	*Muellera*	25–40*	1				1	1	
20	*Camptostemon*	10–25*	2		1	1			
21	*Barringtonia*	10–25*	1		1	1			
22	*Scyphiphora*	25–40*	1		1	1			
23	*Osbornia*	10–25*	1		1	1			
24	*Cynometra*	25–40*	1		1	1			
25	*Dolichandrone*	25–40*	1		1	1			
26	*Excoecaria*	25–40*	1		1	1			
27	*Crenea*	25–40*	1					1	
28	*Diospyros*	5–10*	1			1			
29	*Mora*	3–4*	1				1		
30	*Kandelia*	25–40*	2		2				
31	*Brownlowia*	25–40*	1		1				
32	*Tabebuia*	3–4*	1				1		

Notes: Comparison between major regional discontinuities defined in part by: 1. Malay Peninsula & Wallace's Line; 2. Pacific Ocean; 3. Central American Isthmus; 4. Atlantic Ocean. The barriers dividing these area groupings have different causal factors; see Fig. 2.2; mya = millions of years ago

Mission Bay, around 30–50 years ago (Kay 2007). There were no mangroves extant in this area beforehand. The local native mangrove populations of *Avicennia germinans* are currently limited to a few 100 km further south in northern Mexico, presumably constrained by the temperature limited growth of that species. Of course, this needs to be assessed further, but the south east Australian source stock of introduced *A. marina* has greater cold tolerance than the local *Avicennia* species; this concurs with the unusual natural occurrence of *A. marina* at exceptionally high latitude sites in Australia. This shows notable limitations with the conclusion by Quisthoudt et al. (2012) that high latitude limits might be usefully re-evaluated considering temperature tolerances of individual species and their local genotypes. In any case, as shown for Hawaii, in Mission Bay, artificial and recent occupation of vacant habitat has proven to be suitable for mangroves, previously not occupied by natural means. Could this be evidence of different temperature tolerances of different genotypes, and the current effectiveness of a temperature barrier on native mangrove species along the American Northwest coast?

The effectiveness of land barriers is aptly demonstrated in one prime example (Fig. 2.2, site 3; Table 2.3)—the African land mass. This long-established land barrier works together with the Pacific Ocean water barrier to effectively divide the AEP and IWP global regions. For the African land barrier, only one species, *Acrostichum aureum*, occurs in common between east and west coasts. And, at the genus level, while there are three in common, two other Atlantic genera are unknown in the Indian Ocean, and seven Indian Ocean genera are unknown in the Atlantic.

In another example of a land barrier (Fig. 2.2, site 4), the Central American Isthmus (CAI) has been much less effective in blocking gene flow. While eight genera are shared, there are just two genera not shared across the CAI. The diminished effectiveness of this barrier however corresponds with its recent age, with closure of the isthmus land bridge between the Pacific and Atlantic Oceans, 3–4 mya (Coates et al. 1992; Schmidt 2007). This clearly demonstrates why extant barriers must be assessed in their historical context when evaluating their longer-term effectiveness.

In summary, the map in Fig. 2.2, shows the locations of key extant barriers along with those examples of introduced taxa mentioned earlier. The figure also shows the location of a number of potentially ancient barriers (black bars). These appear no longer to function as currently situated, but each is recognized for its otherwise odd and localized genetic discontinuity at differing taxonomic levels.

2.6 Discontinuities and Deductions Surrounding Relict Barriers

There are notable genetic discontinuities within and amongst present-day barriers (Fig. 2.2, sites 4–7 for example). Our knowledge of these instances is based on prior biogeographical accounts (e.g., Tomlinson 2016; Triest 2008; Spalding et al. 2010) along with other detailed assessments of gene flow amongst specific taxa in particular areas (Duke et al. 2002; Takayama et al. 2013; Lo et al. 2014). Such discontinuities are manifest as unexplained distributional patterns where the range of a taxon might end abruptly, or where sibling (closely related) genotypes exist either in separation or coexisting in overlapping distributional ranges. These instances stand out because they are largely unexplained in the context of the present-day barriers described previously. But, where discontinuities might be artifacts of past geological events over 10 mya, these are better considered as valuable evidence from which we might identify taxa at particular dates and circumstances.

One significant discontinuity exists between Indo-Malesian and Australasian subregions (Fig. 2.2, site 5). The boundary of contact is well-recognized as Wallace's Line, marking the collision juncture of Australian and Asian biotas around 20 mya (e.g., Duke et al. 2002). For mangrove plants, this is marked by 21 genera in common, while 2 are restricted to Asia, and one to the south. But, while 35 species (Appendix 1) are shared between subregions, 7 species are partially or wholly restricted to the south including *Aegialitis annulata, Avicennia integra, Bruguiera exaristata, Camptostemon schultzii, Ceriops australis, Ceriops pseudodecandra,* and *Diospyros littorea*. And, 11 others are restricted to the north, *Aegialitis rotundifolia, Aegiceras floridum, Brownlowia tersa, Camptostemon philippinense, Ceriops decandra, Ceriops zippeliana, Heritiera fomes, Kandelia candel, Kandelia obovata, Sonneratia apetala*, and *Sonneratia griffithii*.

Another recognized discontinuity is that between north and south coasts of New Guinea (Fig. 2.2, site 6). This is part of the same collision event that formed Wallace's Line (Duke 1995). While this discontinuity is much "weaker" (=lower taxonomic level differences), it is marked by sibling taxa replacing each other on either coast like *Avicennia alba* and *A. marina*, *Rhizophora apiculata* (Asian form) and *R. apiculata* (Australian form), and *Sonneratia caseolaris* and *S. lanceolata*.

In recent years, a discontinuity between east and west coasts of the Malay Peninsula (Fig. 2.2, site 7) has been identified in a number of detailed genetic studies (Triest 2008; Guo et al. 2016), but its effectiveness depends on the relevant mangrove's dispersal abilities. For dispersal specialists, like *Rhizophora*, local gene exchanges are governed predominantly by current flows (Lo et al. 2014; Wee et al. 2014), while for species with more limited dispersal abilities, like *Ceriops* and *Lumnitzera* there is a notable ancient land barrier effect (Tan et al. 2004; Liao et al. 2007; Su et al. 2006). The circumstances are explained by the partial submergence of an ancient much larger Sund Peninsula whose remnant outcrops now form the Indonesian Archipelago. The submergence event was the result of the massive collision between Australian and Asian land masses as these continental fragments continue to relocate and adjust. The separation of biotas in the Indian Ocean and the South China Sea therefore appear to have much greater isolation than expected. At least nine species (Appendix 1) affected include: *Aegialtis rotundiolia, Ceriops decandra, Ceriops zippeliana, Ceriops tagal, Heritiera fomes, Kandelia candel, Kandelia obovata, Sonneratia apetala,* and *Sonneratia griffithii*. Much of the genetic differences concern variations within individual species.

2.7 Localized Extinction Events and Recovery

Mangrove distributions have varied considerably over more than 50 million years with both notable range extensions and instances of localized extinction (Duke 1995; Plaziat et al. 2001; Triest 2008). Events most likely to have caused such extreme outcomes have almost certainly been associated with changes in climate, reduced rainfall, and cooling temperatures. A key driver for these events could be associated with shifts in continental fragments as they drifted from one climate zone to another. Or, it could be changes to climate zones from altered oceanic circulation patterns and/or atmospheric conditions.

For north–south orientated sections of coastline in particular, temperature increases have driven poleward shifts, while temperature decreases have driven retreat and extinction from higher latitudes. This is shown in global fossil records (Plaziat et al. 2001) with species also becoming extinct from entire regions of warmer temperatures when rainfall decreased dramatically, like the Middle East and the Caribbean–northern South America areas 20–30 mya. The following case studies illustrate some key processes at play.

Consider the widespread IWP taxon, the mangrove palm, *Nypa fruticans* (Appendix 1). It is restricted naturally today to wet, warm areas within the IWP region. However, fossil evidence shows its past distribution included not only both IWP and AEP regions, but that it also sometimes was in sites of notably higher latitude >40° north and south around 50–55 mya. It is significant that in recent times, the species has been reintroduced into the AEP on multiple occasions (Duke 2013, 2014a), demonstrating the current suitability of such locations and massive changes in local climates. Today, the nearest comparable restrictive climate conditions would be those currently present in the Middle East, west of India. In these locations, *Nypa* is excluded despite its proximity to viable, abundant populations further east. And, the dispersal capacity of *Nypa* is apparently reasonably effective with notable scattered populations spread across islands of the North-West Pacific, and further afield.

A number of mangrove species and genera have shown latitudinal range contractions from past fossil distributions (Appendix 1), including, *Acrostichum, Aegiceras, Avicennia, Bruguiera, Ceriops, Heritiera, Pelliciera,* and *Rhizophora*. However, only *Pelliciera* has shown partly comparable reductions in longitudinal range. Curiously, this relates to similar areas of the AEP in each case. Unlike *Nypa*, *Pelliciera* was not quite extinguished in the AEP, but its broad ancestral range was notably reduced to its western limit (Plaziat et al. 2001). Since then, there has been limited recovery east. As noted for *Nypa*, there are few current-day restrictions in climate, so such climate factors apparently appear not to be limiting the species today. However, *Pelliciera* does seem to have serious limitations for long-distance dispersal because of its poorly buoyant propagules. Furthermore, these are heavily consumed by small crabs (Dangremond 2015). These observations demonstrate the diverse factors to be considered along with the species-specific eco-physiological attributes of each species entity.

2.8 Time Line of Geological Events, Dispersal, and Speciation

Instances of speciation, extinction, and dispersal are driven by internal and external factors where a concurrence of events offers preliminary evidence. As such, ages estimated by carbon dating or genetic dating from rates of mutation and divergence might correspond with dates from fossil records and relevant geological events. In this way, three key instances of diversification seem relevant to the dominant genera of mangrove plants namely *Rhizophora* and *Avicennia*: (1) the diversification and dispersal to both AEP and IWP regions from a more or less common point of origin; (2) secondary diversification separately of each lineage in AEP and IWP occurrences; and (3) subsequent further secondary diversification.

The key global geological changes of continental drift that took place during the last 118 million years are shown in Fig. 2.3. Over this period, there were a series of dramatic and progressive global events with each having profound influences on the plants and animals living during these times. The four most significant circumstances (Fig. 2.3) include, Event 1: the separation of Africa and South America (~100 mya) and progressive opening of the South Atlantic Ocean; Event 2: the closure of the Tethys Sea between Africa and Eurasia 25–35 mya; Event 3: the separation of India (70–75 mya) and Australia (~50 mya) with their subsequent collisions

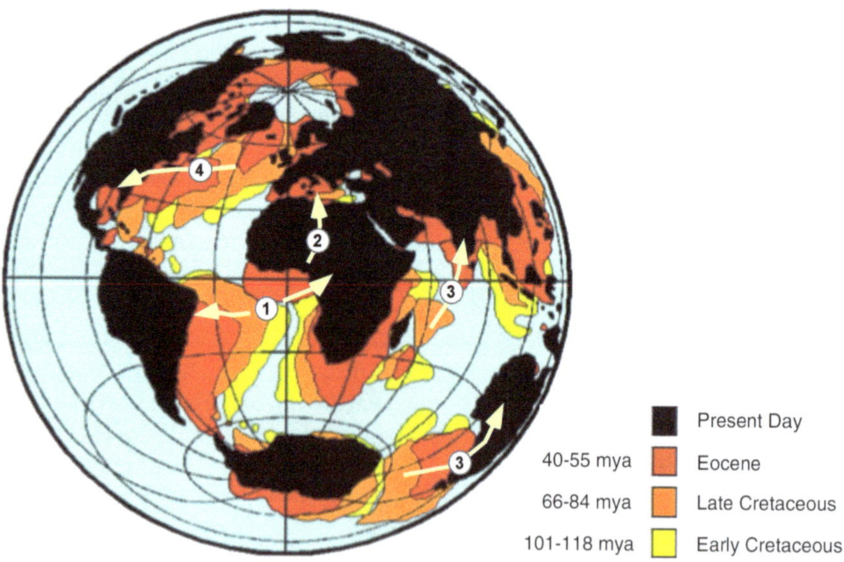

Fig. 2.3 Maps of continental drift in four stages between 118 mya and the present day. Notable events shown with arrows, include: (1) widening of the Southern Atlantic; (2) closure of the Tethys Sea (between Africa and Eurasia); (3) narrowing of the Pacific with northern drifts of India and Australia across the Indian Ocean; and (4) opening of the North Atlantic Ocean

with Asia (~40 mya and ~15 mya, respectively); and Event 4: the opening of the North Atlantic (from ~60 mya onwards).

Fossil evidence (Table 2.3) links mangrove plants with at least the last three circumstances from 55 mya onwards. However, for the first point, it is reasonable to speculate that precursor mangrove flowering plants would almost certainly have benefitted from the creation and progressive widening of the South Atlantic Ocean. The first angiosperm flowering plants appeared nearby around this time. Prior to the Tethys closure, around 40–55 mya, recognized mangrove genera, *Nypa, Acrostichum, Rhizophora, Avicennia, Pelliciera, Sonneratia, Bruguiera, Ceriops, Heritiera,* and *Aegiceras* were present around its shoreline, as well as further afield. It seems likely that these populations were for the most part contiguous. Species like *Nypa* were notably widespread, suggestive of widespread, moist, warm climatic conditions. Then with the closure of the Tethys, widely spread taxa would have been divided as precursors of AEP and IWP regional differences. This was likely to be compounded by populations separated along the northern and southern shorelines of the ancient Tethys Ocean. The subsequent diversification of species was largely dependent on individual circumstances as each genus further diversified and dispersed.

Based on extant and fossil records of *Rhizophora*, a schematic for the phylogeny of genotypes of this genus (Fig. 2.4; Duke et al. 2002) shows likely speciation pathways portrayed as coincident with the geological circumstances described earlier (Fig. 2.3). In consideration of such matters in concordance with known physiological constraints (cs. Figure 2.2), the dispersal and distribution of evolving taxa are described in the following scenario. While AEP and IWP taxa were being separated by the Tethys closure, each was becoming more influenced by region specific events. To better understand these events and circumstances, it is useful to follow likely dispersal paths with known positions of land masses and oceans during periods identified by available fossil evidence (Fig. 2.5; e.g., Plaziat et al. 2001).

In the AEP, *Rhizophora* populations with blunt obtuse leaf tips would have spread along shorelines of the newly forming Atlantic oceans. One group (Fig. 2.5, path #2), as proposed *R. racemosa* progenitors, progressed south from Africa to South America. While the other (Fig. 2.5, path #4), as possible *R. mangle* progenitors may have spread west along the North American coastline before migrating south as conditions progressively cooled during the Eocene. The early progenitor lineages had separated and developed over millions of years, remaining isolated until they were reunited as different species when North and South American land masses merged and formed (Fig. 2.5, point C). The further separation of *R. mangle* and its sibling, *R. samoensis,* is likely to have arisen during this time as populations became isolated on opposing sides of North America which was finally completely isolated after closure of the Caribbean American Isthmus (Fig. 2.5, path #3). During this time, *R. mangle* and *R. racemosa* more or less matched ranges on both sides of the Atlantic.

In the IWP, *Rhizophora* populations with spiked mucronate leaf tips would have spread east along shorelines of the ancestral Indian Ocean as the far western shores of a vast Pacific Ocean. One group (Fig. 2.5, path #10), as proposed *R. apiculata* progenitors, may have progressed eastward to Southeast Asia where founding popu-

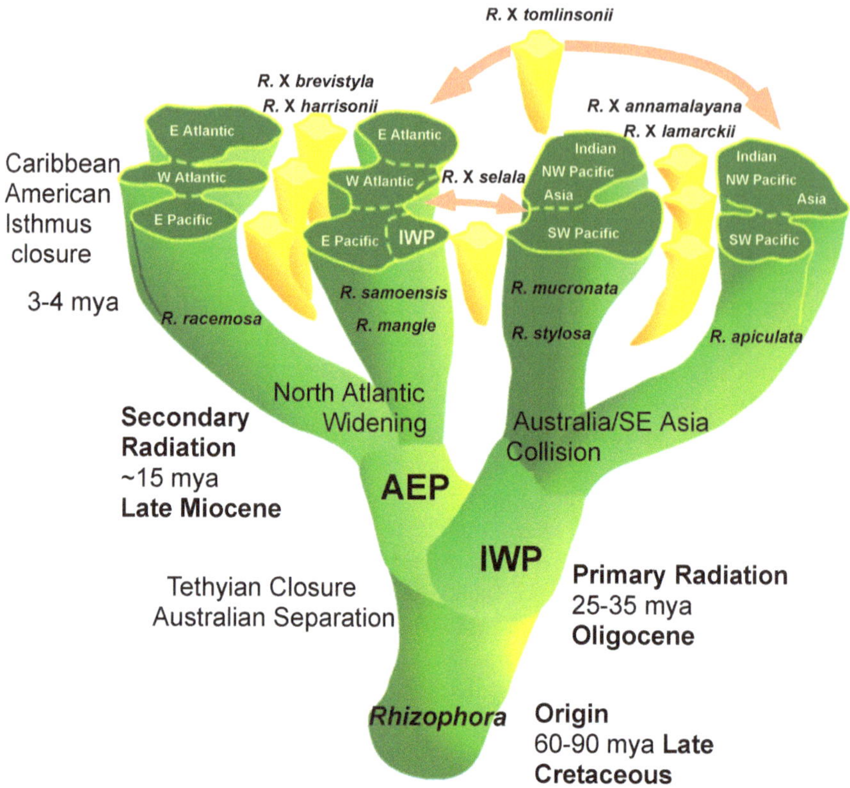

Fig. 2.4 A proposed phylogeny for *Rhizophora* taxa between present day and 118 mya, assuming a single ancestral progenitor possibly created with the emerging Atlantic Oceans (Duke et al. 2002): Seen as a "slice in time," two major radiations with subsequent divergences in extant vicariant populations from global subregions (AEP and IWP) to ocean areas (*green lineages*). Hybrid intermediates (*yellow lineages*) are known between species at each instance of co-existence

lations possibly became isolated on either side of the Sund Peninsula. Meanwhile, the other group (Fig. 2.5, path #6), as *R. mucronata* progenitors, spread south along the East African shoreline where some plants spread to the island of India as it drifted north. The Indian land mass carried populations of *R. mucronata* north to join the Asian shoreline (Fig. 2.5, path #7). Other plants would have spread to the Australian shoreline, and changed slightly to become the *R. mucronata* sibling, *R. stylosa* (Fig. 2.5, path #8). Then, after Australia collided with the Sund Peninsula and Southeast Asia, all diverging IWP species would have been united into the overlapping distributions seen today (Fig. 2.5, point H & I). It is of some interest, that the dispersal and speciation processes were somewhat similar in respective hemispheres where divergent populations appear to be reunited (equalized) after long periods of separation and divergence.

It is of additional great significance, that two distinct and different genera, *Rhizophora* and *Avicennia*, in particular, have comparable global distributions;

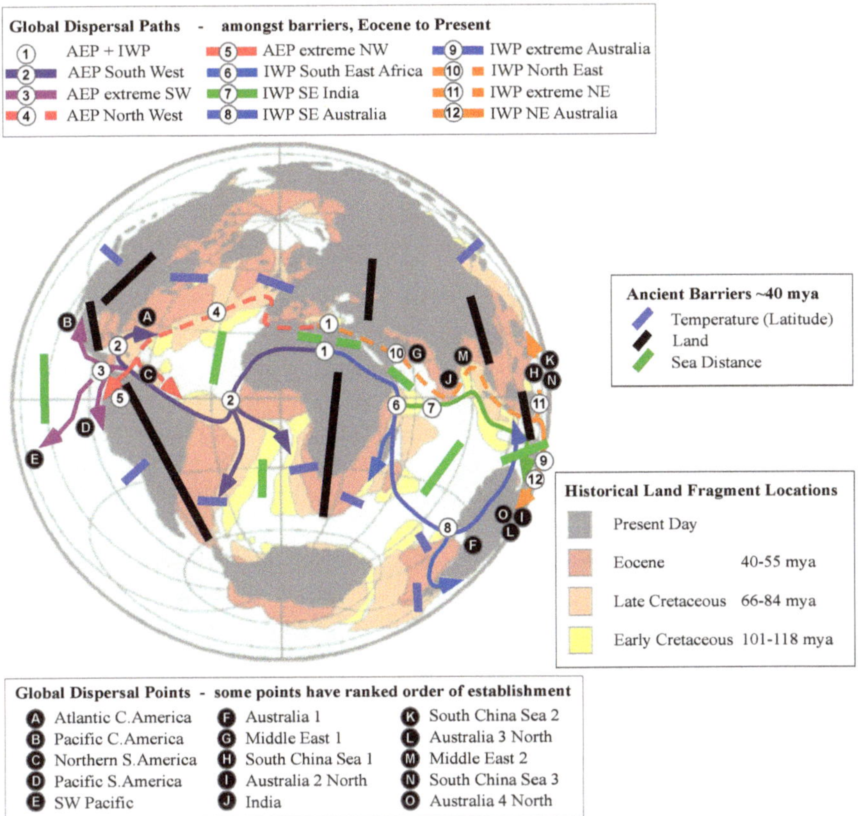

Fig. 2.5 Proposed dispersal pathways taking account of all mangrove species with the help of fossil records; continental locations; land, sea, and temperature barriers; and, overall species affinities

despite acknowledged differences in key functional characteristics, including dispersal ability (cs. Steinke 1986; Drexler 2001). This firmly suggests there might be common overall circumstances influencing dispersal and distribution beyond individual plant differences. So, why do these different genera have similar global distributions? It has been explained earlier that this is despite dispersal being limited by extant land and sea distance barriers. Could it be that the influence of continental drift dominates the evolution of mangrove plants rather than their functional differences? The answer may be explored further by investigating whether common patterns exist in *Avicennia* and other genera, when substituting each genus in the previous scenario for *Rhizophora*. In doing so, some consideration can be made for local influences, particularly, any reported differences in dispersal ability, along with respective levels of taxonomic differentiation.

In the scenario for *Avicennia*, the corresponding AEP species might be as follows: *A. germinans* for *R. racemosa*; *A. schaueriana* for *R. mangle*; and *A. bicolor* for *R. samoensis*. And, matching IWP species would be: *A. marina* for *R. stylosa*; *A. officinalis* for *R. apiculata*; *A. alba* for *R. mucronata*; and two other—*A. integra* appears to a sibling derivative of *A. officinalis* while *A. rumphiana* is a sibling derivative of *A. marina*.

Then, while *Sonneratia* is only recorded in the IWP, the corresponding species for that comparison include *S. alba* for *R. stylosa*; *S. apetala* for *R. apiculata*; *S. caseolaris* for *R. mucronata*; and three others—*S. lanceolata* is a sibling derivative of *S. caseolaris*; but *S. ovata* and *S. griffithii* appear to be outlier derivatives of *S. alba*.

For *Bruguiera* in the IWP only, the corresponding species for that comparison include *B. exaristata* for *R. stylosa*; *B. parviflora* for *R. apiculata*; *B. gymnorhiza* for *R. mucronata*; while *B. sexangula* is a derivative of *B. gymnorhiza*; but *B. cylindrica* and *B. hainesii* are outlier derivatives of *B. parviflora*.

For *Ceriops* in the IWP only, the corresponding species for that comparison include *C. australis* for *R. stylosa*; *C. decandra* for *R. apiculata*; *C. tagal* for *R. mucronata*; while *C. zippeliana* and *C. pseudodecandra* are outlier derivatives of *C. decandra*.

While these comparable groupings and patterns require further validation, they do match the hypothesis of common overall dominance of geophysical influences over individual differences in affected taxa. In making this case, it is not suggested that mangrove taxa have common rates of evolution or common taxonomic diversification, nor that we should underestimate the importance of individual functional differences, like dispersal capabilities, propagule longevity, and phenologies. In fact, it is more likely that with greater awareness of such shared influences, this will allow greater understanding of the deterministic role of individual attributes in the evolution of each taxon.

2.9 Shared Evolutionary Processes and Dispersal Pathways

The first step is to review the initial patterns in distributional occurrences with the known dispersal endpoints for each taxa. All mangrove species (excluding recognized hybrid intermediates) have been assigned to 15 spatiotemporal endpoint groupings, summarized in Table 2.4. Five groups occur in the AEP with two on northern coastlines and three on southern coastlines, notably within its ancient Tethyian context. Similarly, 10 groups occur in the IWP with three on northern coastlines and seven on southern coastlines. Groups with repeated names and number, represent likely overlapping dispersal routes associated with similar extant locations.

To better understand these groupings, each has been located on an Eocene map (Fig. 2.5, starting from red-shaded continents). On this map, ancient dispersal barriers and pathways of the day (~40 mya) have also been tentatively defined (cp. Fig. 2.2). Dominant barriers include temperature (active across latitudes), sea distances (defined by the dispersal and establishment capability of each species), and

Table 2.4 Assignment of 70 extant mangrove taxa (species, subspecies, varieties) to 18 dispersal endpoints grouped west (A–E) and east (F–O) of the ancient Tethys Ocean 40–55 mya (see Figs. 2.5 and 2.6). Each is then further assigned to northern or southern hemispheres, as separating increasingly isolated shorelines. These groups correspond with 12 dispersal pathways, discussed later (see Table 2.5). See Appendix 2 for specific group attributions and estimates of taxa numbers and approximate age (mya)

Site #	Dispersal Endpoints	Description	Taxa	Age mya
A	**Atlantic Central America**	Northern Tethys West	7	25-40
B	**Pacific Central America**	Northern Tethys West	8	3-4
C	**Northern South America**	Southern Tethys West	5	25-40
D	**Pacific South America**	Southern Tethys West	5	3-4
E	**South West Pacific**	Southern Tethys West	1	2
F	**Australia 1**	Southern Tethys East	11	10-25
G	**Middle East 1**	Southern Tethys East	10	25-40
H	**South China Sea 1**	Southern Tethys East	10	25
I	**Australia North 2**	Southern Tethys East	10	5-10
J	**India**	Southern Tethys East	10	40
K	**South China Sea 2**	Southern Tethys East	10	25
L	**Australia North 3**	Southern Tethys East	10	5-10
M	**Middle East 2**	Northern Tethys East	18	25-40
N	**South China Sea 3**	Northern Tethys East	20	10-25
O	**Australia North 4**	Northern Tethys East	19	5-10

land barriers (where they prevent water-borne dispersal). Fossil evidence has helped define the effectiveness and presence of these ancient barriers (cs. Plaziat et al. 2001).

As proposed for the evolution of *Rhizophora* taxa (Fig. 2.4; Duke et al. 2002), these were widely distributed before closure of the Tethys Sea during the latter Eocene, around 45 mya. Using the Eocene map in Fig. 2.5, there were a number of mangrove genera present along the northern and southern shorelines of the Tethys, of Laurasia, and Gondwana, respectively. With the northward movement of Africa, the Tethys closed cutting off western and eastern distributions. The effect would have been to isolate northern and southern populations that separated over time, becoming more isolated entities in AEP and IWP regions. All these observations, and those that follow, are based on known extant and fossil records, in full consideration of individual attributes of each relevant species entity.

Those in the AEP are likely to have dispersed westward in two more or less separate pathways: a southern path along the coasts of West Africa and South America; and, a northern path along the North American east coast, forced south as conditions cooled, and the Atlantic widened.

For the IWP, dispersal eastward had also started out following two isolated pathways: a northern path along the coast to the Middle East and Southeast Asia; and, a southern path along the coasts of East Africa, India, and Australia. To complicate the dispersal outcomes, the latter two land masses with mangrove inhabitants moved rapidly north at different times to collide with the Middle East and Southeast Asia (respectively) during this period.

In both AEP and IWP, there were, at first, partly isolated pathways that independently later reunited toward their respective dispersal extremes. This has led to the consequential genesis of greater diversity and further radiation that now characterizes each region – and, it has led to the creation of the respective hot spots of extant diversity. The greater complexity of geomorphic changes in the IWP is presumed to be the reason for the greater genetic diversity in that region.

Combining these observations, the proposed common dispersal pathways are depicted in 12 key locations of likely isolation (Table 2.5), leading to genetic differentiation and speciation for respective mangrove genera (Figs. 2.5 and 2.6, Appendix 2). These relate also to the 15 extant location endpoints (Table 2.4). The broad assessment of evidence considered for these depictions include, location and age of fossil records; period maps of continental drift; distribution maps of each species; phylogenetic relationships between and among taxa; extant discontinuities; and key barriers of land, water, and climate.

The decision for grouping species in this treatment is based on the idea that instances of genetic divergence and speciation could only have taken place when circumstances resulted in prolonged isolation or during founder events where dispersal from established populations might have reached unoccupied niches suitable for mangrove establishment and diversification. And, that these circumstances would be utilized by all genetic entities present at the time.

2.10 Common Drivers of Diversification and Speciation

While it appears less likely that mutations amongst mangroves taxa have led to genetic differentiation in a process of sympatric speciation (Maynard-Smith 1966), there are other features of these taxa indicative of other recognizable processes of speciation. One is the number of closely related, sibling taxa. While it is notable that some sibling taxa have slightly different ecological preferences, the ranges of these entities often overlap. This implies that the reason for their isolation is no longer applicable, with once separate populations being reunited.

Considering all relevant evidence, it seems the diversification of mangrove plants has largely been driven by three key processes of speciation:

(a) *Allopatric speciation* (Mayr 1970): incidents of isolation by land, sea, or temperature barriers blocking gene flow
(b) *Peripatric speciation* (Provine 2004): founder populations (with their smaller subset of genes) multiplying rapidly into more or less vacant habitat
(c) *Parapatric speciation* (Doebeli and Dieckmann 2003): the reunification of extreme ends of wide, linearly distributed species (as with the broad global shoreline surrounding the Tethys Sea) having notable diversification and genetic drift along its length

The number and diversity of families and genera with representatives specialized for mangrove habitat clearly show that the evolution and speciation of mangrove

Table 2.5 Proposed common dispersal pathways are depicted in 12 key locations of likely isolation, leading to genetic differentiation and speciation for the respective mangrove genera (Figs. 2.5 and 2.6, Appendix 2). These relate to 15 extant location endpoints (Table 2.4). The broad assessment of evidence considered for these depictions include the following: location and age of fossil records; period maps of continental drift; distribution maps of each species; phylogenetic relationships between and among taxa; extant discontinuities; and key barriers of land, water, and climate

Path #	Dispersal pathway	Description	Age mya
1	AEP + IWP	Division of the AEP and IWP with the closure of the Tethys Sea, isolated by the landmass of Africa and Laurasia	40–55
2	AEP South West	AEP South Western pathway from West Africa and across the widening Atlantic to eastern South America; later more or less isolated by the Atlantic but reunited with the North-Western taxa	25–40
3	AEP extreme South West	AEP across northern South America to the western coast, later isolated by the Caribbean American Isthmus	3–4
4	AEP North West	AEP North Western pathway from western Laurasia (Europe) and across the widening North Atlantic to eastern North America; soon isolated by the Atlantic but later reunited with the South-Western taxa	25–40
5	AEP extreme North West	AEP across southern North America to the western coast, later isolated by the Caribbean American Isthmus	3–4
6	IWP South East Africa	IWP South Eastern pathway from northern Africa to eastern Africa into the Indian Ocean; later reunited with Middle Eastern taxa	25–40
7	IWP South East India	IWP from East Africa northward across the Indian Ocean on the Indian landmass to collide with the Middle East; reuniting respective taxa	40
8	IWP South East Australia	IWP from East Africa south and eastward across the Indian Ocean to Australia when breaking away from Gondwana to travel north towards South East Asia to reunite respective taxa	10–25
9	IWP extreme Australia	IWP around Australia from the north coast to isolated founding populations in South East Asia, as Australia collided with the Austral Malesian Archipelago	5–10
10	IWP North East	IWP north eastern pathway from the Middle East to South East Asia	24–40
11	IWP extreme North East	IWP from South East Asia to the South China Sea, to be first isolated by the earlier barrier of the Sund Peninsula (submerged as the Austral Malesian Archipelago), and then later reunited with its submergence	10–25
12	IWP North East Australia	IWP from the Austral Malesian Archipelago to northern Australia, as Australia collided with the Austral Malesian Archipelago	5–10

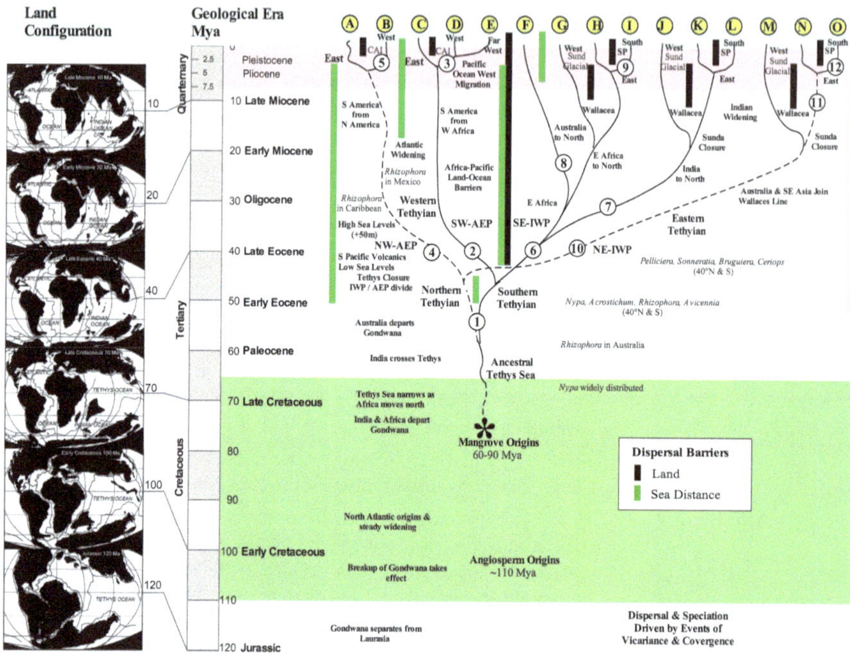

Fig. 2.6 Proposed timeline of common phylogenies of all mangrove species noting global dispersal points (A–O) and dispersal pathways (1–12). Also see Fig. 2.5

plants has taken place on multiple occasions, and independently within various families and genera. As the drivers of diversification and speciation appear strongly influenced by major geological and climatic events, it seems reasonable to conclude that respective phylogenetic pathways of each genetic entity might follow a common structure. The premise is that all taxa have been mostly subjected to the same guiding dominant influences with the same notable events of speciation.

In Fig. 2.6, a common, encompassing phylogenetic tree structure is proposed for the derivation of every mangrove plant taxa. The pathway describes how a small number of progenitors of key plant genera might have diversified from their respective origins during the last 60–90 my. During this time, mangrove taxa appear to have first dispersed east and west, then along northern and southern shores of a diminishing Tethys Sea. The dominant genera had reportedly distributed globally long before closure of the ancient Tethys Sea during the mid-Eocene period, around 45 mya (Plaziat et al. 2001). This is consistent with the presence of dominant mangrove genera in both AEP and IWP regions today. It is notable that similar patterns occur in seagrass and corals (Hoeksema 2007).

Dominant and widespread mangrove genera include *Nypa, Rhizophora, Avicennia*. While *Nypa* has no reported diversification more than its apparently sole species entity, the other two have clear diversification with speciation for common specialized ecological niches, as well as comparable distinct regional forms. This implies these taxa were influenced by a common driver, where the physical environment might have strongly guided and helped form the necessary mangrove traits and features that individually characterize each taxon. While the precise origin of these primary mangrove taxa remains unknown, the separation of South America from Africa would have provided the necessary physical circumstances consistent with the creation of mangrove traits. These circumstances include a slow and progressive increase in salinities as two massive, complex estuarine margins spread apart before becoming the southern Atlantic Ocean. And, this all would have taken place in the formative early years of flowering plants, around 90–110 mya.

It is also curious why so many groups of plants of such antiquity have so few species today, and why these plants have mostly remained unchanged throughout their long fossil record, up to 55 my. A key deduction is that events causing speciation have been rare. And, this concept is consistent with the current proposition of a common phylogenetic structure for the evolution of extant mangrove plants. This is especially the case since each instance of speciation must have been marked by notable and very unusual circumstances. While these events appear to have determined the common phylogenetic tree structure, this does in no way imply that all mangroves originated from the same place, or at the same time. But, it does however provide a means to narrow down the options for determining the individual features of each species and genotype.

For limited range species, like *Mora oleifera* (Duke 2013, 2014a; see Appendix 2), there are a number of important knowledge gaps. This species, restricted to the Pacific coast of Central America, appears to have originated quite recently as there are few if any fossil records. Of course, this observation would be amended if there were fossil evidence, and/or, if this species were found to have unusual dispersal limitations. While there may be other questions about its nearest relatives (nearby, there are 7–9 upland species of *Mora*), it is of some interest that this species has exceptionally large propagules—the largest embryo of any plant. But, it is not really known how buoyant they are, or whether they might be predated upon by small crabs, like those consuming *Pelliciera* propagules (Dangremond 2015). Its limited distributional range however is consistent with either poor dispersal, or that it is newly evolved, or both.

Such questions concern most, if not all, mangrove taxa. So, in the absence of further evidence, it may be concluded that species with limited distributions, might also have recent origins. For example, it was deduced for *Mora oleifera* therefore, that this species was likely to have arisen after closure of the Central American Isthmus, less than 3–4 mya.

Another taxon, the genus *Camptostemon*, appears to have been around only after 10–25 mya, and diversified as two species around 5–10 mya.

Taxa like the genus *Lumnitzera*, are more difficult to place. While the genus is restricted to the IWP, this does imply its post-Tethyian origins, since 40 mya. However, its subsequent diversification and its occurrence in East Africa, does suggest its presence along the Southern Tethys Eastern pathway rather than in the north. There does not appear to be any relevant major dispersal events from north to south.

As such, the two species appear to have diverged with the passage of India north, leaving the other entity to possibly migrate eastward unchanged to Australia.

A summary of estimated ages for mangrove-related representatives from all 32 genera is shown in Table 2.3. While the deductions may be altered and amended in the future as new information comes to hand, the proposed common phylogenetic tree structure does offer a practical working understanding of how and where each mangrove species evolved and dispersed to locations where they exist today.

2.11 Summary and Conclusions

In summary, there are six key deductions and findings raised in this treatment:

1. The mangrove niche is inhabited by around 80 tree and shrub species (including 11 hybrid intermediates) of 32 genera from 17 families worldwide. These mangrove species are widely distributed, but constrained by natural barriers of temperature, water distance, and land masses. For the most part, the effectiveness of each barrier is self-evident, marked by genetic differences or endpoints, as genetic discontinuities.
2. The natural distribution of individual species and genotypes are mostly distinct with few occupying the same range. Taxa are individually constrained by their respective tolerances for key functional factors like temperature, precipitation and evaporation, salinity, soil type, inundation frequency and tidal regimes, as well as individual adaptations in anatomy, physiology, phenology, propagule buoyancy and dispersal longevity, and establishment ability.
3. There are significant instances where extant dispersal barriers have been fully effective in recent times, leaving suitable habitat niches vacant. Because mangrove plants all have buoyant, water-dispersed propagules, the effectiveness of land mass barriers are self-evident. This is complicated for barriers of climate (temperature, rainfall) and water distance because these are dependent on individual capabilities and tolerances of each plant type. But, in all cases, when people intervene by introducing a plant type to habitat previously unoccupied by that genotype, this proves two things: a) that suitable habitat was available; and b) that the introduced entity was unable to disperse and arrive there naturally.
4. There are a number of genetic discontinuities that lack obvious current-day barriers. These can only be explained by past geological circumstances where previous barriers are no longer effective, like Wallace's' line in the IWP. The presence of these occurrences implies there can be appreciable lags in genetic mixing and natural dispersal amongst previously established populations. The relevance and role of each discontinuity is determined by its location; its geological age; and, the level of taxonomic differences in responsible biota.
5. Species distributions worldwide are divided naturally into two global regions of the Indo West Pacific (IWP) and Atlantic East Pacific (AEP), distinguished by relatively high level taxonomic differences in families and genera. Genetic hot

spots are present in each region, but that in the IWP is well recognized for its concurrent and comparable concentrations of mangroves, corals, sea grasses, reef fishes; collectively and individually known for its globally high diversity as the hot spot triangle. These common diversity patterns across different plant and animal taxa demonstrate the overwhelming dominance of geo-physical circumstances over individual functional differences in biota. In this way, plant and animal life appears more to have been shaped and driven by their individual responses to the changing geophysical environment, rather than the other way around.

6. In conclusion, this treatment offers the novel hypothesis that all mangrove species have dispersed, diversified, and evolved following more or less, common, overall phylogenetic pathways formed by the overwhelming geophysical occurrences over the last 100 my. And, it is within these constraints that each taxon displays its individual differences depending on their own circumstances in genesis location and date, as well as their biological capabilities and tolerances in growth and mutation, including their notable attributes of propagule buoyancy, longevity, and establishment success.

Appendix 1

Distribution Maps. Listed are 24 distribution maps (A–X) showing 32 genera with 70 species of mangrove plants. Hybrid species are not included because their ranges for the most part match the overlapping distributions of parental taxa. For more references, specific descriptions, images, and distributional maps of each species, refer to the *World Mangrove* e-book app (Duke 2013, 2014a).

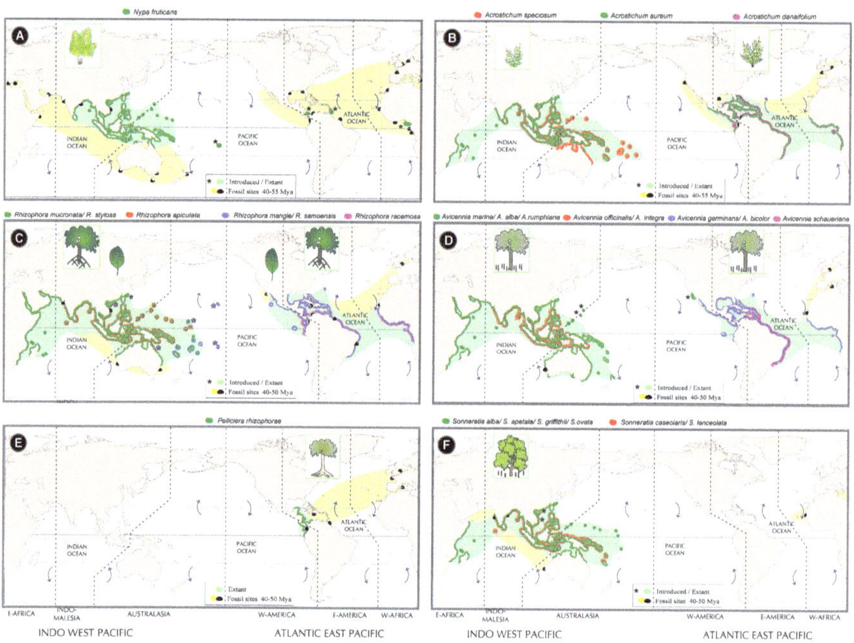

Fig. 2.7 (**A**) Distribution of mangrove species of the genus *Nypa*, showing extant, introduced, and fossil occurrences. Species: *N. fruticans* van Wurmb 1781. (**B**) Distribution of mangrove species of the genus *Acrostichum*, showing extant, introduced, and fossil occurrences. Species: *A. aureum* L. 1753, *A. danaeifolium* Langsdorff and Fischer 1810, *A. speciosum* Willd. 1810. (**C**) Distribution of mangrove species of the genus *Rhizophora*, showing extant, introduced, and fossil occurrences. Species: *R. mucronata* Lam. 1804, *R. stylosa* Griff. 1854, *R. apiculata* Blume 1827, *R. mangle* L. 1753, *R. samoensis* (Hochr.) Salvoza 1936, *R. racemosa* Meyer 1818. (**D**) Distribution of mangrove species of the genus *Avicennia*, showing extant, introduced, and macro fossil occurrences. Species: *A. marina* (Forssk.) Vierh. 1907, *A. alba* Blume 1826, *A. rumphiana* Hallier f. 1918, *A. officinalis* L. 1753, *A. integra* N.C. Duke 1988, *A. germinans* (L.) Stearn 1958, *A. bicolor* Standley 1923, *A. schaueriana* Stapf and Leechman ex Moldenke 1939. (**E**) Distribution of mangrove species of the genus *Pelliciera*, showing extant and fossil occurrences. Species: *P. rhizophorae* Planchon & Triana 1862. (**F**) Distribution of mangrove species of the genus *Sonneratia*, showing extant, introduced, and fossil occurrences. Species: *S. alba* J.Sm. in A.Rees 1816, *S. apetala* Buch.-Ham. 1800, *S. griffithii* Kurz 1871, *S. ovata* Backer 1929, *S. caseolaris* (L.) Engl. 1897, *S. lanceolata* Blume 1851

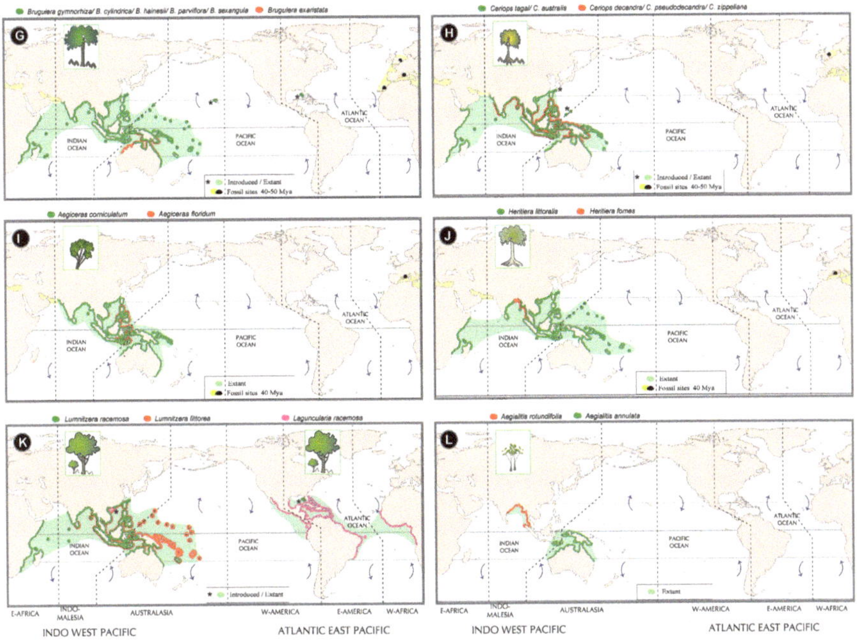

Fig. 2.8 (**G**) Distribution of mangrove species of the genus *Bruguiera*, showing extant, introduced, and fossil occurrences. Species: *B. gymnorhiza* (L.) Savigny ex Lam. & Poiret 1798, *B. cylindrica* (L.) Blume 1828, *B. hainesii* C.G. Rogers 1919, *B. parviflora* (Roxb.) Griff. 1836, *B. sexangula* (Lour.) Poir. 1816, *B. exaristata* Ding Hou 1956. (**H**) Distribution of mangrove species of the genus *Ceriops*, showing extant, introduced, and fossil occurrences. Species**:** *C. tagal* (Perr.) C.B. Robinson 1908, *C. australis* (C.T.White) Ballment, T.J. Sm. & J.A. Stoddart 1988, *C. decandra* (Griff.) Ding Hou 1958, *C. pseudodecandra* Sheue, Liu, Tsai and Yang 2010, *C. zippeliana* Blume 1849. (**I**) Distribution of mangrove species of the genus *Aegiceras*, showing extant and fossil occurrences. Species: *A. corniculatum* (L.) Blanco 1837, *A. floridum* Roemer & Schultes 1819. (**J**) Distribution of mangrove species of the genus *Heritiera*, showing extant and fossil occurrences. Species: *H. littoralis* Aiton 1789, *H. fomes* Buch.-Ham. 1800. (**K**) Distribution of mangrove species of the genus *Laguncularia and Lumnitzera*, showing native and introduced occurrences. Species: *Laguncularia racemosa* (L.) Gaertn. f. 1805; *Lumnitzera racemosa* Willd. 1803, *Lumnitzera littorea* (Jack) Voigt 1845. (**L**) Distribution of mangrove species of the genus *Aegialitis*. Species: *A. annulata* R.Br. 1810, *A. rotundifolia* Roxburgh 1824

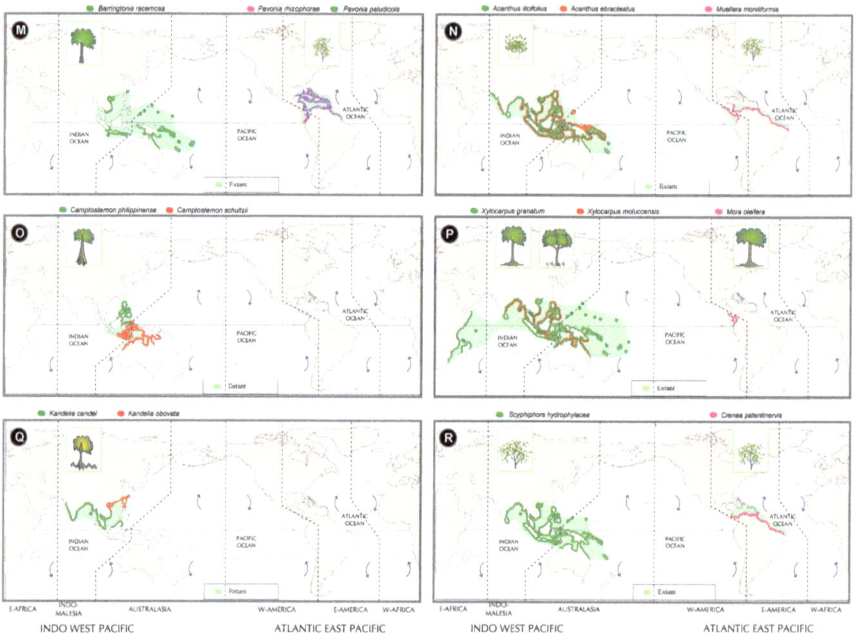

Fig. 2.9 (**M**) Distribution of mangrove species of the genus *Barringtonia and Pavonia*. Species: *B. racemosa* (L.) Spreng. 1826; *P. paludicola* Nicolson ex Fryxell 1989, *P. rhizophorae* Killip. ex Kearney 1954. (**N**) Distribution of mangrove species of *Acanthus* and *Muellera* genera. Species: *A. ebracteatus* Vahl 1791, *A. ilicifolius* L. 1753; *M. moniliformis* L.f. 1781. (**O**) Distribution of mangrove species of the genus *Camptostemon*. Species: *C. philippinense* (Vidal) Becc. 1898, *C. schultzii* Mast. 1872. (**P**) Distribution of mangrove species of the genus *Mora* and *Xylocarpus*. Species: *M. oleifera* (Triana) Ducke 1925; *X. granatum* J.König 1784, *X. moluccensis* (Lam.) M. Roemer 1846. (**Q**) Distribution of mangrove species of the genus *Kandelia*. Species: *K. candel* (L.) Druce 1914, *K. obovata* C.R. Sheue, H.Y. Liu & W.H. Yong 2003. (**R**) Distribution of mangrove species of the genus *Crenea* and *Scyphiphora*. Species: *C. patentinervis* (Koehne) Standl. 1947; *S. hydrophylacea* C.F. Gaertn. 1791

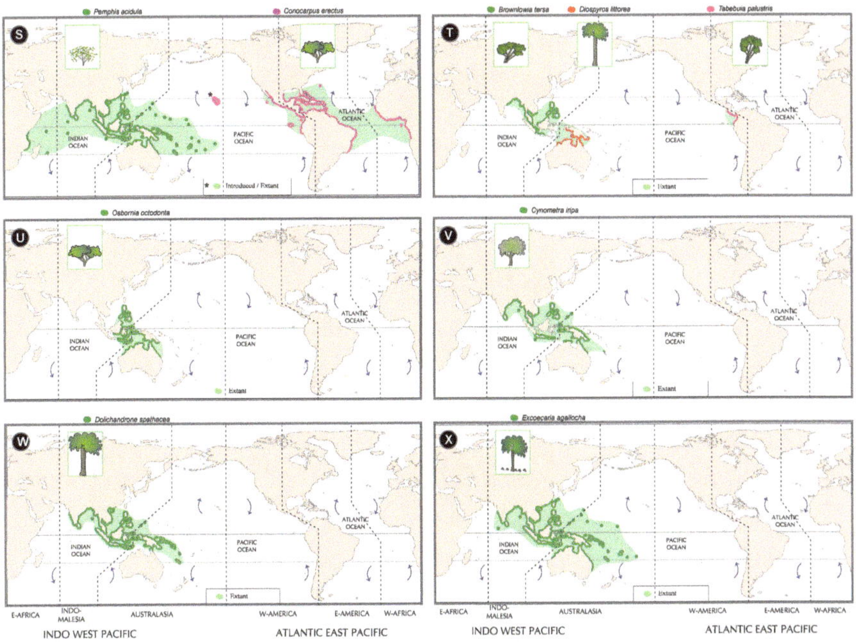

Fig. 2.10 (**S**) Distribution of mangrove species of the genus *Conocarpus* and *Pemphis*, showing native and introduced occurrences. Species: *C. erectus* L. 1753; P. acidula J.R. Forst. & G. Forst. 1775. (**T**) Distribution of mangrove species of the genus *Brownlowia*, *Diospyros*, and *Tabebuia*. Species: *B. tersa* (L.) Kosterm. 1959; *D. littoralis* (R.Br.) Kosterm. 1977; *T. palustris* Hemsl. 1882. (**U**) Distribution of mangrove species of the genus *Osbornia*. Species: *O. octodonta* F.Muell. 1863. (**V**) Distribution of mangrove species of the genus *Cynometra*. Species: *C. iripa* Kostel. 1835.(**W**) Distribution of mangrove species of the genus *Dolichandrone*. Species: *D. spathacea* (L.f.) K. Schum. 1889. (**X**) Distribution of mangrove species of the genus *Excoecaria*.
Species: *E. agallocha* L. 1759

Appendix 2

Table of Species Groupings. Presented are tables showing proposed phylogenetic groupings of 32 mangrove genera (A: with 8 genera, B: with 11 genera, and C: with 13 genera) with selections of key species in each of the 15 global spatiotemporal groupings discussed in the chapter.

Table 6a Proposed phylo-genetic groupings of 8 of 32 mangrove genera and all respective species in 15 spatio-temporal groups

A	B	C	D	E		F	G	H	I	J	K	L	M	N	O
Northern	Tethys West	Southern	Tethys West		← Tethys →	Southern	Tethys East						Northern Tethys East		
Atlan_CAmer	Pacif_CAmer	Nor_SAmer	Pacif_SAmer	SW_Pacific	←AEP-IWP→	Austral 1	Mid_East 1	S_China_S 1	N_Austral 2	India	S_China_S 2	N_Austral 3	Mid_East 2	S_China_S 3	N_Austral 4
25-40 mya >	3-4 mya >	25-40 mya >	3-4 mya >	2 mya >	< 40-55 mya >	< 10-25 mya	< 25-40 mya	25 mya	< 5-10 mya	< 40 mya	25 mya	< 5-10 mya	< 25-40 mya	< 10-25 mya	< 5-10 mya
<Extinct/ ReIntrod>	<Extinct>	<Extinct/ ReIntrod>	<Extinct>		Nypa precursor?	Nypa fruticans	<Extinct>						Nypa fruticans	Nypa fruticans	Nypa fruticans
Acrostichum danaeifolium	Acrostichum danaeifolium	Acrostichum aureum	Acrostichum aureum		Acrostichum precursor	Acrostichum aureum	Acrostichum aureum	Acrostichum aureum	Acrostichu m aureum	Acrostichum aureum	Acrostichum aureum	Acrostichum aureum	Acrostichum speciosum	Acrostichum speciosum	Acrostichum speciosum
Rhizophora racemosa	Rhizophora racemosa	Rhizophora mangle	Rhizophora samoensis	Rhizophora samoensis	Rhizophora precursor?	Rhizophora stylosa	Rhizophora mucronata	Rhizophora mucronata	Rhizophora mucronata	Rhizophora mucronata	Rhizophora mucronata	Rhizophora mucronata	Rhizophora apiculata	Rhizophora apiculata	Rhizophora apiculata southern
Avicennia germinans	Avicennia germinans	Avicennia schaueriana	Avicennia bicolor		Avicennia precursor?	Avicennia marina varieties	Avicennia marina varieties	Avicennia rumphiana	Avicennia rumphiana	Avicennia alba varieties	Avicennia alba varieties	Avicennia alba varieties	Avicennia officinalis	Avicennia officinalis	Avicennia integra
Pelliciera rhizophorae white	Pelliciera rhizophorae red				Pelliciera rhizophorae										
		<Extinct>			Sonneratia precursor?	Sonneratia alba	Sonneratia alba	Sonneratia griffithii	Sonneratia ovata	Sonneratia caseolaris	Sonneratia lanceolata	Sonneratia lanceolata	Sonneratia apetala		
		<Extinct>			Bruguiera precursor?	Bruguiera exaristata	Bruguiera gymnorhiza	Bruguiera gymnorhiza	Bruguiera gymnorhiza	Bruguiera sexangula	Bruguiera sexangula	Bruguiera sexangula	Bruguiera parviflora	Bruguiera cylindrica	Bruguiera cylindrica
					Ceriops precursor?	Ceriops australis	Ceriops tagal	Ceriops tagal	Ceriops tagal	Ceriops tagal	Ceriops tagal	Ceriops tagal	Ceriops decandra	Ceriops zippeliana	Ceriops pseudodecandra

Table 6b Further proposed phylogenetic groupings of 11 of 32 mangrove genera and all respective species in 15 spatiotemporal groups

A	B	C	D	E		F	G	H	I	J	K	L	M	N	O
Northern	Tethys West	Southern	Tethys West		← Tethys →	Southern	Tethys East						Northern Tethys East		
Atlan_CAmer	Pacif_CAmer	Nor_SAmer	Pacif_SAmer	SW_Pacific	← AEP-IWP →	Austral 1	Mid_East 1	S_China_S 1	N_Austral 2	India	S_China_S 2	N_Austral 3	Mid_East 2	S_China_S 3	N_Austral 4
25-40 mya >	3-4 mya >	25-40 mya >	3-4 mya >	2 mya >	< 40-55 mya >	< 10-25 mya	< 25-40 mya	25 mya	< 5-10 mya	< 40 mya	25 mya	< 5-10 mya	< 25-40 mya	< 10-25 mya	< 5-10 mya
					Heritiera precursor?	Heritiera littoralis	Heritiera littoralis	Heritiera littoralis	Heritiera littoralis	Heritiera littoralis	Heritiera littoralis	Heritiera littoralis	Heritiera fomes	Heritiera globosa	
						Lumnitzera racemosa	Lumnitzera racemosa	Lumnitzera racemosa	Lumnitzera racemosa	Lumnitzera littorea	Lumnitzera littorea	Lumnitzera littorea			
						Pemphis acidula	Pemphis acidula	Pemphis acidula	Pemphis acidula	Pemphis acidula	Pemphis acidula	Pemphis acidula			
						Xylocarpus granatum	Xylocarpus granatum	Xylocarpus granatum	Xylocarpus granatum	Xylocarpus rumphii	Xylocarpus rumphii	Xylocarpus rumphii	Xylocarpus moluccensis	Xylocarpus moluccensis	Xylocarpus moluccensis
					Aegiceras precursor?								Aegiceras floridum	Aegiceras corniculatum	Aegiceras corniculatum
		Conocarpus erectus	Conocarpus erectus		><								Brownlowia tersa	Brownlowia tersa	Brownlowia argentata
		Laguncularia racemosa	Laguncularia racemosa		><								Dolichandrone spathacea	Dolichandrone spathacea	Dolichandrone spathacea
Crenea patentinervis					><								Kandelia candel	Kandelia obovata	

Table 6c Further proposed phylo-genetic groupings of 13 of 32 mangrove genera and all respective species in 15 spatio-temporal groups

A	B	C	D	E		F	G	H	I	J	K	L	M	N	O
Northern	Tethys West	Southern	Tethys West		←Tethys→	Southern	Tethys East						Northern Tethys East		
Atlan_CAmer	Pacif_CAmer	Nor_SAmer	Pacif_SAmer	SW_Pacific	←AEP-IWP→	Austral 1	Mid_East 1	S_China_S 1	N_Austral 2	India	S_China_S 2	N_Austral 3	Mid_East 2	S_China_S 3	N_Austral 4
25-40 mya >	3-4 mya >	25-40 mya >	3-4 mya >	2 mya >	< 40-55 mya >	< 10-25 mya	< 25-40 mya	25 mya	< 5-10 mya	< 40 mya	25 mya	< 5-10 mya	< 25-40 mya	< 10-25 mya	< 5-10 mya
Pavonia paludicola	Pavonia rhizophorae				><								Aegialitis rotundifolia	Aegialitis annulata	Aegialitis annulata
Muellera moniliformis	Muellera moniliformis				><								Acanthus ilicifolius	Acanthus ebracteatus	Acanthus ebracteatus ebarbatus
	Mora oleifera				><								Cynometra iripa	Cynometra iripa	Cynometra iripa
	Tabebuia palustris				><								Scyphiphora hydrophylacea	Scyphiphora hydrophylacea	Scyphiphora hydrophylacea
													Excoecaria agallocha	Excoecaria agallocha	Excoecaria agallocha ovalis
														Camptostemon philippinense	Camptostemon schultzii
														Barringtonia racemosa	Barringtonia racemosa
														Osbornia octodonta	Osbornia octodonta
															Diospyros littorea

References

Allen JA (1998) Mangroves as alien species: the case of Hawaii. Glob Ecol Biogeogr Lett 7:61–71
Cowman PF, Bellwood DR (2013) The historical biogeography of coral reef fishes: global patterns of origination and dispersal. J Biogeogr 40:209–224
Castillo-Cardenas MF, Toro-Perea N, Cardenas-Henao H (2005) Population genetic structure of neotropical mangrove species on the Colombian Pacific coast: *Pelliciera rhizophorae* (Pellicieraceae). Biotropica 37(2):266–273
Cerón-Souza I, Rivera-Ocasio E, Medina E, Jiménez JA, McMillan WO, Bermingham E (2010) Hybridization and introgression in new world red mangroves, *Rhizophora* (Rhizophoraceae). Am J Bot 97(6):945–957
Churchill DM (1973) The ecological significance of tropical mangroves in the early tertiary floras of southern Australia. Geol Soc Aust 4:79–86
Coates AG, Jackson JBC, Collins LS, Cronin TM, Dowsell HJ, Bybell LM, Jung P, Obando JA (1992) Closure of Panama: the near-shore marine record in Costa Rica and Western Panama. Geol Soc Am Bull 104:814–828
Dangremond EM (2015) Propagule predation by crabs limits establishment of an endemic mangrove. Hydrobiologia 755(1):257–266
Doebeli M, Dieckmann U (2003) Speciation along environmental gradients. Nature 421:259–264
Drexler JZ (2001) Maximum longevities of *Rhizophora apiculata* and *R. mucronata* propagules. Pac Sci 55:17–22
Duke NC (1992) Mangrove floristics and biogeography. In: Robertson AI, Alongi DM (eds) Tropical mangrove ecosystems, Coastal and estuarine studies series. American Geophysical Union, Washington, D.C., pp 63–100. 329 pp.
Duke NC (1995) Genetic diversity, distributional barriers and rafting continents – more thoughts on the evolution of mangroves. Hydrobiologia 295:167–181
Duke NC (2006) In: Duke NC (ed) Australia's mangroves. The authoritative guide to Australia's mangrove plants. The University of Queensland, Brisbane, p 200
Duke NC (2010) Overlap of eastern and western mangroves in the SW Pacific: hybridization of all three *Rhizophora* combinations in New Caledonia. Blumea 55(2):171–188
Duke NC (2011) Mangroves. Encyclopedia of modern coral reefs. Structure, form and process. D. Hopley (ed.). Springer, Dordrecht, p \-663
Duke NC (2013) World Mangrove iD: expert information at your fingertips, App Store Version 1.1 for iPhone and iPad, Dec 2013. MangroveWatch Publication. e-book. https://itunes.apple.com/us/app/mangrove-id/id761487621?mt=8
Duke NC (2014a) World Mangrove iD: expert information at your fingertips, Google Play Store Version 1.1 for Android, Oct 2014. MangroveWatch Publication. e-book. https://play.google.com/store/apps/developer?id=MangroveWatch+Ltd
Duke NC (2014b) In: Harff J, Meschede M, Petersen S, Thiede J (eds) Mangrove coast. Encyclopedia of marine geosciences. Springer, Dordrecht, p 1–17. Online. 978-94-007-6644-0. DOI 10.1007/978-94-007-6644-0_186-1. http://link.springer.com/referenceworkentry/10.1007%2F978-94-007-6644-0_186-1
Duke NC, Ge XJ (2011) *Bruguiera* (Rhizophoraceae) in the Indo West Pacific: a morphometric assessment of hybridization within single-flowered taxa. Blumea 56:36–48
Duke NC, Ball MC, Ellison JC (1998) Factors influencing biodiversity and distributional gradients in mangroves. Glob Ecol Biogeogr Lett 7:27–47
Duke NC, Lo EYY, Sun M (2002) Global distribution and genetic discontinuities of mangroves – emerging patterns in the evolution of *Rhizophora*. Trees Struct Funct 16:65–79
Duke NC, Meynecke J-O, Dittmann S, Ellison AM, Anger K, Berger U, Cannicci S, Diele K, Ewel KC, Field CD, Koedam N, Lee SY, Marchand C, Nordhaus I, Dahdouh-Guebas F (2007) A world without mangroves? Science 317:41–42

Guo Z, Chen Y, Duke NC, Zhong C, Huang Y, Shi S (2016) Genetic discontinuities in a dominant mangrove *Rhizophora apiculata* (Rhizophoraceae) of Indo Malesia. Biogeogr Evol. doi:10.1111/jbi12770

Hoeksema BW (2007) In: Renema W (ed) Delineation of the Indo-Malayan centre of maximum marine biodiversity: the coral triangle. Biogeography, time, and place: distributions, barriers, and islands. Springer, Heidelberg, pp 117–178

Huang L, Li X, Huang Y, Shi S, Zhou R (2014) Molecular evidence for natural hybridization in the mangrove genus *Avicennia*. Pak J Bot 46(5):1577–1584

Kay I (2007) Mangrove eradication program, Kendall-frost marsh. Mission Bay, San Diego. http://nrs.ucop.edu/reserves/kendall_frost/kendall_frost.htm. Accessed Oct 2015

Liao P-C, Havanond S, Huang S (2007) Phylogeography of *Ceriops tagal* (Rhizophoraceae) in Southeast Asia: the land barrier of the Malay peninsula has caused population differentiation between the Indian Ocean and South China Sea. Conserv Genet 8:89–98

Lo EY, Duke NC, Sun M (2014) Phylogeographic pattern of *Rhizophora* (Rhizophoraceae) reveals the importance of both vicariance and long-distance oceanic dispersal to modern mangrove distribution. BMC Evol Biol 14(83):15pp

Maynard-Smith J (1966) Sympatric speciation. Am Nat 100(916):637–650

Mayr E (1970) Populations, species, and evolution. The Belknap Press of Harvard University Press, Cambridge, MA

Mumby PJ, Edwards AJ, Arias-Gonzalez JE, Lindeman KC, Blackwell PG, Gall A, Gorczynska MI, Harborne AR, Pescod CL, Renken H, Wabnitz CCC, Llewellyn G (2004) Mangroves enhance the biomass of coral reef fish communities in the Caribbean. Nature 427:533–536

Nathan R, Schurr FM, Spiegel O, Steinitz O, Trakhtenbrot A, Tsoar A (2008) Mechanisms of long-distance seed dispersal. Trends Ecol Evol 23(11):638–647

Plaziat J-C, Cavagnetto C, Koeniguer J-C, Baltzer F (2001) History and biogeography of the mangrove ecosystem, based on a critical reassessment of the paleontological record. Wetl Ecol Manag 9:161–180

Provine WB (2004) Ernst Mayr: genetics and speciation. Genetics 167(3):1041–1046

Quisthoudt K, Schmitz N, Randin CF, Dahdouh-Guebas F, Robert EMR, Koedam N (2012) Temperature variation among mangrove latitudinal range limits worldwide. Trees 26:1919–1931

Rabinowitz D (1978) Dispersal properties of mangrove propagules. Biotropica 10(1):47–57

Schlanger SO, Jenkyns HC, Premoli-Silva I (1981) Volcanism and vertical tectonics in the Pacific Basin related to global cretaceous transgressions. Earth Planet Sci Lett 52:435–449

Schlanger SO, Premoli-Silva I (1981) Tectonic, volcanic and paleogeographic implications of redeposited reef faunas of late cretaceous and tertiary age from the Nauru Basin and Line Islands. In: Initial Reports of the Deep Sea Drilling Project 61:817–827

Schmidt DN (2007) The closure history of the central American seaway: evidence from isotopes and fossils to models and molecules. In: Deep-Time Perspect Clim Change Marrying Signal Comput Models Biol Proxies. Micropalaeontol Soc. Spec 2: p 429–445

Spalding MD, Kainuma M, Collins L (2010) World Atlas of mangroves. Earthscan, London, p 319

Steinke TD (1986) A preliminary study of buoyancy behaviour in *Avicennia marina* propagules. Sth Afr J Botany 52:559–565

Su G-H, Huang Y-L, Tan F-X, Ni X-W, Tang T, Shi S-H (2006) Genetic variation in *Lumnitzera racemosa*, a mangrove species from the Indo-West Pacific. Aquat Bot 84(4):341–346

Takayama K, Tamura M, Tchcoaateishi Y, Webb EL, Kajita T (2013) Strong genetic structure over the American continents and transoceanic dispersal in the mangrove genus *Rhizophora* (Rhizophoraceae) revealed by broad-scale nuclear and chloroplast DNA analysis. Am J Bot 100(6):1191–1201

Tan F, Huang Y, Ge X, Su G, Ni X, Shi S (2004) Population genetic structure and conservation implications of *Ceriops decandra* in Malay peninsula and North Australia. Aquat Bot 81(2):175–188

Tomlinson PB (2016) The botany of mangroves. Second Edition. Cambridge University Press, Cambridge, p 418

Triest L (2008) Molecular ecology and biogeography of mangrove trees towards conceptual insights on gene flow and barriers: a review. Aquat Bot 89(2):138–154

Wee AKS, Takayama K, Asakawa T, Thompson B, Onrizal Sungkaew S, Tung NX, Nazre M, Soe KK, Tan HTW, Watano Y, Baba S, Kajita T, Webb EL (2014) Oceanic currents, not land masses, maintain the genetic structure of the mangrove *Rhizophora mucronata* lam. (Rhizophoraceae) in Southeast Asia. J Biogeogr 41:954–964

Chapter 3
Biodiversity

S.Y. Lee, E.B.G. Jones, K. Diele, G.A. Castellanos-Galindo, and I. Nordhaus

3.1 Introduction

Mangrove ecosystems are globally structured by a relatively small number of 'true' mangrove tree species and their associates (Tomlinson 1986). The total number of 'true' mangroves varies between authorities, but the maximum is only about 80 including hybrids (Duke 2006; Polidoro et al. 2010). The global distribution of these species, however, is far from even between the two main mangrove

S.Y. Lee (✉)
Simon F S Li Marine Science Laboratory, School of Life Sciences, and Earth System Science Programme, The Chinese University of Hong Kong, Shatin, Hong Kong SAR, China
e-mail: joesylee@cuhk.edu.hk

E.B.G. Jones
Department of Botany and Microbiology, College of Science, King Saud University, P.O. Box: 2455, Riyadh 1145, Kingdom of Saudi Arabia
e-mail: torperadgj@gmail.com

K. Diele
School of Life, Sport and Social Sciences, Edinburgh Napier University, Edinburgh, UK

St Abbs Marine Station, St Abbs, UK
e-mail: K.Diele@napier.ac.uk

G.A. Castellanos-Galindo
Departmento de Biología, Grupo de Investigación en Ecología de Estuarios y Manglares, Universidad del Valle, A. A. 25360 Cali, Colombia

Leibniz Center for Tropical Marine Ecology (ZMT), Fahrenheitstrasse 6, 28359 Bremen, Germany
e-mail: gustavoa80@yahoo.com

I. Nordhaus
Leibniz Center for Tropical Marine Ecology (ZMT), Fahrenheitstrasse 6, 28359 Bremen, Germany
e-mail: inga.nordhaus@zmt-bremen.de

V.H. Rivera-Monroy et al. (eds.), *Mangrove Ecosystems: A Global Biogeographic Perspective*, https://doi.org/10.1007/978-3-319-62206-4_3

biogeographic regions. The total number of mangroves in the Atlantic-East-Pacific (AEP) biogeographic region is <15, whereas >60 occur in the species-rich Indo-West-Pacific (IWP) (Tomlinson 1986; Ricklefs and Latham 1993; Spalding et al. 2010). The ecological implications of this discrepancy in species richness for ecosystem function are poorly known, as meta-analyses at this spatial scale is riddled with confounding factors ranging from the geological or evolutionary history of the two regions to differences in hydrodynamics, nutrient availability and anthropogenic stressors.

Despite the early effort of mangrove research being focused on the floristics and faunistics of these systems, patterns of biodiversity in mangrove ecosystems are still obscure as system-level analyses are almost non-existent. Major gaps occur in taxa as well as region coverage. An apparent pattern based on limited data from the IWP is that despite the low overall global species richness of mangrove trees, the total species richness of biota supported by mangrove ecosystems is usually two orders of magnitude higher. For example, Sandilyan and Kathiresan (2012) reported a total of 4011 species of bacteria, fungi, algae, plants and animals recorded from mangrove forests comprising of 39 tree species in India. Similarly, Wang and Wang (2007) recorded 2305 species of plants and animals from Chinese mangrove communities, which have a total of 25 mangrove species (Table 3.1). These numbers are significantly larger than those reported in earlier surveys (e.g. Saenger et al. 1983). Similar data are, however, virtually unavailable for the AEP, although data from restricted regions or on selected biodiversity groups, e.g. sponges (Diaz 2012; Rutzler et al. 2014), ascidians (Goodbody 2003; Rocha et al. 2005), goby fish (Greenfield and Johnson 1999), and fungi (Schmit and Shearer 2004; Jones and Puglisi 2006) are available. Since ecosystems are not confined by national boundaries, the country data in Table 3.1 by necessity are only from artificial biodiversity units and may not therefore reflect broad trends.

The full picture on biodiversity supported by mangrove ecosystems is difficult to construct. Mangrove forests are generally problematic habitats to study as the soft substratum and the occurrence of disturbing (e.g. biting insects) and even dangerous (e.g. snakes, crocodiles and tigers) wildlife present significant logistical challenges. Many animals associated with mangrove forests are only transient visitors, e.g. fish using the forest during high tide (e.g. Robertson and Duke 1990; Nagelkerken et al. 2008), while many resident species are endobenthic in the spatially complex underground environment that makes effective sampling and replication demanding (Lee 2008). Many early reports on mangrove animal diversity, especially those on the endobenthic and transient species, are therefore sketchy and lack quantitative detail for analysis of distribution patterns. Further, some of the faunal groups present considerable taxonomic challenges. For example, sesarmid crabs (Decapoda: Brachyura: Sesarmidae) are key structural as well as functional components of tropical mangrove ecosystems (Lee 1998, 2008) especially in the IWP, but their high diversity is yet to be fully appreciated due to the challenges in taxonomic classification and field identification. Many biodiversity components, e.g. the meiofauna or the insect fauna, of the mangrove ecosystem are even more understudied, and consequently, often omitted in assessments of overall mangrove

Table 3.1 Number of species (S) belonging to various taxa reported from mangrove ecosystems of various countries or regions

Country/region	Mangrove species	Non-fish vertebrate	Finfish	Fungi	Algae	Insect	Crustacean	Gastropod	Bivalve	Other macro-invertebrate	Total S	References[j]
Country												
IWP												
ASEAN (Indonesia, Malaysia, Singapore, Philippines)	50	211	212	139[a]	ND	ND	54	ND	ND	ND	ND	[1, 2]
Australia	41	242	ND	ND	ND	ND	65	ND	ND	160	ND	[3]
China	25	402	249	ND	494	440	ND	ND	ND	650[b]	2305	[4]
India	39	591	543	103	557	707	193	ND	ND	1250	4011	[5]
Vietnam	32	403[c]	258	ND		ND	ND	ND	ND	306[b]	ND	[6]
AEP												
Brazil	7 [g]	ND	287	ND	ND	ND	60[d]	11[e]	24	ND	ND	[7–18]
French Guiana	4	ND	83	ND	ND	ND	11[d]	ND	ND	ND	ND	
Guyana	5	ND	ND	ND	ND	ND	2[d,e]	ND	ND	ND	ND	
Surinam	4	ND	ND	ND	ND	ND	6[d,e]	ND	ND	ND	ND	
Venezuela	7	ND	45[e]	ND	ND	ND	40[d]	38	39	ND	ND	[19–20]
Colombia	9	ND	282[g]	ND	ND	ND	66[d]	70	48	ND	ND	[21–28]
Ecuador	7	ND	36[e]	ND	ND	ND	1[d,e]	2[e]	2[e]	ND	ND	[17, 29]
Peru	5	ND	97	ND	ND	ND	23 [d,e]	42	31	ND	ND	[30, 31]

(continued)

Table 3.1 (continued)

Country/region	Mangrove species	Non-fish vertebrate	Finfish	Fungi	Algae	Insect	Crustacean	Gastropod	Bivalve	Other macro-invertebrate	Total *S*	References[j]
Region												
IWP												
Shantou, China (23°15′ – 23°34′ N)	11	203	133	ND	ND	ND	ND	ND	ND	198[b]	534	[32]
Quanzhou Bay, China	3	123[h]	63	ND	73	1	60	28	25	91	464	[33]
Deep Bay, China	7	452	31	ND	ND	372	97	72	55	118	1204	[34, 35]
Segara Anakan lagoon, Java, Indonesia	21	ND	ND	ND	ND	ND	54	41	9	16	ND	[36–38]
AEP												
SE Florida (25–30°N)	3	223	216	ND	74	ND	ND	ND	ND	ND	NA	[39]
South America	10	ND	~500	ND	ND	ND	144[d]	145	114	ND	ND	[1–25]
South America Western Atlantic	8	ND	287	ND	ND	ND	68[d]	50	64	ND	ND	[1–14]
South America Caribbean	NA	ND	80[e]	ND	ND	ND	55[d]	55	25	ND	ND	[15, 16, 19–22]

South America East Pacific	9	ND	228[i]	ND	ND	ND	59[d]	53	48	ND	ND	[11, 17, 18, 23–25]

[a]Only data on Malaysia from Alias et al. (2010)
[b]Including crustaceans
[c]Only birds and mammals in the south of Vietnam
[d]Only Brachyura
[e]Number unlikely representative (few or no studies found)
[f]Reference for Mangrove species: FAO (2007)
[g]Species on Caribbean and Pacific coasts
[h]Only birds included
[i]Includes only Peru, Ecuador and Colombia
[j] [1] Wilkinson et al. (1994), [2] Alias et al. (2010), [3] Saenger et al. (1977), [4] Wang and Wang (2007), [5] Sandilyan and Kathiresan (2012), [6] Hong and Hoang (1993), [7] Aderaldo Vidal & Rocha-Barreira (2009), [8] Araujo et al. (1995), [9] Aviz et al. (2009), [10] Barroso and Metthews-Cascon (2009), [11] Beasley et al. (2005), [12] Costa de Melo et al. (2010), [13] Faraco & Lana (2004), [14] Lopes & Narchi (1997), [15] Leonel et al. (2006), [16] Maia & Coutinho (2013), [17] Reid et al. (2010), [18] Simao do Amaral & Simone (2014), [19] Balbas et al. (2014), [20] Cedeno et al. (2010), [21] Blanco & Cantera (1999), [22] Blanco & Castaño (2012), [23] Cantera et al. (1999), [24] Cantera (2010), [25] Padilla & Palacio (2008), [26] Quirós & Arias (2013), [27] Reyes & Campos (1992), [28] Vilardy & Polanía (2002), [29] MacKenzie & Clyde (2001), [30], Paredes et al. (2005), [31] Sanchez (2007), [32] Peng et al. (2011), [33] Huang (2004), [34] Cai (2015), [35] WWFHK (2014), [36] Geist et al. (2012), [37] Nordhaus et al. (2009), [38] Hinrichs et al. (2009), [39] Odum et al. (1982)

Total species richness may include taxa not detailed in the table. Country/region totals (excluding mangrove tree species) are not computed when significant taxa are not reported. *ND* no data. Finfish data for South America from Castellanos-Galindo & Krumme (unpublished data)

ecosystem biodiversity. Many data gaps therefore exist, limiting the usefulness of comparisons or generalisations.

Notwithstanding, this chapter attempts to examine the pattern and ecological implications of various components of biodiversity supported by mangrove ecosystems, with particular emphasis on a comparison between the IWP and AEP biogeographic regions. Focus is on faunal and microbial assemblages that are important components for mangrove ecosystem functioning.

3.2 Components of Mangrove Biodiversity

3.2.1 *Genetic Diversity of Mangroves*

Mangroves are unique in having their genetic diversity determined not only by pollen motility, but also long-distance dispersal of propagules of viviparous species (Lo et al. 2014; Wee et al. 2014). Populations of mangrove species may therefore demonstrate patterns in genetic structure driven by connectivity via oceanic current flow. A wide range of mangrove species has been reported to demonstrate low levels of genetic diversity (e.g. Sun et al. 1998; Ge and Sun 1999; Jian et al. 2010), even in widely occurring species such as *Avicennia marina* (Maguire et al. 2000; Arnaud-Haond et al. 2006). Mangroves seem to have undergone a process of 'genetic downsizing' upon their colonisation of the intertidal habitat, which is a unique and specific niche for higher plants. This generally low level of genetic diversity in mangroves, which may reflect the result of adaptation to a harsh environment by higher plants, has significant implications for their resilience to changing environmental conditions, especially when multiple stressors are present. Global climate change as well as anthropogenic wholesale modifications ('anthropogenic forcing') of tropical estuarine environments, e.g. nutrient availability (Lee 2016), present a suite of selective pressures and stresses not previously associated within the narrow niche of mangroves. There is no global analysis of mangrove genetic diversity to date, but this component of mangrove biodiversity will not be discussed further in this chapter.

3.2.2 *Functional, Taxonomic and Habitat Components of Mangrove Biodiversity*

Mangroves are efficient producers, with most of their organic production entering the detritus-based food chain or stored in the sediment, with generally <10% of leaf biomass consumed by herbivores (Sousa and Dangremond 2011). Traditional emphasis on the detritus-based food chain has resulted in little work on the herbivore assemblages utilising mangrove live biomass. However, the limited data available to date suggest that this assemblage, usually dominated by insects, can be highly diverse (e.g. Murphy 1990; Tong et al. 2006). Despite their functional importance, the diversity of decomposers and detritivores processing mangrove detritus including leaf litter, wood, as well as belowground roots are still yet to be comprehensively documented. Significant gaps of knowledge exist for taxa and feeding guilds other than the fish, macrobenthic leaf-processing crabs and gastropods. Fungi are ubiquitous decomposers of mangrove leaf litter and wood (Hyde and Jones 1988; Jones and Pang 2012; Loilong et al. 2012), despite a lack of data on their exact ecological roles (Hyde and Lee 1995; Hyde et al. 1998; Swe et al. 2009). While there is considerable taxonomic work on decomposers such as fungi in tropical mangrove wetlands, data are again patchy in relation to geographic and biogeographic settings, making the elucidation of patterns difficult. Wood-boring fauna represent another poorly known but functionally important group (Svavarsson et al. 2002; Davidson et al. 2014) of mangrove biodiversity.

Meiofauna (>40 μm and <500 μm) are ubiquitous in marine soft sediments (Coull 1999) but are generally poorly explored. Limited data available suggest that meiofauna are abundant in mangrove sediment (e.g. Dye 1983a, b; Alongi 1987a, b; Somerfield et al. 1998; Olafsson et al. 2000; Armenteros et al. 2006; Xuan et al. 2007; Ostmann 2010), but to clarify their trophic role and diversity pattern still more data from a wider range of environments are required. As species-level identification is taxonomically challenging and time-consuming, the full picture on diversity of this group and its role in mangrove function and services is still obscure.

Nektonic species, especially commercially important species of finfish and crustaceans, represent the better-studied component of mangrove biodiversity. The challenge to understanding the diversity of this group lies, however, in how to define the association with mangroves, e.g. 'seascape nurseries' (Nagelkerken et al. 2015). Adoption of different definitions of association with or dependence on mangroves contributes to variability in diversity estimates.

Many plants are associated with forests primarily structured by mangrove trees. Epiphytes comprising orchids, macroalgae and lichens are common on mangrove tree trunks and aerial roots, but their diversity and ecological role have seldom been examined.

Apart from the inherent values of the associated species, assessment of the biodiversity of mangrove ecosystems also has implications for the function and services derivable from these habitats (Cannicci et al. 2008). Preliminary analyses suggest that tree and faunal species richness in mangrove ecosystems may be positively correlated (Lee 2008). The reason for this correlation may just be consequences of processes in ecological or evolutionary times. Increased tree species diversity may support more chemical and physical niches to accommodate a larger number of animal species. Similarly, concomitant radiation of tree and animal species from the centre of origin of mangroves in the Indo-Malayan region would have resulted in correlations between tree and faunal diversities. Just as different tree species offer specific physical and chemical matrices such as aboveground structural complexity that underpin key services such as abatement of wave impact on coastal communities, a wider range of and larger capacity for ecosystem services may be realised through a larger diversity of associated biota of mangroves. For example, different species of leaf-eating sesarmid crabs may have different leaf preferences, thus promoting overall organic matter processing rate. Again, this relationship between ecosystem function and the species richness of trees as well as the biodiversity they support is virtually unstudied for mangrove ecosystems. Significant correlations between forest characteristics (e.g. tree density, biomass) and crab abundance and biomass have also been reported (e.g. Ashton et al. 2003a).

The available data on the correlation between diversity of mangrove trees and their associated biota have been limited to the IWP (e.g. Lee 2008; Geist et al. 2012). Whether this relationship extends to cover spatial scales at the level of biogeographic regions has not been explored but will be addressed in this analysis using the better-studied ichthyofauna (Sect. 3.2.2.2). The major biodiversity components of mangrove ecosystems are discussed in more detail in the following sections.

3.2.2.1 Plants and Lichens

A diverse 'mangrove associates' assemblage inhabits the landward fringe of mangrove forests. Wang and Chen (2013) reported 175 species of halophytes from the coasts of south China with 25 mangrove tree species. A large proportion of these associated species live in the ecotone with or close to mangrove forests. As with the upper-intertidal mangrove species, these mangrove associates are strongly threatened by the construction of sea walls and human settlement, agriculture and aquaculture (Spalding et al. 2010). Mangrove tree structures such as tree trunks and aerial roots (e.g. prop roots, pneumatophores) also support epiphytic algae, vascular plants and lichens. Comprehensive accounts of these epiphytic assemblages are rare, but the limited data still suggest significant biodiversity supported by these structures. On the east coast of Australia, mangrove forests support >200 species of macroalgae, lichens and vascular plants (ferns, orchids, mistletoes), with the former two groups being most diverse (Stevens 1979; Barlow 1984; Hutchings and Saenger 1987). Many of these macrophytes play important trophic roles (e.g. epiphytic algae) or have unique ecologies, e.g. the ant-house plant *Myrmecodia beccarii*. As with the landward mangrove species, mangrove associates face extreme anthropogenic threats.

3.2.2.2 Marine Macrofauna

Mangrove macrofauna comprise several main components with different life habits and degree of association with the mangrove environment. Transient species, mainly nektonic aquatic species or aerial visitors from the terrestrial environment, use mangroves for food and refuge. Their association with mangroves is usually determined by habitat availability driven by tides (aquatic visitors) or habitat requirement driven by life history events (e.g. migrating birds) or seasons (e.g. shrimps). Resident species can be epibenthic, endobenthic or arboreal. Megafauna such as birds, reptiles and mammals also demonstrate various degrees of residency in mangrove forests.

Marine macrofauna such as fish and decapod crustaceans contribute the largest groups of transient visitors in terms of both abundance and species richness to mangrove wetlands, whereas molluscs, decapod crustaceans and other groups (polychaetes, sipunculids, echiurids) dominate the resident species, with the molluscs predominantly epibenthic while the other groups mainly endobenthic (Table 3.1). While still considerably species-rich, arboreal macrofauna in mangrove ecosystems usually contribute less biomass and lower abundance than epibenthic or endobenthic species. Grazing insects dominate the arboreal assemblage, consuming only a small percentage of mangrove production but perform vital roles in maintaining mangrove communities such as pollination. Herbivory can sometimes exert devastating effects on tree survival and reproductive output during massive outbreaks (e.g. Anderson and Lee 1995).

Deciphering pattern in the biodiversity supported by mangrove ecosystems is hampered by the lack of commonality in surveying approach and methodology. For example, a large variety of sampling gears and techniques is used for studying the mangrove ichthyofauna, each with their own selectivity and biases. Much of the quantitative data are therefore not directly comparable. Many studies on mangrove macrobenthos focus on epibenthic species, whereas others include endobenthic or arboreal species as well. Several frequently used methods underestimate fast-moving non-burrowing species, often also not adapted to capturing smaller species and specimens.

Sampling effort differs not only between groups, but also between countries and regions. In South America, for example, Peru, Ecuador, Guyana and Surinam are under-sampled for mangrove macrobenthos, while the diversity of Crustacea (mainly Brachyura) in Colombia and Brazil are probably representative because of greater sampling efforts (Table 3.1). However, the notable high number of mangrove-associated Brachyura and Gastropoda in the South-American Caribbean does not relate to differences in sampling effort, but to biogeography, the local seascape and hydrodynamics. The clear waters and additional habitat types (e.g. permanently submerged roots) in the microtidal Caribbean mangrove ecosystems and their connectivity to coral reefs explain, for example, the presence of several species of majid and inachid crabs and coral reef fish species that are absent from most other South-American mangrove wetlands.

Species richness of Gastropoda (145) in mangrove forests of South America is similar to that of Brachyura (144) whereas species number of Bivalvia (114) is slightly lower (Table 3.1). Studies on diversity and distribution of minor components of the endobenthos such as polychaetes and sipunculids in mangrove forests are rare. The few studies available show that polychaete species numbers and densities vary considerably among mangrove forests even in the same biogeographic region. The highest species richness with 68 polychaete species was reported from mangrove areas and salt flats in northern Australia with 30 mangrove tree species (Metcalfe and Glasby 2008). However, sampling occurred over 4 years and included

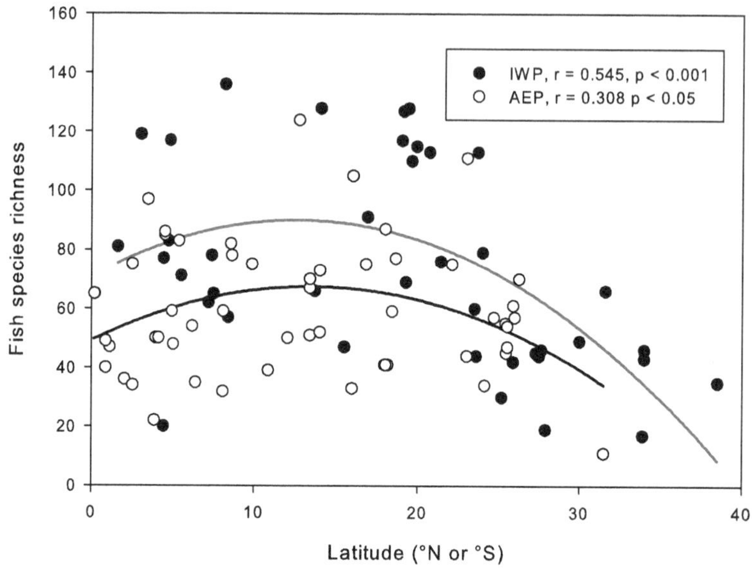

Fig. 3.1 Variation of finfish species richness and latitude for mangroves in the IWP (*grey line*) and AEP (*black line*) biogeographic regions. Both regions seem to peak at around 13–14° latitude, with sharp declines poleward

mangrove sediment and dead wood, making comparison with species richness recorded in other IWP mangrove forests difficult. Lower and variable numbers of polychaete species have been reported from other IWP countries (see Nordhaus et al. 2009 and references therein). The dominant families in terms of species richness and abundance are Capitellidae, Nereididae and Spionidae as reported from Indonesian and Australian mangrove forests (Nordhaus et al. 2009; Metcalfe and Glasby 2008). A mean density of polychaetes (157 individuals m^{-2}, Nordhaus et al. 2009) suggests a potential contribution of this group to bioturbation and processing of organic matter in Indonesian mangrove forests.

Information for other invertebrate phyla is even scarcer. The phyla Sipuncula and Phoronida were only represented by one species each in a mangrove forest in Java, Indonesia; the groups Oligochaeta, Nemertea and Turbellaria were also present (Nordhaus et al. 2009). In northern Australia, one echiuran, three sipunculan, two nemertean, one tubellarian and one oligochaete species were identified (Metcalfe and Glasby 2008). In both studies, sampling effort was high both spatially and temporally, indicating that these groups are of minor importance in terms of taxonomic diversity and abundance.

Due to their commercial value, considerably more data are available on the ichthyofauna of mangrove ecosystems than for other groups. Based on a database comprising 40 studies in the IWP and 54 from the AEP, the species richness of mangrove fish assemblages of the two biogeographic regions seems to follow the same latitudinal trend (Fig. 3.1). Species richness follows a parabolic pattern with a maximum at around 13–14° latitude N or S in either biogeographic regions. The maximum at around 13–14°N or S in both the AEP and IWP is intriguing but may be related to the local extent of mangrove forests or factors controlling their development, e.g. riverine outflow into estuaries. The overall pattern of the data suggests generally higher species richness of mangrove-associated fish assemblages in the IWP than in the AEP. This pattern could be the result of processes occurring over geological and/or ecological timescales and has been previously shown for coral reef fish assemblages (Parravicini et al. 2013) and some mangrove invertebrates (Reid et al. 2010). More work at spatial scales varying from the local forest to the biogeographic region levels is needed to test these hypotheses. The apparent difference may also simply reflect differences in research effort but since there are more studies conducted in the AEP, this is unlikely to be a contributing factor.

Sheaves (2012) used fish assemblages associated with mangrove wetlands to establish ecosystem equivalency between different biogeographic regions. While only two broad biogeographic regions have been recognised for mangrove plants, Castellanos-Galindo and Krumme (2013) commented that distinct fish assemblages exist within these broad regions and that comparisons of equivalency cannot be made without further reference to these detailed differences. Unfortunately, data on other commonly studied taxa such as penaeid prawns are still inadequate for a similar comparison between the two biogeographic regions.

One key issue concerning different levels of faunal diversity in mangrove ecosystems is whether higher diversity may translate to higher capacity for ecosystem services. There is a general paucity of data on how macrofauna contribute to the function of mangrove wetlands, but considerable knowledge is available on how specific taxa, e.g. sesarmid and ucidid crabs, mediate key processes such as carbon dynamics that underpin major ecosystem services.

3.2.2.3 Terrestrial Vertebrates

Mangrove forests are frequently used by terrestrial vertebrates as habitats for foraging, roosting and nesting purposes. Mangrove forests and the associated wetlands such as tidal flats often support large populations of resident and migrant bird species (e.g. Zwarts et al. 2014). Luther and Greenberg (2009) reported a total of 853 species of terrestrial vertebrates (790 birds, 40 mammals, 20 reptiles and 3 amphibians) that are associated with mangrove forests at various levels of permanency. Of these, only 69 species (8.1%) are considered endemic to mangrove ecosystems, i.e. despite occasional use of other habitats, their life cycles are strongly tied to and dependent on the mangrove habitat. Among the endemic species, birds (48 species, 70%) and reptiles (14 species, 20%) contribute the largest groups. Mangrove forests in the Indo-Malayan and Australasian regions support the majority of the mangrove-endemic terrestrial vertebrate fauna (Luther and Greenberg 2009).

3.2.2.4 Decomposers

Mangrove decomposer organisms include bacteria, fungi, fungal-like organisms and crabs, gribble (wood boring isopods) and shipworms (wood boring bivalves) along with other minor groups. Decomposer assemblages in the intertidal mangrove ecosystems vary according to the dominating substrate types: Walls of woody substrates are rich in ligno-cellulose while those of leaves contain less lignin.

3.2.2.4.1 Woody Substrates

Fungi

Jones et al. (2015) listed 1112 marine fungal species (472 genera): Ascomycota with 805 species (352 genera), Basidiomycota with 21 species (17 genera), Chytridiomycota with 26 species (13 genera), Blastocladiomycota with one species (one genus), Zygmomycota with three species (two genera), asexual filamentous morphs with 43 species (35 genera) and marine yeasts with 213 species (61 genera). Hyde and Jones (1988) recognized that mangrove fungi constituted the second largest group of marine fungi, which are widely distributed in Old and New world mangrove ecosystems. Estimates of the number of mangrove fungi vary, with Schmit and Shearer (2003) listing 625 species: 278 (175) Ascomycota, 277 (35) asexual species, 30 (5) Basidiomycota, including those growing on the terrestrial parts of mangrove trees. The number in brackets is a better reflection of the fungi that occur on intertidal or submerged mangrove substrata, a total of 230 fungi and 24 other taxa. Some 287 fungi can be regarded as growing on submerged mangrove substrata (Alias et al. 2010) with the Ascomycota the most numerous and common taxonomic group and the Basidiomycota the least frequently collected (Hyde and Jones 1988; Alias and Jones 2009; Pang et al. 2011).

However, the basidiomycetes play an important role in wood decomposition of aerial parts of mangrove trees (Chalermpongse 1991; Gilbert et al. 2008; Baltazar et al. 2009a, b) and cause brown and white rots of wood and butt rot of the mangrove tree *Xylocarpus granatum* (Sakayaroj et al. 2012).

Extensive surveys of mangrove fungi have been undertaken and these are summarised in Alias and Jones (2009). Biodiversity studies have shown a wide variation in the frequency and abundance of mangrove fungi from one mangrove forest to the next, and this reflects the variation in local conditions, such as availability of substrata, salinity, tidal amplitude, age of mangrove forest, mangrove location (estuarine or fully marine) and sediment type (sand or mud) (Jones 2000; Alias and Jones 2000). One of the key factors is the availability of substrata for colonization by fungi, for example, mature mangrove forests yield a rich species diversity of plants, with senescent leaves, fallen twigs and branches, flowers and fruits. Young mangrove forests, exposed shores, oceanic waters support few fungi due to the lack of suitable substrata for colonization. Also the availability of fungal propagules (zoospores, conidia, ascospores, basidiospores) in the sea is important for colonization of the substrata.

Table 3.2 Core mangrove fungi based on analysis of data from Sarma and Hyde (2001), Raveendran and Manimohan (2007), Alias and Jones (2011), and Loilong et al. (2012)

On woody substrata	On palm fronds and leaves, most only known from *Nypa fruticans*
Halocyphina villosa	*Linocarpon nypae*
Dactylospora haliotrepha	*L. appendiculatum*
Verruculina enalia	*Neolinocarpon nypicola*
Lulworthia grandispora	*Oxydothis nypicola*
Halosarpheia marina	*Anthostomella nypae*
Leptosphaeria australiensis	*Phomatospora nypicola*
Halorosellinia ocenanica	*Tirisporella beccariana*
Kallichroma tethys	*Arecophila nypae*
Hydea pygmea	*Astrosphaeriella nypae*
Savoryella lignicola	*Carinospora nypae*
Netpunella longirostris	*Facsiatispora nypae*
Rhizophila marina	*Ligníncola nypae*
Aigialus grandis	*Vibrissea nypicola*
Periconia prolifica	*Helicorhoidion nypicola*
Morosphaeria velataspora	*Astrosphaeriella striatispora*
Morosphaeria ramunculicola	*Trichocladium nypae*
Marinosphaera mangrovei	*Linocarpon bipolaris*
Matsusporium tropicalis	*Oxydothis nypae*
Savoryella paucispora	*Aniptodera nypae*
Aagaromyces ratnagiriensis	
Halojulella avicenniae	

Studies of tropical and subtropical mangrove ecosystems have demonstrated a core group of fungi (Table 3.2). Many mangrove fungi are unique to this ecosystem and are not found in temperate or other intertidal locations. Many of these are host specific to the brackish water palm *Nypa fruticans* (Loilong et al. 2012).

Geographically, mangrove fungi have been found in every location surveyed (Jones et al. 2013). Schmit and Shearer (2004) analysed the geographical distribution data published on lignicolous mangrove fungi, and found that different oceans supported varying number of species: Atlantic Ocean: 12–46 per site (14 sites; mean 25.6); Indian Ocean: 12–64 (14; 42.9) and the Pacific Ocean: 17–87 (16; 44). This would appear to indicate that more are to be found in the Indo-West-Pacific Ocean where mangrove tree diversity is greatest, but this is more likely to reflect the intensity and frequency of sampling (Jones and Puglisi 2006; Alias and Jones 2009).

Temperature seems to be a major driver in mangrove fungi diversity and activity. Panebianco (1994) examined the temperature requirements of 19 marine fungi, 10 of which were either tropical or cosmopolitan species. The four tropical/mangrove species (*Halorosellinia oceanica, Kallichroma tethys, Saagaromyces glitra, Periconia prolifica*) all grew well at 30 °C, but not at 10 °C. Hughes (1974), Boyd and Kohlmeyer (1982) and Jones (2000) indicated that temperature is one of the most important parameters regulating the geographic distribution of marine fungi in oceans and estuaries.

Marine fungi also contribute to the diverse chemical matrix of mangrove environments. These fungi have been shown to yield over 1100 natural products representing all the major chemical structures (Ebel 2012). All major groups of chemical structures, e.g. polyketides, alkaloids, peptides and terpenoids, are documented for marine-derived fungi (Rateb and Ebel 2011). Mangrove fungi have been extensively screened for their ability to produce new chemical structures that have bioactive properties (Jones et al. 2008; Pan et al. 2008; Pang et al. 2015).

Fungal-Like Organisms

Fungal-like organisms include the Opisthokonta (Cryptomycota, Mesomycetozoea), Chromoalveolata (Oomycota, Hyphochytriomycota, Labyrinthulomycota) and the Rhizaria (Phytomyxea) (Neuhauser et al. 2012). None of these are active in the decomposition of woody material but are prevalent in mangroves, and play a vital role in the enrichment of decomposing leaves and particulate matter (see section on leaf substrates). Although widespread in mangrove ecosystems, comparative data on their geographic distribution is fragmentary.

Prokaryotes

Bacteria are present in all mangrove ecosystems, but few studies enumerate and document their occurrence on different substrata, although their role in the ecosystem is well known, especially their role in bio-mineralization and biotransformation of substances. They are essential food for protists and invertebrates, forming the base of benthic food webs. Whitman et al. (1998) estimated the total number of bacterial and archaeal cells in the marine environment (including the top 10 cm of sediment) and indicated they occur in the range 10^{28}–10^{29}. Bacteria and fungi degrade cellulosic material by producing the enzyme cellulase which hydrolyse the β-1, 4-glycosidic bonds in the polymer to release glucose units.

Wood exposed in the marine environment undergoes sequential colonization by a wide range of bacteria, which are the primary colonizers of wood and lignocellulolytic materials in the sea. These bacteria form a primary film on the wood and subsequently penetrate and digest the wood, in particular the cellulose. Bacteria form burrows in the wood and different forms have been characterized (Holt and Jones 1983; Holt et al. 1980; Venkatasamy 1986; Mouzouras et al. 1987).

Animal Wood Borers

Animal wood borers play a key role in the decomposition of woody material in mangrove habitats. They are more efficient than microorganisms as they can penetrate to a greater depth within the wood (Becker 1971; Turner and Johnson 1971). Two groups are dominant in such habitats: molluscs (Teredinidae and Pholadidae) and crustaceans (Limnoriidae and Sphaeromatidae), the shipworms penetrate to the core of large pieces of wood, while gribbles (limoriids) burrow superficially (Cookson et al. 2012).

Wood-boring molluscs are represented by the Teredinidae and Pholadidae (shipworms and piddocks). Molluscan wood borer's ability to drill into and destroy wood is well documented and they are considered an economic pest because of the extensive damage that their actions have on wooden marine structures. While much is known about molluscan wood borers, few studies have examined their role in mangrove ecosystems (Southwell and Bultman 1971; Cragg 1993). Species dominance and abundance vary widely between locations (e.g. Singh and Sasekumar 1994; Kohlmeyer et al. 1995; Crockett 2014). Twenty-seven molluscan borer species have been reported on wood panels exposed in Australian waters, but at the Burrup Peninsula, Western Australia only one pholad and five teredinid species were collected from mangrove habitats (Marshall and Turner 1974; Brearley et al. 2008). Both pholadid and teredinid species showed distinct habitat differences. Such preferences have also been noted for limnoriid species, with leachate from *Bruguiera gymnorhiza* causing high mortality rates (Borges et al. 2008). Han et al. (2004) suggested that this toxicity might be attributed to terpenoids present in the timber. Data on the geographical distribution of molluscan borers is dependent on where researchers have been based, with observations by Turner (1971) being the most comprehensive. Although Turner (1971) lists the geographic locations of all the species in her monograph, few refer to whether they are to be found in mangrove habitats. Our knowledge of mangrove molluscan borer occurrence remains fragmentary. For example, *Dicyathifer manni* has been found throughout the tropical to subtropical, Indo-Pacific in marine and brackish water and was the most abundant species found in a Western Australian study (Brearley et al. 2008). *Bankia rochi* has an Indo-Pacific tropical to subtropical distribution (Turner 1971), while *Lyrodus pedicellatus* and *Myadora striata* have a worldwide distribution in temperate and tropical waters (Turner 1971).

Most studies of wood-boring limnoriid isopods have been conducted in temperate habitats and focused on the decomposition and preservation of wood (Borges et al. 2008; Cookson 1999; Becker 1971; Kühne 1971). Not all taxa are found in mangrove environments, with *Limnoria insulae*, *Limnoria pfefferi*, *Limnoria sellifera* and *Limnoria unicornis* being the most common in Indonesian mangrove forests (Cookson et al. 2012). Other species recorded from mangrove habitats include: *Limnoria clarkae* (Kensley and Schotte 1987) and *Limnoria cristata* (Cookson and Cragg 1991). Sphaeromid wood borers include three genera: *Sphaeroma* (three species commonly reported as wood borers), *Exosphaeroma* (three species with uncertain status as wood borers) and *Cymodoce* with 50 species (Shiino 1957). *Chelura* species (Cheluridae) are also found on wood in the tropics. *Sphaeroma* species are warm water species, often in brackish waters, while *Chelura* is primarily found in temperate habitats. *Limnoria* and *Sphaeroma* species are of great economic importance due to their high abundances on suitable substrates. Damage of the timber is continuous and proceeds more slowly than that brought about by marine molluscs. However, limnoriid and teredinid may occur simultaneously on mangrove wood (partitioning of resource), teredinids deep in the wood, while the limnoriid occupy the surface layers (Cookson et al. 2012). Ultimately, the wood is destroyed by the teredinids. While limnoriid burrowing into wood is well reported, their ability to eat and digest wood particles has been the source of much speculation (Eltringham and

Jones 1971). Recent studies demonstrate the presence of glycosyl hydrolase (GH7) cellobiohydrolases in *Limnoria*, enabling them to digest lignocellulose even with no microorganisms found in their gut (King et al. 2010; Kern et al. 2013). Others have suggested that fungi condition the wood prior to *Limnoria* burrowing the wood (Ray and Julian 1952).

Wood borers can affect the fecundity, performance and morphology of mangroves. Thus, extensive damage to mangrove trees by *Sphaeroma terebrans* has been reported in Taiwan. Individuals of *Rhizophora stylosa* that were more infested by isopods had significantly fewer propagules, fewer ground roots stabilizing the tree, smaller leaves, and more non-foliated twigs. Similarly, *Avicennia marina* with more burrows had fewer pneumatophores and lenticels (used for gaseous exchange), and pneumatophores with more necrotic tissue (Davidson et al. 2014).

3.2.2.4.2 Leaf Substrates

Macrofauna

The role of gastropods and sesarmid crabs for the initial processing of mangrove leaf litter is a topic of significant attention (Lee 1998, 2008) and will not be covered here. Species richness as well as density of these initial consumers of leaf litter are significantly higher in the IWP than the AEP, but difference in litter processing activity may not be as significant, e.g. the large-bodied *Ucides cordatus* could consume >80% of the leaf litter and propagules produced by *Rhizophora mangle* in north Brazil (Nordhaus et al. 2006), similar to the impact of sesarmids in the IWP (Lee 2008).

Fungi

Fungi also colonize mangrove and seagrass leaves but species diversity is not great compared with the number documented for woody substrates (see above section). The lack of fungal diversity on mangrove leaves may be due to the earlier colonization by fungal-like organisms and bacteria, as leaf material is less lignified and more easily degraded. Many of the microbial organisms listed (Table 3.2) produce lignolytic and cellulolytic enzymes that are responsible for the breakdown of the complex polymers in wood and leaf material, converting it to smaller particulate material which is then acted on by bacteria and other organisms. Thus, setting up the food web on which small animals can survive on (Fig. 3.2). Although fungal-like organisms and bacteria are widely reported from both IWP and AEP little data is available on their geographical frequency and diversity.

Fungal-Like Organisms

Fungal-like organisms belong to the Chromoalveolata and are heterotrophic stramenopiles and include: Oomycota (Saprolegniales, Pythiales, Haliphthorlaes) Labyrinthulomycota (Thraustochytrids, Labyrinthuloids Aplanochytrids) (Neuhauser

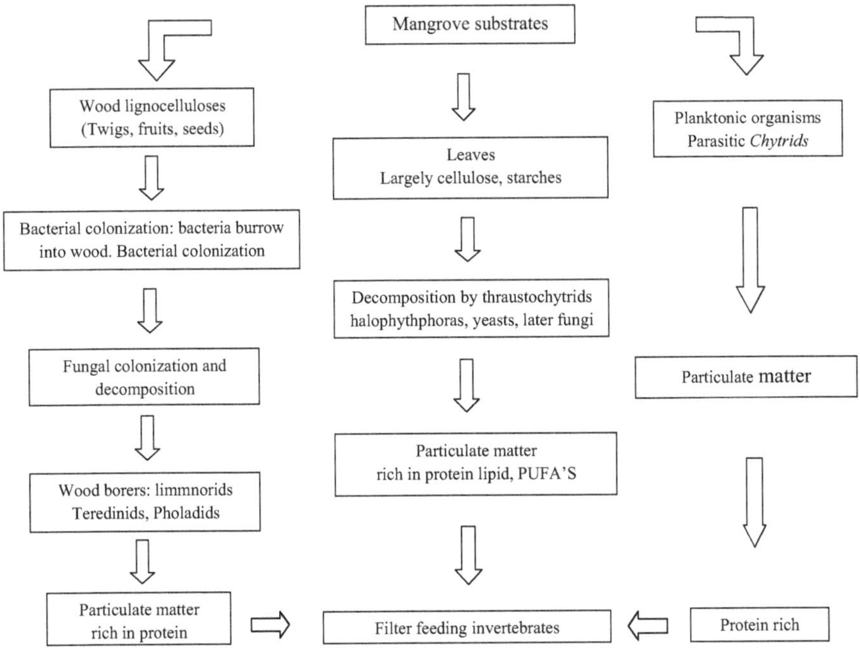

Fig. 3.2 Sequential colonization and decomposition of mangrove substrata by macrofaunal (excluding sesarmid crabs) and microbial decomposers and contribution to filter-feeding animals

et al. 2012; Kearney and Gleason 2014). There are two main groups in the class: (1) labyrinthulids characterized by the production of an ectoplasmic net within which somatic cells can move and absorb nutrients, and (2) thraustochytrids characterized by globose to subglobose vegetative cells with an external ectoplasmic net, scaled cell wall and production of biflagellate zoospores (Leaño and Damare 2012). Some of these organisms are primarily pathogens of animals, algae and microalgae (Haliphthorlaes, Saprolegniales, Pythiales), while the thraustochytrids are best known for their ability to produce omega-3 PUFAS, enriching the particulate matter in mangrove ecosystems (Leaño and Damare 2012).

The earliest colonizers of senescent fallen leaves are the thraustochytrids and only asexual reproduction has been observed, during which zoospores are formed in zoosporangia and upon release, they are attracted to and swim towards organic matter (Fan et al. 2002b). Three groups are recognised based on their mode of growth, types of spores produced: the first group includes thraustochytrids, aplanochytrids and labyrinthulids (Leander et al. 2004) with a total of 53 species (Alias and Jones 2011). Currently, there are 10 thraustochytrid genera (Leaño and Damare 2012) and not all these genera have been reported from the tropics or in mangrove environments. The second group of thraustochytrids is the genus *Aplanochytrium* characterised by aplanospore production and eight species are known (Leaño and Damare 2012). The third group is the labyrinthulids with one genus *Labyrinthula* and characterised by the formation of slime nets with 16 described species (Honda et al. 1999).

These taxa occur on a wide range of substrata in full saline to brackish water, are worldwide in distribution and are an important part of the marine microbial communities (Schärer et al. 2007). Some species are parasitic, but most are saprobes on decaying mangrove leaves and animal detritus (Bremer and Talbot 1995; Bongiorni et al. 2005; Velmurugan and Lee 2012), but with particularly high abundance in plant detritus (Raghukumar et al. 1992). Alias and Jones (2011) list 12 species that have been recorded from mangrove habitats, from India, Hong Kong, Malaysian Philippines, and South America. Thraustochyrids are particularly common on decaying mangrove leaves (Alias and Jones 2011). As a result of their degradative activities, they provide food for filter feeders in mangrove food webs (Wong et al. 2005; Neuhauser et al. 2012; Pang et al. 2015).

Thraustochytrids have attracted particular interest because of their ability to produce high amounts of polyunsaturated omega-3 fatty acids (PUFA) (Bajpaj et al. 1991; Yaguchi et al. 1997; Fan et al. 2000, 2001; Leano et al. 2003; Tanaka et al. 2003), including docsahexaenoic acid (DHA; 22:6ω3). DHA is of great importance ecologically by enriching the diet of juvenile fish, crabs and shrimps and confers an advantage to survival and growth in mangrove habitats (Jarikhuan 2002; Fan and Chen 2006; Jarikhuan et al. 2007).

Another important group of fungal-like organisms in the colonization and degradation of mangrove leaf material are *Halophytophthora* species (Leaño et al. 1998; Hulvey et al. 2010) that appear later in the succession. These organisms are widely reported from Australia, Bahamas, Costa Rica, Japan, Malaysia, Philippines, Singapore, Taiwan, and USA (Marano et al. 2012). *Halophytophthora* spp. are mycelial and produce biflagellate zoospores that show a chemotactic response to decaying mangrove leaves (Leaño et al. 1998). They have been documented from the leaves of a diverse range of mangroves (Marano et al. 2012). *Halophytophthora* species contribute to the enrichment of mangrove detritus, particularly with EPS, for consumption by organisms at higher tropic level (Leaño et al. 1998; Marano et al. 2012).

3.2.2.4.3 Meiofauna in Sediment

Meiofauna represent an abundant, but understudied, biodiversity component of mangrove communities (Table 3.3). A whole ecoregion study of east Africa revealed higher meiofaunal abundance but lower diversity in mangrove forests compared with sandy beaches (Barnes et al. 2011). Again, studies are strongly biased towards IWP assemblages, with only a few from the AEP. Difficulties in extraction, enumeration and identification hamper research on mangrove meiofauna. Diversity at the species level is largely unknown, as most studies only report diversity at higher taxonomic levels due to difficulties in species identification. Little is known about mangrove meiofaunal species to date – for example, the first kinorhynch species of an Indonesian mangrove forest has only just recently been described (Ostmann et al. 2012). The lack of uniformity in taxonomic resolution makes diversity patterns difficult to be established or compared. Almost all studies, however, suggest numerical

Table 3.3 Meiofaunal assemblages associated with mangrove sediment

Location	Density (ind. per 10 cm^2)	Biomass (mg per 10 cm^2)	Identified drivers of assemblage structure	Remarks	Reference
IWP					
Bowling Green Bay, Australia	60–608	NA	Density of harpacticoids correlated with sediment type	No seasonal fluctuation in density; nematodes numerically dominant	Hansen and Alongi (1991)
Gazi Bay, Kenya, Africa	1976–6707	NA	Particle size (−) Oxygenation (+) Fiddler crab burrows (+)	Nematodes as dominant group; significant different density among five mangrove forests	Vanhove et al. (1992)
River Mahanadi, India	1192–1437	NA	Organic content (+)	Higher density at low water mark	Sarma and Wilsanand (1994)
West Coast, Peninsular Malaysia	407–1109	0.96–7.54 (wet wt)	Habitat instability (salinity and tidal inundation fluctuation, −)	Nematodes as dominant group; significant different density among three mangrove forests	Sasekumar (1994)
Zanzibar, Africa	1493	NA	Salinity (−)	Nematodes as dominant group	Olafsson (1995)
Zanzibar, Africa	271–656	NA	Sediment grain size and organic content	28 species in total	Olafsson et al. (2000)
Transkei, S Africa	840–5300	0.37–2.33 (dry wt)	Redox potential (Eh)	Nematodes dominant	Dye (1983a,b)
Cochin, India	508 (max)	NA	NA	Nematodes as dominant group	Chinnadurai and Fernando (2006)
Chuuk lagoon, Micronesia	1671–2967	1.06–1.18 (dry wt)	NA	Nematodes most abundant group; density higher but biomass lower in mangroves compared to seagrass	Min et al. (2006)
Pichavaram and Parangipetti, India	890 (max)	NA	NA	Nematodes most dominant group	Chinnadurai and Fernando (2007)

(continued)

Table 3.3 (continued)

Location	Density (ind. per 10 cm^2)	Biomass (mg per 10 cm^2)	Identified drivers of assemblage structure	Remarks	Reference
Can Gio, Ho chi Minh City, Vietnam	1156–2082	NA	Tidal position (−)	Higher abundance towards the low tide mark	Xuan et al. (2007)
Segara Anakan lagoon, Java, Indonesia	27–3057	NA	Salinity (−), sediment $\delta^{15}N$ (+); no sign. Correlation with sediment C_{org}, N_{tot}, median grain size, chlorophyll a	Nematodes numerically dominant, followed by copepods	Ostmann (2010)
Be River estuary, Nha Trang, Vietnam	244–735	NA	NA	90% of meiofauna found in top 1 cm of sediment	Mokievsky et al. (2011)
Chorao Island, Goa, India	166–519	NA	NA	Nematodes as dominant group; significant different density among four mangrove forests	Sahoo et al. (2013)
Southeast Queensland, Australia	100–300	NA		Nematodes as dominant group, followed by harpacticoid copepods	Abdullah and lee (2016)
AEP					
South Carolina, USA	~3000	NA	Ray excavation activity	Nematodes as dominant group	Cross and Curran (2000)
Island of Santa Catarina, Brazil	77–1589	NA	Detritus abundance (+)	Nematodes as dominant group	Netto and Gallucci (2003)
Gulf of Batabano, Cuba	101	NA	No significant influence of abiotic factors but salinity fluctuation may be important (−)		Armenteros et al. (2006)

+ positive influence, − negative influence, *NA* not available

dominance by nematodes, usually followed by harpacticoid copepods (Table 3.3), confirming earlier observations (Nagelkerken et al. 2008). Large variations up to two orders of magnitude exist among abundance data from different studies, but generally density of meiofauna approaches or exceeds 1000 individuals per 10 cm^2. Biomass information is generally lacking but conversion from published data (0.44 μg per individual, Dye 1983a) suggests this density would equate to ≥0.44 g dry wt m^{-2}.

Drivers determining meiofauna abundance and assemblage structure include the organic content and grain size of the sediment, and general stability of the environment (e.g. fluctuations in salinity, tidal inundation), which may be mediated also by biotic forces such as bioturbation or predation (e.g. Fleeger et al. 2006; Abdullah and Lee 2016). The impact of environmental fluctuations on faunal assemblages including the meiofauna has long been suspected, with Attrill (2002) proposing that the salinity range may determine diversity along an estuary, which has been demonstrated by a study of the nematode assemblages along the Thames estuary (Ferrero et al. 2008). Higher abundance is associated with higher organic content and smaller particle size. This is not surprising, as finer sediments generally have higher organic content (Lopez et al. 1989).

The ecological role played by the meiofauna in mangrove ecosystems is obscure, as few studies have gone beyond simple enumeration and assemblage structure. Some data suggest that meiofauna may support the nutrition of juvenile fish (Coull et al. 1995; Street et al. 1998) and small crustaceans (Dittmann 1993) but their trophic importance is unknown. The small size of meiofauna present significant challenges in conventional approaches to assessing trophic significance, such as gut contents analysis or stable isotope analysis. With the advance in sensitivity of stable isotope analysis, however, new light may be shed on the trophic role of meiofauna to the mangrove food web.

3.3 Mangrove Biodiversity and Ecosystem Function

An intriguing feature of mangrove ecosystems is the high level of functionality, based on natural assemblages of an overall low level of diversity when compared to terrestrial forests both at the global (i.e. biogeographic regions) and local (i.e. forest) scales. For example, mangrove ecosystems support high primary productivity and carbon recycling rates, as well as resilience to species invasion (Lugo 1998), which are general characteristic of species-rich ecosystems. The impact of shifts in diversity of the habitat building, e.g. terrestrial forests (Lohbeck et al. 2012; Sugiura et al. 2013; Kusumoto et al. 2015), or process-mediating, e.g. soil microbes influencing soil carbon cycling (Nielsen et al. 2011) species on ecosystem function have been reported for many ecosystems, mostly with baseline diversity levels much greater than those in mangrove forests. No report to date is available on the nexus between mangrove tree diversity and ecosystem function. Much investment has been made on restoring mangrove habitats through planting projects but simplistic approaches to restoring mangrove wetlands through monoculture plantations, sometimes even using exotic species, could pose serious threats to the diversity of mangrove forests at both the local and global levels. Mangrove forests earmarked for development are often not sufficiently assessed before simple ex situ replanting being accepted as offset or mitigation measures. Without suitable evaluation of the conservation value of mangrove forests (e.g. 'Redlisting' of ecosystems rather than just of species, Keith et al. (2015)), a 'no net loss' approach to maintain overall mangrove area may result in diverse forests being replaced by plantations of low species or structural diversity. Restoration approaches with a narrow focus on establishing the trees are less likely able to restore biodiversity to levels similar to undisturbed forests (Ellison 2000). This is exacerbated by the fact that the return of faunal and other components of biodiversity or function in the planted mangrove forests is often not evaluated or done inadequately (Bosire et al. 2008). However, if suitably monitored, these monoculture plantations may provide comparison sites for rigorous investigations on the relationship between species diversity and ecosystem function (e.g. Macintosh et al. 2002). Limited data on multiple-species plantations suggest there may be some complementarity among species in determining forest traits such as aboveground biomass (Kirui et al. 2012).

As discussed earlier, the two main mangrove biogeographic regions differ not only in mangrove tree diversity but also diversity of their associated biotic assemblages (Bowen et al. 2013). More comparative studies on the function of the species-poor AEP and species-rich IWP mangrove ecosystems will shed important light on the nexus between mangrove diversity and ecosystem function. It would be of fundamental theoretical as well as practical management interest to know if and when functional redundancy occurs.

3.4 The Future of Mangrove Biodiversity

One key finding emerging from an analysis of the total biodiversity of mangrove ecosystems is the disproportionately large number of species supported by a small number of mangrove tree species. Recent advances in biodiversity assessment such as molecular tools in metagenomics could provide more cost-effective approaches to assessing biodiversity in mangrove ecosystems. The present data suggest that total species richness supported by mangrove ecosystems is usually two orders of magnitude greater than the number of mangrove tree species. Rapid loss of mangrove habitat, particularly in species-rich biogeographic regions such as the Indo-West-Pacific, has significant implications for the future of mangrove biodiversity. Specific components of the mangrove habitat are particularly at risk. One of the biodiversity components that would be most endangered because of mangrove loss is species that are endemic to mangrove habitats. A global survey conducted by Luther and Greenberg (2009) found 69 species of terrestrial vertebrates (48 bird, 14 reptile, 1 amphibian and 6 mammal species) that are endemic to mangrove ecosystems, particularly in Asia and Australia where 47 out of the 69 (68%) species occur. Further, 40% of these mangrove-endemic vertebrates are globally threatened according to the IUCN Red List. As the largest-bodied animals depend on mangrove forests as habitat, these mangrove-endemic vertebrates will likely be the first species to become extinct if the present trend of mangrove loss, particularly in the Indo-Malayan region, is not reversed. Most mangrove fungi are endemic to the marine environment with 135 species reported on the brackish water palm *Nypa fruticans*, 24% known only from this mangrove palm (Loilong et al. 2012). Further, this component of mangrove biodiversity is least likely going to return to restored systems, as the very environmental fabric, e.g. connectivity and corridors with adjacent habitats that support their existence, is usually already lost around plantations established for restoration.

The high-intertidal region of tropical coastlines is extensively 'reclaimed' for human settlement, aquaculture and agriculture. For example, >80% of mangrove forests in China are now backed by concrete sea walls (W. Wang, personal comm.), which covers a total length of 11,000 km or 61% of the country's coastline (Ma et al. 2014), seriously jeopardising the biota in the high intertidal, e.g. semi-terrestrial sesarmid crabs and mangrove associates plant species. The high-zoned mangrove species such as *Heritiera* spp. or *Bruguiera gymnorhiza*, for example, are now rare in Chinese mangrove forests. This threat is certainly not unique to China but ubiquitous in many developing countries in both the IWP and AEP. Conversion of upper intertidal mangrove forests into aquaculture ponds significantly reduces species richness, even when managed extensively (Lee 1992). Similar to many anthropogenic large-scale alterations of natural habitats for development (e.g. the recent 53,000 km 'development corridors' in Africa (Laurance et al. 2015)), conversion of the high intertidal mangrove habitat not only harms biodiversity but also results in little net long-term sustainable benefit to coastal communities, as exemplified by the short productive life span of shrimp ponds in SE Asia.

Efforts to generate practical methods to evaluate the conditions ('health') of mangrove ecosystems including their value in biodiversity conservation are urgently needed to identify and protect dwindling mangrove biodiversity resources and the services they support. Mangrove ecosystems that support high levels of plant and animal diversity should be identified and protected, e.g. the Merbok mangrove in Western Peninsular Malaysia supports ~50% of the world mangrove tree species (J.E. Ong, pers. comm.). Systems in the AEP are naturally less species-rich but there is no evidence to date that functionality is compromised to warrant less attention to their conservation. Studies on sesarmid crab assemblages indicate that management practice and history (e.g. rehabilitation) also strongly influence crab diversity and abundance (Macintosh et al. 2002; Ashton et al. 2003b). However, the common practice of using monoculture plantations, sometimes employing exotic species, as a means of mangrove restoration should be discouraged. This approach promotes the continual loss of high-intertidal forests and their associated biodiversity by falsely assuming that their replacement by monocultures results in no net loss of functionality or biodiversity. Incidentally, high-intertidal mangrove forests with little tidal flushing are also hotspots for carbon storage, further arguing for their urgent conservation. Conservation and management of mangrove biodiversity require efforts at various levels of socio-political organisation as well as a multi-pronged approach employing legal and economic instruments (Macintosh and Ashton 2002).

Acknowledgements Gareth Jones is supported by the International Research Group Program (IRG-14-27), Deanship of Scientific Research, King Saud University, Saudi Arabia. We thank Prof. Simon Cragg for assistance with literature on marine borers. GA Castellanos-Galindo is supported by the Alexander von Humboldt Foundation. Karen Diele received funding from the MASTS pooling initiative (The Marine Alliance for Science and Technology for Scotland) and its support is gratefully acknowledged. MASTS is funded by the Scottish Funding Council (grant references HR09011) and contributing institutions. Miss Monica Dayarathane is thanked for preparing Fig. 3.2. We also thank the reviewers for constructive comments and additional references.

References

Abdullah MM, Lee SY (2016) Meiofauna and depositing-feeding crabs: is the interaction physical or trophic? J Exp Mar Biol Ecol 479:69–75

Aderaldo Vidal JM, Rocha-Barreira CDA (2009) Shipworms (Mollusca: Bivalvia; Teredinidae) from a Brazilian northeast estuary. Arquivos Cien Mar 42:43–49

Alias SA, Jones EBG (2000) Vertical distribution of marine fungi on *Rhizophora apiculata* at Morib mangrove, Selangor, Malaysia. Mycoscience 41:431–436

Alias SA, Jones EBG (2009) Marine fungi from mangroves of Malyasia. UniversityMalaysia, IOES. Monogr Ser 8:1–109

Alias SA, Zainuddin N, Jones EBG (2010) Biodiversity of marine fungi in Malaysian mangroves. Bot Mar 53:545–554

Alongi DM (1987a) The influence of mangrove-derived tannins on intertidal meiobenthos in tropical estuaries. Oecologia 71:537–540

Alongi DM (1987b) Intertidal zonation and seasonality of meiobenthos in tropical mangrove estuaries. Mar Biol 95:447–458

do Amaral VS, Simone LRL (2014) Revision of genus *Crassostrea* (Bivalvia: Ostreidae) of Brazil. J Mar Biol Assoc UK 94:811–836

Anderson C, Lee SY (1995) Defoliation of the mangrove *Avicennia marina* in Hong-Kong - cause and consequences. Biotropica 27:218–226

Araujo FV, Soares CAG, Hagler AN, Mendonca-Hagler LC (1995) Ascomycetous yeast communities of marine invertebrates in a Southeast Brazilian mangrove ecosystem. Antonie Van Leeuwenhoek 68:91–99

Armenteros M, Martin I, Williams JP, Creagh B, Gonzalez-Sanson G, Capetillo N (2006) Spatial and temporal variations of meiofaunal communities from the western sector of the Gulf of Batabano, Cuba. I. Mangrove systems. Estuar Coasts 29:124–132

Arnaud-Haond S, Teixeira S, Massa SI, Billot C, Saenger P, Couplands G, Duarte CM, Serrao EA (2006) Genetic structure at range edge: low diversity and high inbreeding in Southeast Asian mangrove (*Avicennia marina*) populations. Mol Ecol 15:3515–3525

Ashton EC, Macintosh DJ, Hogarth PJ (2003a) A baseline study on the diversity and community ecology of crustacean and molluscan macrofauna in the Sematan mangrove forest, Sarawak, Malaysia. J Trop Ecol 19:127–142

Ashton EC, Hogarth P, Macintosh DJ (2003b) A comparison of brachyuran crab community structure at four mangrove locations under different management systems along the Melaka Straits-Andaman Sea Coast of Malaysia and Thailand. Estuaries 26:1461–1471

Aviz D, Ferreira de Mello C, da Silva PF (2009) Macrofauna associada às galerias de Neoteredo reynei (Bartsch, 1920) (Mollusca: Bivalvia) em troncos de Rhizophora mangle Linnaeus durante o período menos chuvoso, em manguezal de São Caetano de Odivelas, Pará (costa norte do Brasil). Bol Mus Para Emílio Goeldi Ciên Nat Belém 4:47–55

Bajpaj P, Bajpaj PK, Ward OP (1991) Production of docosahexaenoic acid by *Thraustochytrium aureum*. Appl Micobiol Biotech 35:706–710

Balbas VA, Tineo RB, Arcas AP (2014) Estructura comunitaria de bivalvos y gasterópodos en raíces del mangle rojo *Rhizophora mangle* (Rhizophoraceae) en Isla Larga, Bahía de Mochima, Venezuela. Rev Biol Trop 62:551–565

Baltazar JM, Trierveiler-Pereira L, Loguercio-Leite C (2009a) A checklist of xylophilous basidiomycetes (Basidiomycota) in mangroves. Mycotaxon 107:221–224

Baltazar JM, Trierveiler-Pereira L, Loguercio-Leite C, Ryvarden L (2009b) Santa Catarina Island mangroves 3: a new species of Fuscoporia. Mycologia 101:859–863

Barlow BA (1984) Loranthaceae. In: George AS (ed) Flora of Australia, vol 22. AGPS, Canberra, pp 68–131

Barnes N, Bamber RN, Bennell G, Cornelius N, Glassom D, Conroy-Dalton Henderson S, Jiddawi N, Lee CNW, Macia A, Msangameno DJ, Paula J, Yahya S, Ferrero TJ (2011) Assessment of regional and local biodiversity in tropical and subtropical coastal habitats in the East African Marine Ecoregion. Biodivers Conserv 20:2075–2109

Barroso CX, Matthews-Cascon H (2009) Occurrence of the exotic freshwater snail *Melanoides tuberculatus* (Mollusca: Gastropoda: Thiaridae) in an estuary of north-eastern Brazil. Mar Biodiv Rec 2:1–4

Beasley CR, Fernandes CM, Gomes CP, Brito BA, Lima dos Santos SM, Tagliaro CH (2005) Molluscan diversity and abundance among coastal habitats of northern Brazil. Ecotropica 11:9–20

Becker G (1971) On the biology, physiology and ecology of marine wood-boring crustaceans. In: EBG J, Eltringham SK (eds) Marine borers, fungi and fouling organisms of wood. Organisaion for Economic Co-operation and Development, Paris, pp 303–326

Blanco JF, Cantera JR (1999) The vertical distribution of mangrove gastropods and environmental factors relative to tide level at Buenaventura Bay, Pacific coast of Colombia. Bull Mar Sci 65:617–630

Borges LMS, Cragg SM, Bergot J, Williams JR, Shayler B, Sawyer GS (2008) Laboratory screening of tropical hardwoods for natural resistance to the marine borer *Limnoria quadripunctata*: the role of leachable and non-leachable factors. Holzforschung 62:99–111

Bosire JO, Dahdouh-Guebas F, Walton M, Crona BI, Lewis RR, Field C, Kairo JG, Koedam N (2008) Functionality of restored mangroves: a review. Aq Bot 89:251–259

Bowen BW, Rocha LA, Toonen RJ, Karl SA, Laboratory TB (2013) The origins of tropical marine biodiversity. Trends Ecol Evol 28:359–366

Boyd PE, Kohlmeyer J (1982) The influence of temperature on the seasonal and geographic distributions of three marine fungi. Mycologia 74:894–902

Brearley A, Kendrick GA, Walker DI (2008) How does burrowing by the isopod *Limnoria agrostisa* (Crustacea: Limnoriidae) affect the leaf canopy of the southern Australian seagrass *Amphibolis griffithii*? Mar Biol 156:65–77

Bremer GB, Talbot G (1995) Cellulolytic enzyme activity in the marine protis *Schizochytrium aggregatum*. Bot Mar 38:37–41

Cai LZ (2015) Zoobenthic ecology in Shenzhen Bay. Xiamen University Press, Xiamen, China, p 303

Cannicci S, Burrows D, Fratini S, Smith TJ, Offenberg J, Dahdouh-Guebas F (2008) Faunal impact on vegetation structure and ecosystem function in mangrove forests: a review. Aq Bot 89:186–200

Cantera KJR (2010) Bivalvos perforadores de madeira (Mollusca: Teredinidae, Pholadidae) en la costa Pacifica Colombiana. Rev Acad Colomb Cienc 34:277–288

Cantera KJR, Thomassin BA, Arnaud PM (1999) Faunal zonation and assemblages in the Pacific Colombian mangroves. Hydrobiologia 413:17–33

Castellanos-Galindo GA, Krumme U (2013) Mangrove fish assemblages from data-sparse regions and the measurement of ecological equivalence: comment on Sheaves (2012). Mar Ecol Prog Ser 474:299–302

Cedeño J, Prieto MJ, Pereda L, Allen T (2010) Abundancia y riqueza de moluscos y crustáceos asociados a las raíces sumergidas del mangle rojo (*Rhizophora mangle*) en la laguna de Bocaripo, Sucre, Venezuela. Rev Biol Trop 58:213–226

Chalermpongse A (1991) Fungal diseases in mangrove ecosystem. In: Proceeding, the 5th silviculture seminar in Thailand, Division of Silviculture, Royal Forest Department, Bangkok, pp 307–338

Chinnadurai G, Fernando OJ (2006) Meiobenthos of Cochin mangroves (southwest coast of India) with emphasis on free-living marine nematode assemblages. Rus. J Nematol 14:127–137

Chinnadurai G, Fernando OJ (2007) Meiofauna of mangroves of the southeast coast of India with special reference to the free-living marine nematode assemblage. Estuar Coast Shelf Sci 72:329–336

Cookson LJ (1999) Twenty year marine trial of single and double preservative treated timber specimens in Australia. Mat Org 33:65–79

Cookson LJ, Cragg SM (1991) *Limnoria cristata* (Isopoda: Limnoriidae), a new species of marine wood-borer from Singapore. Raffles Bull Zool 39:87–97

Cookson LJ, Cragg SM, Hendy IW (2012) Wood-inhabiting limnorids (Crustacea, Isopoda) including a new species from mangrove forests of the Tukang Besi Archipelago, Indonesia. Zootaxa 3248:25–34

Costa de Melo AG, Sousa Varela E, Beasley CR, Schneider H, Sampaio I, Gaffney PM, Reece KS, Tagliaro CH (2010) Molecular identification, phylogeny and geographic distribution of Brazilian mangrove oysters (*Crassostrea*). Genet Mol Biol 33:564–572

Coull BC (1999) Role of meiofauna in estuarine soft-bottom habitats. Aust J Ecol 24:327–343

Coull BC, Greenwood JG, Fielder DR, Coull BA (1995) Subtropical Australian juvenile fish eat meiofauna - experiments with winter whiting *Sillago maculata* and observations on other species. Mar Ecol Prog Ser 125:13–19

Cragg SM (1993) Wood breakdown in mangrove ecosystems: a review. PNG Journal Agric For Fish 36:30–39

Crockett MJ (2014) Marine wood borers of Barbados, West Indies: spatial and temporal variation in recruitment. M. Phil., University of Portsmouth, 99p

Cross RE, Curran MC (2000) Effects of feeding pit formation by rays on an intertidal meiobenthic community. Estuar Coast Shelf Sci 51:293–298

Davidson TM, De Rivera CE, Hsieh HL (2014) Damage and alteration of mangroves inhabited by a marine wood-borer. Mar Ecol Prog Ser 516:177–185

Diaz MC (2012) Mangrove and coral reef sponge faunas: untold stories about shallow water Porifera in the Caribbean. Hydrobiologia 687:179–190

Dittmann S (1993) Impact of foraging soldier crabs (Decapoda, Mictyridae) on meiofauna in a tropical tidal flat. Rev Biol Trop 41:627–637

Duke N (2006) Australia's mangroves. University of Queensland, Brisbane, Australia, p 200

Dye AH (1983a) Composition and seasonal fluctuations of meiofauna in a southern African mangrove estuary. Mar Biol 73:165–170

Dye AH (1983b) Vertical and horizontal distribution of meiofauna in mangrove sediments in Transkei, southern Africa. Estuar Coastal Shelf Sci 16:591–598

Ebel R (2012) Natural products from marine-derived fungi. In: Jones EBG, Pang KL (eds) Marine fungi and fungal-like organisms. De Gruyter, Berlin, pp 411–440

Ellison AM (2000) Mangrove restoration: do we know enough? Restor Ecol 8:219–229

Eltringham SK, Jones EBG (1971) Marine borers, fungi and fouling organisms of wood. OECD, Paris, p 367

Fan KW, Chen F (2006) Production of high-value products by marine microalgae thraustochytrids. In: Yang ST (ed) Bioprocessing for value-added products from renewable resources. Elsevier, Amsterdam, pp 293–323

Fan KW, Chen F, Jones EBG, Vrijmoed LLP (2000) Utilization of food processing waste by Thraustochytrids. In: Hyde KD, Ho WH, Pointing SB (eds) Aquatic Mycology across the Millennium. Fungal Div 5:185–194

Fan KW, Chen F, Jones EBG, Vrijmoed LLP (2001) Eicosapentaenoic and docosahexaenoic acids production by and okara-utilizing potential of thraustochytrids. J Ind Microbiol Biotechnol 227:199–202

FAO (2007) The World's mangroves 1980–2005. FAO, Rome, Italy, p 77

Faraco LFD, Lana PC (2004) Macrobenthic recolonization processes in mangroves of southern Brazil. J Coatal Res 39:1853–1858

Ferrero TJ, Debenham NJ, Lambshead PJD (2008) The nematodes of the Thames estuary: assemblage structure and biodiversity, with a test of Attrill's linear model. Estuar Coastal Shelf Sci 79:409–418

Fleeger JW, Tita G, Carman KR, Millward RN, Moser EB, Portier RJ, Gambrell RP (2006) Does bioturbation by a benthic fish modify the effects of sediment contamination on saltmarsh benthic microalgae and meiofauna? J Exp Mar Biol Ecol 330:180–194

Ge XJ, Sun M (1999) Reproductive biology and genetic diversity of a cryptoviviparous mangrove *Aegiceras corniculatum* (Myrsinaceae) using allozyme and intersimple sequence repeat (ISSR) analysis. Mol Ecol 8:2061–2069

Geist SJ, Nordhaus I, Hinrichs S (2012) Occurrence of species-rich crab fauna in a human-impacted mangrove forest questions the application of community analysis as an environmental assessment tool. Estuar Coast Shelf Sci 96:69–80

Gilbert GS, Gorospe J, Ryvarden L (2008) Host and habitat preferences of polypore fungi in Micronesian tropical flooded forests. Mycol Res 112:674–680

Goodbody I (2003) The ascidian fauna of Port Royal, Jamaica - I. Harbor and mangrove dwelling species. Bull Mar Sci 73:457–476

Greenfield DW, Johnson RK (1999) Assemblage structure and habitat associations of western Caribbean gobies (Teleostei: Gobiidae). Copeia 1999:251–266

Han L, Huang XS, Sattler I, Dahse HM, HZ F, Lin WH, Grabley S (2004) New diterpenoids from the marine mangrove *Bruguiera gymnorrhiza*. J Nat Prod 67:1620–1623

Hansen JA, Alongi DM (1991) Bacterial productivity and benthic standing stocks in a tropical coastal embayment. Mar Ecol Prog Ser 68:301–310

Hinrichs S, Nordhaus I, Geist S (2009) Status, diversity and distribution patterns of mangrove vegetation in the Segara Anakan Lagoon, Java, Indonesia. Reg Environm Change 9:275–289

Holt DM, Jones EBG (1983) Bacterial degradation of lignified wood cell walls in anaerobic aquatic habitats. Appl Environ Microbiol 46:722–727

Holt DM, Jones EBG, Furtado SEJ (1980) Bacterial breakdown of wood in aquatic habitats. Record Annual Convention, British Wood Preserving Association, Huntingdon, UK, p 13–22

Honda D, Yokochi T, Nakahara T, Raghukumar S, Nakagiri A, Schaumann K, Higashira T (1999) Molecular phylogeny of labyrinthulids and thraustochytrids base on the sequencing of 18S ribosomal RNA gene. J Eukaryot Microbiol 46:637–647

Hong PN, Hoang TS (1993) Mangroves of Vietnam. IUCN, Bangkok, p 193

Huang ZG (2004) Biodiversity on marine estuarine wetland. Ocean Press, Beijing

Hughes GC (1974) Geographical distribution of the higher marine fungi. Veröffentlichungen Inst Meeresforschung Bremerhaven 5:419–441

Hulvey JP, Telle S, Nigrelli L, Lamour K, Thines M (2010) Salisapiliaceae - a new family of Oomycetes from marsh grass lietter of Southeastern North Aerica. Persoonia 25:109–116

Hutchings P, Saenger P (1987) Ecology of mangroves. University of Queensland Press, Brisbane, p 388

Hyde KD, Jones EBG (1988) Marine Mangrove fungi. Mar Ecol Pub Staz Zool Napoli 9:15–33

Hyde KD, Lee SY (1995) Ecology of mangrove fungi and their role in nutrient cycling - what gaps occur in our knowledge. Hydrobiologia 295:107–118

Hyde KD, Jones EBG, Leano E, Pointing SB, Poonyth AD, Vrijmoed LLP (1998) Role of fungi in marine ecosystems. Biodivers Conserv 7:1147–1161

Jian SG, Ban JW, Ren H, Yan HF (2010) Low genetic variation detected within the widespread mangrove species *Nypa fruticans* (Palmae) from Southeast Asia. Aq Bot 92:23–27

Jones EBG (2000) Marine fungi: some factors influencing biodiversity. Fungal Div 4:53–73

Jones EBG, Pang KL (2012) Tropical aquatic fungi. Biodivers Conserv 21:2403–2423

Jones EBG, Stanley SJ, Pinruan U (2008) Marine endophytes sources of new chemical natural products: a review. Bot Mar 51:163–170

Jones EBG, Alias SA, Pang K (2013) Distribution of marine fungi and fungus-like organisms in the South China Sea and their potential use in industry and pharmaceutical application. Malay J Sci 32:119–130

Jones EBG, Suetrong S, Sakayaroj J, Bahkali AH, Abdel-Wahab MA, Boekhout T, Pang KL (2015) An updated classification of marine Ascomycota, Basidiomycota, Blastocladiomycota and Chytridiomycota. Fungal Divers 73:1–72

KearneyR GFH (2014) Phylogeny of fungus-like organisms. In: Jones EBG, Hyde KD, Pang KL (eds) Freshwater fungi and fungal-like organisms. De Gruyter, Berlin, pp 155–176

Keith DA, Rodríguez JP, Brooks TM, Burgman MA, Barrow EG, Bland L, Comer PJ, Franklin J, Link J, McCarthy MA, Miller RM, Murray NJ, Nel J, Nicholson E, Oliveira-Miranda MA, Regan TJ, Rodríguez-Clark KM, Rouget M, Spalding MD (2015) The IUCN red list of ecosystems: motivations, challenges, and applications. Conserv Lett 8:214–226

Kensley B, Schotte M (1987) New records of isopod crustacea from the Caribbean, the Florida Keys, and the Bahamas. Proc Biol Soc Wash 100:216–247

Kern M, McGeehan JE, Streeter SD, Martin RNA, Besser K, Elias L, Eborall W, Malyon GP, Payne CM, Himmel ME, Schnorr K, Beckham GT, Cragg SM, Bruce NC, McQueen-Mason SJ (2013) Structural characterization of a unique marine animal family 7 cellobiohydrolase suggests a mechanism of cellulase salt tolerance. Proc Nat Acad Sci USA 110:10189–10194

King AJ, Cragg SM, Li Y, Dymond J, Guille MJ, Bowles DJ, Bruce NC, Graham IA, McQueen-Mason SJ (2010) Molecular insight into lignocellulose digestion by a marine isopod in the absence of gut microbes. Proc Nat Acad Sci USA 107:5345–5350

Kirui BYK, Kairo JG, Skov MW, Mencuccini M, Huxham M (2012) Effects of species richness, identity and environmental variables on growth in planted mangroves in Kenya. Mar Ecol Prog Ser 465:1–10

Kühne H (1971) The identification of wood boring crustaceans. In: Jones EBG, Eltringham SK (eds) Marine borers, fungi and fouling organisms of wood. Organisaion for Economic Co-operation and Development, Paris, pp 66–83

Kusumoto B, Shiono T, Miyoshi M, Maeshiro R, Fujii S, Kuuluvainen T, Kubota Y (2015) Functional response of plant communities to clearcutting: management impacts differ between forest vegetation zones. J Appl Ecol 52:171–180

Laurance WF, Sloan S, Weng L, Sayer JA (2015) Estimating the environmental costs for Africa's massive "development corridors". Curr Biol 25:1–7

Leaño EM, Damare V (2012) Labyrinthulomycota. In: Jones EBG, Pang KL (eds) Marine fungi and fungal-like organisms. De Gruyter, Berlin, pp 215–244

Leaño EM, Vrijmoed LLP, Jones EBG (1998) Physiological studies on *Halophytophthora vesicula* (Staminipilous fungi) isolated from fallen mangrove leaves from Mai Po, Hong Kong. Bot Mar 41:411–441

Lee SY (1992) The management of traditional tidal ponds for aquaculture and wildlife conservation in Southeast Asia: problems and prospects. Biol Conserv 63(113):118

Lee SY (1998) Ecological role of grapsid crabs in mangrove ecosystems: a review. Mar Freshw Res 49:335–343

Lee SY (2008) Mangrove macrobenthos: Assemblages, services, and linkages. J Sea Res 59:16–29

Lee SY (2016) From blue to black: anthropogenic forcing of carbon and nitrogen influx to mangrove-lined estuaries in the South China Sea. Mar Pollut Bull 109:682–690

Leonel RMV, Lopes SGBC, de Moraes DT, Aversari M (2006) The interference of methods in the collection of teredinids (Mollusca, Bivalvia) in mangrove habitats. Iheringia, Sér Zool Porto Alegre 96:25–30

Lo EYY, Duke NC, Sun M (2014) Phylogeographic pattern of *Rhizophora* (Rhizophoraceae) reveals the importance of both vicariance and long-distance oceanic dispersal to modern mangrove distribution. BMC Evol Biol 14:83

Lohbeck M, Poorter L, Paz H, Pla L, van Breugel M, Martinez-Ramos M, Bongers F (2012) Functional diversity changes during tropical forest succession. Pers Plant Ecol Evol Syst 14:89–96

Loilong A, Sakayaroj J, Runjindamai N, Ckoeyklin R, Jones EBG (2012) Biodiversity of fungi on *Nypa fruticans*. In: Jones EBG, Pang KL (eds) Marine fungi and fungal-like organims. De Gruyter, Berlin, pp 273–290
Lopes SGBC, Narchi W (1997) Recrutamento larval e crescimento de Teredinidae (Mollusca - Bivalvia) em região entremarés de manguezais. Rev Brasil Oceanogr 45:77–88
Lopez G, Taghon G, Levinton J (1989) Ecology of marine deposit feeders. Springer, New York
Lugo AE (1998) Mangrove forests: a tough system to invade but an easy one to rehabilitate. Mar Pollut Bull 37:427–430
Luther DA, Greenberg R (2009) Mangroves: a global perspective on the evolution and conservation of their terrestrial vertebrates. Bioscience 59:602–612
Ma Z, Melville DS, Liu J, Chen Y, Yang H, Ren W, Zhang Z, Piersma T, Li B (2014) Rethinking China's new great wall. Science 346:912–914
Macintosh DJ, Ashton EC (2002) A review of mangrove biodiversity conservation and management. Centre for Tropical Ecosystem Research, University of Aarhus, Denmark, p 86
Macintosh DJ, Ashton EC, Havanon S (2002) Mangrove rehabilitation and intertidal biodiversity: study of the Ranong mangrove ecosystem, Thailand. Estuar Coast Shelf Sci 55:331–345
MacKenzie J, Clyde L (2001) The fisheries for mangrove cockles, Anadara spp., from Mexico to Peru, with descriptions of their habitats and biology, the fishermen's lives, and the effects of shrimp farming. Mar Fish Rev 63:1–39
Maguire TL, Saenger P, Baverstock P, Henry R (2000) Microsatellite analysis of genetic structure in the mangrove species *Avicennia marina* (Forsk.) Vierh. (Avicenniaceae). Mol Ecol 9:1853–1862
Maia CR, Coutinho R (2013) The influence of mangrove structure on the spatial distribution of *Melampus coffeus* (Gastropoda: Ellobiidae) in Brazilian estuaries. Pan-Am J Aquat Sci 8:21–29
Marano AV, Pires-Zottarelli CLA, Souza JL, Glockling SL, Leano EM, Gachon CMM, Strittmatter M, Gleason FH (2012) Hyphochytriomycota, Oomycota, and Perkinsozoa (Super-group Chhromalveolata). In: Jones EBG, Pang KL (eds) Marine fungi and fungal-like organisms. De Gruyter, Berlin, pp 167–214
Marshall JV, Turner RD (1974) Survey of marine borers 1970–1972. The family teredinidae in Australian waters. CSIRO Project P5–11. The University of New South Wales Project 12–045–15. p 158
Melcalfe K, Glasby T (2008) Diversity of Polychaeta (Annelida) and other worm taxa in mangrove habitats of Darwin Harbour, northern Australia. J Sea Res 59:70–82
Min W-G, Kim D, Sang KM (2006) Meiobenthic community structures on the sediment of seagrass bed and mangrove forest in Chuuk Lagoon, Micronesia. Ocean Pol Res 28:13–24
Mokievsky VO, Tchesunov AV, Udalov AA, Nguen Duy T (2011) Quantitative distribution of meiobenthos and the structure of the free-living nematode community of the mangrove intertidal zone in Nha Trang Bay (Vietnam) in the South China Sea. Rus J Mar Biol 37:272–283
Mouzouras R, Jones EBG, Venkatasam R, Moss ST (1987) Decay of wood by microorganisms in marine environments. Record Annual Convention, British Wood Preserving Association, Cambridge, pp 27–45
Murphy DH (1990) The natural history of insect herbivory on mangrove trees in and near Singapore. Raffles Bull Zool 38:119–204
Nagelkerken I, Blaber SJM, Bouillon S, Green P, Haywood M, Kirton LG, Meynecke JO, Pawlik J, Penrose HM, Sasekumar A, Somerfield PJ (2008) The habitat function of mangroves for terrestrial and marine fauna: a review. Aq Bot 89:155–185
Nagelkerken I, Sheaves M, Baker R, Connolly RM (2015) The seascape nursery: a novel spatial approach to identify and manage nurseries for coastal marine fauna. Fish Fish 16:362–371
Netto SA, Gallucci F (2003) Meiofauna and macrofauna communities in a mangrove from the Island of Santa Catarina, South Brazil. Hydrobiologia 505:159–170
Neuhauser S, Glocking SJ, Leano EM, Lilje O, Marano AV, Gleason FH (2012) An introduction to fungus-like organisms. In: Jones EBG, Phag KL (eds) Marine fungi and fungal-like organisms. DeGruyter, Berlin, pp 137–152

Nielsen UN, Ayres E, Wall DH, Bardgett RD (2011) Soil biodiversity and carbon cycling: a review and synthesis of studies examining diversity-function relationships. Eur J Soil Sci 62:105–116

Nordhaus I, Wolff M, Diele K (2006) Litter processing and population food intake of the mangrove crab *Ucides cordatus* in a high intertidal mangrove forest in northern Brazil. Estuar Coast Shelf Sci 67:239–250

Nordhaus I, Hadipudjana FA, Janssen R, Pamungkas J (2009) Spatio-temporal variation of macrobenthic communities in the mangrove-fringed Segara Anakan Lagoon, Indonesia, affected by anthropogenic activities. Reg Environm Change 9:291–313

Odum WE, McIvor CC, Smith TJIII (1982) The ecology of the mangroves of south Florida: a community profile. US Fish and Wildlife Service, Washington DC, p 144

Olafsson E (1995) Meiobenthos in mangrove areas in eastern Africa with emphasis on assemblage structure of free-living marine nematodes. Hydrobiologia 312:47–57

Olafsson E, Carlstrom S, Ndaro SGM (2000) Meiobenthos of hypersaline tropical mangrove sediment in relation to spring tide inundation. Hydrobiologia 426:57–64

Padilla JMG, Palacio J (2008) Macroinvertebrados asociados a las raíces sumergidas del Mangle Rojo (Rhizophora mangle) en las bahías Turbo y El Uno, Golfo de Urabá (Caribe Colombiano). Gestión y Ambiente 11:55–66

Pan JY, Jones EBG, She ZY, Ling YC (2008) Review of bioactive compounds from fungi in the South China Sea. Bot Mar 51:179–190

Panebianco C (1994) Temperature requirement of selected marine fungi. Bot Mar 37:157–161

Pang KL, Jheng JS, Jones EBG (2011) Marine mangrove fungi of Taiwan. National Taiwan Ocean University Press, Keelung, p 131

Pang KL, Tsui CKM, Jones EBG, Vrijmoed LLP (2015) Bioprospecting fungi and the Labyrinthulomycetes in ocean-land interface. In: Baker BJ (ed) Marine pharmacognosy–from beach to bedside. CRC Press, Boca Raton, pp 77–389

Paredes C, Indacochea A, Cardoso F, Ortega K (2005) Family Ellobiidae (Gastropoda: Archaeopulmonata) in the Peruvian coast. Rev Peru Biol 12:69–76

Parravicini V, Kulbicki M, Bellwood DR, Friedlander AM, Arias- Gonzalez JE, Chabanet P, Floeter SR, Myers R, Vigliola L, D'Agata S, Mouillot D (2013) Global patterns and predictors of tropical reef fish species richness. Ecography 36:1254–1262

Peng YS, Chen GZ, Lin JJ (2011) Research on wetlands of the South China Sea - a case study of the coastal wetland ecosystem of Shantou. Sun Yat Sen University Press, Guangzhou, p 218

Polidoro BA, Carpenter KE, Collins L, Duke NC, Ellison AM, Ellison JC, Farnsworth EJ, Fernando ES, Kathiresan K, Koedam NE, Livingstone SR, Miyagi T, Moore GE, Vien NN, Ong JE, Primavera JH, Salmo SG, Sanciangco JC, Sukardjo S, Wang YM, Yong JWH (2010) The loss of species: mangrove extinction risk and geographic areas of global concern. Plos One 5(4):e10095

Quirós JA, Arias JE (2013) Taxocenosis of mollusks and crustaceans on roots of *Rhizophora mangle* (Rhizophoraceae) at Cispatá Bay, Córdoba, Colombia. Acta Biol Colomb 18:329–340

Raghukumar C, Nagarkar S, Raghukumar S (1992) Association of thraustochytrids and fungi with living marine algae. Mycol Res 96:542–546

Rateb ME, Ebel R (2011) Secondary metabolites of fungi from marine habitats. Nat Prod Rep 28:290–344

Raveendran K, Monnimohan P (2007) Marine fungi of Kerala- a prelimnary floristic and ecological study. Malabar Nat Hist Soc, Calicaut, p 270

Ray DL, Julian JR (1952) Occurrence of cellulase in *Limnoria*. Nature 169:32–33

Reid DG, Dyal P, Williams ST (2010) Global diversification of mangrove fauna: a molecular phylogeny of *Littoraria* (Gastropoda: Littorinidae). Mol Phylogenet Evol 55:185–201

Reyes R (1992) Moluscos, anelidos y crustaceos asociados a las raices de Rhizophora mangle Linnaeus, en la region de Santa Marta, Caribe Colombiano. CAL 17:133–148

Robertson AI, Duke NC (1990) Mangrove fish-communities in tropical Queensland, Australia - spatial and temporal patterns in densities, biomass and community structure. Mar Biol 104:369–379

Rocha RM, Faria SB, Moreno TR (2005) Ascidians from Bocas del Toro, Panama. I. Biodiversity. Caribb J Sci 41:600–612
Rutzler K, Piantoni C, van Soest RWM, Diaz MC (2014) Diversity of sponges (Porifera) from cryptic habitats on the Belize barrier reef near Carrie Bow Cay. Zootaxa 3805:1–129
Saenger P, Specht MM, Specht ML, Chapman VJ (1977) Mangal and coastal salt marsh communities in Australasia. Elsevier
Saenger P, Hegerl EJ, Davie JDS (1983) Global status of mangrove ecosystems. The Environmentalist 3:1–88
Sahoo G, Suchiang SR, Ansari ZA (2013) Meiofauna-Mangrove interaction: a pilot study from a tropical mangrove habitat. Cah Biol Mar 54:349–358
Sakayaroj J, Preedanon S, Suetrong S, Klaysuban A, Jones EBG, Hattori T (2012) Molecular characterization of basidiomycetes associated with the decayed mangrove tree *Xylocarpus granatum* in Thailand. Fungal Div 56:145–156
Sanchez BCL (2007) Estudio de la Diversidad Hidrobiológica en Tumbes. Instituto del Mar del Perú - IMARPE, Sede Regional de Tumbes, p 36
Sandilyan S, Kathiresan K (2012) Mangrove conservation: a global perspective. Biodivers Conserv 21:3523–3542
Sarma ALN, Wilsanand V (1994) Littoral meiofauna of bhitarkanika mangroves of River Mahanadi system, east-coast of India. Ind. J Mar Sci 23:221–224
Sasekumar A (1994) Meiofauna of a mangrove shore on the west-coast of peninsular Malaysia. Raffles Bull Zool 42:901–915
Schärer L, Knoflach D, Vizoso DB, Rieger G, Peintner U (2007) Thraustochytrids as novel parasitic protists of marine free-living flatworms: *Thraustochytrium caudivorum* sp nov parasitizes *Macrostomum lignano*. Mar Biol 152:1095–1104
Schmit JP, Shearer CA (2003) A checklist of mangrove associated fungi. Mycotaxon 80:423–477
Schmit JP, Shearer CA (2004) Geographical and host distribution of lignicolous mangrove microfungi. Bot Mar 47:496–500
Sheaves M (2012) Ecosystem equivalence and the ability to generalise: insights from global consistencies in mangrove fish assemblages. Mar Ecol Prog Ser 461:137–149
Shiino SM (1957) The marine wood-boring crustaceans of Japan, II. (Sphaeromidae and Cheluridae). Wasmann J Biol 15:161–197
Singh HR, Sasekumar A (1994) Distribution and abundance of marine wood borers on the west coast of Peninsular Malaysia. Hydrobiologia 285:111–121
Somerfield PJ, Gee JM, Aryuthaka C (1998) Meiofaunal communities in a Malaysian mangrove forest. J Mar Biol Assoc UK 78:717–732
Sousa WP, Dangremond EM (2011) Trophic interactions in coastal and estuarine mangrove forest. In: Wolanski E, McLusky DS (eds) Treatise on Estuarine and Coastal Science. Elsevier, Waltham, pp 43–93
Southwell CR, Bultman JD (1971) Marine borer resistance of untreated woods over long periods of immersion in tropical waters. Biotropica 3:81–107
Spalding M, Kainuma M, Collins L (2010) World atlas of mangroves. Earthscan, London, p 319
Stevens GN (1979) Distribution and related ecology of macrolichens on mangroves on the east Australian coast. Lichenologist 11:293–305
Street GT, Coull BC, Chandler GT, Sanger DM (1998) Predation on meiofauna by juvenile spot *Leiostomus xanthurus* (Pisces) in contaminated sediments from Charleston Harbor, South Carolina, USA. Mar Ecol Prog Ser 170:261–268
Sugiura S, Tanaka R, Taki H, Kanzaki N (2013) Differential responses of scavenging arthropods and vertebrates to forest loss maintain ecosystem function in a heterogeneous landscape. Biol Conserv 159:206–213
Sun M, Wong KC, Lee JSY (1998) Reproductive biology and population genetic structure of *Kandelia candel* (Rhizophoraceae), a viviparous mangrove species. Am J Bot 85:1631–1637
Svavarsson J, Osore MKW, Olafsson E (2002) Does the wood-borer *Sphaeroma terebrans* (Crustacea) shape the distribution of the mangrove *Rhizophora mucronata*? Ambio 31:574–579

Swe A, Jeewon R, Pointing SB, Hyde KD (2009) Diversity and abundance of nematode-trapping fungi from decaying litter in terrestrial, freshwater and mangrove habitats. Biodivers Conserv 18:1695–1714

Tomlinson PB (1986) The botany of mangroves. Cambridge University Press, Cambridge, p 413

Tong Y, Lee SY, Morton B (2006) The herbivore assemblage, herbivory and leaf chemistry of the mangrove *Kandelia obovata* in two contrasting forests in Hong Kong. Wetl Ecol Manag 14:39–52

Turner RD (1971) Identification of marine wood-boring mollusks. In: Jones EBG, Eltringam SK (eds) Marine borers, fungi and fouling organisms of wood. Organisation for Economic Co-operation and Development, Paris, pp 17–64

Turner RD, Johnson AC (1971) Biology of marine wood boring molluscs. In: EBG J, Eltringam SK (eds) Marine borers, fungi and fouling organisms of wood. Organisaion for Economic Co-operation and Development, Paris, pp 283–329

Vanhove S, Vincx M, Vangansbeke D, Gijselinck W, Schram D (1992) The meiobenthos of five mangrove vegetation types in Gazi Bay, Kenya. Hydrobiologia 247:99–108

Velmurugan N, Lee YS (2012) Enzymes from marine fungi: current rsearch and future propsects. In: Jones EBG, Pang KL (eds) Marine fungiand fungal-like organisms. De Gruyter, Berlin, pp 441–474

Venkatasamy R (1986) Mechanisms of bacterial deterioration of wood. CNAA, Portsmouth Polytechnic, UK

Vilardy S, Polanía J (2002) Mollusc fauna of the mangrove root-fouling community at the Colombian Archipelago of San Andrés and Old Providence. Wetl Ecol Managem 10:273–282

Wang WQ, Chen Q (2013) Salt-tolerant plant resources from coastal areas of South China. Xiamen University Press, Xiamen, p 444

Wang WQ, Wang M (2007) The Mangroves of China. Science Press, Beijing

Wee AKS, Takayama K, Asakawa T, Thompson B, Onrizal SS, Tung NX, Nazre M, Soe KK, Tan HTW, Watano Y, Baba S, Kajita T, Webb EL (2014) Oceanic currents, not land masses, maintain the genetic structure of the mangrove *Rhizophora mucronata* Lam. (Rhizophoraceae) in Southeast Asia. J Biogeogr 41:954–964

Whitman WB, Coleman DC, Wiebe WJ (1998) Prokaryotes: The unseen majority. Proc Nat Acad Sci USA 95:6578–6583

Wilkinson CR, Sudara S, Chou LM (1994) Proceedings of the third ASEAN-Australia symposium on living coastal resources, vol 1. Status reviews Australian Institute of Marine Science, Townsville, p 454

Wong MKM, Vrijmoed LLP, DWT A (2005) Abundance of thraustochytrids on fallen decaying leaves of *Kandelia candel* and mangrove sediments in Futian National Nature Reserve, China. Bot Mar 48:374–378

WWFHK (2014) Mai Po birds. WWFHK, Hong Kong, p 18

Xuan QN, Thanh NV, Smol N, Vanreusel A (2007) Biodiversity of meiofauna in the intertidal Khe Nhan Mudflat, Can Gio mangrove forest, Vietnam with special emphasis on free living nematodes. Ocean Sci J 42:135–152

Yaguchi T, Tanaka S, Yokochi T, Nakahara T, Higashihara T (1997) Production of high yields of docosahexaenoic acid by *Schizochytrium* sp. J Am Oil Chem Soc 74:1431–1434

Zwarts L, van der Kamp J, Klop E, Sikkema M, Wymenga E (2014) West African mangroves harbour millions of wintering European warblers. Ardea 102:121–130

Chapter 4
Spatial Ecology of Mangrove Forests: A Remote Sensing Perspective

Richard Lucas, Alma Vázquez Lule, María Teresa Rodríguez, Muhammad Kamal, Nathan Thomas, Emma Asbridge, and Claudia Kuenzer

4.1 Introduction

Over the past few decades, remote sensing data from ground, airborne and spaceborne instruments have been used extensively to characterize, map and monitor mangrove forests across their geographical range, at scales ranging from individual trees

R. Lucas (✉) • E. Asbridge
Centre for Ecosystem Science, School of Biological, Earth and Environmental Science, The University of New South Wales, High Street, Kensington, NSW 2052, Australia
e-mail: richard.lucas@unsw.edu.au; e.asbridge@unsw.edu.au

A.V. Lule
Department of Plant and Soil Sciences, University of Delaware, 531 S. College Avenue, 152 Townsend Hall, Newark, DE 19716, Mexico
e-mail: lule684@gmail.com

M.T. Rodríguez
Geomatics, National Commission for the Knowledge and Use of Biodiversity (CONABIO), 14010 Mexico City, Mexico
e-mail: mrodrig@conabio.gob.mx

M. Kamal
Remote Sensing Research Centre, School of Geography, Planning and Environmental Management, The University of Queensland, Brisbane, QLD 4072, Australia
e-mail: m.kamal@uq.edu.au

N. Thomas
Geography and Earth Sciences, Aberystwyth University, Aberystwyth SY23 3DB, UK
e-mail: nmt8@aber.ac.uk

C. Kuenzer
German Remote Sensing Data Centre, DFD of the German Aerospace Centre, DLR, Oberpfaffenhofen, D-82234 Wessling, Germany
e-mail: Claudia.kuenzer@dlr.de

V.H. Rivera-Monroy et al. (eds.), *Mangrove Ecosystems: A Global Biogeographic Perspective*, https://doi.org/10.1007/978-3-319-62206-4_4

to the global and in two or three dimensions (Hardisky et al. 1986; Kuenzer and Tuan 2013). As examples, Green et al. (1998) mapped the spatial (two dimensional) extent of mangrove forests in the Turks and Caicos Islands using a combination of Landsat Thematic Mapper (TM), Système Pour l'Observation de la Terre (SPOT) multispectral and Compact Airborne Spectrographic Imager (CASI) data whilst Giri et al. (2007) generated mangrove maps for the tropics based on classifications of Landsat sensor data. Lucas et al. (2002) derived canopy height models (CHMs) for mangroves from historical stereo aerial photography whilst Fatoyinbo and Simard (2013) achieved this at a regional scale by combining Shuttle Radar Topographic Mission (SRTM) and Ice, Cloud and Land Elevation Satellite (ICESAT) Geoscience Laser Altimeter System (GLAS) data. Time series of both optical and radar data have also been used to quantify changes in mangrove forests across their range (Thomas et al. 2014; Asbridge et al. 2015). Given this capability, remote sensing provides attractive options for understanding spatial mangrove ecology; their spatial patterns and how these relate to changes driven by natural and human-induced events and processes, including those associated with climate fluctuation. Remote sensing is also providing unique access, albeit indirectly and often in retrospect, to a large proportion of mangrove forests that are typically remote and difficult to access (Davis and Jensen 1998; Kuenzer et al. 2011). This chapter highlights how the range of remote sensing data has been used to describe the multiple dimensions of mangrove forests, focusing primarily on spatial scales, temporal frequencies, spectral responses and the three-dimensional state. The ways in which these data have been used to describe the structural complexity of mangrove environments and their connectivity with other habitats across a range of scales are also conveyed. Finally, the means by which these data can contribute to mangrove management programs are outlined.

4.2 Dimensions of Measurement

4.2.1 Coverage and Spatial Resolution

Mapping the areal extent of mangrove forests and their features (e.g. species types or growth stages) over time requires remote sensing data of varying spatial extent and resolution. For global observations, orbital optical and multispectral sensors such as the Moderate Resolution Imaging Spectroradiometer (MODIS) can be useful, but the spatial resolution (250 m to 1 km) is generally too coarse given that mangrove forests often occupy small areas and occur along the margins of coasts and estuaries. Despite this limitation, Vázquez-Lule (2012) was able to evaluate changes in Mexican mangrove forests over an 8-year period using the MOD13Q1 (Normalized Difference Vegetation Index and Enhanced Vegetation Index) and MOD15A2 (Leaf Area Index) products. In most cases, however, focus has generally been on moderate (<30 m) spatial resolution sensors such as Landsat (Giri et al. 2010) and the Sentinels. The advantage of using data from these sensors is that mosaics can be generated at country, regional and global levels. The benefits of using Very High Resolution (VHR) sensors (<5 m) are increasingly been recognized although their use is often restricted to local areas because of their limited spatial coverage.

Knowing the most suited spatial resolution for mapping mangroves features is useful when selecting sensors, particularly when long-term change is to be considered. To investigate this further, Kamal et al. (2014) resampled optical WorldView-2 data at six different spatial resolutions (0.5, 1, 2, 4, 8 and 10 m) and detected the dominant mangrove features by applying a semi-variogram analysis to each image. A gradual gain of mangrove information at smaller resolutions was observed, with a pixel size of ≤2 m being most suitable for mapping canopy and inter-canopy features (e.g. crowns of individual trees, shrubs and canopy gaps). Pixel sizes ≥4 m were considered more appropriate for mapping species and communities, mangrove extent and vegetation formations/zonations. The ability to map mangrove extent and features also depends on their spatial pattern and geometric arrangement. Where mangrove vegetation occurs in narrow strips (e.g. along creeks) or as isolated trees, their detection may only be possible using VHR data. Mangrove forests on the landward and seaward margins are also often difficult to discern particularly at lower spatial resolutions where mixed pixel effects are greater. In many cases, contextual information as well as remote sensing data can increase the accuracy of classification. This might include distance from the shoreline and elevation (as determined from digital terrain models (DTM); (Chadwick 2011; Liu et al. 2008; Zhang et al. 2006; Rodríguez-Zúñiga et al. 2013) Fig. 4.1). A detailed overview of the optical, multispectral and hyperspectral spaceborne sensors that have been used for mapping of mangrove areas is presented in Kuenzer et al. (2011), with those on board the Landsat, SPOT, ASTER, the Chinese-Brazilian satellite CBERS, the WorldViews, IKONOS, RapidEye and QuickBird being the most commonly used at the time of this review. More recent sensors include the Sentinel optical and radar sensors. The majority of studies have focused on mangrove ecosystems in Brazil, Asia and Southeast Asia, the USA and Australia.

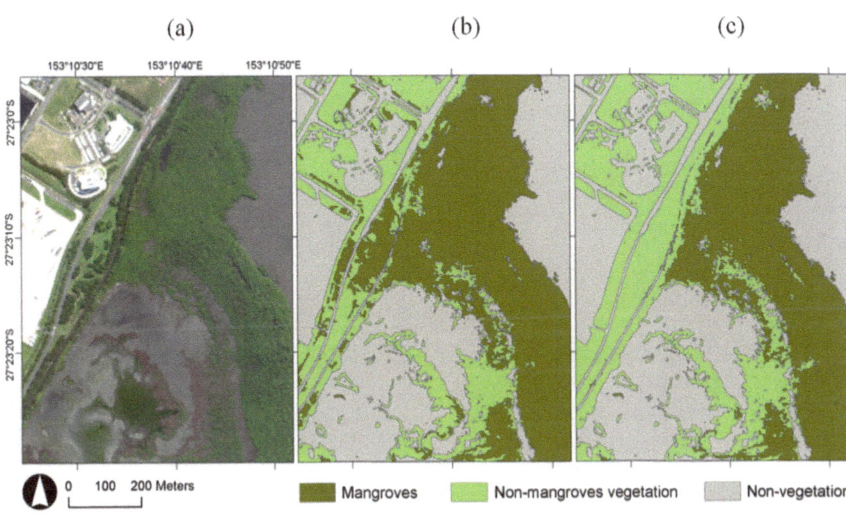

Fig. 4.1 (**a**) True colour composite of a WorldView-2 image of Fisherman Island, southeast Queensland, (**b**) mangrove areas mapped using spectral information only and (**c**) a combination of spectral information, a DTM (derived from LiDAR) and distance from water. The accuracy in the delineation increased from 85% in (**b**) to 94% in (**c**) (Kamal et al. 2015)

4.2.2 Changes over Time

Mangrove forests are dynamic ecosystems, responding to natural and anthropogenic events and processes (Giri et al. 2007). In many regions, mangrove forests have occupied the same area (or 'core zone') for decades, although natural turnover occurs internally through successional processes. However, dieback or colonization is commonplace, particularly at the margins, and occurs over varying time frames (M'rabu et al. 2012; Vogt et al. 2012; Saintilan et al. 2014). Dieback is often slow and progressive and occurs as a result of changes in, for example, salinity and sediment dynamics and often because of alterations in hydrological patterns. However, rapid dieback can occur as a consequence of storm damage, which has a differential impact on some species (e.g. *Rhizophora* spp.; Asbridge et al. 2015), and adverse conditions relating to climate. Colonization by mangroves typically requires increasing sedimentation or a change in the local environment (e.g. increased salinity inland) (Wolanski 1995; Ellison 1999; Anthony et al. 2013). Human-induced change is often rapid (Thomas et al. 2014) and may be associated with logging activities, aquaculture, agriculture development, pollution (e.g. from mine waste) or changing hydrological regimes (Fig. 4.2). The areas affected may be small (e.g. selective logging) to extensive (e.g. when areas are cleared for aquaculture).

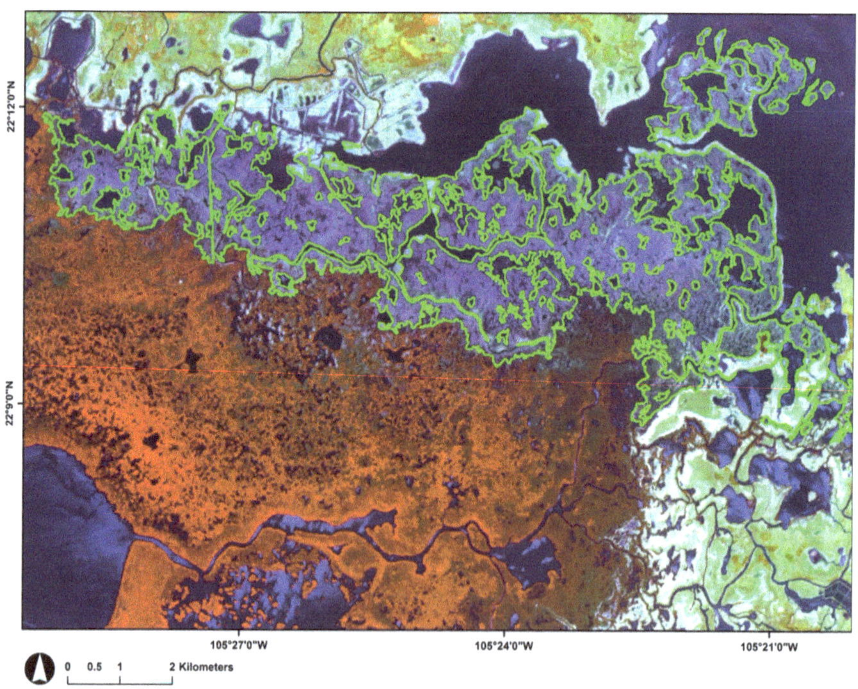

Fig. 4.2 SPOT-5 HRG image (near-infrared, shortwave infrared and *red* in RGB) of Marismas Nacionales, Nayarit, Mexico showing mangrove dieback (within *green* polygons; 3,384 ha) as a consequence of changes in hydrological regimes

The observational time frames for remote sensing observations need to mirror that of the changes that are occurring in the areas occupied by mangroves. The longest continuous time series, from the 1970s to the present, has been captured by the Landsat sensors. These data chronicle the changing extent of mangrove forests in most regions, albeit at 30 m spatial resolution. Unfortunately, the persistence of cloud cover, particularly in tropical areas, reduces the number of usable observations. For this reason, observations are often restricted to dry seasons and even to the morning periods where cloud cover is often minimal. An alternative has been to use spaceborne Synthetic Aperture Radar (SAR) operating at X-, C- and L-band as mangrove forests can be observed regardless of cloud cover and illumination conditions (Lucas et al. 2007), although a shorter time series is available.

Whilst systematic global coverages of high-frequency X-band and C-band SAR are only just becoming available, the Japanese Aerospace Exploration Agency (JAXA) has provided coverages at lower frequency L-band since the early 1990s. The Japanese Earth Resources Satellite (JERS-1) SAR regularly observed the Earth's surface from 1992 until 1998. The Advanced Land Observing Satellite (ALOS) Phased Arrayed L-band SAR (PALSAR; 2007–2010) and ALOS-2 PALSAR-2 (2015 onwards) were the successors and comparisons with JERS-1 SAR data for the same locations (Lucas et al. 2014; Thomas et al. 2014) have highlighted the benefit of combining these sensor data for mangrove monitoring over decadal periods. Changes in mangrove distribution (particularly in the seaward direction but also within mangrove areas) were readily detected, but confusion with terrestrial woody vegetation (plantations, forests) on the landward side compromised the detection of change. As examples of change, areas of mangrove gain (blue) and loss (red) are shown in composites of JERS-1 SAR from the mid-1990s and ALOS PALSAR data from 2007 and 2010 in South East Asia (Fig. 4.3). The comparison of L-band SAR data over time allows for the detection of losses and gains in mangrove extent. Based on a global analysis of these data (UNEP 2014), losses were associated primarily with conversion to agriculture/aquaculture, coastal development, pollution, extreme weather events or industrial logging. Mangrove expansion was the result of sedimentation and recovery from previous deforestation. The extent of losses and gains were highly variable depending upon the causative processes. However, the Mahakam Delta in East Kalimantan (Fig. 4.3b) represents an extreme example where loss of 21,000 ± 152 ha over a period of 11 years (47% by 2000 and 75% by 2010) was reported by Rahman et al. (2013). Whilst the extent of changes could be detected, the establishment of baseline maps of mangrove extent from the SAR data alone is difficult because of confusion with proximal land covers. Hence, some reference to optical (e.g. Landsat) sensor data is necessary.

Changes within mangrove forests (e.g. deforestation, selective logging and regrowth) can also be detected using these L-band SAR time series, even at moderate resolutions (Thomas et al. 2014; Asbridge et al. 2015). For example, in the Matang Forest in Perak State, Malaysia, mangroves arc logged in coupes resulting in discrete areas of forests dominated primarily by *Rhizophora* species at varying stage of regeneration. In the early stage of logging, cut stumps are left and these exhibit a high L-band HH (horizontally transmitted and received; referring to the

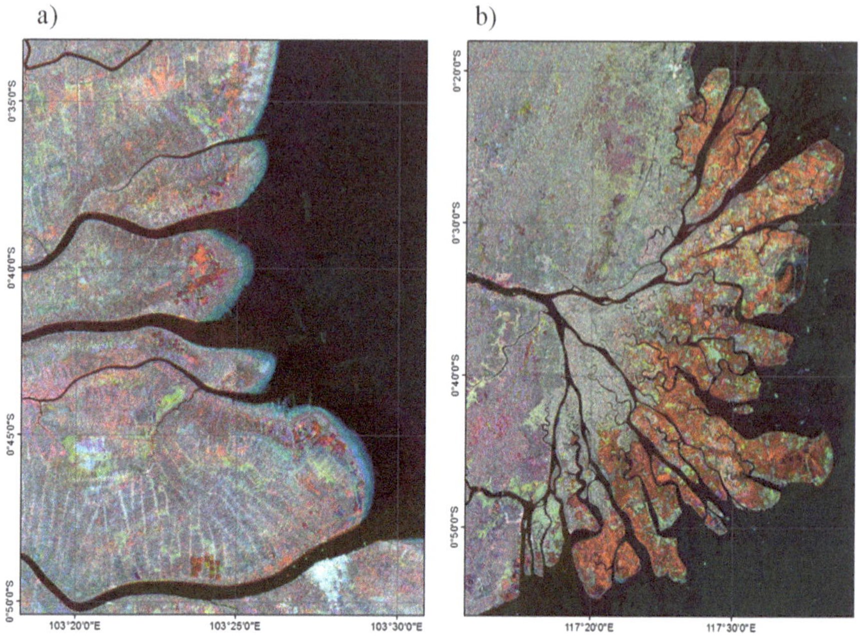

Fig. 4.3 JERS-1 SAR (mid-1990s) and ALOS PALSAR data (2007 and 2010) in RGB showing mangrove gains (*blue*) and losses (*red*) in South East Asia over the observation period near (**a**) Tembilahan, Sumatra and (**b**) the Mahakam Delta, East Kalimantan

orientation of the microwaves within the electric field), and to a lesser extent, HV (vertically received) return several years after the event (Fig. 4.4). However, affected areas cannot be differentiated from the more intact forests following regeneration and hence a time series is necessary for their detection. Time series can also be used to track the sequence of colonization by mangroves and quantify age class distributions and logging histories prior to recovery of the forest.

The capacity for continued monitoring of mangrove forests at L-band has been increased by the ALOS-2 PALSAR-2. As well as providing 10 m observations that complement those of its predecessor, the PALSAR-2 ultrafine stripmap mode (3 m resolution) provides capacity for detecting subtle changes in mangrove areas, such as storm damage (Asbridge et al. 2015). Systematic observations using the European Space Agency's (ESA) Sentinel-1 C-band SAR and the German TanDEM-X (TerraSAR-X add-on for Digital Elevation Measurements) complement the L-band SAR acquisitions. As well as providing observations at commensurate resolutions, these sensors provide higher resolution images that allow changes in mangrove forests at the scale of clusters of trees. This then facilitates differentiation of stands of

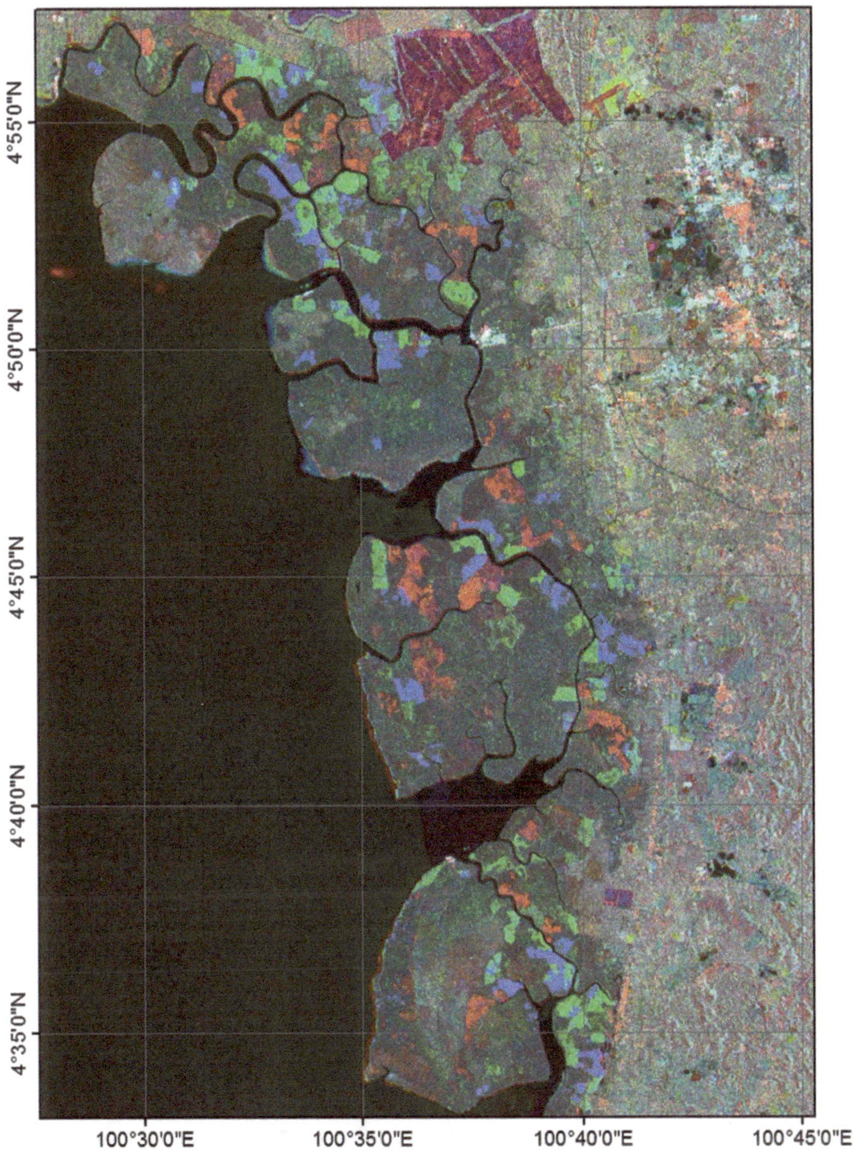

Fig. 4.4 RGB composite of JERS-1 SAR from 1996 and ALOS PALSAR from 2007 and 2008 (in RGB) covering Matang Forest in Perak State. *Red* represents areas that were forested in 1996 but cleared by 2007, *green* represents non-forest in 1996 and regenerating forest in 2007 and 2008 and *blue* represents forests regenerating in 2008 only

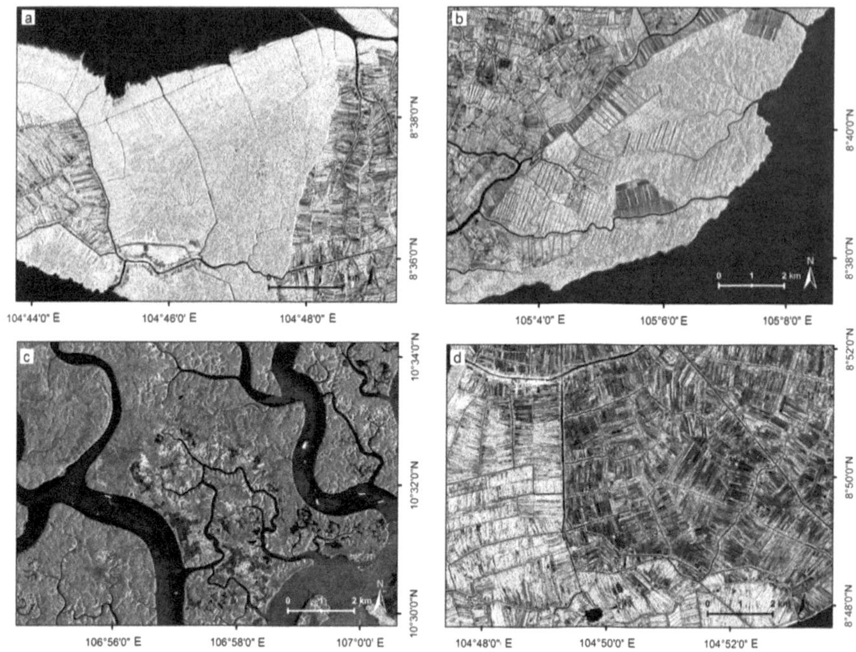

Fig. 4.5 TerraSAR-X SAR strip map SAR data (HH polarization, 3 m resolution) showing different mangrove conditions. The centre of each image indicates (**a**) undisturbed mangrove forests in Ca Mau National Park, Mekong Delta, Vietnam (31 December 2009), (**b**) afforestation of mangroves in rows along the coast of Ca Mau Province Vietnam (24 February 2010), (**c**) clear cuts in mangroves in preparation for aquaculture, Can Gio Mangrove Biosphere Reserve south of Saigon (3 May, 2009) and (**d**) varying densities of mixed mangrove-aquaculture parcels, Vietnam (3 May, 2009)

different tree densities and also disturbance patterns ((Kuenzer et al. 2011); Fig. 4.5). Digital surface models (DSMs) depicting mangrove canopy height can be retrieved from these data (Lee and Fatoyinbo 2015). The sensitivity of the X-band and C-band SAR data to small trees further renders these well suited for detecting early growth of mangroves. As such, these data can support the ongoing monitoring and management of mangrove forests, including that undertaken by community groups or larger restoration programs.

Despite the increasing capability of SAR, optical sensors are often preferred for detecting change partly because of the greater ease of interpretation. VHR data are available to detect changes at the individual tree level but time series are often not available as many sensors are relatively recent (Kuenzer et al. 2011). For this reason, reference is often made to historical aerial photography, particularly as these data can be readily interpreted to obtain the past extent of mangrove forests and to detect change. As an example, the National Commission for the Knowledge and Use of Biodiversity (CONABIO is the Spanish acronym) produced three maps of Mexican mangrove forests as part of the Mexican Mangrove Monitoring System (SMMM) for the period 1970–1980 (based on 1,505 panchromatic aerial photographs) and the years 2005 and 2010 (based on both aerial and SPOT imagery;

Rodríguez-Zúñiga et al. (2013); Valderrama et al. (2014)). Gaps in the coverage representing 17% of the area were in-filled using Landsat sensor data. The study found that the overall decline in mangrove area (through deforestation and degradation) in 17 states and along 12,000 km of coastline was 9.6% in 2005 and 1.2% in 2010 (Valderrama et al. 2014). According to the 2010 map, 60.3% of the mangrove area was found within federal or state natural protected areas and 56 Ramsar sites (wetlands of international importance), highlighting that vast areas were unprotected and hence more vulnerable to change.

4.2.3 Spectral Information

Mangrove forests have distinct spectral reflectance characteristics, particularly in the near-infrared (NIR; 700–1200 nm) and shortwave infrared (SWIR; 1200–2500 nm) wavelength regions, that distinguish them from many adjacent environments (e.g. terrestrial forests, saltmarshes, mudflats; Spalding et al. 2010). Many studies have therefore used spectral data to map the extent of mangrove forests at scales ranging from individual trees/stands (Jensen et al. 2007; Green and Mumby 2000) to entire regions (Giri et al. 2010). Many studies have further used spectral data to distinguish vegetation communities (Rasolofoharinoro et al. 1998; Murray et al. 2003; Vaiphasa et al. 2006; Demuro and Chisholm 2003) and/or species types (Held et al. 2003; Rakotomavo and Fromard 2010; Kamal and Phinn 2011). Separability of many species types is generally high in the NIR spectrum because of differences in internal leaf structure and leaf/canopy geometry (Gausman et al. 1969; Knipling 1970; Kuenzer et al. 2011). The SWIR wavelengths are also well suited because the invariability of pigments across mangrove species (for which the visible wavelengths are most sensitive) is compensated for by different levels of other leaf (including structural) components such as salt, sugar, water, protein, oil, lignin, starch and cellulose (Vaiphasa et al. 2005).

Individual mangrove species can also be distinguished across a range of spatial resolutions because they often occur as distinct zones parallel to the coast. These zones are the result of a differential response to physio-chemical gradients that vary across the intertidal area including topography, geomorphic setting, tidal regime and sediment properties such as salinity, water content, organic matter content, nutrient concentration, texture and chemical composition (Smith 1992; Da Cruz et al. 2013; Yang et al. 2013; Prasad 2011; Ellison et al. 2000). Often, however, the zones are too narrow to allow discrimination at moderate spatial resolution, particularly where a mix of species occurs. Identifications from remote sensing are generally more successful when a single species dominates a zone and where less species occur, as in the Neotropics (Tomlinson 1986; Luther and Greenberg 2009).

Hyperspectral sensors offer greater capacity for discriminating species as these capture spectral information at a much higher resolution and provide often in excess of 200 image bands. Such data allow the user to interrogate the near full spectral response of the land surface. The benefits of such data are the ability to capture information across a greater breadth of the electromagnetic spectrum than that afforded by multispectral imagers. This strategy enables subtle differences in the spectral response of various mangrove species to be identified. Hyperspectral remote sensing data have been shown to provide detailed information on different mangrove species, and also growth stages and stressors. As an example from Kakadu National Park in Australia's Northern Territory, Fig. 4.6 shows a CASI image within which four zones dominated by different species can be distinguished, including a zone of recent seaward expansion.

Whilst spectral data have proved useful for mapping the extent of mangrove forests and their associated species (Heenkenda et al. 2014), there are several challenges particularly when mapping over large areas and at different tidal states. For example, the spectral reflectance of mangroves is strongly influenced by tidal effects on the soil, resulting in mixed pixels (Blasco et al. 1998). Other site-specific factors such as background reflectance and differences in leaf area index (LAI) and leaf inclination also contribute to the variability in the spectral reflectance of mangroves when observed by airborne and spaceborne sensors (Baret et al. 1994; Díaz and Blackburn 2003).

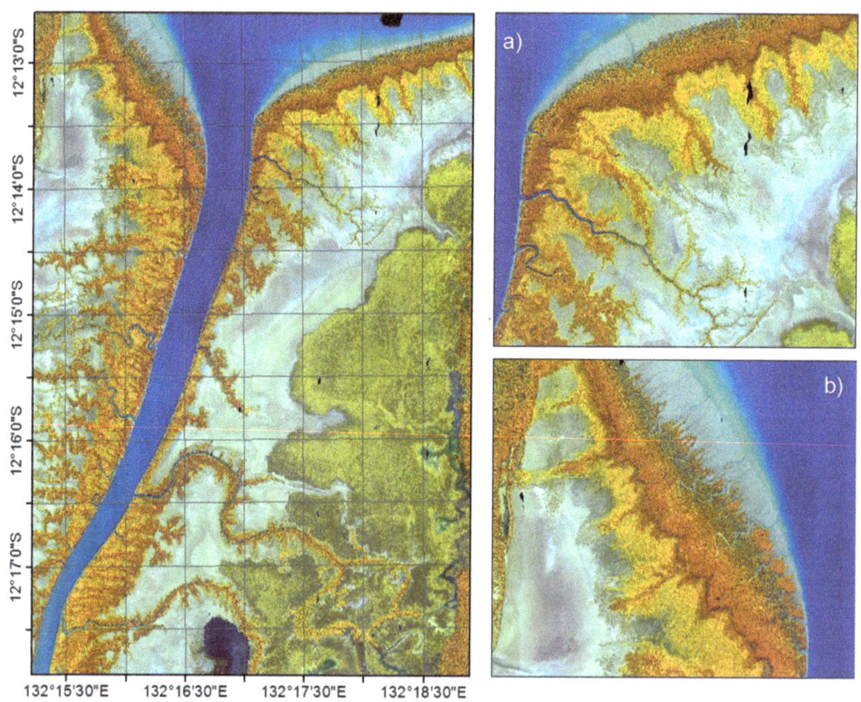

Fig. 4.6 Hyperspectral CASI data, West Alligator River, Australia's Northern Territory showing (**a**) areas of inland intrusion by *Avicennia marina* (*yellow*) and (**b**) areas of seaward expansion of *Sonneratia alba* (*bright orange*) near the river mouth. The central zone is occupied by *Rhizophora stylosa* (*red*) and *S. alba* (*olive*)

The use of textural measures derived from these data can be used as an additional measure to differentiate mangrove species type and growth stages (Ramsey and Jensen 1996). Compared to natural terrestrial forests, the canopy of mangrove forests is generally smoother as most occur in zones that have been established at the same time; hence mangrove forests are often of the same age and height (and hence growth form) and of similar species composition. Image texture can therefore be used and is often measured using first- and second-order metrics computed from a grey-level co-occurrence matrix within a given window, lag distance and direction (Kayitakire et al. 2006). Such information has been included with spectral data to increase the accuracy of mangrove maps (Myint et al. 2008; Wang et al. 2004, 2008).

4.2.4 The Third Dimension

There are numerous descriptors of the three-dimensional structure of mangrove forests, with the most common being maximum (top) and mean height. Others relate to the distribution of plant material (foliage, wood) within the vertical profile, the roughness of the upper canopy and the height of prop roots. Changes in the three-dimensional structure also indicate damage (e.g. storms) or growth (e.g. succession or colonization).

Over small areas and at very high spatial resolution, airborne LiDAR provides a detailed representation of the forest volume and also the underlying surface. LiDAR emits optical (visible, NIR or SWIR) wavelengths of the electromagnetic spectrum, and the return time from the illuminated object surface relates to its distance from the sensor. LiDAR instruments are typically mounted on airborne platforms, including Unmanned Airborne Vehicles (UAVs), although the ICESAT GLAS (operating between 2003 and 2010) is an example of a spaceborne instrument. LiDAR systems can attain resolutions of up to tens of centimetres and large volumes of data are typically generated. Such information can be used to establish the relative growth of mangrove forests, including within the understory when the point cloud is of sufficient density. A wide range of metrics can be obtained from the point cloud, such as canopy openness, as well as mean and maximum height, and canopy height models (CHMs) (Fig. 4.7). These metrics and derived surfaces allow the complex structure of the forest to be described (Wannasiri et al. 2013). A limitation is that the tree trunks can often not be discerned. Thus, a locally based solution is to use ground-based Terrestrial Laser Scans (TLS) within plots, which allows a detailed and very high (in the order of cm) resolution point cloud to be generated (Feliciano et al. 2012, 2014).

An alternative approach to deriving CHMs is to use stereo images, with these generally acquired at more local scales as aerial photographs or by VHR sensors on board satellites or, more recently, UAVs. As an example, Lucas et al. (2002) generated CHMs of mangrove forests in northern Australia from historical (1991) stereo photography, with these being accurate to within a few meters. To quantify change, CHMs derived from optical data can be compared against those generated using photography, LiDAR or radar interferometry acquired on different dates. Across

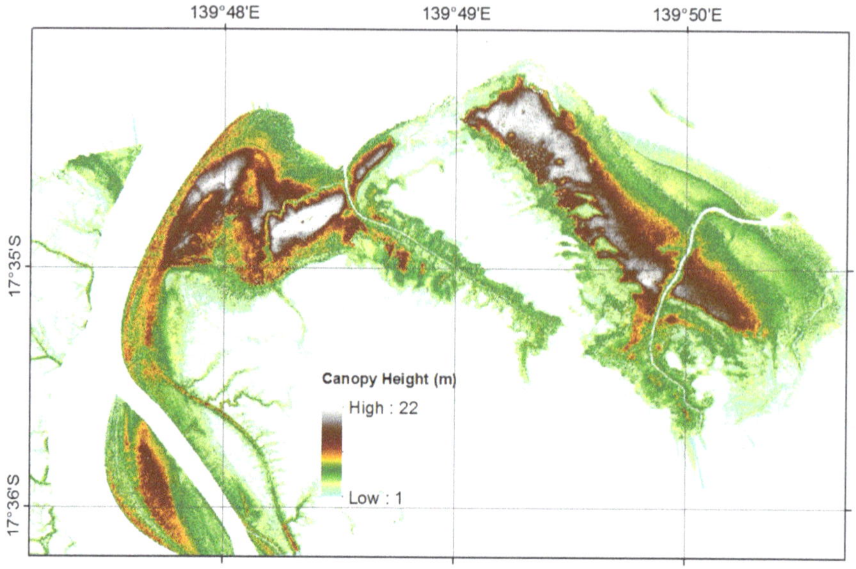

Fig. 4.7 LiDAR-derived CHM of mangroves in the Gulf of Carpentaria, northern Queensland, Australia. The tall mangroves in the 'core zone' are dominated by *Rhizophora stylosa* and *Ceriops tagal*. *Avicennia marina* dominates the lower stature mangroves either side. Note the different height levels on the north-eastern edge depicting the progressive colonization of mudflats

wider areas, CHMs for mangrove forests have been generated from interferometric SAR, with usable missions being the SRTM and the TanDEM-X. Using SRTM data, for example, Fatoyinbo and Simard (2013) generated CHMs calibrated against ICESAT/GLAS data for all mangroves in Africa.

The three-dimensional structure of mangrove forests can also be quantified using single or multi-frequency polarimetric SAR. Microwaves at X- and C-band interact primarily with the foliage and small branches whilst those at lower frequency L and P-band interact with the larger branches, trunks and prop roots. To a certain extent, the degree of tidal inundation can be best established at these lower frequencies, as water under the canopy leads to strong double-bounce interactions at HH polarizations between the water surface and the trunks. The ability to detect inundation depends though upon the frequency of observation, the amount of overtopping canopy material, and the openness of the canopy. The backscattered signal is, however, reduced at lower frequencies where large prop root systems occur because of disruption to the double bounce scattering mechanism (Lucas et al. 2007).

Another approach to retrieve detailed three-dimensional structures of forests is SAR tomography (Reigber and Moreira 2000). This technique requires at least two SAR acquisitions, which are combined to form an additional synthetic aperture in the elevation direction. When used in combination with polarimetric data, better separation between different components of complex volumetric media is achieved and the different scattering mechanisms as a function of height can be retrieved. These can then be used to describe the three-dimensional structures of forests.

Tomographic analysis can be conducted using high-resolution imagery provided by the TerraSAR-X and Italy's Cosmo-SkyMed, but is particularly effective when using lower frequency SAR. As part of the European Space Agency's (ESA's) BIOMASS mission, this method is proposed for the retrieval of above-ground biomass (AGB) from P-band data, whereby contributions to the radar signal from volume and double bounce scattering can be discerned and the height of the forest quantified. Furthermore, as a better understanding of microwave interaction with different components of the forest volume is achieved, inversion models that use intensity and Pol-InSAR data can be better developed to improve estimation of AGB (Le Toan et al. 2011).

4.2.5 *Above-ground Biomass (AGB)*

For estimating the AGB and also the trunk and branch biomass of mangroves from VHR optical data, the Fourier-based textural ordination (FOTO) method (Couteron 2002) has been proposed. This method computes textural indices of canopy grain where no saturation in the relationships with AGB was observed. Ploton et al. (2012) also demonstrated, for tropical forests, how the technique could be applied to Google Earth images, hence widening its broader applicability and accessibility. The advantage of using textural measures is that the structural diversity of tree crowns as a function of growth stage and species can be better captured (Eckert 2012).

Other studies have estimated the AGB of mangrove forests using relationships established with CHMs, as generated using airborne or spaceborne stereo imaging, LiDAR or SAR interferometry. The canopy height retrieved from these data assumes that the underlying topography is flat, although variations in elevation are often observed between the seaward and landward margins. With airborne LiDAR, the increased spatial resolution allows the underlying topography to be better discerned and the retrieved heights are generally more reliable (Wannasiri et al. 2013). Height is then related to AGB, which is often determined by referencing allometric relationships with diameter at breast height (DBH). As an example, Simard et al. (2006) used a stand-level relationship between mean stem height and AGB, which was applied to a CHM obtained from the Shuttle Radar Topographic Mission (SRTM) interferometric SAR at 30 m spatial resolution.

An alternative approach is to use the polarimetric information associated with SAR data to retrieve AGB. Mougin et al. (1999), Proisy et al. (2000) and Proisy et al. (2002) found the lower frequency L- and P-band to be more sensitive to AGB in mangrove forests of French Guiana. However, in two areas with contrasting physiognomy in México, Vázquez-Lule (2012) found ALOS PALSAR L-band HH and HV backscatter to be highest in mangrove forests with low structural development or experiencing degradation. A significant correlation was also observed with ground estimated tree height, density and areal AGB. In mangrove forests with greater structural development, the highest coefficient of determination was between the HH polarization and tree height ($r^2 = 0.82$; $p < 0.001$) and, in forests with lower structure, between HV polarization and areal AGB ($r^2 = 0.75$, $p < 0.001$). The AGB can also be estimated using combinations of height and backscatter. For example, Asbridge et al. (2015) demonstrated a decrease in L-band HH and HV backscatter with increasing biomass (above 100 Mg ha^{-1}) for mangroves >10 m tall and with prop root systems. Significant improvements in estimating the AGB of mangrove forests changes over time, particularly at higher biomass levels (approximately 60–100 Mg ha^{-1}, which is the saturation level at L-band), are anticipated following launch of the European Space Agency's (ESA's) P-band BIOMASS P-band SAR in 2020. However, potential hindrances to the use of these data are the relatively coarse spatial resolution (100–200 m) that may limit retrieval in areas where mangrove forests are fragmented and occupy small areas and where the AGB is very low (as in the case of forests undergoing colonization or restoration). The integration of these data with, for example, L-band SAR or spaceborne LiDAR is more likely to lead to better estimates.

4.3 Towards Characterization of Mangrove Habitats and Their Management

4.3.1 Habitat Structural Complexity at Local Scales

Mangrove forests can be remarkably homogenous in terms of their structure, particularly where they consist of one tree species that have colonized and grown at a similar pace. Even so, there are often noticeable structural gradients (e.g. from the seaward to the landward margins) as a result of the different response of species to environmental conditions (e.g. salinity, tidal regimes). Management practices such as logging within discrete and contiguous areas can also lead to homogenization of structures and species composition. More structurally heterogeneous forests occur where species diversity is higher, strong gradients (e.g. in inundation) exist or where disturbance has occurred. In general, and as examples, decreases in the structural complexity of mangrove forests often lead to lower provision of nursery habitats (de la Morinière et al. 2004) and fewer refuges for faunal species (e.g. fish), with the latter affecting predator–prey relationships (e.g. invertebrates).

An overview of how remote sensing data can be used to quantify the structural complexity of mangrove habitats is outlined in Fig. 4.8, where the focus is on links with faunal diversity and ecological interactions. Through classification of optical imagery, the areal extent of mangrove forests can be delineated with mapped units related directly to, for example, catches of invertebrates (e.g. prawns) and/or inshore fish production (Nagelkerken et al. 2008). Landscape (connectivity) metrics such as perimeter or closest distance to other habitats (including saltmarshes, mud/sandflats, seagrass beds and coral reefs) can also be used to quantify interactions between static and migratory faunal communities. By differentiating mangrove tree species and growth forms, the distribution of fauna such as gastropods, sesarmid crabs and other fish can be ascertained. Knowledge of the extent of mangroves with prop root

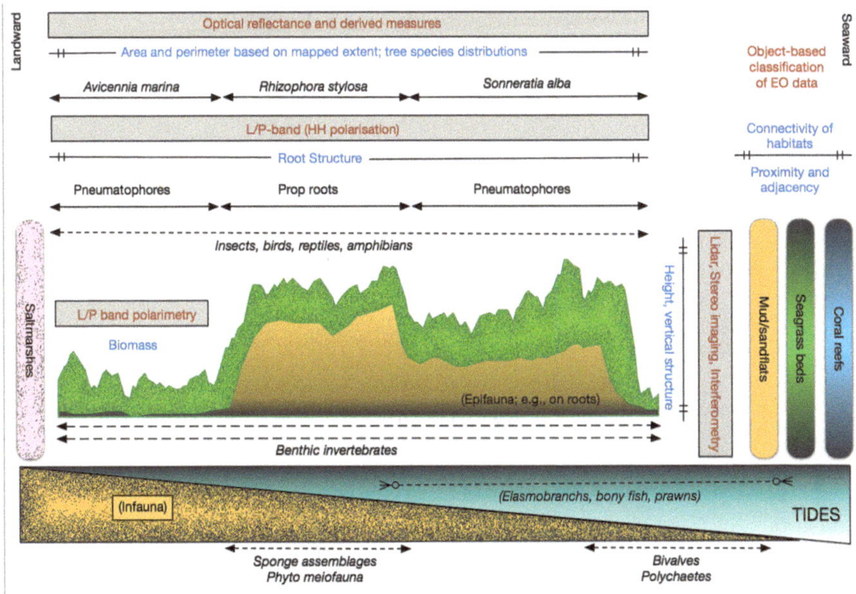

Fig. 4.8 Diagrammatic link between remote sensing observations of the tree species composition structure, biomass and geometric arrangement of mangrove forests in northern Australia (see Fig. 4.5) and potential relationships with the distribution of fauna

systems (e.g. *Rhizophora*, *Ceriops*) or pneumatophores obtained using L- or P-band SAR data in combination with CHMs can be used to infer faunal distributions. For example, a greater number of epifauna are associated with prop root systems, and the number of fish species differs in areas where mangroves have prop roots or pneumatophores (Rönnbäck et al. 1999). Some species assemblages (e.g. sponges) favour prop roots as they protect from fish predation. Together with pneumatophores, these prop roots also serve as a substratum for phyto-meiofauna, which are a dietary component of many commercially important fish and crustaceans.

The distribution of plant material within the vertical profile of mangrove forests, as described using LiDAR data, stereo imaging and/or interferometric SAR, can also be related to the distribution of fauna within the forest. In particular, the trees and roots support benthic epifauna (including crabs, bivalves and gastropods). This habitat pattern is in contrast to the benthic invertebrates (infauna; crabs, polychaetes, pistol prawns and sipunculids) that inhabit the surface or subsurface of the sediment (Leung 2015; Nagelkerken et al. 2008). The extent of mud or sand flats on the landward and seaward margins is also important to map from optical and/or SAR data in order to provide an insight into the distribution of epifaunal and infaunal communities (Dissanayake and Chandrasekara 2014). For example, bivalves and polychaetes often occur in a narrow zone on the seaward margins because of optimal conditions for feeding and larval settlement (Nagelkerken et al. 2008). Moreover, by delimiting regions of mangrove loss, the associated impacts on epifauna and infauna can be predicted. For example, mangrove removal in the Mangawhai Estuary, New Zealand, was followed by an increase in the accumulation of sand, which increased the number of snails, crabs and bivalves (Alfaro 2010).

Information regarding tidal inundation is useful as many fish that inhabit shallow waters migrate into the mangrove swamps at high tide where they feed on, for example, invertebrates. Crustaceans (particularly prawns) also make extensive use of mangrove forests during high tide. The tidal inundation within the forest can be mapped indirectly through reference to tidal data and knowledge of the underlying terrain surface as obtained, for example, from airborne LiDAR. Such information can be used as input to models of faunal population dynamics within mangrove forests.

4.3.2 Connectivity at Different Scales

Mangrove forests occur in proximity to other habitats, including mudflats, seagrasses and coral reefs. These ecosystems are functionally interlinked with synergistic relationships. The mangrove forests provide benefits such as trapping heavy metals, stabilizing sediment and filtering water-borne pollutants, which improves water quality and clarity for seagrasses and coral reefs. In turn, coral reefs reduce wave action, allowing mangroves to thrive in the more sheltered environments. Mangrove forests also act as nursery grounds for many species of reef fish, because the network of dense roots and reduced wave action provide a refuge for juveniles (Mumby et al. 2004; Lee et al. 2014). The fish primarily migrate between these habitats through larval recruitment and ontogenetic migration. Prawn species also often spawn offshore in seagrass and algal beds, and their larvae subsequently migrate into the mangrove forests where they develop into juveniles. The mangrove habitat is preferred due to the greater availability of food and structural complexity, high turbidity and soft sediment providing protection from predators (e.g. large fish). Relating the extent of these different habitats at multiple scales, their interactions and how they are changing over time is therefore important, but represents a significant challenge that has not been adequately addressed using remote sensing data.

When considering habitat connectivity, classification schemes based on object-oriented analysis of remote sensing data are often beneficial (Conchedda et al. 2008; Blaschke 2010; Lucas et al. 2014). In these approaches, groups of pixels within an image are amalgamated into objects and then assigned statistical values (e.g. the mean and variation of bands or derived measures such as vegetation indices, but also shape (e.g. roundness, length, area, perimeter) and contextual measures (e.g. adjacency, proximity)). Such information can be used in a classification approach to differentiate the main habitats surrounding or in proximity to mangrove forests. VHR (<2 m) imagery is often preferred as more detail can be resolved. The spatial connectivity between habitats can then be quantified in more detail. In more extensive mangrove systems with offshore reefs and seagrass beds, the use of moderate spatial resolution Landsat and SAR data would be necessary.

A limitation of such approaches is that temporal components are rarely included because of the availability of imagery and often only a static product (i.e. a habitat map) is generated. However, mangrove environments are very dynamic and hence the usefulness of these classifications will diminish over time. Along the coast of French Guiana, for example, rapid losses and gains in mangrove distribution occur as a consequence of erosion and sediment accretion (Fromard et al. 2004; Fromard 1998), and the changing connectivity of habitats (mudflats, forests) is difficult to quantify without a high temporal frequency of observation. Nevertheless, the release of the Landsat archive is providing new opportunities for quantifying changes in mangrove forests and associated habitats (Thomas et al. 2014).

Quantifying the level of mangrove fragmentation and landscape connectivity is a critical component of modern strategies aimed at biodiversity conservation, and many tools are available to address this using data from remote sensing. For example, in Mexico, CONABIO used the open source software package Conefor (www.conefor.org/index.html; Pascual-Hortal and Saura 2006; Saura et al. 2011; Saura and Torné 2009) with time series of aerial photography and Landsat/SPOT sensor data to describe the loss and fragmentation of mangrove habitats. The study established that in 2005 and since 1970–1980, mangrove forests in 16 out of 17 states with a coastline had experienced fragmentation and connectivity losses.

4.3.3 Challenges for Coastal Zone Management

With increasing coastal populations, considerable pressure has been placed upon mangrove ecosystems, resulting in clearance for resource use, industrialisation, urbanization and aquaculture. Often, secondary effects are noted from pollution and upstream land use. Mangrove wetlands are also affected by climate change, which include changes in sea level, temperature, precipitation, frequency and intensity of storm/cyclonic events, ocean circulation and atmospheric carbon dioxide (CO_2) concentration ((Asbridge et al. (2015); see Chap. 7). For example, in many cases such changes lead to alterations in carbon budgets and biodiversity and exert an influence on the social, political and economic aspects of many countries. Managing such change to maintain ecosystem services has largely been undertaken at the ground level but is a role that is increasingly addressed by remote-sensing observations.

As an example, a key ecosystem service provided by mangrove forests is the maintenance or enhancement of carbon stocks (Chmura et al. 2003; Bird et al. 2004; Manson et al. 2005; Polidoro et al. 2010; Lovelock et al. 2010; Sanders et al. 2010; McLeod et al. 2011). As indicated earlier, remote sensing estimates of carbon stocks are typically based on surrogates such as height which have been generated at a continental level using the 2000 SRTM supported by ICESAT GLAS data (Fatoyinbo and Simard 2013) and more locally using TanDEM-X data (Lee and Fatoyinbo 2015), airborne LiDAR (Wannasiri et al. 2013) or historical aerial photography (Mitchell et al. 2007). Quantifying changes in carbon stocks as a function of height is, however, more problematic. In particular, updates often use data acquired in modes that are different from those used for the baseline generation; hence, changes in height may be artificial. The data used for updating may also not be acquired at the most opportune time. As a complementary source, the dense time series of

Landsat sensor data can be used to establish the patterns of deforestation, regeneration or colonisation by mangroves and hence assist interpolation of standing biomass estimates. Time series of Landsat sensor data linked with mangrove growth models may be used to estimate actual or potential carbon sequestration but these approaches have not been undertaken to date.

Changes in carbon stocks may also be quantified using time series of L-band SAR data. To support this, the JAXA Global Mangrove Watch (GMW) was initiated in 2001 as part of the Kyoto and Carbon (K&C) Initiative (Lucas et al. 2014). The main objectives were to a) generate revised baseline maps of mangrove extent in the tropics and subtropics using ALOS PALSAR data from 2010, b) map changes relative to existing (2000) baselines (Giri et al. 2010) using JERS-1 SAR and ALOS PALSAR data acquired in the mid-1990s and annually between 2007–2010, and c) provide routine monitoring of mangrove areas post-2014 using ALOS-2 PALSAR-2 data. Using maps of canopy height and biomass generated from a combination of SRTM and ICESAT GLAS data, the potential for quantifying changes in carbon is being realized. The causes and consequences of change can also be discerned from these data.

Many other ecosystem services associated with mangroves have been assessed by different authors (Vo et al. 2012), with several employing higher resolution remote sensing data and (Vo et al. 2015, 2013) analyzing these data with household surveys. Kuenzer and Tuan (2013) also used remote sensing observations to quantify ecosystem services for the Can Gio Mangrove Biosphere Reserve in Vietnam, including timber provision, mangrove-related fish catch and protection from storms. By providing information on the spatial distribution of mangrove forests, their structure, species composition and biomass, and changes over time, remote sensing data at all scales can play a key role in quantifying ecosystems services and provide a contribution to integrated coastal zone management.

4.4 6. Summary

Remote sensing data have allowed observations of mangrove environments at global, biogeographic and local scales (Table 4.1). At the global level, studies have focused mainly on areal extent but increasingly there is capacity to quantify attributes that, until recently, were the domain of the local scale (Giri et al. 2010; Spalding et al. 2010; Valderrama et al. 2014). An example of this is the mangrove height and biomass maps for Africa derived using a combination of SRTM and ICESAT data (Fatoyinbo and Simard 2013). The use of SAR data is also becoming more widespread with the public release of ALOS PALSAR mosaics and Sentinel-1 data. Regional studies have also become more common, with many countries addressing the need for baseline maps of mangrove forests against which to quantify and monitor change. For example, Hay et al. (2005) generated baselines of mangrove species composition for northern and eastern Australia using a combination of aerial photography and Landsat sensor data. Local characterization of

Table 4.1 Examples of scale in relation to information retrieved

Scale	Spatial ecology of mangroves variations	Remote sensing and/or spatial analysis sources	Spatial resolution	Temporal resolution	Relevance	Reference examples
Global	Extent	Landsat	25/30 m	Decadal to potentially annual	Knowledge of changing distribution	Giri et al. (2010)
Biogeographic	Species, zonation height	Landsat SRTM TanDEM-X	12–30 m	Annual to monthly	More detailed knowledge of mangrove characteristics	Hay et al. (2005), Fatoyinbo and Simard (2013), Lee and Fatoyinbo (2015)
Local	Tree patches, canopy gaps, species, individual tree, 3D structure	Aerial photos LIDAR, UAV, WorldView-2, Airborne SAR	<2.5 m	Variable, from monthly to weekly	Knowledge of tree and stand-level characteristics and dynamics Monitoring of tree health disturbance and dieback	Lucas et al. (2002), Verheyden et al. (2002), Simard et al. (2006), Wannasiri et al. (2013), Jiali et al. (2015), Heenkenda et al. (2014), Held et al. (2003)

mangrove forests has focused primarily on the use of VHR airborne and spaceborne optical (e.g. WorldView), hyperspectral and LiDAR data with increasing opportunities for repeat coverage. As such, these data are collectively providing information, which is contributing to our understanding of spatial characteristics and dynamics of mangrove forests in relation to their ecology.

Despite significant advances in mangrove observations, however, remote sensing data need to be better linked with ecological and process models that are specific to this ecosystem. To achieve this, there needs to be greater comparison of observations, data and modelled outputs involving teams of scientists from different biogeographical regions. In Fig. 4.9, we underscore the overlap between remote sensing and ecological perspectives of mangrove forests in the spatial and temporal domain.

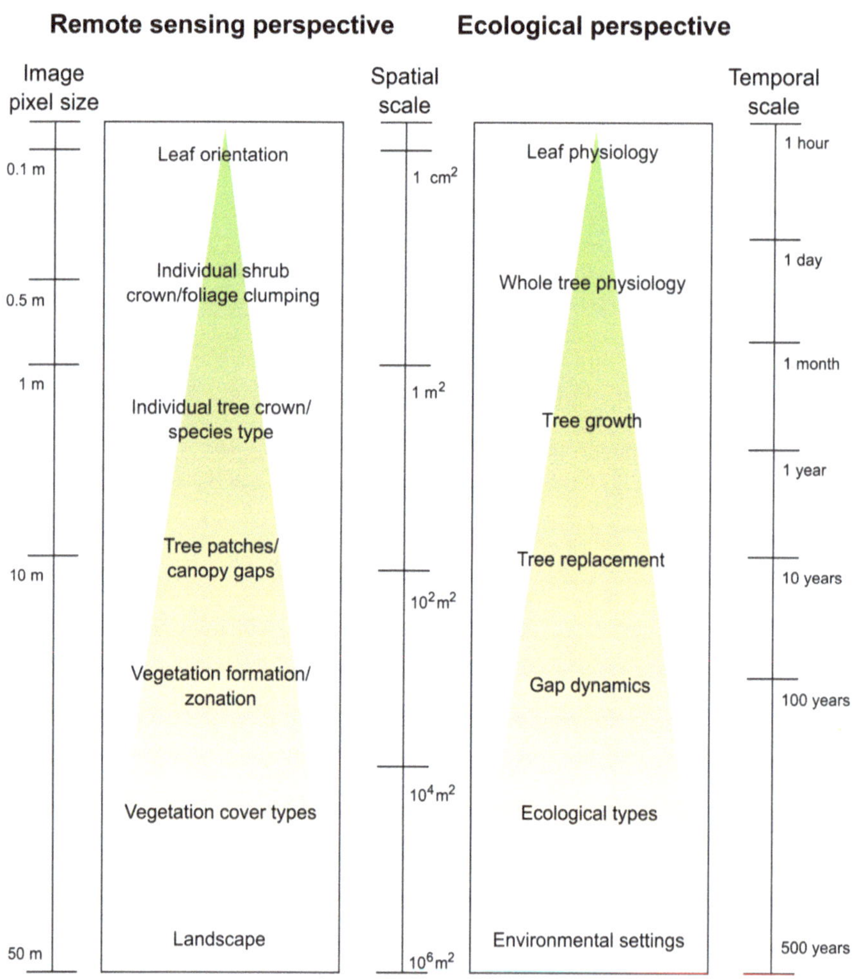

Fig. 4.9 Relationship between the temporal and structural hierarchy of mangrove forests and the size of features that are able to be detected using remote sensing data (Modified from Kamal (2015), p. 17)

Remote sensing products tend to be static and include leaf attributes (e.g. orientation), individual crown delineations by species, the extent of tree patches and canopy gaps, mangrove zones by species and growth stage and broader vegetation types. However, by increasing the spatial and temporal resolution of biophysical retrieval or mapping, these data can be transferred to the ecological domain and used to quantify leaf and whole tree physiology, growth and turnover, gap dynamics and ecological function. The context of mangrove forests within the wider landscape also needs to be considered, particularly given their dependence of, for example, river discharge in several coastlines around the world. By integrating this information, there are significant opportunities for species distribution and growth modelling that can be used to understand habitat use by fauna, fluxes of greenhouse gases and hydrological processes. Such knowledge can then lead to a better understanding of how best to manage mangrove forests in a more sustainable way and to establish their response to change, both now and into the future. Ultimately, this can assist with their long-term conservation.

Acknowledgement The authors would like to thank Jorg Hacker of Airborne Research Australia for the provision of the LIDAR mosaic for the Leichardt River.

References

Alfaro AC (2010) Effects of mangrove removal on benthic communities and sediment characteristics at Mangawhai Harbour, northern New Zealand. ICES J Mar Sci 67(6):1087–1104

Anthony EJ, Gardel A, Proisy C, Fromard F, Gensac E, Peron C, Walcker R, Lesourd S (2013) The role of fluvial sediment supply and river-mouth hydrology in the dynamics of the muddy, Amazon-dominated Amapá–Guianas coast, South America: a three-point research agenda. J S Am Earth Sci 44:18–24. doi:10.1016/j.jsames.2012.06.005

Asbridge E, Lucas R, Accad A, Dowling R (2015) Mangrove response to environmental changes predicted under varying climates: case studies from Australia. Curr Forest Rep 1(3):178–194

Baret F, Vanderbilt VC, Steven MD, Jacquemoud S (1994) Use of spectral analogy to evaluate canopy reflectance sensitivity to leaf optical properties. Remote Sens Environ 48(2):253–260

Bird M, Fifield L, Chua S, Goh B (2004) Calculating sediment compaction for radiocarbon dating of intertidal sediments. Radiocarbon 46(1):421–436

Blaschke T (2010) Object based image analysis for remote sensing. ISPRS J Photogramm Remote Sens 65(1):2–16

Blasco F, Gauquelin T, Rasolofoharinoro M, Denis J, Aizpuru M, Caldairou V (1998) Recent advances in mangrove studies using remote sensing data. Mar Freshw Res 49:287–296

Chadwick J (2011) Integrated LiDAR and IKONOS multispectral imagery for mapping mangrove distribution and physical properties. Int J Remote Sens 32(21):6765–6781

Chmura GL, Anisfeld SC, Cahoon DR, Lynch JC (2003) Global carbon sequestration in tidal, saline wetland soils. Glob Biogeochem Cycles 17:1111

Conchedda G, Durieux L, Mayaux P (2008) An object-based method for mapping and change analysis in mangrove ecosystems. ISPRS J Photogramm Remote Sens 63(5):578–589

Couteron P (2002) Quantifying change in patterned semi-arid vegetation by Fourier analysis of digitized aerial photographs. Int J Remote Sens 23:3407–3425

Da Cruz CC, Mendoza UN, Queiroz JB, Berrêdo JF, Neto SVDC, Lara RJ (2013) Distribution of mangrove vegetation along inundation, phosphorus, and salinity gradients on the Bragança Peninsula in Northern Brazil. Plant Soil 370(1–2):393–406

Davis BA, Jensen JR (1998) Remote sensing of mangrove biophysical characteristics. Geocarto Int 13(4):55–64

Demuro M, Chisholm L (2003) Assessment of Hyperion for characterizing mangrove communities. In: Proceedings of the International Conference the AVIRIS 2003 Workshop, pp 18–23

Díaz BM, Blackburn GA (2003) Remote sensing of mangrove biophysical properties: evidence from a laboratory simulation of the possible effects of background variation on spectral vegetation indices. Int J Remote Sens 24(1):53–73

Dissanayake N, Chandrasekara U (2014) Effects of mangrove zonation and the physicochemical parameters of soil on the distribution of macrobenthic fauna in Kadolkele mangrove forest, a tropical mangrove forest in Sri Lanka. Adv Ecol 2014

Eckert S (2012) Improved forest biomass and carbon estimations using texture measures from WorldView-2 satellite data. Remote Sens 4:810–829

Ellison JC (1999) Impacts of sediment burial on mangroves. Mar Pollut Bull 37(8):420–426

Ellison AM, Mukherjee BB, Karim A (2000) Testing patterns of zonation in mangroves: scale dependence and environmental correlates in the Sundarbans of Bangladesh. J Ecol 88(5):813–824. doi:10.1046/j.1365-2745.2000.00500.x

Fatoyinbo TE, Simard M (2013) Height and biomass of mangroves in Africa from ICESat/GLAS and SRTM. Int J Remote Sens 34(2):668–681

Feliciano E, Wdowinski S, Potts M (2012) Vegetation structure, tree volume and biomass estimation using terrestrial laser scanning remote sensing: a case study of the mangrove forests in the Everglades National Park. In: AGU Fall Meeting Abstracts, p 0358

Feliciano E, Wdowinski S, Potts M (2014) Assessing mangrove above-ground biomass and structure using terrestrial laser scanning: a case study in the Everglades National Park. Wetlands 34(5):955–968. doi:10.1007/s13157-014-0558-6

Fromard F (1998) Mangrove research discussion list communication.

Fromard F, Vega C, Proisy C (2004) Half a century of dynamic coastal change affecting mangrove shorelines of French Guiana. A case study based on remote sensing data analyses and field surveys. Mar Geol 208(2):265–280

Gausman H, Allen W, Cardenas R (1969) Reflectance of cotton leaves and their structure. Remote Sens Environ 1(1):19–22

Giri C, Pengra B, Zhu Z, Singh A, Tieszen LL (2007) Monitoring mangrove forest dynamics of the Sundarbans in Bangladesh and India using multi-temporal satellite data from 1973 to 2000. Estuar Coastal Shelf Sci 73(1):91–100

Giri C, Ochieng E, Tieszen LL, Zhu Z, Singh A, Loveland T, Masek J, Duke N (2010) Status and distribution of mangrove forests of the world using earth observation satellite data. Glob Ecol Biogeogr 20:154–159

Green EP, Mumby PJ (2000) Mapping mangroves. Remote sensing handbook for tropical coastal management. UNESCO Publishing, Paris

Green EP, Clark CD, Mumby PJ, Edwards AJ, Ellis A (1998) Remote sensing techniques for mangrove mapping. Int J Remote Sens 19(5):935–956

Hardisky M, Gross M, Klemas V (1986) Remote sensing of coastal wetlands. Bioscience 36:453–460

Hay T, Gribble N, de Vries C, Danaher K, Dunning M, Hearnden M, Caley P (2005) Methods for monitoring the abundance and habitat of the Northern Australian mud crab Scylla serrata, Fishery report no. 80. Northern Territory Department of Business, Industry and Resource Development, Darwin

Heenkenda MK, Joyce KE, Maier SW, Bartolo R (2014) Mangrove species identification: comparing WorldView-2 with aerial photographs. Remote Sens 6(7):6064–6088

Held A, Ticehurst C, Lymburner L, Williams N (2003) High resolution mapping of tropical mangrove ecosystems using hyperspectral and radar remote sensing. Int J Remote Sens 24(13):2739–2759

Jensen R, Mausel P, Dias N, Gonser R, Yang C, Everitt J, Fletcher R (2007) Spectral analysis of coastal vegetation and land cover using AISA+ hyperspectral data. Geocarto Int 22(1):17–28

Jiali F, Kai L, Yuanhui Z, Yong L, Liu L, Meng L (2015) Application of unmanned aerial vehicles to mangrove resources monitoring. Trop Geogr 35:35–42
Kamal M (2015) Remote sensing for multi-scale mangrove mapping. The University of Queensland, Brisbane
Kamal M, Phinn S (2011) Hyperspectral data for mangrove species mapping: a comparison of pixel-based and object-based approach. Remote Sens 3(10):2222–2242
Kamal M, Phinn S, Johansen K (2014) Characterizing the spatial structure of mangrove features for optimizing image-based mangrove mapping. Remote Sens 6(2):984–1006
Kamal M, Phinn S, Johansen K (2015) Object-based approach for multi-scale mangrove composition mapping using multi-resolution image datasets. Remote Sens 7(4):4753–4783
Kayitakire F, Hamel C, Defourny P (2006) Retrieving forest structure variables based on image texture analysis and IKONOS-2 imagery. Remote Sens Environ 102(3):390–401
Knipling EB (1970) Physical and physiological basis for the reflectance of visible and near-infrared radiation from vegetation. Remote Sens Environ 1(3):155–159
Kuenzer C, Tuan VQ (2013) Assessing the ecosystem services value of Can Gio Mangrove Biosphere Reserve: combining earth-observation-and household-survey-based analyses. Appl Geogr 45:167–184
Kuenzer C, Bluemel A, Gebhardt S, Quoc TV, Dech S (2011) Remote sensing of mangrove ecosystems: a review. Remote Sens 3(5):878–928
Le Toan T, Quegan S, Davidson M, Balzter H, Paillou P, Papathanassiou K, Plummer S, Rocca F, Saatchi S, Shugart H (2011) The BIOMASS mission: mapping global forest biomass to better understand the terrestrial carbon cycle. Remote Sens Environ 115(11):2850–2860
Lee S-K, Fatoyinbo TE (2015) TanDEM-X Pol-InSAR inversion for mangrove canopy height estimation. IEEE J Sel Top Appl Earth Obs Remote Sens 8(7):3608–3618
Lee SY, Primavera JH, Dahdouh-Guebas F, McKee K, Bosire JO, Cannicci S, Diele K, Fromard F, Koedam N, Marchand C (2014) Ecological role and services of tropical mangrove ecosystems: a reassessment. Glob Ecol Biogeogr 23(7):726–743
Leung JY (2015) Habitat heterogeneity determining the macrobenthic infaunal community in a mangrove swamp in South China: implication for plantation and plant invasion. J Coast Res 31(3):624–633
Liu K, Li X, Shi X, Wang S (2008) Monitoring mangrove forest changes using remote sensing and GIS data with decision-tree learning. Wetlands 28(2):336–346
Lovelock CE, Sorrell BK, Hancock N, Hua Q, Swales A (2010) Mangrove forest and soil development on a rapidly accreting shore in New Zealand. Ecosystems 13:437–451
Lucas RM, Ellison J, Mitchell A, Donnelly B, Finlayson M, Milne A (2002) Use of stereo aerial photography for quantifying changes in the extent and height of mangroves in tropical Australia. Wetl Ecol Manag 10(2):159–173
Lucas RM, Mitchell AL, Rosenqvist A, Proisy C, Melius A, Ticehurst C (2007) The potential of L-band SAR for quantifying mangrove characteristics and change: case studies from the tropics. Aquat Conserv Mar Freshwat Ecosyst 17(3):245–264
Lucas R, Rebelo L-M, Fatoyinbo L, Rosenqvist A, Itoh T, Shimada M, Simard M, Souza-Filho PW, Thomas N, Trettin C (2014) Contribution of L-band SAR to systematic global mangrove monitoring. Mar Freshw Res 65(7):589–603
Luther DA, Greenberg R (2009) Mangroves: a global perspective on the evolution and conservation of their terrestrial vertebrates. Bioscience 59(7):602–612
M'rabu E, Bosire JO, Cannicci S, Koedam N, Dahdouh-Guebas F (2012) Mangrove die-back due to massive sedimentation and its impact on associated biodiversity. In: Meeting on Mangrove Ecology, Functioning and Management (MMM3) 2–6 July 2012, Galle, Sri Lanka, p 104
Manson F, Loneragan N, Harch B, Skilleter G, Williams L (2005) A broad-scale analysis of links between coastal fisheries production and mangrove extent: a case-study for northeastern Australia. Fish Res 74(1):69–85
McLeod E, Chmura GL, Bouillon S, Salm R, Björk M, Duarte CM, Lovelock CE, Schlesinger WH, Silliman BR (2011) A blueprint for blue carbon: toward an improved understanding of the role of vegetated coastal habitats in sequestering CO_2. Front Ecol Environ 9(10):552–560

Mitchell AL, Lucas RM, Donnelly BE, Pfitzner K, Milne AK, Finlayson M (2007) A new map of mangroves for Kakadu National Park, Northern Australia, based on stereo aerial photography. Aquat Conserv Mar Freshwat Ecosyst 17(5):446–467

de la Morinière EC, Nagelkerken I, Van Der Meij H, Van Der Velde G (2004) What attracts juvenile coral reef fish to mangroves: habitat complexity or shade? Mar Biol 144(1):139–145

Mougin E, Proisy C, Marty G, Fromard F, Puig H, Betoulle JL, Rudant JP (1999) Multifrequency and multipolarization radar backscattering from mangrove forests. IEEET Trans Geosci Remote Sens 37(1):94–102

Mumby P, Edwards A, Arlas-Gonzalez J, Lindeman K, Blackwell P, Gall A, Gorczynska M, Harbone A, Pescod C, Renken H, Wabnitz C, Llewellyn G (2004) Mangroves enhance the biomass of coral reef fish communities in the Caribbean. Nature 427:533–536

Murray MR, Zisman SA, Furley PA, Munro DM, Gibson J, Ratter J, Bridgewater S, Minty CD, Place CJ (2003) The mangroves of Belize: part 1 Distribution, composition and classification. For Ecol Manag 174:265–279

Myint SW, Giri CP, Wang L, Zhu Z, Gillette SC (2008) Identifying mangrove species and their surrounding land use and land cover classes using an object-oriented approach with a lacunarity spatial measure. GISci Remote Sens 45:188–208

Nagelkerken I, Blaber S, Bouillon S, Green P, Haywood M, Kirton L, Meynecke J-O, Pawlik J, Penrose H, Sasekumar A (2008) The habitat function of mangroves for terrestrial and marine fauna: a review. Aquat Bot 89(2):155–185

Pascual-Hortal LY, Saura S (2006) Comparison and development of new graph-based landscape connectivity indices: towards the priorization of habitat patches and corridors for conservation. Landsc Ecol 21:959–967

Ploton P, Pélissier R, Proisy C, Flavenot T, Barbier N, Rai SN, Couteron P (2012) Assessing above ground tropical forest biomass using Google earth canopy images. Ecol Appl 22(3):993–1003

Polidoro BA, Carpenter KE, Collins L, Duke NC, Ellison AM, Ellison JC, Farnsworth EJ, Fernando ES, Kathiresan K, Koedam NE (2010) The loss of species: mangrove extinction risk and geographic areas of global concern. PLoS One 5(4):e10095

Prasad NV (2011) Remote sensing of mangrove wetlands: concept, methods, ecology, zonation, dynamics. VDM Verlag Dr. Müller, Saarbrucken

Proisy C, Mougin E, Fromard F, Karam M (2000) Interpretation of polarimetric radar signatures of mangrove forests. Remote Sens Environ 71(1):56–66

Proisy C, Mougin E, Fromard F, Trichon V, Karam MA (2002) On the influence of canopy structure on the polarimetric radar response from mangrove forest. Int J Remote Sens 23(20): 4197-4210

Rahman AF, Dragoni D, Didan K, Barreto-Munoz A, Hutabarat JA (2013) Detecting large scale conversion of mangroves to aquaculture with change point and mixed-pixel analyses of high-fidelity MODIS data. Remote Sens Environ 130:96–107

Rakotomavo A, Fromard F (2010) Dynamics of mangrove forests in the Mangoky River delta, Madagascar, under the influence of natural and human factors. For Ecol Manag 259(6):1161–1169. doi:10.1016/j.foreco.2010.01.002

Ramsey EW III, Jensen JR (1996) Remote sensing of mangrove wetlands: relating canopy spectra to site-specific data. Photogramm Eng Remote Sens 62:939–948

Rasolofoharinoro M, Blasco F, Bellan MF, Aizpuru M, Gauquelin T, Denis J (1998) A remote sensing based methodology for mangroves studies in Madagascar. Int J Remote Sens 10(19):1873–1886

Reigber A, Moreira A (2000) First demonstration of airborne SAR tomography using multi-baseline L-band data. IEEE Trans Geosci Remote Sens 38(5):2142–2152

Rodríguez-Zúñiga MT, Troche-Souza C, Vázquez-Lule AD, Márquez-Mendoza JD, Vázquez-Balderas B, Valderrama-Landeros L, Velázquez-Salazar S, Cruz-López MI, Ressl R, Uribe-Martínez A, Cerdeira-Estrada S, Acosta-Velázquez J, Díaz-Gallegos J, Jiménez-Rosenberg R, Fueyo-Mac Donald LY, Galindo-Leal C (2013) Manglares de México/Extensión, distribución y monitoreo. Comisión Nacional para el Conocimiento y Uso de la Biodiversidad, México

Rodríguez-Zúñiga MT, Troche-Souza C, Vázquez-Lule AD, Márquez-Mendoza JD, Vázquez-Balderas B, Valderrama-Landeros L, Velázquez-Salazar S, Uribe-Martínez A, Acosta-Velázquez J, Díaz-Gallegos J, Cruz-López MI, Ressl R (2012) Los Manglares de México: Estado Actual y Establecimiento de un Programa de Monitoreo a Largo Plazo: 2ª y 3era etapas. Informe final del proyecto GQ004. Comisión Nacional para el Conocimiento y Uso de la Biodiversidad. México

Rönnbäck P, Troell M, Kautsky N, Primavera JH (1999) Distribution pattern of shrimps and fish among Avicennia and Rhizophora microhabitats in the Pagbilao mangroves, Philippines. Estuar Coastal Shelf Sci 48:223–234

Saintilan N, Wilson N, Rogers K, Rajkaran A, Krauss KW (2014) Mangrove expansion and salt marsh decline at mangrove poleward limits. Glob Chang Biol 20(1):147–157

Sanders CJ, Smoak JM, Naidu AS, Sanders LM, Patchineelam SR (2010) Organic carbon burial in a mangrove forest, margin and intertidal mud flat. Estuar Coastal Shelf Sci 90(3):168–172. doi:10.1016/j.ecss.2010.08.013

Saura S, Torné J (2009) Conefor Sensinode 2.2: a software package for quantifying the importance of habitat patches for landscape connectivity. Environ Model Softw 24(1):135–139. doi:10.1016/j.envsoft.2008.05.005

Saura S, Estreguil C, Mouton C, Rodríguez-Freire M (2011) Network analysis to assess landscape connectivity trends: application to European forests (1990–2000). Ecol Indic 11(2):407–416. doi:10.1016/j.ecolind.2010.06.011

Simard M, Zhang K, Rivera-Monroy VH, Ross MS, Ruiz PL, Castañeda-Moya E, Twilley RR, Rodriguez E (2006) Mapping height and biomass of mangrove forests in everglades national park with SRTM elevation data. Photogramm Eng Remote Sens 3(72):299–311

Smith TJI (1992) Forest structure. Tropical mangrove ecosystems. American Geophysical Union, Washington, DC

Spalding M, Kainuma M, Collins L (2010) World atlas of mangroves. Earthscan, London

Thomas N, Lucas R, Itoh T, Simard M, Fatoyinbo L, Bunting P, Rosenqvist A (2014) An approach to monitoring mangrove extents through time-series comparison of JERS-1 SAR and ALOS PALSAR data. Wetl Ecol Manag:1–15. doi:10.1007/s11273-014-9370-6

Tomlinson PB (1986) The botany of mangroves. Cambridge University Press, Cambridge

UNEP (2014) The importance of mangroves to people: a call to action. United Nations Environment Programme World Conservation Monitoring Centre, Cambridge

Vaiphasa C, Ongsomwang S, Vaiphasa T, Skidmore AK (2005) Tropical mangrove species discrimination using hyperspectral data: a laboratory study. Estuar Coastal Shelf Sci 65:371–379

Vaiphasa C, Skidmore AK, de Boer WF (2006) A post-classifier for mangrove mapping using ecological data. ISPRS J Photogramm Remote Sens 61:1–10

Valderrama L, Troche C, Rodriguez MT, Marquez D, Vázquez B, Velázquez S, Vázquez A, Cruz I, Ressl R (2014) Evaluation of mangrove cover changes in Mexico during the 1970–2005 period. Wetlands 34:747–758

Vázquez-Lule AD (2012) Exploración de parámetros biofísicos con series de tiempo de productos modis y estimación de biomasa con imágenes de radar en los manglares de laguna Pom Atasta. Tesis Maestria, Campeche y laguna Agua brava, Nayarit, México

Verheyden A, Dahdouh-Guebas F, Thomaes K, De Genst W, Hettiarachchi S, Koedam N (2002) High resolution vegetation data for mangrove research as obtained from aerial photography. Remote sensing and GIS in the sustainable management of tropical coastal ecosystems. Environ Dev Sustain 4:113–133

Vo QT, Kuenzer C, Ming VQ, Oppelt N (2012) Review of valuation methods for mangrove ecosystem services. J Ecol Indic 23:431–446

Vo QT, Oppelt N, Kuenzer C (2013) Remote sensing in mapping mangrove ecosystems – an object-based approach. Remote Sens 5(1):183–201

Vo QT, Kuenzer C, Oppelt N (2015) How remote sensing supports mangrove ecosystem service valuation: a case study in Ca Mau Province, Vietnam. Ecosyst Serv 14:67–75

Vogt J, Skóra A, Feller IC, Piou C, Coldren G, Berger U (2012) Investigating the role of impoundment and forest structure on the resistance and resilience of mangrove forests to hurricanes. Aquat Bot 97(1):24–29

Wang L, Sousa WP, Gong P, Biging GS (2004) Comparison of IKONOS and QuickBird images for mapping mangrove species on the Caribbean coast of Panama. Remote Sens Environ 91:432–440

Wang L, Silván-Cárdenas JL, Sousa WP (2008) Neural network classification of mangrove species from multi-seasonal Ikonos imagery. Photogramm Eng Remote Sens 74:921–927

Wannasiri W, Nagai M, Honda K, Santitamnont P, Miphokasap P (2013) Extraction of mangrove biophysical parameters using airborne LiDAR. Remote Sens 5:1787–1808

Wolanski E (1995) Transport of sediment in mangrove swamps. Hydrobiologia 295:31–42

Yang S-C, Shih S-S, Hwang G-W, Adams JB, Lee H-Y, Chen C-P (2013) The salinity gradient influences on the inundation tolerance thresholds of mangrove forests. Ecol Eng 51:59–65

Zhang K, Houle P, Ross M, Ruiz P, Simard M (2006) Airborne laser mapping of mangroves on the Biscayne Bay Coast, Miami, Florida. In: Proceedings of the Geoscience and Remote Sensing Symposium IGARSS 2006, July 31, 2006–August 4, 2006, pp 3750–3754

Chapter 5
Productivity and Carbon Dynamics in Mangrove Wetlands

Robert R. Twilley, Edward Castañeda-Moya, Victor H. Rivera-Monroy, and Andre Rovai

5.1 Introduction

The focus of this chapter is the net primary productivity and carbon (C) dynamics of mangrove wetlands as related to the potential to sequester atmospheric C in tropical and subtropical coastal ecosystems (Fig. 5.1; Table 5.1) (Donato et al. 2011; Kauffman et al. 2011). Mangrove wetlands produce organic carbon well in excess of ecosystem respiration and are considered important sites for C burial (~10%) and C export (~40%) to adjacent coastal waters, indicating their significant contribution to C biogeochemistry in the coastal zone (Fig. 5.1) (Twilley et al. 1992; Duarte and Cebrian 1996; Bouillon et al. 2008; Komiyama et al. 2008). Whole C storage estimates using structural data (tree size, density), dead wood biomass, soil C content, and soil depth in mangrove wetlands of the Indo-Pacific region have estimated an average storage of 1023 MgC ha^{-1} ± 88, which exceeds 2.5–5 times the mean C

R.R. Twilley (✉) • V.H. Rivera-Monroy
Department of Oceanography and Coastal Sciences, School of the Coast and the Environment, Louisiana State University, Baton Rouge, LA 70803, USA
e-mail: rtwilley@lsu.edu; vhrivera@lsu.edu

E. Castañeda-Moya
Department of Oceanography and Coastal Sciences, School of the Coast and the Environment, Louisiana State University, Baton Rouge, LA 70803, USA

Southeast Environmental Research Center, Florida International University, Miami, FL 33133, USA
e-mail: ecastane@fiu.edu

A. Rovai
Departamento de Ecologia e Zoologia Universidade Federal de Santa Catarina, Florianópolis, SC 88040-900, Brazil

Department of Oceanography and Coastal Sciences, College of the Coast and Environment, Louisiana State University, Baton Rouge, LA 70803, USA
e-mail: asrovai@gmail.com

V.H. Rivera-Monroy et al. (eds.), *Mangrove Ecosystems: A Global Biogeographic Perspective*, https://doi.org/10.1007/978-3-319-62206-4_5

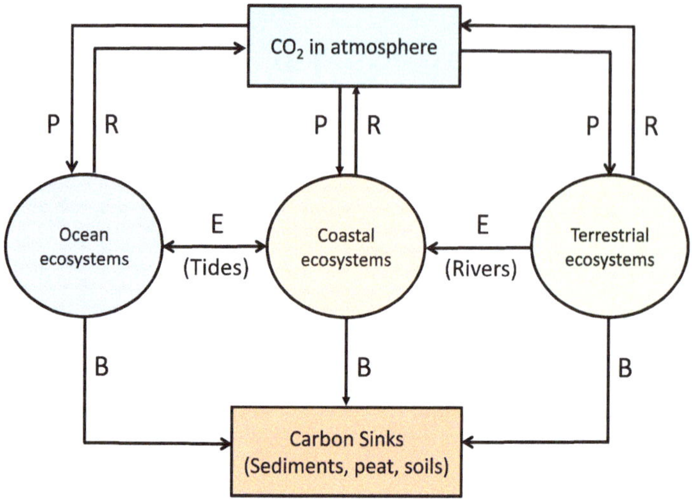

Fig. 5.1 A simple model for the cycling of carbon in the coastal ocean including coastal ecosystems, dominated in the tropics and subtropics by mangrove ecosystems. The model demonstrates significance of mangrove NEP to exchange with atmosphere, but also exchanges across boundaries with terrestrial ecosystems and ocean ecosystems. *D* diffusion, *P* photosynthesis (gross productivity), *R* respiration, *E* exchange, *B* burial (Modified from Suratman 2008, Twilley et al. 1992)

Table 5.1 Summary of global scale estimates of organic carbon burial based on local carbon sequestration rates from Breithaupt et al. (2012)

Authors	Local burial rate (g m^{-2} yr^{-1})	Study's mangrove areal extent (km^2)	Global burial rate (Tg C yr^{-1})	Global burial rate standardized to 160,000 km^2 (Tg C yr^{-1})
Twilley et al. (1992)	100	240,000	24.0	16.0
Jennerjahn and Ittekkot (2002)	115	200,000	23.0	18.4
Chmura et al. (2003)	210	181,000	38.0	33.6
Duarte et al. (2005)	139	200,000	27.8	22.2
Bouillon et al. (2008)	115	160,000	18.4	18.4
Alongi (2009)	181	160,000	29.0	29.0
Mcleod et al. (2011)	226	137,760	31.1	36.2
		152,361	34.4	
Breithaupt et al. (2012)	163	137,760	22.5	26.1
		152,361	24.9	

stock in tropical upland, temperate, and boreal forests (200–400 MgC ha^{-1}) (Donato et al., 2011). Thus, this biomass value suggests that mangrove wetlands are "*among the most C-rich forests in the tropics*" (Donato et al. 2011). However, current studies show that the actual mangrove global burial rates range is wider and uncertain

(18.4–38 TgC y^{-1}) and the aerial C storage estimates depend on the extent used in these estimations of both biomass and production (Bouillon et al. 2008). An updated review of mangrove C dynamics indicates that global mangrove primary production is ~218 ± 72 TgC yr^{-1}, yet approximately >50% (~112 ± 85 TgC yr^{-1}) of the C fixed by mangrove vegetation is unaccounted (Bouillon et al. 2008). We propose that the inability to constrain the productivity and C sequestration numbers include (1) the complexity of C dynamics in forest ecosystems that inhabit the intertidal zone, (2) the inability to account for variation in ecosystem processes among geomorphological settings, and (3) inability to account for net ecosystem dynamics in highly disturbed coastal landscapes.

We will utilize a systems ecology approach of mangrove productivity and C dynamics to scale the large variation in these ecosystem properties across dynamic coastal environmental settings where the interaction between hydrology (e.g., tides, frequency and duration of inundation) and geomorphology defines a wide range of mangrove ecotypes with highly distinctive ecosystem attributes (Lugo and Snedaker 1974). In addition, we will pay particular attention to global patterns of these ecosystem processes by comparing the Atlantic-East-Pacific (AEP) and Indo-West-Pacific (IWP) biogeographic regions to understand the distribution of global observations that have been used to constrain C budgets of mangrove wetlands (Fig. 5.2). The high net primary production (e.g., NPP = 17.7 ± 7 Mg dry mass ha^{-1} yr^{-1}, $N = 20$) and aboveground biomass (169 ± 112 Mg dry mass ha^{-1}) observed in these diverse mangrove ecotypes are regulated by key interactions across gradients among resources (e.g., nitrogen, phosphorus), regulators (e.g., salinity), and hydroperiod (e.g., duration of flooding) (Twilley and Rivera-Monroy 2005, 2009). This review will synthesize existing information on above- and belowground productivity and will provide a description of how net ecosystem productivity (NEP) relates to net ecosystem carbon exchange (NECE). Models will be introduced to make preliminary estimates of the global scale of these ecosystem processes across biogeographic regions and coastal environmental settings. We propose that using the energy signature hypothesis (see below, Twilley 1995) improves understanding of the variation with ecogeomorphology, which should improve attempts to scale from local and regional settings to global budgets.

It is a challenge to scale up a variety of ecosystem attributes and processes from local to global scales given the variability of rate processes and the high biodiversity of both coastal environmental settings and biogeographic regions of mangrove wetlands (Rovai et al. 2016). In addition, the combination between geographical sampling bias (e.g., scarce and uneven distributed sampling points across biogeographic regions), the lack of consistency in robust global sampling designs (Brown and Lugo 1984), and the choice of environmental predictors used in contemporary global mangrove C assessments (Hutchison et al. 2013; Jardine and Siikamäki 2014) have been hindering the development of models capable of delivering more precise mangrove C global estimates. Global estimates of C storage in mangroves colonizing environments ranging from deltas to estuaries and to oceanic islands are based on annual increments of mangrove biomass (above- and belowground) and C

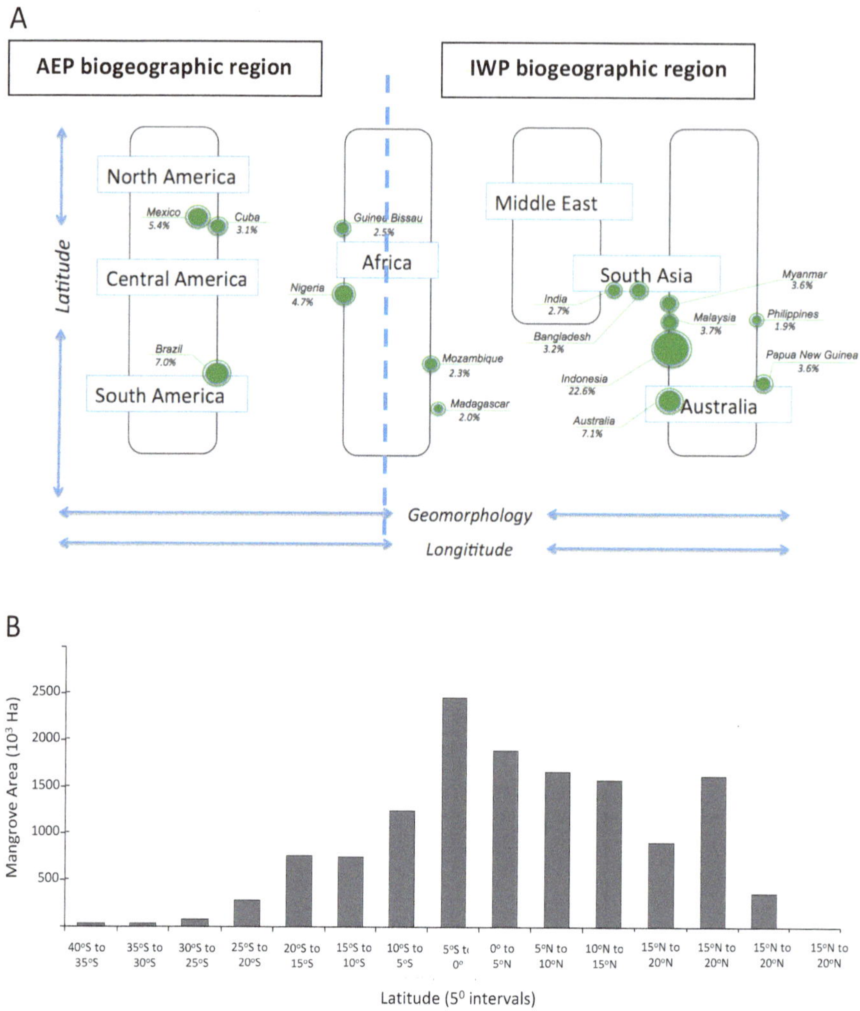

Fig. 5.2 (**a**) The biogeographical and regional dimensions of mangrove distribution considered in comparative analysis of ecological attributes discussed in this chapter. The *circles* represent the 15 countries contributing 75.3% of the global mangrove area in 2000 (Data from Giri et al. 2011). (**b**) Latitudinal distribution of mangrove forests in the world from Giri et al. (2011) used to develop statistics in (**a**)

stocks in soil and account for large exchanges across complex boundaries between forests and the sea (Bouillon et al. 2008; Breithaupt et al. 2012; Alongi 2014). Published global estimates on central components of the mangrove C budget (i.e., net primary productivity, herbivory, mineralization, burial, organic carbon export) are based on limited C budgets that integrate processes at a single site and therefore represent relatively small data sets (Table 5.1). We will do preliminary assessments of new global budgets of biomass, C content, and sequestration by classifying the physical characteristics to coastal environmental settings.

The role of C sequestration by mangrove wetlands in mitigating the C feedback to climate conditions is also a very complex argument. Mangrove C sequestration and storage may provide global benefits and economic value because this ecosystem service may mitigate the effects of climate change (Hopkinson et al. 2012). However, global estimates of benefits from C sequestration are limited by not accounting for diversity of environmental settings, and presently no global estimate of C sinks accounts for worldwide loss of mangroves through extensive degradation and over exploitation (Caldeira 2012; Gedan et al. 2011; Giri et al. 2008). According to the Intergovernmental Panel on Climate Change, deforestation and land-use change account for 8–20% of the total global anthropogenic carbon dioxide (CO_2) emissions (Allen et al. 2014). Former global assessments estimated that a third of mangrove area has been lost over the last 50 years as a result of land use change and degradation (Valiela et al. 2001; Alongi 2002). However, these estimates are as much a result of variation in mapping of historical mangrove cover as real changes in global mangrove area. In addition, recent estimates indicate that global losses during the last decade (2000–2012) were around 10% (Friess and Webb 2014). Despite occupying only 0.7% of tropical forest area, deforestation of mangrove biomass is apparently responsible for 10% (0.02–0.12 Pg C per year) of global CO_2 emissions (Siikamäki et al. 2012; Donato et al. 2011). Nonetheless, reforestation of mangroves may stimulate C storage processes, as observed in terrestrial forests. In addition, many mangrove wetlands occur in coastal regions with high frequency of disturbance from cyclones, and the effects of these pulsing events on regenerating C sequestration during forest reorganization may be a global consideration in C mitigation techniques. These discussions of net C sinks with disturbances are important considering the significance of C along tropical and subtropical shorelines. However, we need to improve estimates of how these ecosystems may contribute to achieve desired climate change mitigation goals (Canadell and Raupach 2008).

5.1.1 Coastal Environmental Settings

The energy signature hypothesis for tropical coastal ecosystems states that geophysical forces (river discharge, tides, waves), along with regional climate (temperature, precipitation, evapotranspiration), control soil biogeochemistry, ultimately determining the structure (e.g., basal area, tree height) and function (e.g., net primary productivity) of mangrove ecosystems (Fig. 5.3; Twilley 1995). Energy signature hypothesis is a readily applicable concept to other flood pulse-dominated systems (Lugo et al. 1988; Junk et al. 2014), but it also underscores that adopting geomorphic-related controls into predictive models is critical when scaling from local and regional to global patterns of C dynamics (Rovai 2016). The regional scale description of coastal environments using geomorphology and geophysical processes can be further separated into ecological classification systems (Fig. 5.3). The ecological classification of mangroves as fringe, basin, or scrub forests describes the

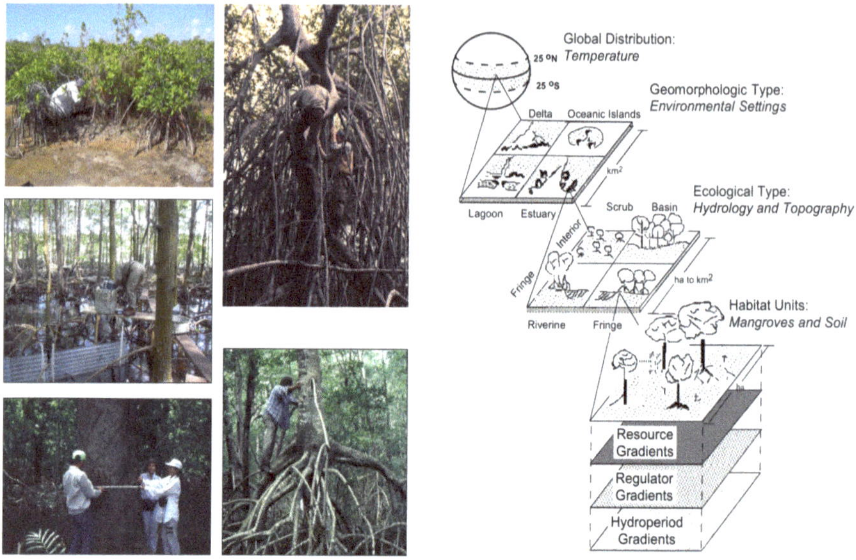

Fig. 5.3 Hierarchical classification system to describe patterns of mangrove structure and function based on global, geomorphological (regional), and ecological (local) factors that control the concentration of nutrients resources and regulators in soil along gradients from fringe to more interior locations from shore (Modified from Twilley et al. 1998 and Twilley and Rivera-Monroy 2005). Pictures of mangrove sites describing the structure of mangrove forests across the ecogeomorphological types

microtopographic effects of hydrology on the formation of forest ecotypes (Lugo and Snedaker 1974). Resources, regulators, and hydroperiod are very distinct along the environmental gradient of the intertidal zone defined by microtopography. This sets up the ecologically significant attributes of fringe versus interior zones of mangroves (Ewel et al. 1998). Fringe mangroves colonize the boundary of mangroves along a coastal water body, and their ecological attributes are established by the relative influence of river, tides, and waves. Interior mangrove wetlands are less frequently inundated and as found for most upper regions of intertidal zone, soil salinities are dominated by soil moisture as a result of precipitation, evapotranspiration, and upland flow. Distinct soil characteristics in this interior intertidal zone can be regulated by regional climate as in the case for drier environments (e.g., Pacific coast of Honduras, Castaneda-Moya et al. 2006). We will review some of the test hypotheses across different spatial scales, given that mangrove macroecology remains as an emergent science, with only a handful of studies addressing ecological trends across global biogeographic regions (e.g. Twilley et al. 1992; Ellison 2002; Lovelock et al. 2007; see also Chap. 11). In addition, the broad distribution along the world's tropical and subtropical shorelines (Giri et al. 2011) and over a variety of costal environmental settings, from deltaic to karstic landforms (Woodroffe et al. 2016), enables such investigations.

5.1.2 Conceptual Model of Productivity and Carbon Dynamics

The fixation of CO_2 from the atmosphere by mangrove wetlands represents potentially significant global C exchanges that influence the net C budget of the atmosphere (Figs. 5.1 and 5.4). Organic carbon budgets for mangrove wetlands based on what is fixed into forest biomass from the atmosphere or deposited by rivers and tides into soils require operational definitions on what processes constitute net

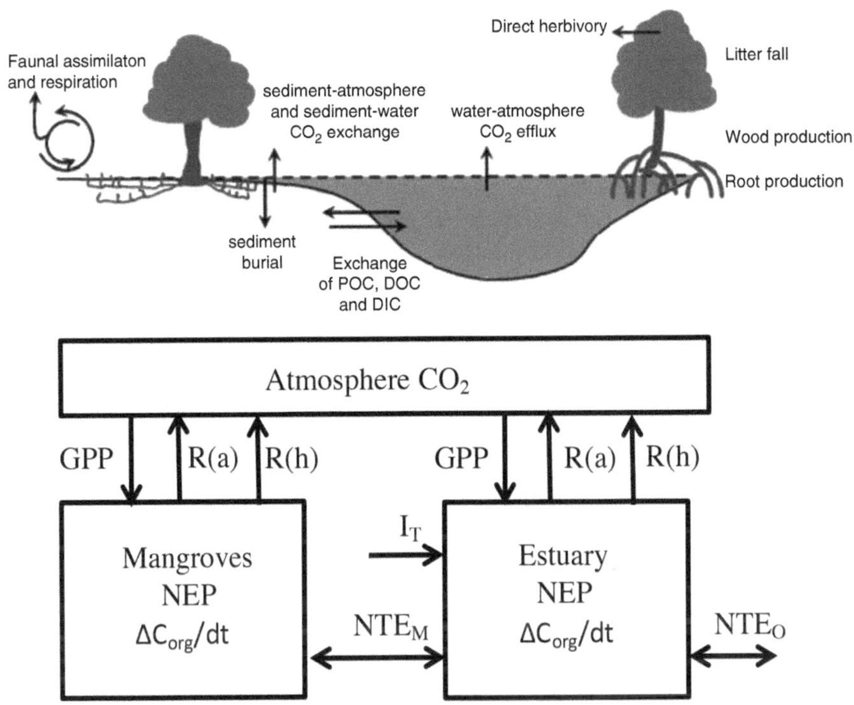

$$NEP\ (mangroves) = (GPP) - (Ra + Rh) \pm NTE_M)$$
$$I_T = DIC + DOC + POC\ (river\ inflow)$$
$$NTE = DIC + DOC + POC\ (inflow - outflow\ of\ tides)$$
$$NEP\ (estuary) = (GPP + I_T) - (Ra + Rh) \pm NTE_M \pm NTE_O$$

Fig. 5.4 (*Top panel*) Summary of the major components in the mangrove carbon budgets normally measured in field observations including production (litter fall, wood, and root production) and various sinks (sediment burial) along with exchange with coastal waters (carbon forms including particulate organic (*POC*), dissolved organic (*DOC*) and dissolved inorganic carbon (*DIC*). From Bouillon et al. 2008). (*Lower panel*) Mass balance of organic carbon using concepts of net ecosystem production (*NEP*) used in the text for mangroves (*GPP* gross primary production, *Ra* respiration autotrophs, *Rh* respiration heterotrophs, NTE_M is the net exchange of tidal inflow-outflow). For estuary NEP, additional terms are IT for input from rivers, NTE_O is exchange with coastal ocean

fluxes across the boundaries of coastal forests (Twilley et al. 1992; Bouillon et al. 2008). As will be described in this section, some approaches use the net accumulation of wood and soil to estimate the net effects of all these processes in the coastal zone compared to more detailed accounting of all forms of C to derive a net C sink (Table 5.1). Current estimates indicate that litter fall (NPP_L), along with wood (NPP_W) and belowground production (NPP_B), account for ~31, 31, and 38%, respectively, of the total net productivity (NPP_T) on a global basis (Bouillon et al. 2008). These estimates underscore the significant contribution of NPP_W and NPP_B to NPP_T of mangrove wetlands worldwide. Recent summaries indicate there are few examples of simultaneous measurements of both the aboveground net primary productivity ($NPP_A = NPP_L + NPP_W$) and NPP_B to accurately estimate NPP_T of mangrove wetlands. There are few long-term studies that test the temporal and spatial variation in C allocation above and within the soil compartment, limiting our understanding of how these processes may be stimulated by disturbance. In addition, estimates of export to coastal waters and respiration by all consumers must be subtracted to indicate what may be a net sink of organic carbon in these coastal forests.

Our review to account for the biomass and productivity of mangroves to assess C dynamics will build upon the methods recently reviewed for mangrove ecosystems (Rivera-Monroy et al. 2013). We will use the ecogeomorphology models described for biogeochemistry of mangroves (Twilley and Rivera-Monroy 2009) and apply these methods to C accounting and productivity across different coastal environmental settings to scale estimates at the global level. There are significant efforts to measure the blue carbon of mangroves and all these efforts will be reviewed within the context of our conceptual model and methodologies to try and explain patterns from local sites to biogeographical regions. There are still some estimates of key processes influencing fate of C in mangroves that are lacking global coverage such as belowground components (see Castañeda-Moya et al. 2011; Donato et al. 2011; Rivera-Monroy et al. 2013) and C export to coastal waters (Cai 2011). We will define landscape patterns of mangrove NPP_T and C allocation by applying models of these processes for specific coastal environmental settings using the energy signature hypothesis.

NECE involves closer attention to C fluxes including dissolved inorganic carbon, organic vapors, and C gases other than carbon dioxide (e.g., methane) to assess the role of mangrove wetlands as C sinks or sources in subtropical and tropical latitudes. Attempts to estimate the fate of organic carbon by evaluating production, biomass allocation, respiration, soil accumulation, and export do not account for all the forms of C that are exchanged with the atmosphere. It is very important to clearly define which processes are actually measured and how those measures contribute to different approaches to account for C budgets to define the impact of such processes on mitigating C dynamics at the continent–atmosphere–ocean interfaces (e.g., Komiyama et al. 2008; Alongi 2009, 2014). We will provide guidance on our approach to account for these different techniques to define NEP compared to NECE estimates for mangrove ecosystems.

The relationships among gross primary production (GPP), NPP, and NEP are critical to understanding how much organic carbon is accumulated in a mangrove wetland by comparing net inputs from photosynthesis to losses by respiration and export (Fig. 5.4). The equations for mangrove wetlands, as for other forested ecosystems, represent C exchanges between the forest canopy and the atmosphere as measures of net primary productivity of aboveground biomass (AGB). In this review, we will focus on the following equation for aboveground net primary productivity as follows:

$$\mathrm{NPP_A} = \mathrm{NPP_L} + \mathrm{NPP_W} \tag{5.1}$$

where $\mathrm{NPP_L}$ is litter production and $\mathrm{NPP_W}$ is wood production. $\mathrm{NPP_B}$ is measured by changes in biomass of live roots over time intervals. Thus, our estimates of total net primary productivity of mangroves are as follows:

$$\mathrm{NPP_T} = \mathrm{NPP_L} + \mathrm{NPP_W} + \mathrm{NPP_B} \tag{5.2}$$

where $\mathrm{NPP_T}$ is the gain in organic carbon by the formation of plant tissues (GPP) minus the respiration of autotrophs in both above- and belowground compartments (Fig. 5.4).

The net productivity of mangrove wetlands can be evaluated by measuring the net metabolism across a defined ecosystem boundary, defined above as NEP. The net exchange of C with the atmosphere is controlled by the transformation of inorganic to organic carbon assimilated into mangrove biomass, balanced by the relative return of inorganic carbon associated with total ecosystem respiration (R_T) as the sum of respiration of autotrophs (Ra) and heterotrophs (Rh). $\mathrm{NPP_T}$ of mangrove wetlands accounts for losses associated with Ra, and Rh can be tracked by measuring soil respiration rates in mangroves (Rs) based on soil CO_2 efflux rates. Thus, the atmosphere exchange of NEP can be accounted for using the following:

$$\mathrm{NEP} = \left(\mathrm{NPP_L} + \mathrm{NPP_W} + \mathrm{NPP_B}\right) - \left(\mathrm{Rs}\right). \tag{5.3}$$

However, open ecosystems such as mangrove wetlands can be associated with large organic carbon exchanges by tidal and river inundation, including extreme weather events (Twilley et al. 1992; Bouillon et al. 2008) (Fig. 5.4). Given the bidirectional flux of tides in mangroves, this exchange has to be considered a net tidal exchange ($\mathrm{NTE_M}$), with negative values indicating a loss of organic carbon from mangrove wetlands to coastal waters. Thus, the net balance of organic carbon in mangrove wetlands as defined by NEP is the following:

$$\mathrm{NEP} = \left(\mathrm{NPP_L} + \mathrm{NPP_W} + \mathrm{NPP_B}\right) - \left(\mathrm{Rs} \pm \mathrm{NTE_M}\right). \tag{5.4}$$

C gains in NEP depend on how much of the net balance between $\mathrm{NPP_T}$ and Rs ($\mathrm{NPP_T} - \mathrm{Rs}$) is lost at the coastal water boundary by $\mathrm{NTE_M}$. In this analysis, a negative $\mathrm{NTE_M}$ represents organic carbon export from mangrove wetlands to coastal waters, reducing the organic carbon storage potential of mangrove wetlands as a C sink. $\mathrm{NTE_M}$ becomes an import of C to the coupled coastal system (e.g. estuary, lagoon, delta), where it may contribute to ecosystem respiration or be buried in soil. Therefore, we do not consider NTE_M as part of the blue carbon potential of mangrove wetlands, which is the ecosystem that is the focus of this review.

The accumulation rate of organic carbon in mangrove soils (ΔS_{org}) may integrate several of the processes that contribute to NEP (Chen and Twilley 1999). For example, the net amount of organic carbon in mangrove soils integrates NPP_T by including net C balance of root production and litter fall ($NPP_B + NPP_L$), losses due to Rs, including soil CO_2 efflux linked to leaching of organic carbon from canopy. ΔS_{org} also integrates the net effect of NTE. Thus, NEP can largely be determined by focusing on the sum of ΔS_{org} and the NPP_W based on the following substitutions:

$$\mathrm{NEP} = \left(\mathrm{NPP_L} + \mathrm{NPP_W} + \mathrm{NPP_B}\right) - \left(\mathrm{Rs} \pm \mathrm{NTE_M}\right) \quad (5.5)$$

$$\Delta S_{org} = \left(\mathrm{NPP_L} + \mathrm{NPP_B}\right) - \left(\mathrm{Rs} \pm \mathrm{NTE_M}\right) \quad (5.6)$$

$$\mathrm{NEP} = \mathrm{NPP_W} + \Delta S_{org} \quad (5.7)$$

This establishes the significance of organic carbon accumulation in mangrove soils and wood production to define the rates that mangroves may store organic carbon from atmosphere, as suggested originally by Lugo and Snedaker (1974). The accumulation of organic carbon in mangrove soils can be measured directly, using a variety of techniques in vertical accretion multiplied by the C density of soils (Lynch et al. 1989; Chmura et al. 2003; Rivera-Monroy et al. 2013; Breithaupt et al. 2012). Equation (5.4) suggests that NEP can be determined by measuring the components of NPP_T ($NPP_L + NPP_W + NPP_B$, Eq. 5.2), account for total soil CO_2 efflux rates (Rs), and include NTE_M. We will review the literature on both direct soil accumulation rates and soil CO_2 efflux rates to compare with various components of NPP_T and NTE_M to resolve the C balance by a variety of mass balance techniques. By comparing a combination of approaches to mass balance the C flux across the atmosphere boundary and at the river/coastal interface, we can resolve how well our efforts have been at deriving the role of mangrove wetlands as a C sink.

5.2 Aboveground Biomass

AGB represents a significant portion of the total organic carbon reservoir in mangrove wetlands that is found in vegetation and soils (Lugo and Snedaker 1974). The annual incremental change in AGB is a traditional approach to estimate NPP_A, considering that annual C stored in trees via wood growth, along with litter fall productivity from the canopy, represents the residual of gross production remaining after autotrophic respiration (see Eq. 5.1). The variation in annual C storage in AGB may be a function of climatic regimes coupled with distinct geophysical and geomorphological variables (Twilley and Rivera-Monroy 2009). It has been proposed that mangrove NPP_A and biomass accumulation are high in coastal regions where tidal ranges are high and significant run-off provides sediment and nutrients to mangrove soils, in contrast to harsh environments with low moisture, frequent frosts, periodic droughts, and hypersalinity that restrict forest biomass accumulation (low AGB) (Schaeffer-Novelli et al. 1990; Twilley 1995; Castañeda-Moya et al. 2006). The close relationship between mangrove NPP_A and biomass and these environmental drivers has seldom been directly tested. A literature review to assemble a data set containing information on published mangrove AGB and forest structure data is summarized in a review by Rovai et al. (2016). This search included 134 studies encompassing 1047 sites that were aggregated into 0.25° cells, producing 135 AGB values, which were used to develop regression analyses with coastal environmental settings.

Previous attempts to predict continental-scale mangrove AGB include latitude (Twilley et al. 1992; Saenger and Snedaker 1993) and climate-based models (Hutchison et al. 2014). Although latitude-based models can indirectly encompass critical climatic and geophysical variables, their individual contribution to explain AGB value spatial patterns is unknown since their explanatory power is not explicitly weighted in the statistical analysis. And although a climatic modeling approach explicitly includes climate variables such as temperature (mean temperatures of warmest and coldest quarters) and precipitation (precipitation of wettest and driest quarters) to explain mangrove AGB at the global scale (Hutchison et al. 2014), this analysis is limited not only by the number of climatic variables included in the model, but also by the lack of other environmental variables that directly influence mangrove structural and functional properties at regional and local scales (Twilley 1995; Twilley and Rivera-Monroy 2009). Most estimates of mangrove AGB also assume maturity of stand age, but very few mangroves represent old growth stands and thus represent some younger stage of biomass accumulation (Lugo 1997). The inclusion of other geophysical variables in the climatic-geophysical model of Rovai et al. (2016) significantly improves AGB estimates at the latitudinal scale. This model had values ranging from 16.6 to 627.0 t ha^{-1} (mean = 88.7 t ha^{-1}) and demonstrated that climate-based and latitude-based models overestimated mangrove AGB by 25.3% (Hutchison et al. 2013), 34.3% (Saenger and Snedaker 1993), and 44.4% (Twilley et al. 1992) in the neotropics.

Coastal environmental settings represent a major constraint on mangrove wetlands spatial distribution and realized maximum biomass, particularly considering the diversity of mangrove environmental settings and associated ecogeomorphic dynamics (Thom 1982; Twilley 1995; Woodroffe 1992). This dynamic change is strongly influenced by the local tidal range, a critical geophysical variable explaining a significant percentage of the AGB total variance (Rovai et al. 2016). Indeed, tidal amplitude, a component of the hydroperiod regime in coastal regions, significantly influenced mangrove structural development by promoting nutrient exchange and aeration of soil layers, which reduces sulfide accumulation, allowing higher growth rates and forest development (Castañeda-Moya et al. 2013; Lugo and Snedaker 1974). The model by Rovai et al. (2016) supports the environmental signature hypothesis approach for mangrove dominated ecosystems suggesting that AGB allocation at the continental scale is not only dependent on climatic variables, but also on local and regional geophysical forcings such as tidal amplitude. Their estimates of AGB are ecologically significant because current published models to estimate AGB do not include or test the relative contribution of these variables in explaining mangrove AGB spatial distribution. Further, the Rovai et al. (2016) model is composed of a set of environmental drivers that are ecologically meaningful and closely associated to observed AGB spatial distribution patterns at larger geographical scales; thus, it could potentially be applied to other continental coastal regions worldwide.

The global data set from 135 sites with observations of AGB, using the local models of biomass estimates based on forest structure statistics, presents a comparison of IWP and AEP biogeographic regions (Figs. 5.5 and 5.6). Nearly half of the observations have AGB values <100 t ha^{-1}. There are only 10 of the 135 sites with AGB >300 t ha^{-1} (Fig. 5.5, left panels). There are no clear patterns distinguishing the two biogeographic regions, except the majority of highest AGB values are in IWP region (Fig. 5.5, left panels). When applied to Giri's global mangrove coverage mask (Giri et al. 2011), using 25 km^2 degree cells at the equator ($N = 6850$ grids), the model by Rovai et al. (2016) gives a somewhat different perspective of the global distribution of AGB (Fig. 5.5, right panels). Nearly half of the grids have AGB between 50 and 140 t ha^{-1}, and very few sites have values >300 t ha^{-1} (Fig. 5.5, right panels). In addition, there does seem to be a trend between IWP and AEP, with higher AGB values in IWP region in the more frequent AGB categories. This is particularly true for AGB values >300 t ha^{-1}. This similarity is more evident when comparing observed and simulated results for IWP and AEP (Fig. 5.6). AEP and IWP have average AGB of about 140 t ha^{-1} using observations, with larger variation around observations for AEP (Fig. 5.6, left panels). For the modeled observations, the mean AGB for IWP and AEP is about 80 t ha^{-1} (Fig. 5.6 right panels). Using both observed and simulated results, there does not appear to be any difference in AGB between the IWP and AEP regions. But there does seem to be an overestimate of what may be considered an average value of mangrove biomass based on aboveground compartments when using the observed plots. This result could be an artifact of the simulation or represent the lack of balance (e.g. sample size) in coverage in mangrove types in coastal environmental settings on a global scale.

The analytical approach by Rovai et al. (2016) across different neotropical latitudes also underscores the bias generally found in current AGB global estimates. This bias is based on the historical selection of study sites representing mostly forests with typically high AGB values (e.g., riverine, fringe), and the absence of data from forests of lower stature (e.g., basin, scrub) and consequently low AGB values. Current mangrove AGB models do not consider this skewness to avoid such bias in the estimation, extrapolation, and ecological interpretation of AGB values at the global scale. Average AGB value reported for the neotropics using a climate based-model was 165.5 t ha^{-1} (Hutchison et al. 2014) compared to 82.9 t ha^{-1} using the geophysical-climate model (Rovai et al. 2016). The difference in output of the two

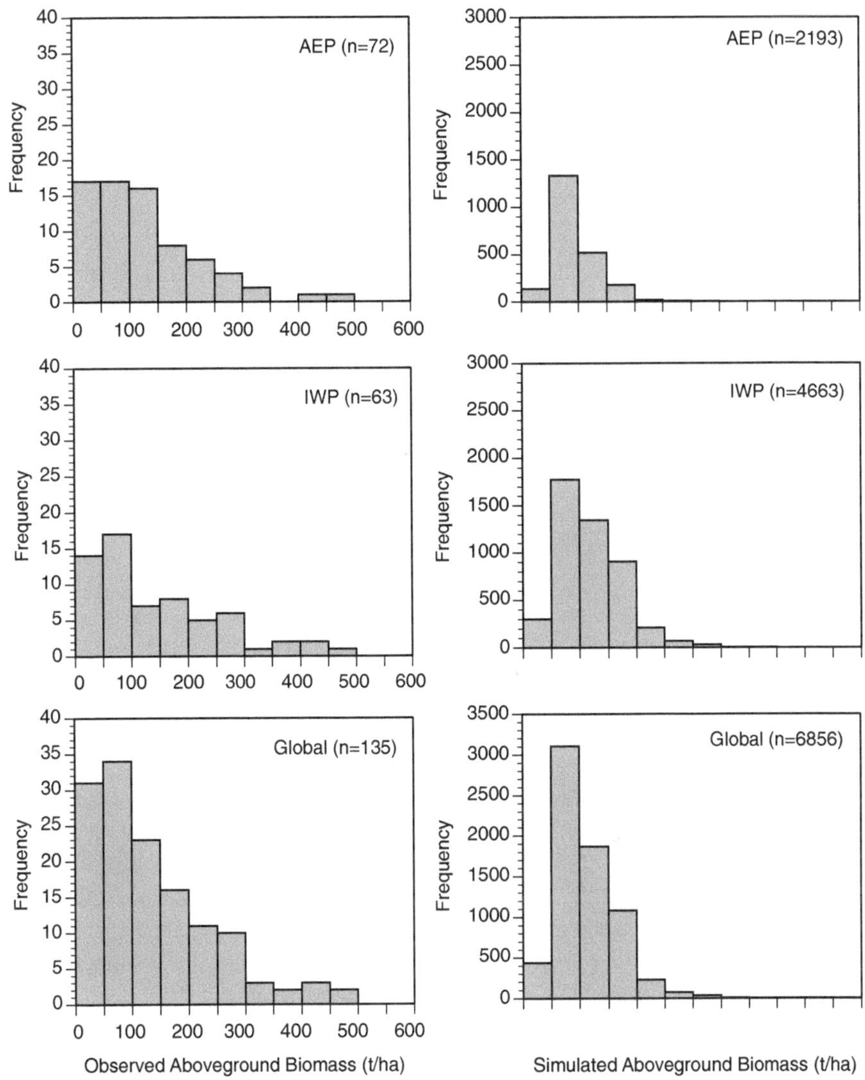

Fig. 5.5 *Left panels* are frequencies of AGB (t/ha) based on field observations from 135 sites. *Right panels* are frequencies of AGB based on the model of by Rovai et al. (2016) using frequency of grids with mangrove cover in Giri et al. (2008). Results distributed by AEP and IWP biogeographic regions

models (54%) is not only related to the type and number of variables included in the continental-scale model by Rovai et al. (2016) but also to the sample size and spatial coverage of real observations used to produce this model. The biases associated with sampling protocols were highlighted in global estimates of AGB in tropical forests (Brown and Lugo 1984). The interest in continuing to develop global

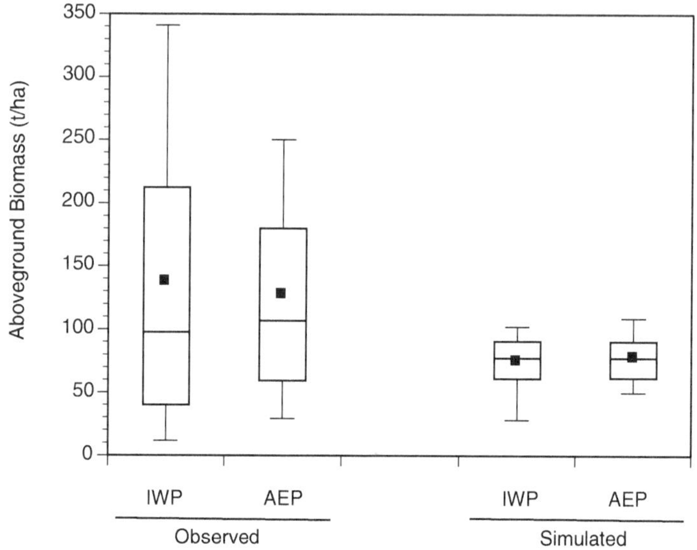

Fig. 5.6 AGB between the IWP and AEP biogeographic regions based on observations (*left panels*) and simulations of AGB based on Rovai et al. 2016 (*right panels*). For each statistic, the *upper*, *middle*, and *lower bar* in the box is the 75th, 50th, and 25th percentile, the smaller *black box* the mean, and the whiskers range from 10th to 90th percentile of the data for each category

estimates of C stored in AGB suggests that these issues need special attention by the mangrove research community.

As discussed earlier in this chapter, the relative influence of regional and local factors determines not only species-specific mangrove spatial distribution (Crase et al. 2013), but also both AGB and belowground biomass (BGB) allocation patterns (Castañeda-Moya et al. 2013). The database of 132 sites with estimates of AGB was catalogued into one of seven ecogeomorphic settings and analyzed to test the energy signature hypothesis for mangroves (Fig. 5.7). Deltaic coasts have higher AGB compared to other continental settings, but the highest average AGB was found in high oceanic islands. These are based largely on the observations from Micronesia (Cole et al. 1999) where values are much higher (nearly 200 t ha^{-1}) compared to the average of about 100 t ha^{-1} discussed above. Fringe mangroves have AGB just >100 t ha^{-1} compared to just <100 t ha^{-1} for interior mangroves (Fig. 5.7). By splitting the fringe and interior classifications across coastal setting type, there is some evidence of zonation for estuarine and composite delta/estuarine shorelines (Fig. 5.7). Interior mangroves for high oceanic islands have higher AGB than fringe mangroves in these coastal settings. Using the energy signature hypothesis as an approach provides some additional power in integrating global AGB variance by accounting for the interaction of regional and local drivers (Twilley et al. 1992; Rovai et al. 2016). Grid models that can define these coastal setting characteristics may prove to improve understanding global AGB patterns.

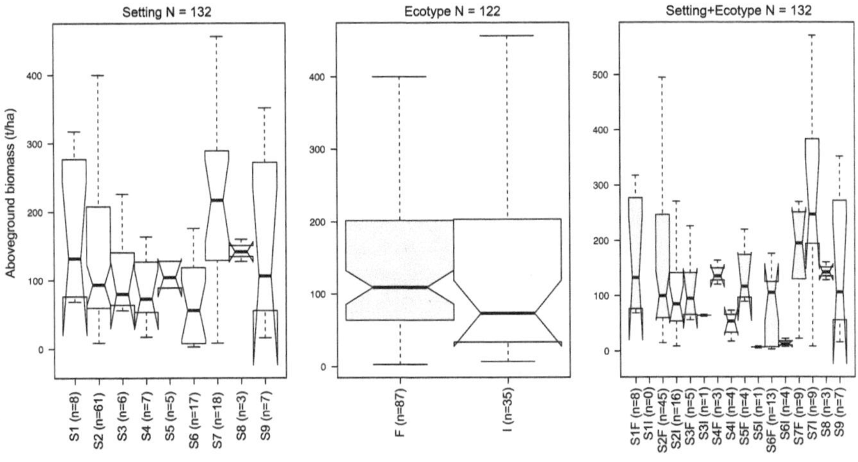

Fig. 5.7 AGB based on observations in mangroves grouped by coastal environmental setting. (*Left panel*) Average measures for mangrove sites classified as one of nine ecogeomorphic settings (*S1*, deltas, *S2*, estuaries (embayment perpendicular to shore, *S3*, lagoon (embayment horizontal to shore), *S4*, composite river and wave (delta and lagoon), *S5*, drowned bedrock valley, *S6*, low Oceanic islands—low islands with carbonate settings, *S7*, high Oceanic islands—with significant runoff from interior; *S8*, disturbed; *S9*, restoration site). (*Central panel*) Measures for 190 mangrove sites classified as either fringe or interior mangroves as ecological type. (*Right panel*) The 190 mangrove sites are classified into one of the nine ecogeomorphic types that are subdivided into fringe and interior ecotypes. *Light-gray* shading indicates fringe mangrove ecotypes

5.3 Aboveground Net Primary Productivity

5.3.1 Litter Fall

Litter fall from a mangrove canopy is dominated by leaf fall with the remainder fraction consisting of wood (stems from the canopy), grass, and reproductive parts. Leaf fall is continuous year-round but with seasonal rates that, in the Caribbean, vary during fall months from September to November. Reproductive input from the canopy is highest in late summer, just before peak leaf fall from the canopy. A mean leaf fall rate for mangrove wetlands in the Caribbean region is about 2 g dry mass (gdm) m^{-2} d^{-1} or about 750 gdm m^{-2} yr^{-1}. Regional rates in litter production in mangrove wetlands are generally a function of the production envelope described above, and rank among the ecological types as follows: riverine > fringe > interior > scrub (Pool et al. 1975; Twilley et al. 1986; Ewe et al. 2006). The fate of leaf litter on the forest floor in most upland forests is controlled by rates of decomposition, thus the residence time of litter will vary with local environment together with chemical characteristics (e.g., substrate quality) of leaf tissue. Mangrove leaf litter decomposes much faster than most upland temperate forests and even tropical moist forests, with average residence time about 0.6 years. *Rhizophora* has slower decomposition rate than *Avicennia* and *Sonneratia* with corresponding residence time of

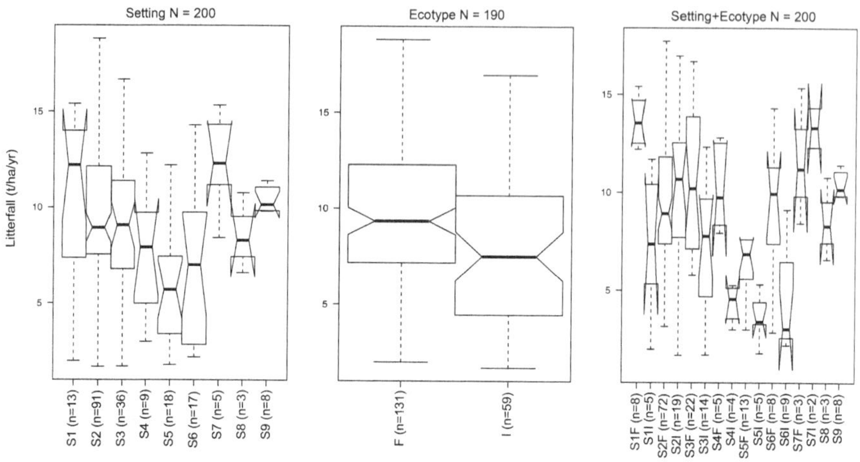

Fig. 5.8 Litter fall rates based on observations in mangroves grouped by coastal environmental setting. (*Left panel*) Average measures for mangrove sites classified as one of nine ecogeomorphic settings [*S1*, deltas, *S2*, estuaries (embayment perpendicular to shore, *S3*, lagoon (embayment horizontal to shore), *S4*, composite river and wave (delta and lagoon), *S5*, drowned bedrock valley, *S6*, low Oceanic islands—low islands with carbonate settings, *S7*, high Oceanic islands – with significant runoff from interior; *S8*, disturbed; *S9*, restoration site]. (*Central panel*) measures for 190 mangrove sites classified as either fringe or interior mangroves as ecological type. (*Right panel*) The 190 mangrove sites are classified into one of the nine ecogeomorphic types that are subdivided into fringe and interior ecotypes. *Light-gray* shading indicates fringe mangrove ecotypes

2–3 months (Twilley et al. 1997, 1986; Woodroffe et al. 1988; Lee 1989b; Sessegolo and Lana 1991; Mfilinge et al. 2002, 2005; Bosire et al. 2005; Chen et al. 2009; Alongi 2011; Coronado-Molina et al. 2012).

The energy signature hypothesis was applied to testing global patterns in litter fall using 164 sites where field observations are available. There is a strong gradient in litter fall productivity from deltas to estuaries to oceanic islands (Fig. 5.8), with rates ranging from 4.5 to 8 t ha^{-1} yr^{-1}. Fringe mangroves have litter fall rates about 7 t ha^{-1} yr^{-1} compared to 5.8 t ha^{-1} yr^{-1} for interior mangroves (Fig. 5.8). There are some clear differences in fringe and interior mangrove litter productivity in certain coastal settings, such as deltas, estuaries, and delta/estuary complex, and particularly for low oceanic islands (Fig. 5.8). However, no such pattern is evident for lagoons or bedrock valley settings. Again, these patterns indicate some important considerations in scaling mangrove productivity based on litter fall estimates.

Grapsid crabs can process significant amounts of mangrove leaf litter (Lee 2005), especially in the Indo-west-Pacific region, and can serve as ecological engineers in mangrove ecosystems (Kristensen 2008). While grapsid crabs can consume up to ~80% of the mangrove leaf litter production in some forests (Robertson et al. 1992), assimilation rates are generally low (<50%), and most of the organic material (about 60%) is egested as fecal material (Lee 1997). Grapsid crabs in mangrove wetlands have been described as playing a role similar to shredders in freshwater environments, i.e., processing detritus by enriching leaf litter with nitrogen-enriched microbial populations, increasing the trophic quality of leaf litter (Peckarsky 1980; Lee 2005; Werry and Lee 2005). The trophic value of this conversion of leaf material, via fecal material, to benefit other invertebrate consumers has been demonstrated as a significant component of mangrove tropho-dynamics (Lee 1997; Werry and Lee 2005). This role of grapsid crabs as grazers in mangrove wetlands has also been observed foraging on microalgae and bacteria (Bouillon et al. 2002a; Cannicci et al. 2008; Kristensen 2008). They disturb the microbenthic primary producers and meiofauna communities in the top sediment and change the physico-environmental conditions of the sediment by increasing sediment drainage, soil redox potential, translocate sediment, increase the sediment surface area, and stimulate decomposition and nutrient cycling processes (Kristensen and Alongi 2006; Kristensen 2008; Kristensen 1988; Andersen and Kristensen 1992; Olafsson and Ndaro 1998; Lee 2008). Such modifications to respiratory processes may shift metabolism from sulfate reduction to more aerobic processes in otherwise anaerobic conditions as oxygen is transported with water that exchanges in animal burrows (Penha-Lopes et al. 2010; Kristensen et al. 2011; Araújo et al. 2012; Ólafsson et al. 2002; Ferreira et al. 2007; Nielsen et al. 2003). These modifications to the bioturbation of leaf material and organic matter in mangrove wetlands enhance rates of decomposition and reduce tidal export of organic matter to adjacent coastal waters. But as will be discussed later, the bioturbation of organic matter may enhance the transport of dissolved inorganic carbon from mangrove wetlands with tidal exchange.

5.3.2 *Wood Production*

Woody production or woody biomass increment measurements in mangrove wetlands typically involve repeated measurements of tree diameters and the application of allometric equations to estimate changes in biomass (Cintrón and Schaffer-Novelli 1984; Clark et al. 2001; Komiyama et al. 2008). For a forest stand, increments are summed for all trees surviving the interval to estimate wood productivity (Clark et al. 2001). Synthesis of wood productivity estimates (Table 5.2) range from 1.1 to 14.6 t ha^{-1} yr^{-1}. This is actually a wider range of net productivity than that observed for litter fall. Also, the number of global estimates for wood production (only about 60 sites) is much less than observed for litter fall; and much less than for estimates of AGB (135 sites). The average wood production is about 5.8 t ha^{-1} yr^{-1}, which is very similar to global average of litter fall productivity. Wood production can range from one-third to nearly double the rate of litter fall productivity (Table 5.2). However, a general pattern is that litter fall and wood production are about 6 t ha^{-1} yr^{-1} for a NPP_A of about 12 t ha^{-1} yr^{-1} as a global average.

Table 5.2 Synthesis of wood production data (and litter fall, if available) in mangroves from Bouillon et al. (2008)

Site	Wood production (t ha yr^{-1})	Litter fall (t ha yr^{-1})	Wood:litter production	Reference
0–10 S or N				
Malaysia	11.8			Ong et al. (1979)
Malaysia	24.1			Ong et al. (1979)
Phuket (Thailand)	20			Christensen (1978)
Malaysia	6.7	11	0.61	Putz and Chan (1986)
Java (Indonesia)	13.96	8.2	1.70	Sukardjo and Yamada (1992)
Java (Indonesia)	13.75	7.37	1.87	Sukardjo and Yamada (1992)
Java (Indonesia)	13.75	7.1	1.95	Sukardjo and Yamada (1992)
Java (Indonesia)	14.18	8.2	1.72	Sukardjo and Yamada (1992)
Java (Indonesia)	14.60	10.4	1.40	Sukardjo and Yamada (1992)
Kala Oya (Sri Lanka)	6.76	6.23	1.09	Amarasinghe and Balasubramaniam (1992)
Kala Oya (Sri Lanka)	5.62	5.52	1.02	Amarasinghe and Balasubramaniam (1992)
Erumathivu (Sri Lanka)	4.34	4.41	0.98	Amarasinghe and Balasubramaniam (1992)
Erumathivu (Sri Lanka)	1.40	3.74	0.37	Amarasinghe and Balasubramaniam (1992)
10–20 S or N				
Puerto Rico	3.07			Golley et al. (1962)
Estero Pargo	7.72	8.34	0.93	Day et al. (1987)
Boca Chica (Mexico)	12.06	12.52	0.96	Day et al. (1987)
Dominican Republic	16.3		0.58	Sherman et al. (2003)
Dominican Republic	11.8		0.58	Sherman et al. (2003)
Dominican Republic	6.6		0.58	Sherman et al. (2003)
Laguna de Terminos	1.96	4.96	0.40	Day et al. (1996)
Laguna de Terminos	1.11	3.01	0.37	Day et al. (1996)
Laguna de Terminos	1.99	4.14	0.48	Day et al. (1996)
20–30 S or N				
Florida (USA)	13.9	12.2	1.14	Ross et al. (2001)
Hong Kong	13.3	6.87	1.94	Lee (1990)
Florida (USA)	7.31			Sell (1977)
Florida (USA)	13.33			Sell (1977)
Shark River, Florida (USA)	12.57	8.46	1.49	Ewe et al. (2006)
Shark River, Florida (USA)	7.01	8.03	0.5	Ewe et al. (2006)
Shark River, Florida (USA)	11.9	1.11	1.11	Ewe et al. (2006)

Data previously compiled by Twilley et al. (1992)

Average value for wood:litter production mentioned by Sherman et al. (2003) for different vegetation types

5.4 BGB and Root Productivity

BGB allocation in forest and wetland ecosystems is considered significant to soil formation and surface elevation gain (Chen and Twilley 1999; McKee et al. 2007), making significant contributions to ecological processes associated with C budgets due to the high proportion of biomass allocated belowground relative to aboveground tissues (Nadelhoffer et al. 1985; Nadelhoffer and Raich 1992; Jackson et al. 1997; Chmura et al. 2003; Bouillon et al. 2008). Our understanding of the contributions of BGB and productivity to soil formation and C dynamics in mangrove wetlands is limited by the difficulties associated with measuring productivity, mortality, and longevity of complex root systems (Santantonio and Hermann 1985; Cuevas and Medina 1988; Majdi and Kangas 1997). The low number of studies, with only 48 considered in this review, currently limits the estimates of biomass and productivity contributions to C budgets by belowground components of NEP (Bouillon et al. 2008; Castañeda-Moya et al. 2011). Roots of mangrove wetlands are phenotypically plastic in their ability to exploit nutrients in response to environmental change (Feller 1995; Feller et al. 2010; Reef et al. 2010) and mangrove species can allocate up to 40–60% of their total biomass to belowground roots (Komiyama et al. 1987; Lugo 1990; Snedaker et al. 1995; Khan et al. 2009). The BGB and net primary productivity of mangroves is thought to be a very important contribution to ΔS_{org} and thus NEP (Eq. 5.5; McKee and Faulkner 2000; Lovelock et al. 2006; Bouillon et al. 2008; Lovelock 2008; Alongi 2011; Donato et al. 2011; Mcleod et al. 2011; Chen and Twilley 1999).

Global trends in BGB estimates in mangroves are confusing given that different methodological approaches can cause varying results (Vogt et al. 1998; Clark et al. 2001; Bouillon et al. 2008; see review by Rivera-Monroy et al. 2013). Comparing root biomass across studies shows a wide range of biomass values reported for mangrove locations around the world, even when similar sampling techniques and sampling depths are used (Fig. 5.9, left panels). Some of these differences may be due to sampling methods (Rivera-Monroy et al. 2013), but also other factors such as forest age, species composition, history (planted vs. natural), and local climate variation could influence biomass allocation patterns contributing to observed differences (Kairo et al. 2008; Tamooh et al. 2008). Nearly half of the biomass estimates reviewed are <50 t/ha, and 13 of the 48 sites have BGB >100 t/ha (Fig. 5.9, left panel). For the majority of observations, there is no distinct difference in IWP and AEP (Fig. 5.10a). The only difference is that the two greater estimates of BGB occur in IWP (>200 t/ha).

Root necromass is seldom specified in describing root biomass but can be a substantial portion of total biomass. Values in Micronesia mangroves range from 63.6 to 345.6 t/ha (sampled to a depth of 45 cm) in sites sampled by Cormier et al. (2015), which is similar to root necromass values of 220–230 t/ha in the top 30 cm in the nearby Okat and Utwe mangroves on Kosrae (Gleason and Ewel 2002). In contrast to fine root biomass, these estimates of root necromass are considerably greater than those found in Gazi Bay, Kenya (10.3–32.6 t/ha in cores 0–60 cm; Tamooh et al. 2008)

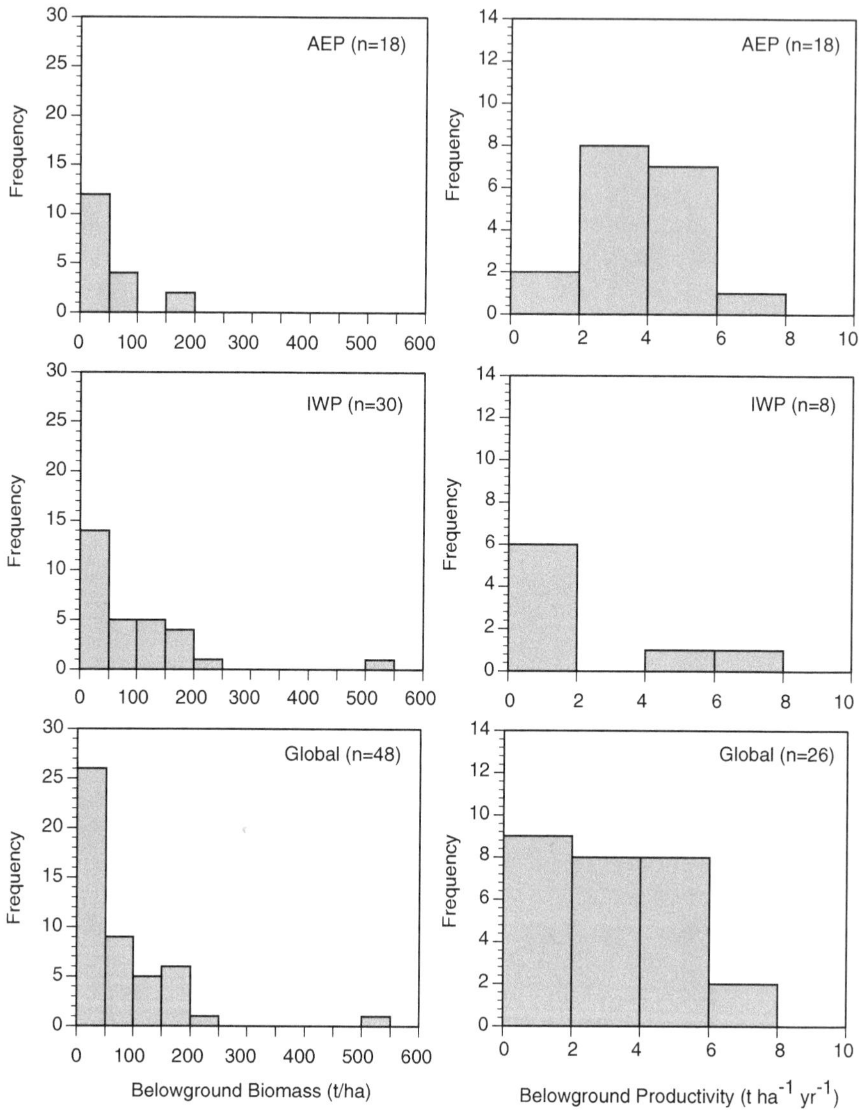

Fig. 5.9 Frequencies of BGB (*left panels*) based on observations from 48 sites. Frequencies of belowground productivity (*right panels*) based on 26 sites. Distribution of observations is distinguished between the IWP and AEP biogeographic regions

and in Thailand (15–133 t/ha in cores 0–30 cm; Chalermchatwilai et al. 2011), indicating a particularly strong capacity for C storage in Micronesian mangroves through root production. This partition contributes to the high overall rates of soil C storage attributed to mangroves in places such as Yap, Federated States of Micronesia (Kauffman et al. 2011).

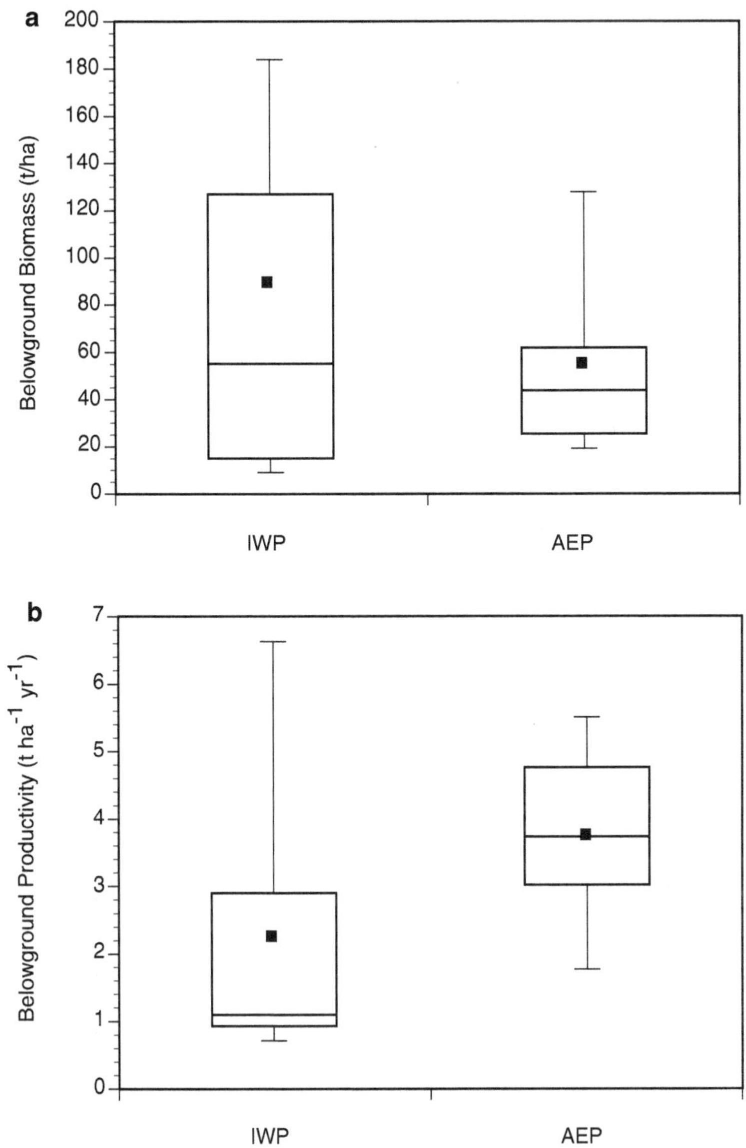

Fig. 5.10 BGB between the IWP and AEP biogeographic regions based on observations (**a**) and belowground productivity between the IWP and AEP biogeographic regions based on observations (**b**). For each statistic, the *upper*, *middle* and *lower bar* in the box is the 75th, 50th, and 25th percentile, the smaller *black box* the mean, and the whiskers range from 10th to 90th percentile of the data for each category

Most values of root productivity in the literature are based on root ingrowth bag technique, which may underestimate (Rivera-Monroy et al. 2013) or stimulate (Graham and Mendelssohn 2016) root production rates. However, this measure does provide a relative comparison of root productivity among sites (McKee and Faulkner 2000; Li et al. 2012). Most of these measures focus on the productivity and turnover of fine roots, which may again describe a portion of the total belowground processes of mangrove ecosystems. Highest root production (57–78%) occurs in the shallow root zone of mangrove sites that is comparable to other studies in terrestrial forests and wetland ecosystems. These patterns have been reported for mangrove wetlands in Florida (McKee and Faulkner 2000), Honduras (Cahoon et al. 2003), Belize (McKee et al. 2007), and in other subtropical and tropical latitudes where indirect methods were applied to measure root productivity in scrub and tall mangrove forms (Lovelock 2008; Fig. 5.9, right panel).

Root productivity based on observations at 26 sites varies from about 0.5 to 7.5 t ha^{-1} yr^{-1}, with no clear pattern of frequency among rates (Fig. 5.9, right panel). Most of the lower rates are observed in IWP region, and rates measured in AEP region range across the entire distribution of observations (Fig. 5.9, right panel). The present trend, again based on only a few observations (8 in IWP and 18 in AEP), is that the mean rates of root productivity are much higher in AEP (Fig. 5.10b). The overall mean of belowground productivity, based on these 36 observations, is about 3 t ha^{-1} yr^{-1}, or 30% of the NPP_A estimate (12 t ha^{-1} yr^{-1}) described above. This is consistent with most distributions of productivity between NPP_A and NPP_B in forest ecosystems.

Direct measurements of fine root productivity in the mangrove literature range from 0.46 t ha^{-1} yr^{-1} to 7.5 t ha^{-1} yr^{-1} in Micronesia, at depth of about 30–50 cm (Gleason and Ewel 2002; Cormier et al. 2015), in plots dominated by *S. alba*. Fine root growth ranged from 0.43 to 1.97 t ha^{-1} yr^{-1} in Belizean mangroves (McKee et al. 2007), from 1.83 to 2.10 t ha^{-1} yr^{-1} in the Florida Coastal Everglades, USA (Castañeda-Moya et al. 2011), and from 1.40 to 2.80 t ha^{-1} yr^{-1} in restored mangrove sites near Rookery Bay, Florida, USA (McKee and Faulkner 2000). Moreover, fine root productivity significantly contributed (21–50%) to the total (0–90 cm) belowground allocation in mangrove sites along the Florida Coastal Everglades (USA). For comparison, carbon allocation to fine root production accounts for approximately 30% of the total annual C allocation compared to belowground in forest ecosystems (Nadelhoffer and Raich 1992). The higher biomass allocation to coarse roots (~70%) and the fact that fine root production is a significant component of NPP_B supports the hypothesis that belowground root allocation is a significant contribution to soil C storage in mangrove wetlands (Chen and Twilley 1999; Chmura et al. 2003).

5.5 Allocation Ratios of Biomass and Production

Mangrove wetlands show adaptive mechanisms when plants are growing in benign soil conditions of coastal environmental settings where regulators and hydroperiod allow for the expression of phenotypic plasticity to nutrient resource availability. However, when stress conditions develop due to soil conditions that reduce plant growth, regulators and hydroperiod can stimulate plant allocation patterns that may not reflect solely the root:shoot ratio expected based on nutrient resource gradients. Using the BGB data published by Castañeda-Moya et al. (2011), the proportion of BGB allocation (BGB:AGB ratios) is 17 times higher in mangrove wetlands under stressed soil conditions located in southeastern Florida Everglades, USA, compared to those with daily tidal exchanges in estuaries in southwest Florida Everglades. Also, BGB:AGB ratios increased with increasing sulfide concentrations across this coastal environmental setting ($r^2 = 0.78$, $p < 0.001$), suggesting the regulatory effect of hydroperiod on root biomass allocation (Krauss et al. 2006). Differences in total production between above- and belowground components can be distinct; for example, riverine mangroves in higher soil phosphorus (P) concentrations and more frequent tidal inundations allocate most of the total NPP to above ground (69%), whereas scrub mangroves in lower-P and long duration of flooding showed the highest allocation to below ground (58%) (Castañeda-Moya et al. 2013). The environmental setting of Florida Coastal Everglades in USA demonstrates that mangroves allocate a large proportion of their total biomass to below ground in response to P limitation and soil stress conditions (Komiyama et al. 2000; Sherman et al. 2003; Lopez-Hoffman et al. 2007; Naidoo 2009).

These findings are consistent with global budget estimates of NPP_T for mangrove wetlands, suggesting the significant contribution of NPP_B (38%) to the overall production (Bouillon et al. 2008). There is evidence that biomass production increases with increasing nutrient availability in mangroves (Lovelock et al. 2004, 2006; Feller et al. 2007; Naidoo 2009), and in benign soil conditions more of this production is associated with wood production. This pattern was observed in Florida Coastal Everglades with higher wood:litterfall production ratios in riverine mangroves (0.35 ± 0.08) along Shark River (higher-P and more frequent tides), compared to considerably lower ratios in scrub mangroves (0.22 ± 0.03) in Taylor River (lower-P and longer flood durations) (see Table 5.2 for review). The production to biomass (P:B) ratios also confirm patterns of C partitioning to biomass and productivity between benign and stressed soil conditions. The lower P:B ratios observed in all Shark River sites of Florida Coastal Everglades (higher-P and more frequent tides) suggest that more of the energy flux from primary production is converted into plant biomass, in contrast to nearby Taylor River sites (lower-P and longer flooding durations) where most of the available energy is used to offset the higher soil stress conditions (i.e., sulfide and permanent flooding) and lower P availability in these scrub forests. Similarly, higher P:B ratios have been reported for scrub forests in Biscayne Bay, Florida (0.36 yr^{-1}; Ross et al. 2001) compared to lower ratios for riverine mangroves in Terminos Lagoon, Mexico (0.18 yr^{-1}; Day et al. 1987). These patterns support the hypothesis that scrub mangrove wetlands allocate more BGB (i.e., high BGB:AGB ratios) and belowground production relative to aboveground components (litter fall and wood production) in response to P limitation and high soil stress conditions, at expenses of aboveground growth and development (Lovelock 2008). These trade-offs indicate a strong link between biomass and NPP allocation patterns and the phenotypic plasticity of mangrove species in response to environmental gradients to maximize resource use efficiency and conserve the most limiting nutrient (Chapin et al. 1986). Accordingly, the relative contribution of belowground allocation to soil C storage and wood production to NPP_T in mangrove forested wetlands have significant implications to net C exchange in these coastal forested wetlands (Eq. 5.7; Chmura et al. 2003; Khan et al. 2007; Donato et al. 2011; Kauffman et al. 2011; Mcleod et al. 2011).

5.6 Soil CO_2 Efflux and Accumulation Rates

A cohort model (NUMAN) of soil biogeochemistry demonstrates the balance of processes that control soil organic carbon accumulation including the inputs by both root productivity and litter fall, compared to losses due to decomposition (soil CO_2 efflux) and net estuarine exchange (Chen and Twilley 1999). The model can simulate the balance of organic carbon in soils based on these inputs and losses, accounting for the differential decomposition of labile and refractory organic matter, and the depth distribution of root productivity to soil column. The key to resolving organic carbon mass balance of soils, as an index of NEP, is to compare soil CO_2 efflux rates as consumption of NPP_T. As described above in Eq. (5.5), litter fall and belowground production rates are included in NUMAN. Heterotrophic respiration losses to the atmosphere occur through the decomposition of organic matter in soil, fine leaf litter, coarse woody debris, and surface water (Kristensen et al. 2000, 2008, 2011; Troxler et al. 2015; Alongi et al. 1993; Lovelock 2008; Bulmer et al. 2015; Leopold et al. 2013, 2015). The assumption here is that root productivity techniques described above measure NPP_B, while soil CO_2 efflux includes the autotrophic respiration by roots. Because NPP_B is net root production, some of the soil CO_2 efflux component of the Rs term may be considered double accounting. This consideration gets very complicated for the root system of mangroves, especially in the case of aerial root structures used for plant aeration in hypoxic or anoxic soils (i.e., pneumatophores; Kristensen 2007). Partitioning these sources of soil CO_2 from soil CO_2 efflux is particularly challenging in mangrove wetlands, and determining how this measure of soil CO_2 efflux should be compared to balance NEP (see review in Alongi 2009).

Simulation models of mangrove soils adjusting for relative rates of litter turnover and root productivity suggest that root production is a critical process in controlling organic carbon accumulation rates in mangrove soils in the neotropics (Chen and Twilley 1999). These models also predicted that variations in root turnover have a more significant effect on these soil characteristics than variation in litter fall. For example, observations of deposition and slow degradation of mangrove roots have shown to contribute more to organic matter accumulation and vertical building of mangrove islands in Belize than total litter fall (Middleton and McKee 2001; McKee et al. 2007, McKee and Faulkner 2000, Krauss et al. 2008). Root data used in the south Florida model, however, were based on very few direct observations of mangrove biomass in relation to belowground nutrient concentrations, relying mostly on derived estimates of root:shoot ratios. Another key coefficient in determining rates of organic carbon accumulation in mangrove soils is the relative proportion of labile and refractory organic carbon. Several studies have found that decay of below-ground material is slower than leaf litter (McKee and Faulkner 2000; Middleton and McKee 2001; Poret et al. 2007). A large part of sedimentary organic matter in mangrove wetlands is derived from root organic matter (Alongi et al. 2001) and in many forest systems can be the principle source of organic matter in the deeper soil layers (Ludovici et al. 2002).

Root productivity and BGB accumulation contribute to soil volume and consequently soil elevation change in mangrove wetlands (Cahoon et al. 2003; McKee et al. 2007; McKee 2011), particularly in carbonate settings (Lynch et al. 1989; Parkinson et al. 1994). Mangroves are touted as highly efficient C sinks in the wet tropics due to relatively high primary productivity and low rates of decomposition (Komiyama et al. 2008). One of the most comprehensive reviews of C sequestration of mangrove soils by Breithaupt et al. (2012) demonstrates that rates can range from 4 to 1000 gC m^{-2} yr^{-1} (Fig. 5.11). However, the largest proportion of observations, 41 of the 65 sites, occur from 0 to 200 gC m^{-2} yr^{-1}, with an average of 163 gC m^{-2} yr^{-1} for all sites. The importance of the contributions of mangrove wetlands to global C sequestration reinforces the need for a better understanding of biomass allocation in these systems.

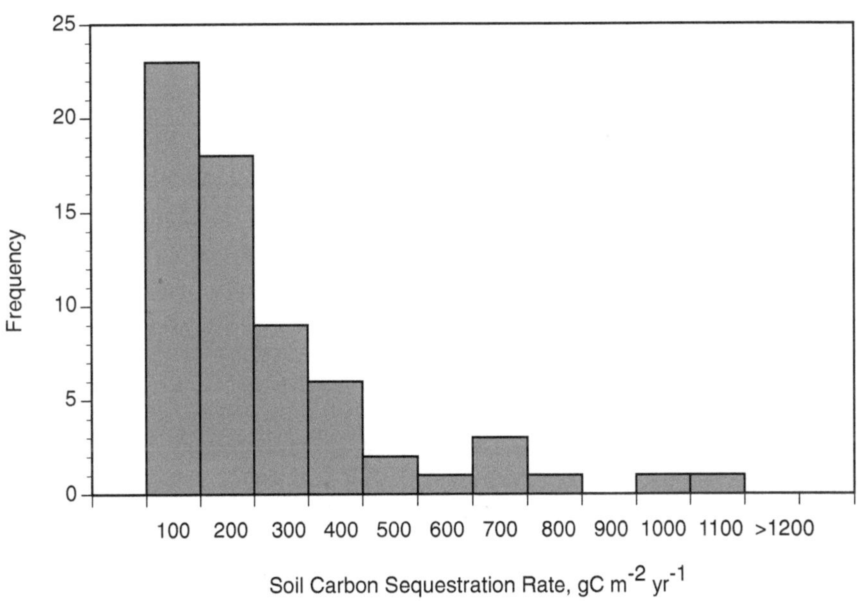

Fig. 5.11 Distribution of soil carbon sequestration rates (gC m^{-2} yr^{-1}) for mangrove forests based on review by Breithaupt et al. (2012)

5.7 Net Estuarine Exchange with Mangrove Wetlands (NTE_M)

Several authors have suggested that mangrove-derived organic matter is of global significance in the coastal zone (Robertson et al. 1993; Dittmar and Lara 2001; Kristensen et al. 2008; Alongi 2009). Estimates indicate that mangrove wetlands could be responsible for ~10% of the global export of terrestrial particulate and dissolved organic carbon (POC and DOC) to the coastal zone (Jennerjahn and Ittekkot 2002; and Dittmar et al. 2006, respectively), and for ~10% of the global organic carbon burial along with seagrasses in the coastal ocean (Duarte et al. 2005). The exchange of C between tidal wetlands such as mangrove wetlands or salt marshes and the coastal ocean and its ultimate fate in the ocean is therefore increasingly recognized as a potentially critical component in the ocean C budget (Twilley et al. 1992; Bouillon et al. 2008). This may be particularly evident in river-dominated mangrove systems such as muddy coasts and deltas where organic material exchanged is greater at the boundary of the forest compared to other coastal settings (Twilley 1985), as observed for major river systems such as Fly River, Papua New Guinea, or systems with large tidal exchanges in Australia (Alongi 2009). Recent reviews on C dynamics in these coupled wetland/estuarine ecosystems describe organic matter from wetlands as contributing to the heterotrophic nature of estuaries (Cai 2011; Hopkinson et al. 2012). The assumption is that C exported from mangrove wetlands, which may be a significant loss of mangrove NEP, may stimulate respiration in coastal waters and not contribute significantly to burial of C in coastal ecosystems. This is still a major question in the C budget of coastal ecosystems.

Resolving estimates of mangrove NEP have benefited from an expanded literature on biomass and productivity of both NPP_A and NPP_B, but estimates of C export in the literature are still very limited. Carbon export from mangrove wetlands ranges from 1.86 to 401 gC m^{-2} yr^{-1}, with an average rate of about 210 gC m^{-2} yr^{-1} (Twilley and Rivera-Monroy 2009; see Table 5.3 from Bouillon et al. 2008). Carbon export from mangrove wetlands is nearly double the rate of average C export from salt marshes (Nixon 1980), which may be associated with the more buoyant mangrove leaf litter, higher precipitation in tropical wetlands, and greater tidal amplitude in mangrove ecosystems (Twilley 1988). The fate of mangrove primary production has been a major topic of debate in the literature during the past decades (Bouillon et al. 2008). In particular, the outwelling hypothesis, first proposed by Odum and Heald (1975), suggested that a large fraction of organic matter produced by mangrove trees is exported to the coastal ocean, where it would form the basis of a detritus food chain and thereby support coastal fisheries. Despite the large number of case studies dealing with various aspects of organic matter cycling in mangrove systems (Kristensen et al. 2008), there is still no consensus on the ecological fate of organic matter produced in mangroves on the metabolism and food webs of the coastal zone (Bouillon et al. 2008).

The function of mangrove wetlands as habitat and food source to estuarine-dependent fisheries is one of the most important ecosystem services of coastal wetlands (Fig. 5.4). The outwelling hypothesis of mangrove wetlands has been revised

Table 5.3 Summary of literature estimates on organic carbon export from mangroves defined by particulate (POC), dissolved (DOC), and total (TOC) organic constituents from Bouillon et al. (2008) (Adopted from Lee 1995)

Site	POC export ($gC\ m^{-2}\ yr^{-1}$)	DOC export ($gC\ m^{-2}\ yr^{-1}$)	TOC export ($gC\ m^{-2}\ yr^{-1}$)	Data source
Australia	420			Boto and Bunt (1981)
New Zealand	110			Woodroffe (1985)
Australia	340			Robertson (1986)
Hong Kong	2			Lee (1989a)
Hong Kong	5			Lee (1990)
Zanzibar	65	230	295	Machiwa (1999)
Florida (USA)	16	48	64	Twilley (1985)
Brazil		44		Dittmar and Lara (2001)
Florida (USA)		56		Romigh et al. (2006)
Brazil		144		Dittmar and Lara (2001)
Florida (USA)		381		Davis et al. (2001)
Malaysia			176	Gong and Ong (1990)
Florida (USA)			186	Heald (1969)
Florida (USA)			292	Odum and Heald (1972)
Florida (USA)			91	Lugo and Snedaker (1974)
Australia			332	Woodroffe et al. (1988)
Australia			−7	Boto and Wellington (1988)
Australia			994	Alongi et al. (1989)
Papua New Guinea			343	Robertson and Alongi (1995)
Florida (USA)			7.1	Sutula et al. (2003)

from the original paradigms described by Odum and Heald (1972) and reevaluated by Odum (2002), Lee (1995, 2004, 2005), Sousa and Dangremond (2011), and Saintilan (2004). Trophic links between mangrove production and higher trophic level consumers depend on microbial decomposition of litter fall and consumption by a variety of small detritivores before energy is available to higher trophic level organisms (see review by Sousa and Dangremond 2011). The role of detritus export from mangrove wetlands for coastal foodwebs is substantiated by a correlation of nearshore catch of shrimp or fish to mangrove area in the vicinity of those harvests (Macnae 1974; Sasekumar et al. 1992; Primavera 1996). Many of the recent reviews described above warn that such correlations do not mean causal relationships and that isotope signatures in diets are a better suited for testing the food composition of coastal fish.

There are several well-documented cases where the strength of the mangrove C signal in tissues of consumers in estuaries and coastal waters is related to the location of the organism when sampled relative to the dominance of mangrove detritus production relative to estuary area (or volume). For example, sharp declines in mangrove detrital signature with distance from the mouth of a mangrove-dominated creek have been found in a Kenyan estuary (Hemminga et al. 1994), where the signal was absent >2 km from mangroves. In the Godavari delta mangrove ecosystem in India, there was a much stronger mangrove C signal in detritivores of mangrove creeks relative to an adjacent bay (Bouillon et al. 2002a, 2002b, 2004a, 2004b). For Shark River estuary in southern Florida (USA) mangrove-derived C supported up to 60% of the nutrition of filter feeders half-way up the estuary (Fry and Smith 2002). In Bocas del Toro, Panama, organic matter from mangroves declined by about 40–50% over the first 250–300 m from the forest edge, but was found in reef organisms living >10 km from a mangrove forest (Granek et al. 2009). In a Malaysian mangrove and adjacent water bodies, the isotopic signature of mangrove-derived C in organisms was evident among mangrove stands or in adjacent mudflats, but decreased to low levels in coastal inlets <2 km from the coast, and was negligible in offshore water (2–18 km from the coast, Rodelli et al. 1984). And finally similar gradients were found in Biscayne Bay Florida (USA) where mangroves contribute C to heterotrophic organisms within or immediately adjacent to forests (Fleming et al. 1990), as was found off the coast of Colombia (Pineda 2003).

These isotope observations support the dual gradient hypothesis of organic matter in estuaries as proposed by Odum (1984) that explains the complex patterns of organic detritus vs phytoplankton and seagrasses in the diets of estuarine food webs. Nekton and sessile organisms in small tidal creeks with high amounts of mangrove area relative to water volume have diets with higher influence of mangrove detritus than similar organisms in open water at distance from the mangrove wetland (Odum 1984). As stream orders of tidal creeks increase and salinity increases, phytoplankton and seagrasses progressively dominate the diets of estuarine food webs. Such statistics support the claim that the significance of mangrove wetlands to fisheries depends on the total wetland (mangrove) area compared to the area and/or volume of water habitats in the region (Twilley 1995, Cifuentes et al. 1996), similar to the original idea for salt marshes (Nixon 1980). In regions with low ratio of wetland to water area, mangrove C contribution can range from 2% to 52% of the total available C pool for secondary productivity (Twilley 1988; Wafar et al. 1997; see review by Sousa and Dangremond 2011 and Saintilan 2004). Evaluations of different C budgets suggest that detritus is an important component of the energy budget of coastal ecosystems depending on the mangrove area:open water area ratio. Other estimates show that large fraction of organic matter produced in mangrove wetlands is buried or consumed by residents of mangrove food webs (Lee 1989a, b; Robertson et al. 1992; Twilley et al. 1997).

5.8 Net Ecosystem Carbon Exchange

During the last 20 years, forest-level C exchange with the atmosphere has been measured using the Eddy-Covariance method (ECM) (see review by Rivera-Monroy et al. 2013). ECM is a micrometeorological method implementing instrumentation along towers above vegetation that measures turbulent motions (i.e., "eddies") of upward and downward moving air parcels that transport trace gases (e.g., CO_2, H_2O, CH_4, N_2O) across the biosphere atmosphere interface (Moncrieff et al. 2000; Baldocchi 2003, 2008) (Fig. 5.12). Fluxes of C measured by ECM are net fluxes of the whole system with the atmosphere and thus represent net ecosystem exchange (NEE) (Fig. 5.12). There are few examples of where an ECM, and thus NEE, have been applied to measure CO_2 fluxes to evaluate ecosystem productivity at large spatial scales in mangrove wetlands (reviewed in Rivera-Monroy et al. 2013). Two towers over the Sundarbans mangrove wetlands on the northern Bay of Bengal, India (Ganguly et al. 2008), measure a significant seasonality in CO_2 exchange over the year with a mean net CO_2 flux of −0.48 t ha^{-1} yr^{-1}. Integrative approaches to resolving the net C budgets of mangroves have been suggested using the combination of NEE based on ECM and NEP based on biometric methods that define organic carbon accumulation. It is important that the inclusion of soil CO_2 efflux (Rs) and net tidal export (NTE_M) with biometric methods properly accounts for NEP when comparing with NEE using ECM. In addition, the biometric methods focus on organic carbon exchange, whereas ECM can measure all forms of C given the instrumentation that is used in measuring NEE.

NECE of a mangrove wetland can be described by the following:

$$\mathrm{NECE} = \Delta\mathrm{C} / \mathrm{d}t \tag{5.8}$$

Because the focus of this review is on C exchange in mangrove wetlands (given that the boundary is defined by the location of forested wetlands in the intertidal zone), a positive sign of NECE indicates an increase in the C pool of mangrove wetlands (C sink) and negative sign is a loss from the C pool of mangrove wetlands (C source). This is in contrast to the conventions of positive and negative signs used for NEE to denote changes relative to the C balance of the atmosphere (positive is increase in atmospheric C and negative is net loss of atmospheric C). It is important that we keep the focus of this analysis relevant to the role of mangrove wetlands in mitigating the global C exchange to the atmosphere.

This use of positive and negative signs to denote storage and loss to NECE of mangrove wetlands, respectively, allows ECM and biometric methods to resolve NEE and NEP. Given this clarification,

$$\mathrm{NECE} = \sum \mathrm{NEE} + F_{\mathrm{CO}} + F_{\mathrm{CH4}} + F_{\mathrm{VOC}} + F_{\mathrm{DIC}} + F_{\mathrm{DOC}} + F_{\mathrm{PC}} \tag{5.9}$$

where NEE is net ecosystem exchange of CO_2 (i.e., the net CO_2 flux from the atmosphere to the ecosystem, or net CO_2 uptake in mangrove wetlands = positive sign); F_{CO} is net carbon monoxide (CO) absorption (or efflux = negative sign); F_{CH4} is net methane (CH_4) consumption (or efflux = negative sign); F_{VOC} is net volatile organic C (VOC) absorption (or efflux = negative sign); F_{DIC} is net dissolved inorganic C (DIC) input to the ecosystem (or net DIC leaching loss = negative sign); F_{DOC} is net dissolved organic C (DOC) input (or net DOC leaching loss = negative sign); and F_{PC} is the net lateral transfer of particulate (nondissolved, nongaseous) C into the ecosystem (or out of = negative sign, by fluxes represented by soot emission during

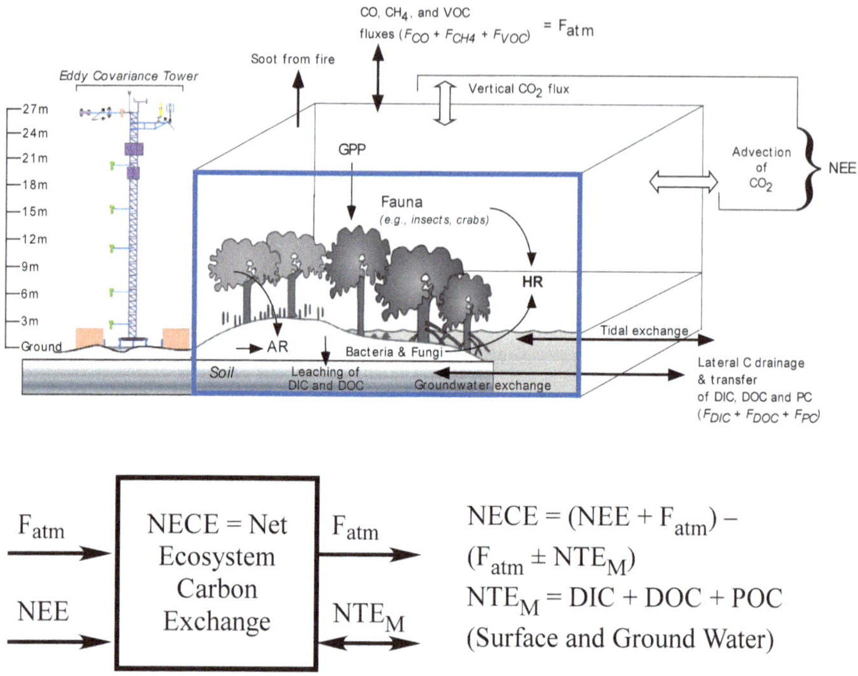

Fig. 5.12 Carbon fluxes identified as the main variables determining the NECE as related to the net ecosystem exchange (*NEE*) measurement obtained with the eddy covariance method. The solid *blue box* represents the mangrove forest interacting with coastal waters. Fluxes (F) contributing to NECE are emissions to or uptake from the atmosphere of CO_2 (i.e., NEE), CH_4, CO, and VOC, along with lateral leaching and drainage fluxes of DOC and DIC. Lateral or vertical movement processes of particulate C (PC) (nongaseous, undissolved) are influenced by animal movement, soot emission during fires, water and wind deposition and erosion, and anthropogenic transport or harvest. Net ecosystem production is regulated by C fluxes due to gross primary production (*GPP*), autotrophic respiration (*AR*), and heterotrophic respiration (*HR*) (Modified from Chapin et al. 2006, as reported in Rivera-Monroy et al. 2013). The *bottom panel* is a diagram converting the measures into inputs and outputs of mass balance of carbon resulting in estimates of NECE

fires, water and wind deposition and erosion, animal movement, and anthropogenic transport or harvest).

The fluxes of C other than NEE are very important to understand and to compare NECE with NEP. These fluxes in addition to NEE, which we will denote as F_T, include the following:

$$F_T = F_{CO} + F_{CH4} + F_{VOC} + F_{DIC} + F_{DOC} + F_{PC} \tag{5.10}$$

Methane fluxes in mangroves are minor compared to carbon balance (F_{CH4}) (Purvaja and Ramesh 2001; Kreuzwieser et al. 2003; Alongi et al. 2005), and very little data are available for most mangrove wetlands to estimate F_{CO} and F_{VOC}. Thus, these F terms will not be included in our calculations, and thus a modified total flux is as follows:

$$F_T = F_{DIC} + F_{DOC} + F_{PC} \tag{5.11}$$

These fluxes of C (F_{DIC} + F_{DOC} + F_{PC}) can occur across the mangrove boundary with the atmosphere (F_A) and across the boundary with coastal waters (F_W). Fluxes with the atmosphere, in the form of soot in response to burning, will be denoted as $F_A = F_{DIC} + F_{DOC} + F_{PC}$. Fluxes associated with tidal exchange with coastal waters will be denoted as $F_W = F_{DIC} + F_{DOC} + F_{POC}$ (note that F_{POC} is used here assuming that all the particulate C exchange in water is organic). Again, assuming that F_{CO}, F_{CH4}, and F_{VOC} are negligible,

$$\text{NECE} = \sum \text{NEE} + (F_W + F_A) \tag{5.12}$$

Again, the fluxes of F_W and F_A are relative to C budget of mangrove wetlands and thus negative flux denotes net loss from mangroves and positive numbers a net gain of C across the mangrove boundary with respective to the atmosphere (F_A) or coastal waters (F_W).

If we assume that F_A is negligible (lack of burning or other atmosphere exchange) and that all the F_T terms are exchanges with coastal waters (F_W), then we can compare NECE with NEP and include appropriate estimates with biometric methods.

$$\text{NECE} = \sum \text{NEE} + (F_W) = \text{NEP} = \text{NPP}_W + (\Delta S_{org}) \tag{5.13}$$

$$\text{NECE} = \sum \text{NEE} + (F_W) = \text{NEP} = (\text{NPP}_L + \text{NPP}_W + \text{NPP}_B) - (\text{Rs}) \pm \text{NTE}_M) \tag{5.14}$$

The problem is that NEP in both cases does not account for some of the inorganic fluxes in F_W or F_A, and that although NEE includes total ecosystem respiration, NEP has to include soil respiration.

We will use the summary provided in Rivera-Monroy et al. (2013) for biometric and ECM estimates of C exchange for a riverine mangrove forest in the Shark River Estuary of Florida Coastal Everglades (SRS-6 will be used to denote station identity). The NEE value estimated for SRS-6 in 2004, based on the sum of monthly NEE values measured during that year, was 1170 ± 127 $\text{gCm}^{-2}\ \text{yr}^{-1}$. Thus, assuming that $F_A = 0$, the NECE can be estimated with the following:

$$\text{NECE} = 1170 + (F_W) \tag{5.15}$$

We then need to estimate F_W ($F_W = F_{DIC} + F_{DOC} + F$) to determine NECE. Barr et al. (2010) constrained the values with direct and indirect measurements of DOC and POC fluxes in the study site where tidal activity generally results in significant lateral fluxes of particulate and dissolved carbon (Whelan et al. 2005; Krauss et al. 2006). A flume study in SRS-6 (Romigh et al. 2006) estimated a net DOC export rate of 56 gC m^{-2} yr^{-1}, and ratios of inorganic to organic carbon flux (Bouillon et al. 2008) were used to estimate a F_W value of −550 ± 260 gC m^{-2} yr^{-1}. Thus,

$$\mathrm{NECE} = 1170 - 550 = 620\,\mathrm{gC\,m^{-2}yr^{-1}} \tag{5.16}$$

NECE can also be estimated using biometric information to generate NEP by using the difference between NPP and soil CO_2 efflux (Barr et al. 2010; Eq. 5.14). Average maximum NPP_A estimated for SRS-6 (Rivera-Monroy et al. 2013) is 1150 ± 29 gC m^{-2} yr^{-1}. Average maximum belowground production (i.e., fine and coarse roots, <20 mm diameter) (NPP_B) (2004–2006) is 311 ± 24 gC m^{-2} yr^{-1} (Castañeda-Moya et al. 2011). Soil CO_2 efflux in SRS-6 based on the sum of heterotrophic and autotrophic components (that also includes root respiration estimates) is about 492 gC m^{-2} yr^{-1} for Rs (Troxler et al. 2015). Thus, we can estimate:

$$\mathrm{NEP} = \left(\mathrm{NPP_A} + \mathrm{NPP_B}\right) - \left(\mathrm{Rs}\right) \pm \mathrm{NTE_M}) = 1150 + 311 - 492 - 550 = 419\,\mathrm{gC\,m^{-2}yr^{-1}} \tag{5.17}$$

This comparison of NECE and NEP is very similar given the gross estimates of NTE_M (Fw) and the fact that NEE in SRS-6 ranged from 1040 to 1290 gC m^{-2} yr^{-1} (Barr et al. 2010).

Finally, the use of wood production and soil C accumulation method can also be compared to NECE. Carbon accumulation (burial) in soils (ΔS_{org}) for SRS-6 is about 125 ± 3 gC m^{-2} yr^{-1} (Castañeda-Moya unpublished results; see Rivera-Monroy et al. 2013). Wood production estimated for the period 2001–2004 was 456 ± 33 gC m^{-2} yr^{-1}. Thus, on an annual basis,

$$\mathrm{NEP} = \mathrm{NPP_W} + \Delta S_{\mathrm{org}} = \left(456 + 125\right) = 581\,\mathrm{gC\,m^{-2}yr^{-1}} \tag{5.18}$$

This does not include inorganic carbon accumulation. In summary, the three techniques to resolve the net C sink in this mangrove wetlands (Eqs. 5.16, 5.17 and 5.18) provide values for SRS-6 in the Shark River Estuary of Florida Coastal Everglades within a narrow range of 419 to 620 gC m^{-2} yr^{-1}.

The differences described in these three approaches to estimate C sinks in mangrove wetlands are within 200 gC m^{-2} yr^{-1}. The closest estimates are those that use NEE and estimate for both organic and inorganic exchange (expanding NTE_M to include dissolved inorganic carbon) with the estuary, compared to just focusing on the two C accumulation rates in wood and soil. These differences are only 40 gC m^{-2} yr^{-1}. This result gives some confidence that this C exchange with the estuary, which is much higher than values summarized above, captures all the different C species that are coupled to the estuary. Because NEE is the CO_2 flux from mangrove wetlands to the atmosphere by definition (Chapin et al. 2006), NEE is different from NEP and NECE when C enters or leaves the forest as DIC in the aquatic phase rather than through atmospheric change (Twilley et al. 1992; Chapin et al. 2006). For example, DIC leaching of groundwater (Call et al. 2015), DIC exchange at different depths with the adjacent estuarine waters (Dittmar and Lara

2001; Bouillon et al. 2008) and transfers of respiration-derived DIC from the mangrove to the estuary, cause NEE to be greater than mangrove NEP or NECE. Given the frequent tidal exchange between Shark River mangroves and the estuary, these differences in fluxes can be significant (Romigh et al. 2006). Indeed, Bouillon et al. (2008) pointed out that DIC export is a major component of unaccounted C in mangrove wetlands. Unfortunately, understanding spatial and temporal organic carbon flux trends in mangrove wetlands has been difficult due to the scarcity of studies (particularly on DIC fluxes), long-term data sets, and the high variability in experimental approaches used in mangrove studies (Romigh et al. 2006; Bouillon et al. 2008; Barr et al. 2010; Adame and Lovelock 2011; Call et al. 2015).

Crab burrows can greatly enhance the surface area of the sediment–air or sediment–water interface where exchange of CO_2 or DIC can take place, serving as significant conduits for enhancing CO_2 exchange between mangrove sediments and atmosphere (Kristensen 2008, see also Chap. 6). The process of tidal exchange enhanced by bulk hydraulic permeability associated with burrows built by animals such as crabs and other crustaceans is known as tidal pumping of DIC. This process is also well established in mangrove creeks (Borges et al. 2003; Kristensen et al. 2008; Maher et al. 2013; Stieglitz et al. 2013; Zablocki et al. 2011) and it has been shown to dominate resulting pCO_2 distributions and CO_2 emissions to air (Bouillon et al. 2007; Linto et al. 2014; Koné and Borges 2008). The impact of these burrows, and possibly mangrove root formations, on the biochemical pathways of anaerobic and aerobic processes is defined by the relative distribution of lower redox zones described in the geochemical model above. What is critical is to define the relative volumes of water via subsurface advective exchange with tidal pumping that transports DIC produced within these redox zones (Maher et al. 2013).

These updated estimates of C sequestration based on NEP and NECE for mangrove wetlands have implications to the debate on how coastal ecosystems influence the global C budget (Twilley et al. 1992; Saenger and Snedaker 1993; Lee 1995; Duarte et al. 2005; Jennerjahn and Ittekkot 2002). A reassessment of global mangrove C budgets by Bouillon et al. (2008) estimates that more than 50% of the C fixed by mangrove vegetation, estimated ~217 ± 72 TgC yr^{-1}, appears to be unaccounted for based on estimates of various C sinks (organic carbon export, soil burial, and mineralization). This missing C sink is conservatively estimated at ~112 ± 85 Tg C yr^{-1}, equivalent in magnitude to ~30–40% of the global riverine organic carbon input to the coastal zone. The analysis above suggests that inorganic carbon flux from sediments and mangrove waters is severely underestimated and that the majority of C export from mangrove wetlands to adjacent waters occurs as DIC (Fig. 5.4). Using the average rate of DIC flux above, global levels of CO_2 efflux from both soils and the water column can be estimated at 178 ± 165 Tg C yr^{-1}. The magnitude of this process could be similar to that of the missing C sink but may vary in range among different mangrove systems from muddy coasts to carbonate settings. Despite the preliminary nature of *F* (DOC, DIC, POC) fluxes (and associated uncertainties) estimated for Shark River mangroves in our example, it is clear that directly measuring rates at the mangrove–tidal creek/river interface should be a major priority in future C and nutrient studies in mangrove wetlands (Bouillon et al. 2008; Barr et al. 2010). In addition, research emphasis should be placed on measuring C that can be oxidized and consequently reduce NEP (Jaffe et al. 2001; Jaffe et al. 2004; Maie et al. 2008).

Another estimate of how important inorganic carbon exchange is to reconciling NECE for mangroves, compared to NEP, is to compare the analysis in Eqs. (5.16, 5.17 and 5.18) with estimates above for wood production and soil C sequestration ($NPP_W + \Delta S_{org}$). The average NEP from Eqs. (5.16, 5.17 and 5.18) is about 500 gCm^{-2} yr^{-1}. Based on global reviews of observations, wood production is about 600 gdm m^{-2} yr^{-1} or about 240 gC m^{-2} yr^{-1}. The average C sequestration in mangroves soils, again based on observations described above (Fig. 5.11), is about 163 gC m^{-2} yr^{-1}. The sum of these two measures, ($NPP_W + \Delta S_{org}$), is an estimate of NEP at about 400 gC m^{-2} yr^{-1}. This value is very similar to the NEP estimates for Shark River mangroves described above in Eqs. (5.16, 5.17 and 5.18), again using NEE as one of the methods to constrain the NEP estimate. The other significance to this comparison is that the only way to get agreement among these estimates in Eqs. (5.16, 5.17 and 5.18) is to include a very significant flux of inorganic carbon as part of the NTE_M in Eq. (5.17). The flux of inorganic carbon in this estimate is equal to the organic carbon flux, or about 250 gC m^{-2} yr^{-1}. However, a general conclusion based on these different approaches, and reviewing the existing global literature, is that NEP accumulates about 400 gC m^{-2} yr^{-1}, equally distributed between wood production and soil C accumulation. The literature for soil carbon accumulation is more thorough than global estimates of carbon storage in wood production. Estimates of the latter are particularly important to have a better estimate of global carbon storage in mangrove NEP.

5.9 Disturbance and Carbon Exchange

The significance of nonanthropogenic and anthropogenic disturbances on ecosystems associated with land use and pulses such as floods, fires, and cyclones has to be assessed when estimating C mitigation strategies due to increasing C pools in the atmosphere. The NEP of wetlands change with ecosystem development following disturbance and can be further modified by anthropogenic effects such as drainage of organic soils in wetlands (Armentano and Menges 1986). Such changes in NEP will alter the function of wetlands, including mangroves, as sources or sinks of atmospheric C (Fig. 5.13), and thus modifying how ecosystems contribute to the mass balance of C in the atmosphere (Houghton 1994; Houghton 2007). The particular question is whether the net flux of C in coastal ecosystems such as mangrove wetlands may contribute to the unidentified sink of C to balance the amount of C in the atmosphere relative to the anthropogenic inputs that cannot be accounted for based on actual C increases over the last century (Houghton 2007). There has been a focus on whether disturbances to terrestrial ecosystems may enhance C sinks or sources of NEP, but no global analysis has been done to account for role of tropical or subtropical coastal zone to this unidentified atmospheric carbon (Fig. 5.13).

Long-term studies of mangrove wetlands at Shark River Estuary in Florida Coastal Everglades have provided insights into how disturbances may change NEP from sources to sinks during recovery, as forest successional stages change NECE following the passage of hurricane Wilma in October 24, 2005 (Zhang et al. 2008; Smith et al. 2009; Castañeda-Moya et al. 2010). Wilma, a category three hurricane, produced extensive inundation along the southwestern Florida coast, with a maximum storm surge of 3–5 m in some mangrove stands (Krauss et al. 2009; Smith et al. 2009; Castañeda-Moya et al. 2010). Mangrove wetlands lost nearly all canopy and experienced extensive tree damage, increasing woody debris, and reducing NPP_A (Smith et al. 2009; Danielson 2016). Mangrove damage as a result of Wilma included defoliation, tree snapping, and uprooting and was related to storm surge energy and proximity relative to the eye-wall of the storm (Zhang et al. 2008). In certain locations along Shark River estuary (e.g., SRS-6 and mouth of the estuary), Wilma caused severe damage to the mangrove forest, with 99% of the canopy defoliated and several large trees (i.e., tree height > 15 m) downed, broken, or uprooted. Recent studies using light detection and ranging (LIDAR) measurements in the southwestern Everglades region before and after the 2005 hurricane season showed that the total area of canopy gaps in the Shark River estuary increased from 1% to 2% to 12% after the storm (Zhang et al. 2008). Estimates of wood damage (WD) volume across mangroves in the Shark River and Broad Creek areas ranged from 170 to 259 m^3 ha^{-1} following Wilma, with twice the volume of WD in areas closer to Wilma's eye-wall. In contrast, posthurricane mangrove WD studies in high oceanic islands of Micronesia have found considerably lower estimates (35–104 m^3 ha^{-1}) likely due to the lower frequency of hurricanes and also attributed to differences in age and forest structure (i.e., lower canopies) (Allen et al. 2000). These results highlight the significant contribution of C (and nutrients) from downed wood into mangrove soils that may be significant to accurate C budgets for mangrove wetlands, particularly in areas with high recurrence of storms such as the Caribbean Sea in the neotropics (Krauss et al. 2005).

Recovery of the mangrove wetland described above occurred within five years, stimulating NPP and NEP, including both enhanced litter fall and root production (Castañeda-Moya unpublished results; Danielson 2016). This requires a modification of Fig. 5.13 whereby C sinks in mangroves wetlands can be stimulated by disturbance. The impact of Hurricane Wilma on Shark River estuary (at SRS-6, same site used in NEP comparison above) is reflected in the significant reduction (65%) of litter fall production 1 year after the storm. After 2006, NPP_L gradually increased to prestorm levels (4.8 Mg C ha^{-1} yr^{-1}) by 2009. Similar resilience strategies were observed with mangrove root biomass and production after the storm. For instance, total (0–90 cm) root C storage after the storm (2009: 3633 ± 498 gC m^{-2}) was 3.3 times higher compared to values reported before Wilma (2000–2003: 1109 ± 185 gC m^{-2}; Castañeda-Moya et al. 2011), with fine (<2 mm diameter) roots accounted for 51–85% of the total C storage in the shallow (0–45 cm) and deeper (45–90 cm) root zones, respectively. Before the storm, the contribution of fine roots

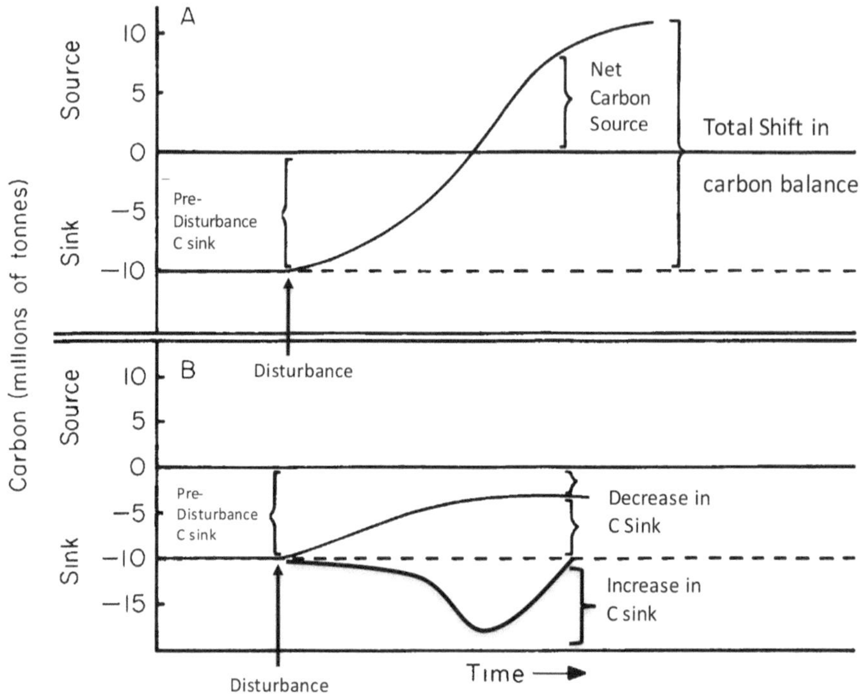

Fig. 5.13 Definitions of carbon sources and sinks as exchanged with the atmosphere using two hypothetical wetland regions subject to disturbance (after Armentano and Menges 1986). (**a**) The original net carbon sink is totally lost by the changes in ecosystem processes in response to disturbance, to a condition where the wetland is a net carbon source to atmosphere. (**b**) The wetland currently functions as a diminished carbon sink because CO_2 release in disturbed wetlands is lower that net CO_2 fixation in undisturbed wetlands; with a modification to demonstrate that carbon sinks can actually increase in some scenarios following a disturbance by increasing NEP (net ecosystem production)

to total C storage ranged from 13–18% for both root zones, whereas coarse (>5 mm) roots contributed up to 60–67% of the total (Castañeda-Moya et al. 2011). The higher biomass allocation to coarse roots (~40–67%) and substantial fine root production before and after the storm in all our mangrove sites support the hypothesis that belowground allocation is a significant contribution to soil C storage in mangrove wetlands (Chmura et al. 2003; Khan et al. 2007). These results demonstrate the resiliency of mangrove wetlands to recover to predisturbance NPP rates within a short (<5 years) period of time as result of disturbances that caused significant changes in mangrove community structure and function. Continued long-term ecological research on mangrove ecosystems in response to disturbance will be critical to understand trajectories of recovery in C dynamics of mangrove wetlands, particularly given the current climate change scenarios and accelerating rates of sea-level rise affecting these forested wetlands and other coastal ecosystems worldwide.

Similar observations have been made on C budgets in mangrove plantations following disturbance (Bosire et al. 2006; Fontalvo-Herazo et al. 2011; Jin-Eong et al. 1995; Kairo et al. 2008, 2009; Ren et al. 2008). These estimates associated with tropical cyclones are very important to any global C exchange estimate of mangroves given the frequency of this disturbance in the subtropics (Fig. 5.14). Mangrove wetlands are known as frequently disturbed ecosystems, particularly associated with tropical cyclones, and this forcing may have a significant impact on NEP mitigating C exchange with the atmosphere. This is similar to the debate for land use of forest ecosystems that may explain the unidentified C sink in the biosphere (Houghton 1994). For instance, mangrove wetlands in the southeastern region of Florida Everglades have encroached inland approximately 1.5 km during the past 50 years (Ross et al. 2000), transforming previous freshwater wetlands dominated by *C. jamaicense*- to brackish *R. mangle*-dominated communities. These changes in community composition and peat depth result from reductions in freshwater drainage into this region along with gradual increase in sea level and temperature, simulating the migration of mangroves under scenarios of climate change. Similar expansion has been observed in the Atlantic coast of the United States and the Gulf of Mexico (Doughty et al. 2016; Osland et al. 2013; Comeaux et al. 2012). There is a global trend of mangroves expanding into the warm tropics (Saintilan et al. 2009, 2014), potentially expanding the effects of mangrove NEP on mitigating atmospheric C.

Finally, the continued contribution of mangrove wetlands as C sinks in the coastal zone is enhanced by accelerated rates of sea level rise, which can be considered an anthropogenic disturbance to coastal zones. Woodroffe (2002) summarized how the organic carbon storage (peat formation) of salt marshes changes with different responses of morphodynamics to sea level rise scenarios. If coastal wetlands accrete with sea level rise (and subsidence), then the NEP of wetlands is nonsteady state and there is constant annual net C sequestration. Accelerated sea level rise will potentially accelerate C sequestration as soil accretion also accelerates. Yet the latter scenario of accelerated rates depends on adequate source of particles that can enhance deposition along with the contribution of NPP_B to soil accretion (Kirwan and Temmerman 2009; Kirwan and Guntenspergen 2012; Kirwan and Mudd 2012). There are critical rates of relative sea level rise (eustatic sea level plus subsidence rates, RSLR) that are predicted to cause the drowning of coastal wetlands because such RSLR rates are above the adaptation potential for wetlands to survive. The stability of intertidal wetlands over decadal time scales in response to accelerated sea level rise (thus accelerated RSLR) relies on sediment concentrations and tidal amplitudes of coastal environmental setting (Kirwan et al. 2010). Thus, ecogeomorphic types of coastal settings from carbonate lagoons to estuaries and river-dominated deltas will determine the capacity of coastal wetlands to adapt to increase RSLR in the next century. This significant interaction between latitude and ecogeomorphology will also determine what coastal environmental settings will continue to sequester C in mangrove soils. This outcome has important implications as to how mangrove wetlands may be able to mitigate anthropogenic increase of CO_2 in the atmosphere as both C sequestration and RSLR also increase.

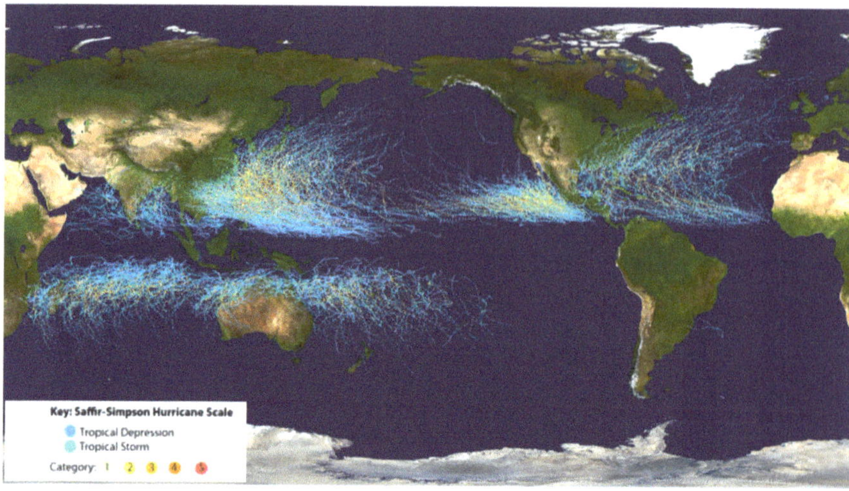

Fig. 5.14 Map of the cumulative tracks of all tropical cyclones during the 1985–2005 period. *The points* show the locations of the storm paths at six-hourly intervals and colors of each path indicate the strength of the storm based on the Saffir–Simpson Hurricane Scale (*bottom left*) (Source: https://en.wikipedia.org/wiki/File:Global_tropical_cyclone_tracks-edit2.jpg)

International trends of disturbance in mangrove area are important to really understand the impact of mangroves wetlands on mitigating C dynamics in the atmosphere. An extensive review of global changes of mangrove area concluded that estimates exhibited high variability depending on the modeled data points (Friess and Webb 2014). The high variability in deforestation rate is evident when comparing Indonesia (331.65 ± 222.26 km^2 yr^{-1}), Nigeria (92.09 ± 188.95 km^2 yr^{-1}) and Cuba 34.82 ± 142.17 km^2 yr^{-1}). The standard deviations were large due to high variability in national estimates, and hence the number of contradictory trends that could be extracted (Friess and Webb 2014). Conflicting trends of long-term mangrove loss and gain could be derived for eight of the 15 countries analyzed, suggesting that estimates of ecosystem NEP effects on C mitigation will suffer by low confidence in global estimates of mangrove area disturbance. As described by Armentano and Menges (1986), global estimates of how wetland ecosystems impact atmosphere C dynamics require precise measures of NEP associated with different responses to disturbance and the global area associated with each of these disturbances. Many of the modeling efforts to improve global estimates of C stocks and NEP are important to improve these types of assessments. Recently, advances in a wealth of remote sensing studies assessing mangrove wetlands spatial distribution (see review in Kuenzer et al. 2011 and Chap. 4) have improved estimates of global C storage in different types of mangroves on regional and local scales (e.g., Simard et al. 2006, 2008). These techniques have provided insights on how to extrapolate values from a few plots/transects to regional estimates of selective attributes of mangrove C storage such as AGB. However, net ecosystem C exchange of mangrove wetlands still requires ability to extrapolate other more complex fluxes of C across both the atmosphere and coastal waters with the forest boundary. This dynamic coupling should be a future research and assessment focus for determining the global role of mangroves and coastal ecosystems to C mitigation.

Acknowledgments This work represents support from several funding sources including the Florida Coastal Everglades Long-Term Ecological Research (FCE-LTER) program funded by the National Science Foundation (Grant #DBI-0620409, Grant #DEB-9910514, and Grant #DEB-1237517), Coastal SEES program of National Science Foundation (EAR-1427389), and Frontiers of Earth Surface Dynamics (OCE-1135427). Support was also provided by Louisiana Sea Grant College Program. ASR was partially supported by CAPES/CNPq Science without Borders Program (grant no. BEX1930/13-3). VHRM was partially funded by NASA-JPL project "Vulnerability Assessment of Mangrove Forest Regions of the Americas" (LSU Subcontract no. 1452878), and the Department of the Interior South Central Climate Science Center through Cooperative Agreement # G12AC00002. We would like to thank Paulo R. Pagliosa from Universidade Federal de Santa Catarina, Florianópolis, Brasil, and Alessandra Larissa Fonseca from Universidade Federal de Santa Catarina, Florianópolis, Brasil in helping to compile some of the mangrove statistics.

References

Adame MF, Lovelock CE (2011) Carbon and nutrient exchange of mangrove forests with the coastal ocean. Hydrobiologia 663(1):23–50

Allen JA, Ewel KC, Keeland BD, Tara T, Smith TJ (2000) Downed wood in Micronesian mangrove forests. Wetlands 20(1):169–176

Allen MR, Barros VR, Broome J, Cramer W, Christ R, Church JA, Clarke L, Dahe Q, Dasgupta P, Dubash NK (2014) IPCC fifth assessment synthesis report-climate change 2014 synthesis report

Alongi D, Pfitzner J, Trott L, Tirendi F, Dixon P, Klumpp D (2005) Rapid sediment accumulation and microbial mineralization in forests of the mangrove Kandelia candel in the Jiulongjiang Estuary, China. Estuar Coast Shelf Sci 63(4):605–618

Alongi D, Wattayakorn G, Pfitzner J, Tirendi F, Zagorskis I, Brunskill G, Davidson A, Clough B (2001) Organic carbon accumulation and metabolic pathways in sediments of mangrove forests in southern Thailand. Mar Geol 179(1):85–103

Alongi DM (2002) Present state and future of the world's mangrove forests. Environ Conserv 29(3):331–349

Alongi DM (2009) The energetics of mangrove forests. Springer, Dordrecht/London

Alongi DM (2011) Patterns of mangrove wood and litter production within a beach ridge-fringing reef embayment, northern great barrier reef coast. Estuar Coasts 34(1):32–44. doi:10.1007/s12237-010-9289-y

Alongi DM (2014) Carbon cycling and storage in mangrove forests. Annu Rev Mar Sci 6:195–219

Alongi DM, Boto KG, Tirendi F (1989) Effect of exported mangrove litter on bacterial productivity and dissolved organic carbon fluxes in adjacent tropical nearshore sediments. Mar Ecol Prog Ser 56:133–144

Alongi DM, Christoffersen P, Tirendi F (1993) The influence of forest type on microbial-nutrient relationships in tropical mangrove sediments. J Exp Mar Biol Ecol 171:201–223

Amarasinghe MD, Balasubramaniam S (1992) Net primary productivity of two mangrove forest stands on the northwestern coast of Sri Lanka. Hydrobiologia 247(1–3):37–47

Andersen FO, Kristensen E (1992) The importance of benthic macrofauna in decomposition of microalgae in a coastal marine sediment. Limnol Oceanogr 37(7):1392–1403

Araújo J Jr, Otero X, Marques A, Nóbrega G, Silva J, Ferreira T (2012) Selective geochemistry of iron in mangrove soils in a semiarid tropical climate: effects of the burrowing activity of the crabs Ucides cordatus and Uca maracoani. Geo-Mar Lett 32(4):289–300

Armentano T, Menges E (1986) Patterns of change in the carbon balance of organic soil-wetlands of the temperate zone. J Ecol 74:755–774

Baldocchi D (2008) 'Breathing' of the terrestrial biosphere: lessons learned from a global network of carbon dioxide flux measurement systems. Aust J Bot 56(1):1–26. doi:http://dx.doi.org/10.1071/BT07151

Baldocchi DD (2003) Assessing the eddy covariance technique for evaluating carbon dioxide exchange rates of ecosystems: past, present and future. Glob Chang Biol 9(4):479–492

Barr JG, Engel V, Fuentes JD, Zieman JC, O'Halloran TL, Smith TJ, Anderson GH (2010) Controls on mangrove forest-atmosphere carbon dioxide exchanges in western Everglades National Park. J Geophys Res Biogeo 115, G02020, doi:10.1029/2009JG001186

Borges AV, Djenidi S, Lacroix G, Theate J, Delille B, Frankignoulle M (2003) Atmospheric CO2 flux from mangrove surrounding waters. Geophys Res Lett 30(11):12-11–12-14

Bosire JO, Dahdouh-Guebas F, Kairo JG, Wartel S, Kazungu J, Koedam N (2006) Success rates of recruited tree species and their contribution to the structural development of reforested mangrove stands. Mar Ecol Prog Ser 325:85–91. doi:10.3354/meps325085

Bosire JO, Dahdouh-Guebas F, Kairo JG, ZKazungu J, Dehairs F, Koedam N (2005) Litter degradation and CN dynamics in reforested mangrove plantations at Gazi Bay, Kenya. Biol Conserv 126:287–295

Boto KG, Bunt JS (1981) Tidal export of particulate organic matter from a northern Australian mangrove system. Estuar Coast Shelf Sci 13:247–255

Boto KG, Wellington JT (1988) Seasonal variations in concentrations and fluxes of dissolved organic and inorganic materials in a tropical, tidally-dominated, mangrove waterway. Mar Ecol Prog Ser 50:151–160

Bouillon S, Borges AV, Castaneda-Moya E, Diele K, Dittmar T, Duke NC, Kristensen E, Lee SY, Marchand C, Middelburg JJ, Rivera-Monroy VH, Smith TJ, Twilley RR (2008) Mangrove production and carbon sinks: a revision of global budget estimates. Glob Biogeochem Cycles 22(2). doi:10.1029/2007gb003052

Bouillon S, Koedam N, Baeyens W, Satyanarayana B, Dehairs F (2004a) Selectivity of subtidal benthis invertebrate communities for local microalgal production in an estuarine mangrove ecosystem during the post-monsoon period. J Sea Res 51(2):133–144

Bouillon S, Koedam N, Raman AV, Dehairs F (2002a) Primary producers sustaining macro-invertebrate communities in intertidal mangrove forests. Oecologia 130:441–448

Bouillon S, Middelburg JJ, Dehairs F, Borges AV, Abril G, Flindt MR, Ulomi S, Kristensen E (2007) Importance of intertidal sediment processes and porewater exchange on the water column biogeochemistry in a pristine mangrove creek (Ras Dege, Tanzania). Biogeosci Discuss 4(1):317–348

Bouillon S, Moens T, Overmeer I, Koedam N, Dehairs F (2004b) Resource utilization patterns of epifauna from mangrove forests with contrasting inputs of local versus imported organic matter. Mar Ecol Prog Ser 278:77–88

Bouillon S, Raman A, Dauby P, Dehairs F (2002b) Carbon and nitrogen stable isotope ratios of subtidal benthic invertebrates in an estuarine mangrove ecosystem (Andhra Pradesh, India). Estuar Coast Shelf Sci 54(5):901–913

Breithaupt JL, Smoak JM, Smith TJ, Sanders CJ, Hoare A (2012) Organic carbon burial rates in mangrove sediments: strengthening the global budget. Glob Biogeochem Cycles 26, GB3011, doi:10.1029/2012GB004375

Brown S, Lugo AE (1984) Biomass of tropical forests: a new estimate based on forest volumes. Science 223:1290–1293

Bulmer RH, Lundquist C, Schwendenmann L (2015) Sediment properties and CO2 efflux from intact and cleared temperate mangrove forests. Biogeosciences 12(20):6169–6180

Cahoon DR, Hensel P, Rybczyk J, McKee KL, Proffitt CE, Perez BC (2003) Mass tree mortality leads to mangrove peat collapse at Bay Islands, Honduras after Hurricane Mitch. J Ecol 91(6):1093–1105. doi:10.1046/j.1365-2745.2003.00841.x

Cai W-J (2011) Estuarine and coastal ocean carbon paradox: CO2 sinks or sites of terrestrial carbon incineration? Annu Rev Mar Sci 3:123–145

Caldeira K (2012) Avoiding mangrove destruction by avoiding carbon dioxide emissions. Proc Natl Acad Sci 109(36):14287–14288

Call M, Maher DT, Santos IR, Ruiz-Halpern S, Mangion P, Sanders CJ, Erler DV, Oakes JM, Rosentreter J, Murray R (2015) Spatial and temporal variability of carbon dioxide and methane fluxes over semi-diurnal and spring–neap–spring timescales in a mangrove creek. Geochim Cosmochim Acta 150:211–225

Canadell JG, Raupach MR (2008) Managing forests for climate change mitigation. Science 320(5882):1456–1457

Cannicci S, Burrows D, Fratini S, Smith TJ, Offenberg J, Dahdouh-Guebas F (2008) Faunal impact on vegetation structure and ecosystem function in mangrove forests: a review. Aquat Bot 89(2):186–200

Castañeda-Moya E, Rivera-Monroy VH, Twilley RR (2006) Mangrove zonation in the dry life zone of the Gulf of Fonseca, Honduras. Estuar Coasts 29(5):751–764

Castañeda-Moya E, Twilley RR, Rivera-Monroy VH (2013) Allocation of biomass and net primary productivity of mangrove forests along environmental gradients in the Florida Coastal Everglades, USA. For Ecol Manag 307:226–241

Castañeda-Moya E, Twilley RR, Rivera-Monroy VH, Marx BD, Coronado-Molina C, Ewe SML (2011) Patterns of root dynamics in mangrove forests along environmental gradients in the Florida Coastal Everglades, USA. Ecosystems 14(7):1178–1195. doi:10.1007/s10021-011-9473-3

Castañeda-Moya E, Twilley RR, Rivera-Monroy VH, Zhang KQ, Davis SE, Ross M (2010) Sediment and nutrient deposition associated with Hurricane Wilma in mangroves of the Florida Coastal Everglades. Estuar Coasts 33(1):45–58. doi:10.1007/S12237-009-9242-0

Chalermchatwilai B, Poungparn S, Patanaponpaiboon P (2011) Distribution of fine-root necromass in a secondary mangrove forest in Trat province, Eastern Thailand. ScienceAsia 37:1–5

Chapin FS III, Woodwell GM, Randerson JT, Rastetter EB, Lovett GM, Baldocchi DD, Clark DA, Harmon ME, Schimel DS, Valentini R (2006) Reconciling carbon-cycle concepts, terminology, and methods. Ecosystems 9(7):1041–1050

Chen L, Zan Q, Li M, Shen J, Liao W (2009) Litter dynamics and forest structure of the introduced Sonneratia caseolaris mangrove forest in Shenzhen, China. Estuar Coast Shelf Sci 85(2):241–246. doi:http://dx.doi.org/10.1016/j.ecss.2009.08.007

Chen R, Twilley RR (1999) A simulation model of organic matter and nutrient accumulation in mangrove wetland soils. Biogeochemistry 44:93–118

Chmura GL, Anisfeld SC, Cahoon DR, Lynch JC (2003) Global carbon sequestration in tidal, saline wetland soils. Glob Biogeochem Cycles 17(4):12. doi:10.1029/2002gb001917

Christensen B (1978) Biomass and primary production of Rhizophora apiculata Bl. in a mangrove in southern Thailand. Aquat Bot 4:43–52. doi:http://dx.doi.org/10.1016/0304-3770(78)90005-0

Cifuentes LA, Coffin RB, Solorzano L, Cardenas W, Espinoza J, Twilley RR (1996) Isotopic and elemental variations of carbon and nitrogen in a mangrove estuary. Estuar Coast Shelf Sci 43(6):781–800

Cintrón G, Schaeffer-Novelli Y (1984) Caracteristicas y desarrollo estructural de los manglares de Norte y Sur America. Cienc Interam 25(1–4):4–15

Clark DA, Brown S, Kicklighter DW, Chambers JQ, Thomlinson JR, Ni J (2001) Measuring net primary production in forests: concepts and field methods. Ecol Appl 11(2):356–370

Cole TG, Ewel KC, Devoe NN (1999) Structure of mangrove trees and forests in Micronesia. For Ecol Manag 117(1–3):95–109. doi:http://dx.doi.org/10.1016/S0378-1127(98)00474-5

Comeaux RS, Allison MA, Bianchi TS (2012) Mangrove expansion in the Gulf of Mexico with climate change: implications for wetland health and resistance to rising sea levels. Estuar Coast Shelf Sci 96:81–95. doi:10.1016/j.ecss.2011.10.003

Cormier N, Twilley RR, Ewel KC, Krauss KW (2015) Fine root productivity varies along nitrogen and phosphorus gradients in high-rainfall mangrove forests of Micronesia. Hydrobiologia 750(1):69–87

Coronado-Molina C, Alvarez-Guillen H, Day JW Jr, Reyes E, Perez BC, Vera-Herrera F, Twilley R (2012) Litterfall dynamics in carbonate and deltaic mangrove ecosystems in the Gulf of Mexico. Wetl Ecol Manag 20(2):123–136. doi:10.1007/s11273-012-9249-3

Crase B, Liedloff A, Vesk PA, Burgman MA, Wintle BA (2013) Hydroperiod is the main driver of the spatial pattern of dominance in mangrove communities. Glob Ecol Biogeogr 22(7):806–817. doi:10.1111/geb.12063

Cuevas E, Medina E (1988) Nutrient dynamics within amazonian forests II. Fine root growth, nutrient availability and leaf litter decomposition. Oecologia 76:222–235

Danielson TM (2016) Everglades mangrove forest response to large-scale disturbance: long-term assessment of structural and functional properties. Louisiana State University, Baton Rouge

Davis SE, Childers DL, Day JW, Rudnick DT, Sklar FH (2001) Wetland-water column exchanges of carbon, nitrogen, and phosphorus in a southern Everglades dwarf mangrove. Estuaries 24(4):610–622

Day JW, Conner WH, Ley-Lou F, Day RH, Navarro AM (1987) The productivity and composition of mangrove forests, Laguna de Terminos, Mexico. Aquat Bot 27:267–284

Day JW, Coronado-Molina C, Vera-Herrera FR, Twilley RR, Rivera-Monroy VH, Alvarez-Guillen H, Day R, Conner W (1996) A 7 year record of above-ground net primary production in a southeastern Mexican mangrove forest. Aquat Bot 55(1):39–60

Dittmar T, Hertkorn N, Kattner G, Lara RJ (2006) Mangroves, a major source of dissolved organic carbon to the oceans. Glob Biogeochem Cycles 20, GB1012, doi:10.1029/2005GB002570

Dittmar T, Lara RJ (2001) Do mangroves rather than rivers provide nutrients to coastal environments south of the Amazon River? Evidence from long-term flux measurements. Mar Ecol Prog Ser 213:67–77

Donato DC, Kauffman JB, Murdiyarso D, Kurnianto S, Stidham M, Kanninen M (2011) Mangroves among the most carbon-rich forests in the tropics. Nat Geosci 4(5):293–297. doi:http://www.nature.com/ngeo/journal/v4/n5/abs/ngeo1123.html - supplementary-information

Doughty CL, Langley JA, Walker WS, Feller IC, Schaub R, Chapman SK (2016) Mangrove range expansion rapidly increases coastal wetland carbon storage. Estuar Coasts 39(2):385–396

Duarte CM, Cebrian J (1996) The fate of marine autotrophic production. Limnol Oceanogr 41(8):1758–1766

Duarte CM, Middelburg JJ, Caraco N (2005) Major role of marine vegetation on the oceanic carbon cycle. Biogeosciences 2(1):1–8

Ellison AM (2002) Macroecology of mangroves: large-scale patterns and processes in tropical coastal forests. Trees-Struct Funct 16(2–3):181–194. doi:10.1007/s00468-001-0133-7

Ewe SML, Gaiser EE, Childers DL, Iwaniec D, Rivera-Monroy VH, Twilley RR (2006) Spatial and temporal patterns of aboveground net primary productivity (ANPP) along two freshwater-estuarine transects in the Florida Coastal Everglades. Hydrobiologia 569:459–474. doi:10.1007/S10750-006-0149-5

Ewel KC, Twilley RR, Eong-Ong J (1998) Different kinds of mangrove forests provide different goods and services. Glob Ecol Biogeogr 7:83–94

Feller IC (1995) Effects of nutrient enrichment on growth and herbivory of dwarf red mangrove (*Rhizophora mangle*). Ecol Monogr 65(4):477–505

Feller IC, Lovelock C, Berger U, McKee K, Joye S, Ball M (2010) Biocomplexity in mangrove ecosystems. Annu Rev Mar Sci 2:395–417

Feller IC, Lovelock C, McKee KL (2007) Nutrient addition differentially affects ecological processes of Avicennia germinans in nitrogen versus phosphorus limited mangrove ecosystems. Ecosystems 10(3):347–359

Ferreira T, Otero X, Vidal-Torrado P, Macías F (2007) Effects of bioturbation by root and crab activity on iron and sulfur biogeochemistry in mangrove substrate. Geoderma 142(1):36–46

Fleming M, Lin G, Sternberg LSL (1990) Influence of mangrove detritus in an estuarine ecosystem. Bull Mar Sci 47:663–669

Fontalvo-Herazo ML, Piou C, Vogt J, Saint-Paul U, Berger U (2011) Simulating harvesting scenarios towards the sustainable use of mangrove forest plantations. Wetl Ecol Manag 19(5):397–407. doi:10.1007/s11273-011-9224-4
Friess DA, Webb EL (2014) Variability in mangrove change estimates and implications for the assessment of ecosystem service provision. Glob Ecol Biogeogr 23(7):715–725
Fry B, Smith TJ (2002) Stable isotope studies of red mangroves and filter feeders from the Shark River estuary, Florida. Bull Mar Sci 70(3):871–890
Ganguly D, Dey M, Mandal S, De T, Jana T (2008) Energy dynamics and its implication to biosphere–atmosphere exchange of CO2, H2O and CH 4 in a tropical mangrove forest canopy. Atmos Environ 42(18):4172–4184
Gedan KB, Kirwan ML, Wolanski E, Barbier EB, Silliman BR (2011) The present and future role of coastal wetland vegetation in protecting shorelines: answering recent challenges to the paradigm. Clim Chang 106(1):7–29. doi:10.1007/s10584-010-0003-7
Giri C, Ochieng E, Tieszen LL, Zhu Z, Singh A, Loveland T, Masek J, Duke N (2011) Status and distribution of mangrove forests of the world using earth observation satellite data. Glob Ecol Biogeogr 20(1):154–159
Giri C, Zhu Z, Tieszen L, Singh A, Gillette S, Kelmelis J (2008) Mangrove forest distributions and dynamics (1975–2005) of the tsunami-affected region of Asia. J Biogeogr 35(3):519–528
Gleason S, Ewel KC (2002) Organic matter dynamics on the forest floor of a Micronesian mangrove forest: an investigation of speices composition shifts. Biotropica 34(2):190–198
Golley FB, Odum HT, Wilson RF (1962) The structure and metabolism of a Puerto Rican red mangrove forest in May. Ecology 43:9–19
Gong W-K, Ong J-E (1990) Plant biomass and nutrient flux in a managed mangrove forest in Malaysia. Estuar Coast Shelf Sci 31(5):519–530. doi:http://dx.doi.org/10.1016/0272-7714(90)90010-O
Graham SA, Mendelssohn IA (2016) Contrasting effects of nutrient enrichment on below-ground biomass in coastal wetlands. J Ecol 104(1):249–260
Granek EF, Compton JE, Phillips DL (2009) Mangrove-exported nutrient incorporation by sessile coral reef invertebrates. Ecosystems 12(3):462–472
Heald EJ (1969) The production of organic detritus in a south Florida estuary. Ph.D. Disseration, University of Miami, Coral Gables, Florida
Hemminga MA, Slim FJ, Kazungu J, Ganssen GM, Nieuwenhuize J, Kruyt NM (1994) Carbon outwelling from a mangrove forest with adjacent seagrass beds and coral reefs (Gazi Bay, Kenya). Mar Ecol Prog Ser 106:291–301
Hopkinson CS, Cai WJ, Hu XP (2012) Carbon sequestration in wetland dominated coastal systems – a global sink of rapidly diminishing magnitude. Curr Opin Environ Sustain 4(2):186–194. doi:10.1016/j.cosust.2012.03.005
Houghton R (2007) Balancing the global carbon budget. Annu Rev Earth Planet Sci 35:313–347
Houghton RA (1994) The worldwide extent of land-use change. Bioscience 44(5):305–313
Hutchison J, Manica A, Swetnam R, Balmford A, Spalding M (2013) Predicting global patterns in mangrove forest biomass. Conserv Lett 7(3):233–240
Jackson RB, Mooney H, Schulze E-D (1997) A global budget for fine root biomass, surface area, and nutrient contents. Proc Natl Acad Sci 94(14):7362–7366
Jaffe R, Boyer JN, Lu X, Maie N, Yang C, Scully NM, Mock S (2004) Source characterization of dissolved organic matter in a subtropical mangrove-dominated estuary by fluorescence analysis. Mar Chem 84(3–4):195–210
Jaffe R, Mead R, Hernandez ME, Peralba MC, DiGuida OA (2001) Origin and transport of sedimentary organic matter in two subtropical estuaries: a comparative, biomarker-based study. Org Geochem 32(4):507–526
Jardine SL, Siikamäki JV (2014) A global predictive model of carbon in mangrove soils. Environ Res Lett 9(10):104013
Jennerjahn TC, Ittekkot V (2002) Relevance of mangroves for the production and deposition of organic matter along tropical continental margins. Naturwissenschaften 89(1):23–30
Jin-Eong O, Khoon GW, Clough BF (1995) Structure and productivity of a 20-year-old stand of Rhizophora apiculata Bl. mangrove forest. J Biogeogr 22(2/3):417–424. doi:10.2307/2845938

Junk W, Piedade M, Lourival R, Wittmann F, Kandus P, Lacerda L, Bozelli R, Esteves F, Nunes da Cunha C, Maltchik L (2014) Brazilian wetlands: their definition, delineation, and classification for research, sustainable management, and protection. Aquat Conserv Mar Freshwat Ecosyst 24(1):5–22

Kairo J, Bosire J, Langat J, Kirui B, Koedam N (2009) Allometry and biomass distribution in replanted mangrove plantations at Gazi Bay, Kenya. Aquat Conserv Mar Freshwat Ecosyst 19(S1):S63–S69

Kairo JG, Lang'at JKS, Dahdouh-Guebas F, Bosire J, Karachi M (2008) Structural development and productivity of replanted mangrove plantations in Kenya. For Ecol Manag 255(7):2670–2677. doi:http://dx.doi.org/10.1016/j.foreco.2008.01.031

Kauffman JB, Heider C, Cole TG, Dwire KA, Donato DC (2011) Ecosystem carbon stocks of Micronesian mangrove forests. Wetlands 31(2):343–352

Khan MNI, Suwa R, Hagihara A (2009) Biomass and aboveground net primary production in a subtropical mangrove stand of Kandelia obovata (S., L.) Yong at Manko Wetland, Okinawa, Japan. Wetl Ecol Manag 17(6):585–599

Khan NI, Suwa R, Hagihara A (2007) Carbon and nitrogen pools in a mangrove stand of Kandelia obovata (S., L.) Yong: vertical distribution in the soil-vegetation system. Wetl Ecol Manag 15:141–153

Kirwan M, Temmerman S (2009) Coastal marsh response to historical and future sea-level acceleration. Quat Sci Rev 28(17):1801–1808. doi:http://dx.doi.org/10.1016/j.quascirev.2009.02.022

Kirwan ML, Guntenspergen GR (2012) Feedbacks between inundation, root production, and shoot growth in a rapidly submerging brackish marsh. J Ecol 100(3):764–770. doi:10.1111/j.1365-2745.2012.01957.x

Kirwan ML, Guntenspergen GR, D'Alpaos A, Morris JT, Mudd SM, Temmerman S (2010) Limits on the adaptability of coastal marshes to rising sea level. Geophys Res Lett 37. doi:10.1029/2010gl045489

Kirwan ML, Mudd SM (2012) Response of salt-marsh carbon accumulation to climate change. Nature 489(7417):550. doi:10.1038/nature11440

Komiyama A, Ogino K, Aksornkoae S, Sabhasri S (1987) Root biomass of a mangrove forest in southern Thailand,. 1. Estimation by the trench method and the zonal structure of root biomass. J Trop Ecol 3:97–108

Komiyama A, Ong JE, Poungparn S (2008) Allometry, biomass, and productivity of mangrove forests: a review. Aquat Bot 89(2):128–137

Koné Y-M, Borges A (2008) Dissolved inorganic carbon dynamics in the waters surrounding forested mangroves of the Ca Mau Province (Vietnam). Estuar Coast Shelf Sci 77(3):409–421

Krauss KW, Doyle TW, Twilley RR, Rivera-Monroy VH, Sullivan JK (2006) Evaluating the relative contributions of hydroperiod and soil fertility on growth of south Florida mangroves. Hydrobiologia 569:311–324. doi:10.1007/S10750-006-0139-7

Krauss KW, Doyle TW, Twilley RR, Smith TJ, Whelan KRT, Sullivan JK (2005) Woody debris in the mangrove forests of South Florida. Biotropica 37(1):9–15

Krauss KW, Lovelock CE, McKee KL, López-Hoffman L, Ewe SM, Sousa WP (2008) Environmental drivers in mangrove establishment and early development: a review. Aquat Bot 89(2):105–127

Kreuzwieser J, Buchholz J, Rennenberg H (2003) Emission of methane and nitrous oxide by Australian mangrove ecosystems. Plant Biol 5(04):423–431

Kristensen E (1988) Benthic fauna and biogeochemical processes in marine sediments. II. Microbial activities and fluxes. In: Blackburn TH, Sørensen J (eds) Nitrogen cycling in coastal marine environments. Wiley, Chichester, pp 275–299

Kristensen E (2007) Carbon balance in mangrove sediments: the driving processes and their controls. In: Greenhouse gas and carbon balances in mangrove coastal ecosystems. Gendai Tosho, Tokyo, pp 61–78

Kristensen E (2008) Mangrove crabs as ecosystem engineers; with emphasis on sediment processes. J Sea Res 59(1):30–43

Kristensen E, Alongi DM (2006) Control by fiddler crabs (Uca vocas) and plant roots (Avicennia marina) on carbon, iron, and sulfur biogeochemistry in mangrove sediment. Limnol Oceanogr 51(4):1557–1571

Kristensen E, Andersen FØ, Holmboe N, Holmer M, Thongtham N (2000) Carbon and nitrogen mineralization in sediments of the Bangrong mangrove area, Phuket, Thailand. Aquat Microb Ecol 22(2):199–213

Kristensen E, Bouillon S, Dittmar T, Marchand C (2008) Organic carbon dynamics in mangrove ecosystems: a review. Aquat Bot 89(2):201–219

Kristensen E, Mangion P, Tang M, Flindt MR, Holmer M, Ulomi S (2011) Microbial carbon oxidation rates and pathways in sediments of two Tanzanian mangrove forests. Biogeochemistry 103(1–3):143–158

Kuenzer C, Bluemel A, Gebhardt S, Quoc TV, Dech S (2011) Remote sensing of mangrove ecosystems: a review. Remote Sens 3(5):878–928

Lee S (1995) Mangrove outwelling: a review. Hydrobiologia 295(1–3):203–212

Lee S (1997) Potential trophic importance of the faecal material of the mangrove sesarmine crab Sesarma messa. Mar Ecol Prog Ser 159:275–284

Lee SY (1989a) The importance of sesarminae crabs Chiromanthes spp. and inundation frequency on mangrove (Kandelia candel (L.) Druce) leaf litter turnover in a Hong Kong tidal shrimp pond. J Exp Mar Biol Ecol 131:23–43

Lee SY (1989b) Litter production and turnover of the mangrove Kandelia candel (L.) Druce in a Hong Kong tidal shrimp pond. Estuar Coast Shelf Sci 29:75–87

Lee SY (1990) Primary productivity and particulate organic matter flow in an estuarine mangrove-wetland in Hong Kong. Mar Biol 106:453–463

Lee SY (2004) Relationship between mangrove abundance and tropical prawn production: a re-evaluation. Mar Biol 145(5):943–949

Lee SY (2005) Exchange of organic matter and nutrients between mangroves and estuaries: myths, methodological issues and missing links. Int J Ecol Environ Sci 31(3):163–176

Lee SY (2008) Mangrove macrobenthos: assemblages, services, and linkages. J Sea Res 59(1):16–29

Leopold A, Marchand C, Deborde J, Allenbach M (2015) Temporal variability of CO_2 fluxes at the sediment-air interface in mangroves (New Caledonia). Sci Total Environ 502:617–626

Leopold A, Marchand C, Deborde J, Chaduteau C, Allenbach M (2013) Influence of mangrove zonation on CO_2 fluxes at the sediment–air interface (New Caledonia). Geoderma 202:62–70

Li X, Zhu J, Lange H, Han S (2012) A modified ingrowth core method for measuring fine root production, mortality and decomposition in forests. Tree Physiol 33(1):18–25

Linto N, Barnes J, Ramachandran R, Divia J, Ramachandran P, Upstill-Goddard R (2014) Carbon dioxide and methane emissions from mangrove-associated waters of the Andaman Islands, Bay of Bengal. Estuar Coasts 37(2):381–398

Lopez-Hoffman L, Anten NP, Martinez-Ramos M, Ackerly DD (2007) Salinity and light interactively affect neotropical mangrove seedlings at the leaf and whole plant levels. Oecologia 150(4):545–556

Lovelock C, Feller IC, Ball M, Engelbrecht B, Ewe ML (2006) Difference in plant function in phosphorous- and nitrgoen- limited mangrove ecosystems. New Phytol 172:514–522

Lovelock C, Feller IC, McKee KL, Engelbrecht B, Ball M (2004) The effect of nutrient enrichment on growth, photosynthesis and hydraulic conductance of dwarf mangroves in Panama. Funct Ecol 18(1):25–33

Lovelock CE (2008) Soil respiration and belowground carbon allocation in mangrove forests. Ecosystems 11(2):342–354

Lovelock CE, Feller IC, Ellis J, Schwarz AM, Hancock N, Nichols P, Sorrell B (2007) Mangrove growth in New Zealand estuaries: the role of nutrient enrichment at sites with contrasting rates of sedimentation. Oecologia 153(3):633–641

Ludovici KH, Zarnoch SJ, Richter DD (2002) Modeling in-situ pine root decomposition using data from a 60-year chronosequence. Can J For Res 32(9):1675–1684

Lugo AE (1990) Fringe wetlands. In: Lugo AE, Brinson M, Brown S (eds) Forested wetlands, Ecosystems of the world, vol 15. Elsevier, Amsterdam, pp 143–169

Lugo AE (1997) Old-growth mangrove forests in the United States. Conserv Biol 11(1):11–20. doi:10.2307/2387271

Lugo AE, Brown S, Brinson MM (1988) Forested wetlands in freshwater and salt-water environments. Limnol Oceanogr 33(4):894–909. doi:10.2307/2837228

Lugo AE, Snedaker SC (1974) The ecology of mangroves. Ann Rev Ecol Syst 5:39–64

Lynch JC, Meriwether JR, McKee BA, Vera-Herrera F, Twilley RR (1989) Recent accretion in mangrove ecosystems based on ^{137}Cs and ^{210}Pb. Estuaries 12:284–299

Machiwa JF (1999) Lateral fluxes of organic carbon in a mangrove forest partly contaminated with sewage wastes. Mangrove Salt Marshes 3(2):95–104

Macnae W (1974) Mangrove forests and fisheries. Food and Agriculture Organization of the United Nations/United Nations Development Programme, Rome

Maher DT, Santos IR, Golsby-Smith L, Gleeson J, Eyre BD (2013) Groundwater-derived dissolved inorganic and organic carbon exports from a mangrove tidal creek: the missing mangrove carbon sink? Limnol Oceanogr 58(2):475–488

Maie N, Pisani O, Jaffe R (2008) Mangrove tannins in aquatic ecosystems: their fate and possible influence on dissolved organic carbon and nitrogen cycling. Limnol Oceanogr 53(1):160–171

Majdi H, Kangas P (1997) Demography of fine roots in response to nutrient applications in a Norway spruce stand in southwestern Sweden. Ecoscience 4(2):199–205

McKee KL (2011) Biophysical controls on accretion and elevation change in Caribbean mangrove ecosystems. Estuar Coast Shelf Sci 91(4):475–483

McKee KL, Cahoon DR, Feller IC (2007) Caribbean mangroves adjust to rising sea level through biotic controls on change in soil elevation. Glob Ecol Biogeogr 16(5):545–556. doi:10.1111/j.1466-8238.2007.00317.x

McKee KL, Faulkner PL (2000) Restoration of biogeochemical function in mangrove forests. Restor Ecol 8(3):247–259

Mcleod E, Chmura GL, Bouillon S, Salm R, Björk M, Duarte CM, Lovelock CE, Schlesinger WH, Silliman BR (2011) A blueprint for blue carbon: toward an improved understanding of the role of vegetated coastal habitats in sequestering CO2. Front Ecol Environ 9(10):552–560

Mfilinge PL, Atta N, Tsuchiya M (2002) Nutrient dynamics and leaf litter decomposition in a subtropical mangrove forest at Oura Bay, Okinawa, Japan. Trees 16:172–180

Mfilinge PL, Meziane T, Bachok Z, Tsuchiya M (2005) Litter dynamics and particulate organic matter outwelling from a subtropical mangrove in Okinawa Island, South Japan. Estuar Coast Shelf Sci 63(1–2):301–313. doi:http://dx.doi.org/10.1016/j.ecss.2004.11.022

Middleton BA, McKee KL (2001) Degradation of mangrove tissues and implications for peat formation in Belizean island forests. J Ecol 89:818–828

Moncrieff JB, Jarvis PG, Valentini R (2000) Canopy fluxes. In: Methods in ecosystem science. Springer, New York/London, pp 161–180

Nadelhoffer KJ, Aber JD, Melillo JM (1985) Fine roots, net primary production, and soil nitrogen availability: a new hypothesis. Ecology 66(4):1377–1390

Nadelhoffer KJ, Raich JW (1992) Fine root production estimates and belowground carbon allocation in forest ecosystems. Ecology 73(4):1139–1147

Naidoo G (2009) Differential effects of nitrogen and phosphorus enrichment on growth of dwarf Avicennia marina mangroves. Aquat Bot 90(2):184–190

Nielsen OI, Kristensen E, Macintosh DJ (2003) Impact of fiddler crabs (Uca spp.) on rates and pathways of benthic mineralization in deposited mangrove shrimp pond waste. J Exp Mar Biol Ecol 289(1):59–81

Nixon SW (1980) Between coastal marshes and coastal waters-a review of twenty years of speculation and research on the role of salt marshes in estuarine productivity and water chemistry. In: Hamilton P, MacDonald KB (eds) Estuarine and wetland processes with emphasis on modeling. Plenum, New York, pp 437–525

Odum EP (2002) Tidal marshes as outwelling/pulsing systems. In: Concepts and controversies in tidal marsh ecology. Dordrecht, pp 3–7

Odum WE (1984) Dual-gradient concept of detritus transport and processing in estuaries. Bull Mar Sci 35(3):510–521
Odum WE, Heald EJ (1972) Trophic analyses of an estuarine mangrove community. Bull Mar Sci 22:671–738
Odum WE, Heald EJ (1975) The detritus-based food web of an estuarine mangrove community. In: Cronin LE (ed) Estuarine research. Academic, New York, pp 265–286
Ólafsson E, Buchmayer S, Skov MW (2002) The East African decapod crab Neosarmatium meinerti (de Man) sweeps mangrove floors clean of leaf litter. AMBIO J Hum Environ 31(7):569–573
Olafsson E, Ndaro S (1998) Impact of the mangrove crabs Uca annulipes and Dotilla fenestrata on meiobenthos. Oceanogr Lit Rev 6(45):1004–1005
Ong JE, Khoon GW, Hoong WC, Dhanarajan G (1979) Productivity of a managed mangrove forest in west Malaysia. In: Noor, YM (ed) Trends in applied biology in southeast Asia. University Sains Press, Penang, pp 1–10
Osland MJ, Enwright N, Day RH, Doyle TW (2013) Winter climate change and coastal wetland foundation species: salt marshes vs. mangrove forests in the southeastern United States. Glob Chang Biol 19(5):1482–1494
Parkinson RW, DeLaune RD, White JR (1994) Holocene sea-level rise and the fate of mangrove forests within the wider caribbean region. J Coast Res 10(4):1077–1086
Peckarsky BL (1980) Influence of detritus upon colonization of stream invertebrates. Can J Fish Aquat Sci 37(6):957–963
Penha-Lopes G, Kristensen E, Flindt M, Mangion P, Bouillon S, Paula J (2010) The role of biogenic structures on the biogeochemical functioning of mangrove constructed wetlands sediments–a mesocosm approach. Mar Pollut Bull 60(4):560–572
Pineda JEM (2003) The contribution of mangrove outwelling to coastal food webs as a function of environmental settings. University of Louisiana at Lafayette, Lafayette
Pool DJ, Lugo AE, Snedaker SC (1975) Litter production in mangrove forests of southern Florida and Puerto Rico. In: Walsh GE, Snedaker SC, Teas HJ (eds) Proceedings of the international symposium on biology and management of mangroves. East-West Center, Honolulu, pp 213–237
Poret N, Twilley RR, Rivera-Monroy VH, Coronado-Molina C (2007) Belowground decomposition of mangrove roots in Florida Coastal Everglades. Estuar Coasts 30(3):491–496
Primavera JH (1996) Stable carbon and nitrogen isotope ratios of Penaeid juveniles and primary producers in a reverine mangrove in Guimaras, Philippines. Bull Mar Sci 58:675–683
Purvaja R, Ramesh R (2001) Natural and anthropogenic methane emission from coastal wetlands of South India. Environ Manag 27(4):547–557
Putz FE, Chan HT (1986) Tree growth, dynamics, and productivity in a mature mangrove forest in Malaysia. For Ecol Manag 17:211–230
Reef R, Feller IC, Lovelock CE (2010) Nutrition of mangroves. Tree Physiol 30(9):1148–1160
Ren H, Jian S, Lu H, Zhang Q, Shen W, Han W, Yin Z, Guo Q (2008) Restoration of mangrove plantations and colonisation by native species in Leizhou bay, South China. Ecol Res 23(2):401–407. doi:10.1007/s11284-007-0393-9
Rivera-Monroy, VH, Castañeda-Moya E, Barr JG, Engel V, Fuentes JD, Troxler TG, Twilley RR, Bouillon S, Smith TJ, O'Halloran TL (2013). Current Methods to Evaluate Net Primary Production and Carbon Budgets in Mangrove Forests. In: DeLaune RD, Reddy KR, Richardson CJ, Megonigal JP (eds) Methods in Biogeochemistry of Wetlands, SSSA Book Ser. 10. SSSA, Madison, WI, pp 243–288. doi:10.2136/sssabookser10.c14
Robertson A, Alongi D (1995) Role of riverine mangrove forests in organic carbon export to the tropical coastal ocean: a preliminary mass balance for the Fly Delta (Papua New Guinea). Geo-Mar Lett 15(3–4):134–139
Robertson A, Alongi D, Boto K (1993) Food chains and carbon fluxes. In: Alongi DM, Robertson AI (eds) Tropical mangrove ecosystems. American Geophysical Union, Washington, D. C, pp 293–326. doi:10.1029/CE041p0293
Robertson AI (1986) Leaf-burying crabs: their influence on energy flow and export from mixed mangrove forests (Rhizophora spp) in northeastern Australia. J Exp Mar Biol Ecol 102:237–248

Rodelli MR, Gearing JN, Gearing PJ, Marshall N, Sasekumar A (1984) Stable isotope ratio as a tracer of mangrove carbon in Malaysian ecosystems. Oecologia 61:326–333

Romigh MM, Davis SE, Rivera-Monroy VH, Twilley RR (2006) Flux of organic carbon in a riverine mangrove wetland in the Florida Coastal Everglades. Hydrobiologia 569:505–516. doi:10.1007/S10750-006-0152-X

Ross MS, Meeder JF, Sah JP, Ruiz PL, Telesnicki GJ (2000) The southeast saline everglades revisited: 50 years of coastal vegetation change. J Veg Sci 11:101–112

Ross MS, Ruiz PL, Telesnicki GJ, Meeder JF (2001) Estimating above-ground biomass and production in mangrove communities of Biscayne National Park, Florida (USA). Wetl Ecol Manag 9(1):27–37

Rovai A, Riul P, Twilley R, Castañeda-Moya E, Rivera-Monroy V, Williams A, Simard M, Cifuentes-Jara M, Lewis R, Crooks S (2016) Scaling mangrove aboveground biomass from site-level to continental-scale. Glob Ecol Biogeogr 25(3):286–298

Saenger P, Snedaker SC (1993) Pantropical trends in mangroves above-ground biomass and annual litterfall. Oecologia 96:293–299

Saintilan N (2004) Relationships between estuarine geomorphology, wetland extent and fish landings in New South Wales estuaries. Estuar Coast Shelf Sci 61(4):591–601

Saintilan N, Rogers K, Mckee K (2009) Salt marsh-mangrove interactions in Australasia and the Americas. In: G.M Perillo E Wolanski, D. Cahoon & M. Brinson (ed) Coastal wetlands: an integrated ecosystem approach. Elsevier, Amsterdam, pp 855–884

Saintilan N, Wilson NC, Rogers K, Rajkaran A, Krauss KW (2014) Mangrove expansion and salt marsh decline at mangrove poleward limits. Glob Chang Biol 20(1):147–157. doi:10.1111/gcb.12341

Santantonio D, Hermann R (1985) Standing crop, production, and turnover of fine roots on dry, moderate, and wet sites of mature Douglas-fir in western Oregon. Ann Sci For 2:113–142

Sasekumar A, Chong V, Leh M, D'cruz R (1992) Mangroves as a habitat for fish and prawns. Hydrobiologia 247(1–3):195–207

Schaeffer-Novelli Y, Cintrón-Molero G, Adaime RR, de Camargo TM (1990) Variability of mangrove ecosystems along the Brazilian coast. Estuar Coasts 13(2):204–218. doi:10.2307/1351590

Sell MG (1977) Modeling the response of mangrove ecosystems to herbicide spraying, hurricanes, nutrient enrichment and economic development. Ph.D. Thesis, University of Florida, Gainesville, Florida

Sessegolo GC, Lana PC (1991) Decomposition of Rhizophora mangle, Avicennia schaueriana and Laguncularia racemosa leaves in a mangrove of Paranagua Bay (southeastern Brazil). Bot Mar 34:285–289

Sherman RE, Fahey TJ, Martinez P (2003) Spatial patterns of biomass and aboveground net primary productivity in a mangrove ecosystem in the Dominican Republic. Ecosystems 6(4):384–398

Siikamäki J, Sanchirico JN, Jardine SL (2012) Global economic potential for reducing carbon dioxide emissions from mangrove loss. Proc Natl Acad Sci 109(36):14369–14374

Simard M, Rivera-Monroy VH, Mancera-Pineda JE, Castaneda-Moya E, Twilley RR (2008) A systematic method for 3D mapping of mangrove forests based on Shuttle Radar Topography Mission elevation data, ICEsat/GLAS waveforms and field data: application to Cienaga Grande de Santa Marta, Colombia. Remote Sens Environ 112(5):2131–2144. doi:10.1016/J.Rse.2007.10.012

Simard M, Zhang KQ, Rivera-Monroy VH, Ross MS, Ruiz PL, Castaneda-Moya E, Twilley RR, Rodriguez E (2006) Mapping height and biomass of mangrove forests in Everglades National Park with SRTM elevation data. Photogramm Eng Remote Sens 72(3):299–311

Smith TJ, Anderson GH, Balentine K, Tiling G, Ward GA, Whelan KR (2009) Cumulative impacts of hurricanes on Florida mangrove ecosystems: sediment deposition, storm surges and vegetation. Wetlands 29(1):24–34

Snedaker SC, Baquer SJ, Behr PJ, Ahmed SI (1995) Biomass distribution in *Avicennia marina* plants in the Indus River Delta, Pakistan. In: Thompson MF, Tirmizi NM (eds) The Arabian sea. Vanguard, Lahore, p 724

Sousa W, Dangremond E (2011) Trophic interactions in coastal and estuarine mangrove forest ecosystems. Treatise Estuar Coast Sci 6:43–93

Stieglitz TC, Clark JF, Hancock GJ (2013) The mangrove pump: the tidal flushing of animal burrows in a tropical mangrove forest determined from radionuclide budgets. Geochim Cosmochim Acta 102:12–22

Sukardjo S, Yamada I (1992) Biomass and productivity of a Rhizophora mucronata Lamarck plantation in Tritih, Central Java, Indonesia. For Ecol Manag 49(3–4):195–209

Sutula MA, Perez BC, Reyes E, Childers DL, Davis S, Day JW, Rudnick D, Sklar F (2003) Factors affecting spatial and temporal variability in material exchange between the Southern Everglades wetlands and Florida Bay (USA). Estuar Coast Shelf Sci 57(5):757–781

Tamooh F, Huxham M, Karachi M, Mencuccini M, Kairo J, Kirui B (2008) Below-ground root yield and distribution in natural and replanted mangrove forests at Gazi bay, Kenya. For Ecol Manag 256(6):1290–1297

Thom BG (1982) Mangrove ecology-a geomorphological perspective. In: Clough BF (ed) Mangrove ecosystems in Australia. Australian National University Press, Canberra, pp 3–17

Troxler TG, Barr JG, Fuentes JD, Engel V, Anderson G, Sanchez C, Lagomasino D, Price R, Davis SE (2015) Component-specific dynamics of riverine mangrove CO2 efflux in the Florida coastal Everglades. Agric For Meteorol 213:273–282

Twilley RR (1985) The exchange of organic-carbon in basin mangrove forests in a southwest Florida estuary. Estuar Coast Shelf Sci 20(5):543–557

Twilley RR (1988) Coupling of mangroves to the productivity of estuarine and coastal waters. In: Janson BO (ed) Coastal-offshore ecosystem interactions. Springer, Berlin, pp 155–180

Twilley RR (1995) Properties of mangrove ecosystems related to the energy signature of coastal environments. In: Hall CAS (ed) Maximum power: the ideas and applications of H. T. Odum. University Press of Colorado, Niwot, pp 43–62

Twilley RR, Chen RH, Hargis T (1992) Carbon sinks in mangroves and theirs implications to carbon budget of tropical coastal ecosystems. Water Air Soil Pollut 64:265–288

Twilley RR, Lugo AE, Patterson-Zucca C (1986) Litter production and turnover in basin mangrove forests in southwest Florida. Ecology 67(3):670–683

Twilley RR, Pozo M, Garcia VH, Rivera-Monroy VH, Zambrano R, Bodero A (1997) Litter dynamics in riverine mangrove forests in the Guayas River estuary, Ecuador. Oecologia 111:109–122

Twilley RR, Rivera-Monroy VH (2005) Developing performance measures of mangrove wetlands using simulation models of hydrology, nutrient biogeochemistry, and community dynamics. J Coast Res 40:79–93

Twilley RR, Rivera-Monroy VH (2009) Ecogeomorphic models of nutrient biogeochemistry for mangrove wetlands. Coastal wetlands: an integrated ecosystem approach. Elsevier, Amsterdam, pp 641–683

Valiela I, Bowen JL, York JK (2001) Mangrove forests: one of the world's threatened major tropical environments. Bioscience 51:807–815

Vogt KA, Vogt DJ, Bloomfield J (1998) Analysis of some direct and indirect methods for estimating root biomass and production of forests at an ecosystem level. Plant Soil 200:71–89

Werry J, Lee S (2005) Grapsid crabs mediate link between mangrove litter production and estuarine planktonic food chains. Mar Ecol Prog Ser 293:165–176

Whelan KRT, Smith TJ, Cahoon DR, Lynch JC, Anderson GH (2005) Groundwater control of mangrove surface elevation: shrink and swell varies with soil depth. Estuaries Coasts 28(6):833–843. doi:10.1007/bf02696013

Woodroffe C (1992) Mangrove sediments and geomorphology. In: Robertson AI, Alongi DM (eds) Coastal and estuarine studies, Chapter 2. American Geophysical Union, Washington, D. C, pp 7–41

Woodroffe CD (1985) Studies of a mangrove basin, Tuff Crater, New Zealand: II. Comparison of volumetric and velocity-area methods of estimating tidal flux. Estuar Coast Shelf Sci 20:431–445

Woodroffe CD (2002) Coasts: form, process and evolution. Cambridge University Press, Cambridge

Woodroffe CD, Bardsley KN, Ward PJ, Hanley JR (1988) Production of mangrove litter in a macrotidal embayment, Darwin Harbour, N.T., Australia. Estuar Coast Shelf Sci 26:581–598
Woodroffe CD, Rogers K, McKee KL, Lovelock CE, Mendelssohn I, Saintilan N (2016) Mangrove sedimentation and response to relative sea-level rise. Annu Rev Mar Sci 8:243–266
Zablocki JA, Andersson AJ, Bates NR (2011) Diel aquatic CO2 system dynamics of a Bermudian mangrove environment. Aquat Geochem 17(6):841–859
Zhang KQ, Simard M, Ross M, Rivera-Monroy VH, Houle P, Ruiz P, Twilley RR, Whelan KRT (2008) Airborne laser scanning quantification of disturbances from hurricanes and lightning strikes to mangrove forests in Everglades National Park, USA. Sensors 8(4):2262–2292

Chapter 6
Biogeochemical Cycles: Global Approaches and Perspectives

Erik Kristensen, Rod M. Connolly, Xose L. Otero, Cyril Marchand, Tiago O. Ferreira, and Victor H. Rivera-Monroy

6.1 Introduction

Mangrove wetlands are intriguing ecosystems because they share biological, geochemical, and ecological properties from both terrestrial and marine environments (Alongi 2009; Mitsch and Gosselink 2015). The mangrove ecosystem is

E. Kristensen (✉)
Department of Biology, University of Southern Denmark,
Campusvej 55, DK-5230 Odense, Denmark
e-mail: ebk@biology.sdu.dk

R.M. Connolly
School of Environment & Australian Rivers Institute - Coast & Estuaries, Griffith University,
Gold Coast Campus, Gold Coast, QLD 4222, Australia
e-mail: r.connolly@griffith.edu.au

X.L. Otero
Dept. de Edafoloxía e Química Agrícola, Facultade de Bioloxía, Universidade de Santiago de Compostela, 15782 Santiago de Compostela, Spain

Departamento de Tecnología de Ciencia de los Alimentos y Biotecnología, Universidad Politécnica Nacional, Quito, Ecuador
e-mail: xl.otero@usc.es

C. Marchand
Institut de Recherche pour le Développement (IRD), UMR 7590, UR 206,
BP A5, 98848 Nouméa, New Caledonia, France
e-mail: cyril.marchand@ird.fr

T.O. Ferreira
Dept. de Ciência do Solo, ESALQ/USP, São Paulo, Brazil
e-mail: toferreira@usp.br

V.H. Rivera-Monroy
Department of Oceanography and Coastal Sciences, College of the Coast and Environment, Louisiana State University, Baton Rouge, LA 70803, USA
e-mail: vhrivera@lsu.edu

V.H. Rivera-Monroy et al. (eds.), *Mangrove Ecosystems: A Global Biogeographic Perspective*, https://doi.org/10.1007/978-3-319-62206-4_6

characterized by dense growth of highly productive trees and shrubs (Tomlinson 1994; see Chap. 1). They provide most of the primary production, although other autotrophs, including pelagic, benthic, and epiphytic algae may also contribute significantly to the organic input (Kristensen et al. 2008a). The export of autochthonous production and import of allochthonous material are strongly dependent on complex spatio-temporal hydrological patterns regulated by large-scale physical and geomorphological processes. From a biogeochemical perspective, the input of organic and inorganic matter from various sources and their mixing within the mangrove ecotone create highly spatially and temporally heterogeneous sediments with microbial processes that are challenging to understand and evaluate. The available biogeochemical information is yet to be integrated into a generalized ecosystem model due to the wide variety of geomorphic settings and mangrove ecotypes within and among biogeographical regions (Twilley and Rivera-Monroy 2005).

Although the ecological functioning of mangrove environments has been described for a variety of climatic regions (e.g., tropical, subtropical) and ecogeomorphic settings (e.g. deltas, lagoons, estuaries, oceanic islands) (e.g. Bouillon et al. 2007; Adame and Lovelock 2011; Alongi et al. 2012), our understanding of how carbon (C), iron (Fe), sulfur (S), and nutrient (e.g. N and P) cycling are controlled and interact in these locally diverse environments is still developing. The research has come predominantly from Australasia and North America, with less from locations in Oceania, Asia, Africa, and Central and South America. Information from the understudied regions is, therefore, required to achieve a full global overview. Additional knowledge on the spatiotemporal patterns of biogeochemical mechanisms and processes will improve the reliability of mangrove C and nutrient budgets as well as estimates of the impact of human activities on global cycles (Bouillon et al. 2008; Pendleton et al. 2012). Organic matter decomposition in mangrove sediments is mediated by microbial processes utilizing a variety of electron acceptors under a wide range of redox conditions (Kristensen and Alongi 2006; Ferreira et al. 2007a; Kristensen et al. 2011). The fraction of mangrove detritus that escapes degradation and export is a significant source of in situ C sequestration via accretion and storage (Bouillon 2011; Donato et al. 2011; Pendleton et al. 2012). The accumulation and residence time of C stocks, however, depends strongly on the interaction among local environmental and biological variables such as hydrology, plant activity, crab foraging, and bioturbation (Lee 1997; Kristensen 2008; Mitsch and Gosselink 2015). It is, therefore, imperative not only to evaluate mangrove biogeochemical patterns among biogeographical regions but also to examine the spatio-temporal variability within each region with focus on anthropogenic impacts.

The main objective of this chapter is to advance our understanding of the biogeochemistry of mangrove wetlands by comparing differences in element cycling at biogeographical scales. Through a comparative literature review, we identify potential sources of variation when applying different methods and techniques and provide an understanding of the small- and large-scale variability as well as complexity of biogeochemical transformations in these productive wetlands. Our goal is, therefore, to identify knowledge gaps, and thus research priorities, in biogeochemical cycling of C and other critical macro (N, P) and micro (e.g., Fe, Mn) elements in mangrove environments across biogeographic regions and latitudes.

6.2 Characteristics of Mangrove Substrata

6.2.1 Terms and Definitions; Sediment or Soil?

The operational differentiation between sediments and soils in coastal and wetland-dominated environments has been an ongoing discussion since the insertion of the "subaqueous soil" term in the Soil Taxonomy classification (Soil Survey Staff 1999; Ferreira et al. 2007a; Kristensen and Rabenhorst 2015). This discussion has its genesis in the different methodological and conceptual approaches historically followed by soils scientists (pedologists) and marine scientists and reflects on the current understanding of biogeochemical processes in mangrove wetlands. The development of the subaqueous soil concept was based on the original work by Demas and colleagues (Demas et al. 1996; Demas and Rabenhorst 1999) who defined sediment layers as a function of pedogenesis and proposed the presence of subaqueous soil horizons in wetlands. Along the same line, Ferreira et al. (2007a) argued that sedimentary material (the parental material) that is permanently colonized by higher vascular plants interacting with fauna and microbial activity leads to substantial changes in the composition and properties of the original substratum. These changes transform the original sediment to a more complex, geochemically contrasting environment, which should lead to soil formation. However, the development of oxic or suboxic mosaics within anoxic layers due to plant–substratum interactions is not only associated with soil formation, but is also common in subtidal sediments affected by bioturbation and roots of submerged vegetation (Kristensen and Rabenhorst 2015).

Demas and Rabenhorst (1999) argued that pedogenic processes leading to horizon differentiation are required to consider estuarine substrata as soils. The formation of soil horizons through pedogenesis includes four generalized processes: additions, losses, transfers (or translocations), and transformations (Simonson 1959). Conversely, the biogeochemical zonation of aquatic sediments is controlled by sediment diagenesis, which includes the composition of deposited material and the involved physical, chemical, and biological processes (Burdige 2006; Aller 2014). The generalized pedogenic processes are, therefore, a subset of the sediment diagenetic processes. Under this perspective, Kristensen and Rabenhorst (2015) pointed out that pedogenic processes identified by pedologists in shallow water environments cannot be distinguished from the diagenetic processes described for sediments by marine scientists and concluded that the terms "sediment" and "soil" to describe the substrata in coastal environments vegetated by mangrove forests are not mutually exclusive.

Thus, under these perspectives, which term should be used? In this chapter, we compromise and use the term sediment as a more general historic term. It is not our objective to prefer one term over the other, but to underscore the context of the arguments currently still under discussion (see Ferreira et al. 2007a; Kristensen and Rabenhorst 2015). Our own research and interdisciplinary approach in writing this chapter has motivated us into an inclusive approach to advance the understanding of mangrove wetland spatiotemporal biogeochemical processes and functions. Particularly, in trying to fulfill one of the objectives of the chapter where "both scientific communities (i.e., pedologists, and marine scientists) will benefit from comparable studies in the same environments, (recognizing that) their different backgrounds may even promote collaborations with the developments of new ideas and revolutionary concepts" (Kristensen and Rabenhorst 2015).

6.2.2 *Litter Fall and Sediment Organic Matter*

Organic matter delivered to microbial decomposers in mangrove sediments is of both autochthonous and allochthonous origin. Litter from mangrove trees (i.e., leaves, propagules, twigs, and wood) is usually considered the most important source of organic C and nutrients to mangrove sediments (Alongi et al. 2005a; Sousa and Dangremond 2011; Murdiyarso et al. 2015). A range of other sources may also provide significant inputs; including below-ground mangrove roots, as well as local production by benthic or epiphytic micro- or macroalgae, phytoplankton in tidal creeks or estuarine waters, and materials imported via rivers (e.g. terrestrial tree litter) or tides (e.g., seagrass)(Kristensen et al. 2008a; Alongi 2009; Twilley and Rivera-Monroy 2009; Adame and Lovelock 2011; Leopold et al. 2015). Because of its important functional role, the quantification of litter fall is crucial for assessing productivity of a mangrove ecosystem and thus forest organic matter contribution to benthic food webs in both the mangrove and its adjacent coastal environment (Imgraben and Dittmann 2008).

Annual litter fall, which is the most widely used proxy of mangrove net productivity, differs substantially within and among forests due to a number of factors including tidal and hydrological gradients (Feller et al. 1999), salinity (Day et al. 1996), anthropogenic influence (Silva et al. 1998), mangrove species composition (Coupland et al. 2005), and latitude (Twilley et al. 1992; Saenger and Snedaker 1993). The global average mangrove litter fall, which is in the order of ~460 g C m^{-2} $year^{-1}$ (range: 48–924 g C m^{-2} $year^{-1}$), does not show a clear biogeographical trend (Twilley et al. 1992; Saenger and Snedaker 1993; Jennerjahn and Ittekkot 2002) due in part to major differences in ecotype dominance and spatial distribution within latitude (Twilley et al. 1998). There are apparently some distinct differences in productivity and litter fall among mangrove species, for example, *Rhizophora* spp. shows about 50% higher litter yield than *Avicennia* spp. (Bunt 1995). It must be stressed, however, that most available estimates of mangrove production do not include wood and below-ground components (Middleton and McKee

2001; Castaneda-Moya et al. 2013). Current below-ground biomass estimates indicate that this component contributes a substantial part (10–55%) of the total mangrove biomass (Twilley et al. 1992; Matsui 1998; Alongi and Dixon 2000; Rivera-Monroy et al. 2013) and probably account for a similar part of the total productivity. The ongoing effort to establish a latitudinal-based network of eddy-covariance towers in mangrove coastal regions in the near future might improve net mangrove ecosystem productivity estimates to include both above- and below-ground production (Rivera-Monroy et al. 2013; see Sect. 6.4).

Regardless of differences in organic matter sources and pathways through food webs among mangrove forests, all organic matter not exported by tidal exchange enters the sediment where it is consumed, degraded, or buried. Mangrove sediments are relatively rich in organic C with an estimated global particulate organic carbon (POC) median value of 2.6% (Kristensen et al. 2008a), ranging from 2.2% in the Indo-west-Pacific (IWP: East Africa, Asia, and Australasia) to 3.1% in the Atlantic-East Pacific region (AEP: West Africa and America) (Fig. 6.1), which is higher than generally observed in marine sediments (Seiter et al. 2004) and terrestrial soils (Donato et al. 2011). Current available information indicates that the fraction of organically enriched mangrove sediments having POC content >10% is higher in America (23%) and Asia (19%) than in East Africa (8%) and Australasia (6%) (Fig. 6.1). These differences in POC among regions are difficult to explain and may be confounded by variations within and among mangrove forests depending on hydrological regimes (i.e., hydroperiod) and mangrove species composition as well as other structural and environmental variables (Alongi 2012). The median sediment molar POC/PN ratios also vary among regions, ranging from 18 to 19 in East Africa and Asia to 24 to 25 in America and Australasia (Fig. 6.1). Most mangrove sediments have POC/PN ratios above 10 (100% in America; 96% in East Africa; 98% in Asia; and 92% in Australasia), whereas POC/PN ratios above 30 are more frequent in America (24%) and Australasia (34%) than in East Africa (9%) and Asia (5%). Although the generally high POC/PN ratios indicate that mangrove sediments contain a significant input of mangrove litter, the large differences among regions may indicate higher inputs of N-rich marine organic matter coupled with global differences in rivers and river flows and more extensive eutrophication in East Africa and Asia than in America and Australasia (Lee 2016).

In addition to the substantial deposition of litter from mangrove canopies, vegetation structure has a profound impact on the magnitude of sedimentation by actively capturing mineral and organic particles (Furukawa et al. 1997). Large trees with complex aerial root systems (e.g. tree height >10 m), such as *Rhizophora* species, facilitate the retention and deposition of particles from tidal currents to a much greater extent than smaller trees with simpler architecture, such as *Avicennia* species. Accordingly, sediments under *Rhizophora* stands are often richer in POC than under *Avicennia* stands (Table 6.1). However, this difference is not always evident and may in some cases be reversed due to location-specific and climatic-driven differences in litter fall rates, litter composition, and hydrological patterns. This is evident from Table 6.1 where two *Rhizophora mangle* locations in the Americas support the lowest sediment POC. Variations in mangrove zonation pattern may

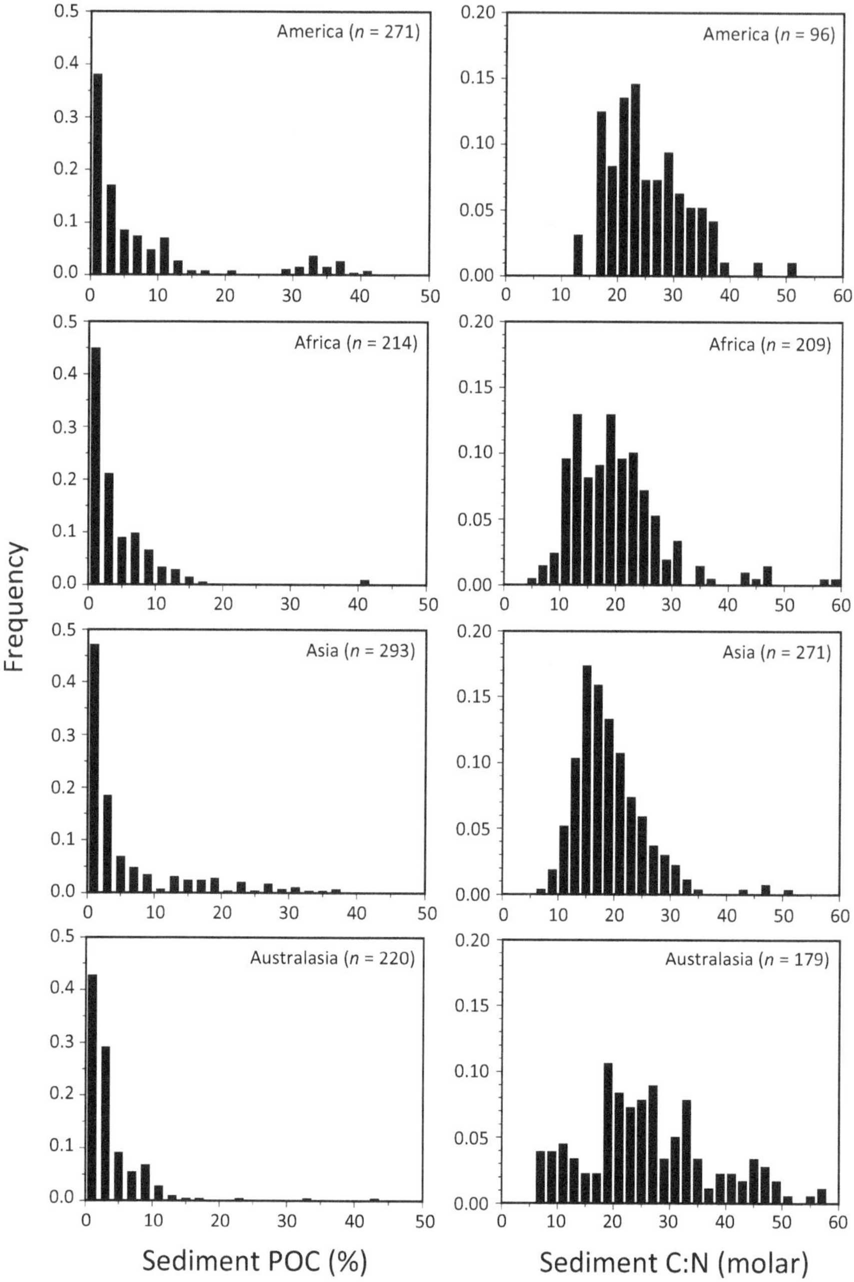

Fig. 6.1 Compilation of (*left*) bulk POC and (*right*) bulk POC:TN ratios of intertidal mangrove sediments from four biogeographical subregions. Data compiled from various sources (Modified from Kristensen et al. 2008a)

Table 6.1 Sediment content of POC and TN in mangrove forests from around the world dominated by *Rhizophora* spp. and *Avicennia* spp. Only data from sediments underlying forests of about the same age (15–30 years) and same intertidal location (mid-intertidal) are included

Location	Tree species	POC (%)	TN (%)	Ref
Pambala, Sri Lanka	*Rhizophora apiculata*	19.1 ± 2.4	1.04 ± 0.13	1
	Avicennia officinalis	10.4 ± 1.7	0.64 ± 0.05	1
Pichavaram, India	*Rhizophora apiculata*	10.0 ± 1.9	0.70 ± 0.01	2
	Avicennia marina	7.0 ± 1.0	0.42 ± 0.01	2
Dampier, W. Australia	*Rhizophora stylosa*	6.5 ± 1.3	0.23 ± 0.03	3
	Avicennia marina	1.4 ± 0.1	0.09 ± 0.01	3
Port Hedland, W. Australia	*Rhizophora stylosa*	2.3 ± 1.1	0.13 ± 0.02	3
	Avicennia marina	1.7 ± 0.2	0.16 ± 0.02	3
Ras Dege, Tanzania	*Rhizophora mucronata*	4.3 ± 1.1	0.18 ± 0.07	4
	Avicennia marina	2.9 ± 0.3	0.11 ± 0.02	4
Gazi Bay, Kenya	*Rhizophora mucronata*	4.4 ± 1.4	0.23 ± 0.06	5
	Avicennia marina	2.2 ± 1.8	0.11 ± 0.05	5
Somone, Senegal	*Rhizophora* sp.	1.7 – 2.1	–	6
	Avicennia sp.	0.3 – 0.6	–	6
Balandra Bay, Mexico	*Rhizophora mangle*	3.7 ± 1.8	0.16 ± 0.08	7
	Avicennia germinans	7.9 ± 4.0	0.32 ± 0.17	7
Itacuruca, Brazil	*Rhizophora mangle*	2.7 ± 0.1	0.17 ± 0.01	8
	Avicennia schaueriana	4.6 ± 1.3	0.26 ± 0.08	8

(1) Bouillon et al. (2003); (2) Alongi et al. (2005b); (3) Alongi et al. (2000a); (4) Kristensen et al. (2011); (5) Andreetta et al. (2014); (6) Sakho et al. (2015); (7) Giani et al. (1996); (8) Lacerda et al. (1995)

partly explain this biogeographic difference in both retention and deposition of POC (Twilley et al. 1992; Chmura et al. 2003). Intertidal position may also affect POC deposition as frequently exposed upper intertidal mangrove areas are often dominated by simple scrub vegetation with limited POC accumulation capacity due to evaporation-driven high porewater salinity (>50) (Adame et al. 2010; Deborde et al. 2015). Conversely, mangrove margins and adjacent intertidal mudflats where tall and dense fringing mangrove stands dominate are often sites of higher POC accumulation (Sanders et al. 2010). However, changes in hydrodynamics, especially turbulent kinetic energy from waves and currents at the edge of the forest, may alter sediment/organic matter deposition and its interaction with different species of mangroves and density of roots (Wolanski et al. 1990; Zhang et al. 2015).

The complex composition of mangrove litter with high content of structural organic polymers and polyphenolic compounds (e.g., cellulose, lignin, and tannins) hampers degradation and promotes long-term preservation of organic C once these organic substrates enter anoxic conditions in waterlogged sediments (Hernes et al. 2001; Marchand et al. 2005; Alongi 2009). Although detrital POC from litter fall is a mixture of more or less refractory biomolecules in various stages of decomposition, it also contains labile components (mainly amino acids, proteins, and sugars). Root exudates are particularly rich in these reactive components and may represent an important source of labile POC in sediments densely vegetated by mangrove trees (Reddy and DeLaune 2008; Weng et al. 2013). Such subsurface sources of reactive POC may be the principal drivers of fast microbial processes deep in mangrove sediment in contrast to oceanic sediments, where partly degraded POC from the water column is deposited at the surface, and slowly buried through active sedimentation and accretion. The reactivity, rather than the quantity of POC, is, therefore, a key factor driving anaerobic respiration in mangrove sediments (Opsahl and Benner 1999; Tremblay and Benner 2006).

6.2.3 Sediment Geochemical Characteristics

Redox processes involving reactive Fe are important for sediment biogeochemistry in most mangrove areas. The actual role of Fe in any mangrove ecotype depends on the availability and delivery of reactive forms and the concentration is typically high in tropical mangrove regions receiving surface runoff and groundwater from adjacent land and watersheds with Fe-rich soils (Souza-Júnior et al. 2007; Sanders et al. 2012; Gonneea et al. 2014; Noel et al. 2014). Much of the Fe delivered to sediments occur as solid phase Fe(III) oxyhydroxides (Table 6.2). These forms are generally very reactive and can be reduced considerably faster than solid phase crystalline Fe(III), such as silicate-bound Fe. The reactivity of Fe(III) forms to undergo reduction by microorganisms typically follows the sequence: iron phosphate tetrahydrate ($FePO_4{\bullet}4H_2O$) > ferrihydrite ($Fe_{10}O_{14}(OH)_2$) > iron hydroxide ($Fe(OH)_3$) > lepidrocrocite (γ-FeO(OH)) > goethite (α-FeO(OH)) (Fischer and Pfanneberg 1984; Roden and Zachara 1996). Once amorphous Fe(III) forms more reactive than goethite has been reduced, sulfate reduction becomes energetically favorable (Canfield et al. 2005). This shift in microbial reactions may be the reason for the presence of goethite in mangrove sediments dominated by sulfate reduction (Otero et al. 2009). Dissolved Fe^{2+} generated by reduction of solid phase Fe(III) oxyhydroxides may either diffuse to oxic layers where it is reoxidized and precipitated as Fe(III) forms,

Table 6.2 Chemical species and characteristics of iron oxides, with information on crystal structure, color, weight-specific area of reactive surfaces, and reactivity (Adapted from Cornel and Schwertmann 1996; Canfield et al. 1992; Poulton et al. 2004)

Mineral name	Crystal structure	Color	Surface area ($m^2\ g^{-1}$)	Reactivity (yr^{-1})
Ferrihydrite ($Fe_{10}O_{14}(OH)_2$)	Trigonal	Reddish-brown	200–400	500–2200
Lepidocrocite (γ-FeO(OH))	Orthorhombic	Orange	15–260	85–557
Goethite (α-FeO(OH))	Orthorhombic	Yellowish-brown	30–90	4–22
Haematite (α-Fe_2O_3)	Trigonal	Bright red	10–36	1–12
Magnetite (Fe_3O_4)	Cubic	Black	20–60	10^{-2}–4

or precipitate in anoxic sediment as carbonate (siderite, $FeCO_3$), phosphate (vivianite, $FePO_4$), or sulfide (mackinawite, FeS and pyrite, FeS_2), depending on the sediment geochemical conditions (Fig. 6.2). As a result, Fe speciation changes dramatically with depth in mangrove sediments from dominance of Fe(III) oxyhydroxides near oxic surfaces to primarily pyrite (FeS_2) in reduced layers. Close spatial coupling of iron and sulfate reduction favors rapid precipitation of Fe(II) sulfides as mackinawite (FeS) that may act as a transient phase in pyrite formation (Holmer et al. 1994; Butler and Rickard 2000; Ferreira et al. 2007b). Analyses using scanning electron microscopy and dispersive X-ray spectroscopy have revealed that pyrite framboids are commonly formed along mangrove roots (Noel et al. 2014). However, newly formed pyrite near sediment interfaces can be rapidly reoxidized to amorphous Fe(III) oxyhydroxides by oxygen intrusion through the action of tides, bioturbation, plant roots, and seasonal changes in hydrology (Noel et al. 2014).

The spatial heterogeneity of redox processes is much more complex and variable in intertidal mangrove sediments with high abundance of roots and burrows than in oceanic sediments (Fig. 6.3; Otero et al. 2006; Ferreira et al. 2010). The upper 60–100 cm of the sediment is normally characterized by a substantial redox variation, in some cases without any clear vertical trend. However, redox models consistently show an oxidized surface zone of variable thickness (upper oxidation zone) overlying a zone with more reducing conditions (upper reduced zone; Clark et al.

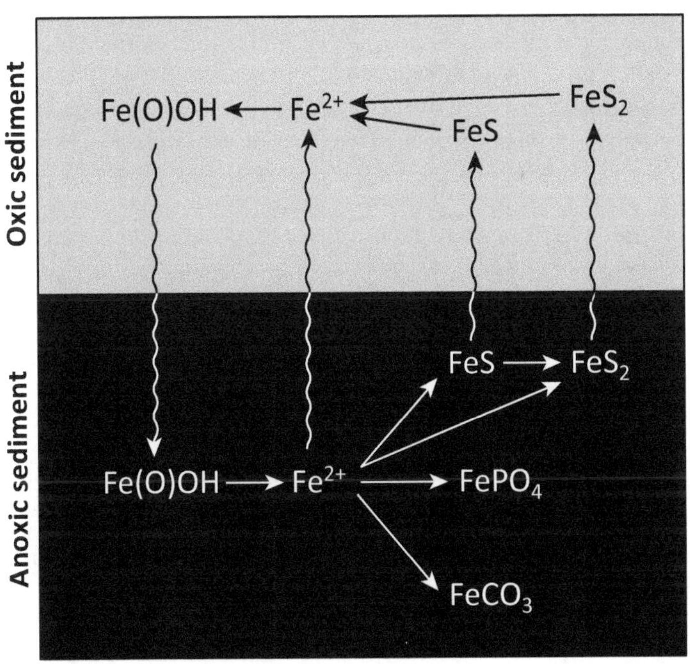

Fig. 6.2 Oxidation and reduction processes involving Fe in mangrove sediment. The *zig-zag arrows* indicate solid phase (Fe(O)OH, FeS, and FeS_2) and solute (Fe^{2+}) transport between oxic and anoxic sediment. *Straight arrows* indicate a chemical reaction or process

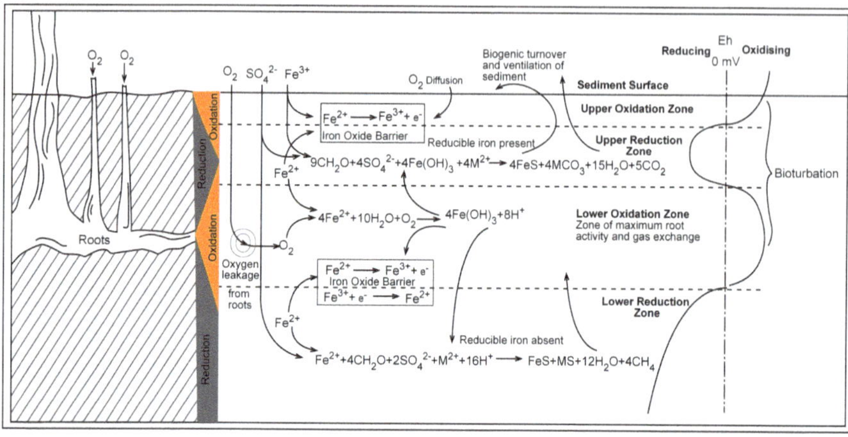

Fig. 6.3 Vertical redox zones in mangrove sediment. The drawing to the left indicates the position of *Avicennia* spp. roots. All redox relevant microbial reactions and transport processes are indicated

1998). The thickness of the upper oxidation zone in mangrove forests typically ranges from <2 cm at the seaward edge to >10 cm near the landward edge (Clark et al. 1998). The upper oxidation zone may vary seasonally and can disappear when intense rain causes prolonged flooding events, or expand after long drought periods. These changes are particularly evident in the high intertidal part of the forest, where tidal inundation normally is infrequent (Marchand et al. 2004, 2006). Below the upper reduction zone, a second oxidizing layer is found (lower oxidation zone) at a depth that can vary between 30 and 60 cm (Clark et al. 1998). The oxidizing effect in this layer is the result of downward translocation of oxygen by bioturbation and via aerenchyma tissue in roots followed by release into the sediment (Fig. 6.3). The lower oxidation zone is actually a mosaic of alternating oxidizing areas near burrows and roots and reduced areas away from biogenic structures (Kristensen and Alongi 2006). The thickness of the lower oxidation zone depends on the forest structure, mangrove species composition, and hydrological regime (Marchand et al. 2006). Thus, this zone is typically less distinct in *Avicennia* spp. than in *Rhizophora* spp. dominated forests due to the restricted root depth of the former (Fiala and Hernandez 1993). Finally, a permanently reduced layer (lower reduced zone) with high pyrite content is found below the root penetration depth (Clark et al. 1998; Marchand et al. 2006; Otero et al. 2009).

The redox depth profiles and spatial distribution also vary among biogeographical regions and types of mangrove forest (basin, fringe, scrub, overwash, and riverine; Lugo and Snedaker 1978). For example, monospecific *Rhizophora mangle* forests in Sao Paulo State (SE Brazil) show redox conditions that vary according to their physiographic position (Ferreira et al. 2007c). Basin forests in this region have strongly reducing conditions throughout the sediment profile, whereas fringe and riverine forests typically have oxidized surface sediments (Table 6.3). Similar to the vegetation zones, the intertidal gradient also affects the redox zonation. The

Table 6.3 Geochemical characteristics of near-surface sediment in various forest types of Sao Paulo State, Brazil (Ferreira et al. 2007b)

Location	Forest type	Depth cm	Sand %	TOC %	Eh mV	Total S %	Total Fe %	Fe oxides %	Pyrite Fe %
Cardoso Island	Basin	0–15	12	8.5	−52	1.85	2.46	0.46	0.57
	Riverine	0–15	86	2.0	102	0.35	0.57	0.15	0.09
Paimatos Island	Fringe	0–15	8	5.2	300	0.80	5.63	0.79	0.22
Baixada Santista	Riverine	0–10	15	23.7	331	2.89	3.23	0.17	0.73

infrequent tidal flooding near the mangrove landward side may result in the formation of numerous mud cracks due to desiccation allowing downward penetration of oxygen causing oxidation of deeper sediment layers (Marchand et al., 2011; Noel et al., 2014; Deborde et al., 2015). Consequently, Fe and S redox cycling intensifies significantly from the landward to seaward section of mangrove forests due to differences in duration and frequency of tidal inundation (Noel et al. 2014). Moreover, the higher tree productivity generally observed toward the seaward front of the forest may lead to strongly reduced sediment conditions because of lower saline stress, higher input of labile POC, and faster microbial activity.

6.3 Factors Affecting Element Cycling in Mangrove Sediments

Rates and pathways of microbial C and nutrient transformation in mangrove sediments are dependent on a number of key factors. The most important are organic matter input, availability of electron acceptors, bioturbation activity, and presence of tree roots, as well as geomorphology and hydrology (Canfield et al. 2005; Kristensen et al. 2008b; Mitsch and Gosselink 2015). The reactivity of organic matter toward microbial degradation using a variety of electron acceptors maintains a delicate redox zonation (Ferreira et al. 2007b). However, this zonation can be interrupted by downward translocation of oxygen via crab burrows and tree roots (Kristensen and Alongi 2006). C oxidation by heterotrophic microbial communities in mangrove environments is also dependent on the interaction between tidal elevation and hydroperiod (frequency, duration, and depth of inundation). Release of CO_2 may in certain cases vary several folds during different tidal conditions with the highest rates observed during low tide promoting degassing through air exposure (Alongi et al. 2004; Kristensen and Alongi 2006). It is not yet fully understood how this CO_2 exchange is controlled, but large biogenic structures (e.g. pneumatophores and crab burrows) may have an important role as conduits for O_2 intrusion and CO_2 release (Kristensen et al. 2008b).

6.3.1 Carbon Oxidation and Partitioning of Electron Acceptors

Sediment C oxidation can be quantified as CO_2 release in the dark, which represents the sum of all aerobic and anaerobic respiration processes and provides a reliable estimate of the total organic matter decomposition occurring within the sediment (Kristensen et al. 2011). To obtain the actual dark CO_2 fluxes in intertidal mangrove environments, measurements should be performed both during inundation and air exposure. The global average dark CO_2 release from inundated mangrove sediments is 62 mmol m^{-2} d^{-1} (range: 8–224 mmol m^{-2} d^{-1}) and 44 mmol m^{-2} d^{-1} (range: 4–156 mmol m^{-2} d^{-1}) for air-exposed sediments (Table 6.4). The variability in fluxes among study sites is undoubtedly a valid indication of regional environmental differences. However, the quite inconsistent differences between inundated and air-exposed rates must partly be caused by the applied methodological approach. Most of the reported CO_2 fluxes are probably underestimated, particularly during air exposure because measurements are typically performed on bare sediment away from trees and burrows. Air-exposed pneumatophores and open crab burrows considerably increase CO_2 release by efficient translocation of CO_2 gas from deeper sediments. For example, measurements of CO_2 efflux in a Tanzanian mangrove forest revealed that 100 pneumatophores per m^2 of the mangrove species *Sonneratia alba* and *Avicennia marina* released about 170 and 60 mmol CO_2 d^{-1}, respectively, whereas 100 crab burrows (*Uca* spp.) per m^2 released about 90 mmol CO_2 d^{-1} (Kristensen et al. 2008b). However, the contribution of biogenic structures to CO_2 exchange may vary among mangrove ecotypes and biogeographical regions, depending on mangrove species composition as well as the abundance of both trees and burrowing crabs (see Sect. 6.3.3; Araújo et al. 2012). Future studies on sediment metabolism in mangrove environments, therefore, need to incorporate the role of aerial roots and crab burrows when quantifying sediment CO_2 exchange and estimating reliable whole-forest C budgets (Rivera-Monroy et al. 2013; Troxler et al. 2015).

Microbial electron acceptor utilization in mangrove sediments follows the same energy yield sequence observed in other marine sediments: O_2, Mn^{4+}, NO_3^-, Fe^{3+}, and SO_4^{2-}(Kristensen et al. 2000; Alongi et al. 2005b). Aerobic microorganisms have the enzymatic capacity for complete oxidation of organic C to CO_2 using oxygen as electron acceptor (Canfield et al. 2005). In contrast, anaerobic heterotrophic processes occur stepwise involving several competitive respiration pathways (Mn respiration, denitrification, Fe respiration, and sulfate reduction) (Canfield et al. 2005). However, the bioavailability of organic macromolecules requires the prior intervention of fermenting microorganisms to generate low molecular organic substrates (e.g. lactate, butyrate, propionate, and acetate) that allow uptake and metabolism by anaerobic respiring microorganisms (Valdemarsen and Kristensen 2010). The specific role of each respiration process depends on the environmental setting

Table 6.4 O_2 and CO_2 exchange by exposed (air) and inundated (water) mangrove sediments in various biogeographical regions. Values in parentheses indicate ± SE (n = 3–4). Locations are within forested areas in the mid- to low-intertidal zone. The forests are older than 10 years since the last clearing. The dominant tree species and the season are indicated. The contribution of sulfate reduction and iron reduction to the daily integrated carbon dioxide efflux is given in percent

Location	Dominant tree	Season	O_2 air	O_2 water	CO_2 air	CO_2 water	SRR	FeR	Salinity	Temp[b]	Ref
			mmol m^{-2} d^{-1}				%	%		°C	
IWP: West Pacific											
Hinchinbrook, Australia	*R. stylosa*	Dry	–	43 (14)	–	21 (13)	79	0[a]	31	9–24	1
Hinchinbrook, Australia	*R. stylosa*	Wet	18 (5)	–	12 (6)	12 (9)	100	0[a]	24	27–36	1
Port Hedland; Australia	*R. stylosa*	Dry	16 (5)	18 (6)	14 (5)	27 (25)	100	0[a]	43	21–33	2
Port Hedland; Australia	*A. marina*	Dry	14 (7)	33 (21)	15 (3)	8 (3)	94	0.1[a]	51	22–32	2
Dampier, Australia	*R. stylosa*	Dry	24 (2)	38 (15)	36 (3)	25 (7)	90	0[a]	50–53	24–29	2
Dampier, Australia	*A. marina*	Dry	12 (1)	39 (15)	4 (0)	20 (4)	58	0[a]	44–54	23–30	2
Bay of Rest, Australia	*R. stylosa*	Dry	18 (5)	17 (8)	25 (12)	44 (7)	100	0.1[a]	42–50	22–28	2
Mangrove Bay, Australia	*A. marina*	Dry	14 (5)	12 (5)	26 (7)	30 (11)	56	0[a]	38–45	20–23	2
Haughton, Australia	*A. marina*	Dry	49 (7)	14 (2)	36 (5)	47 (5)	44	37	35	25–31	3
IWP: SE Asia											
Ao Nam Bor, Thailand	*R. apiculata*	Dry	–	45 (2)	–	70 (10)	100	–	34	29	4
Ao Nam Bor, Thailand	*R. apiculata*	Dry	35 (2)	19 (2)	–	35 (5)	85	–	34	29	5
Bangrong, Thailand	*R. mucronata*	Dry	46 (8)	21 (2)	–	70 (2)	18	–	35	29	6
Bangrong, Thailand	*R. mucronata*	Wet	29 (2)	24 (2)	–	96 (27)	20	–	35	29	6
Bangrong, Thailand	*R. mucronata*	Dry	–	15 (1)	–	43 (4)	19	79	35	29	7
Bangrong, Thailand	*R. mucronata*	Wet	–	24 (2)	–	49 (8)	19	79	35	29	7
Khlong Sawi, Thailand	*R. apiculata*	Dry	45 (17)	28 (2)	17 (4)	26 (26)	59	0[a]	8–10	~30	8
Khlong Sawi, Thailand	*R. apiculata*	Wet	9 (79)	24 (5)	19 (7)	21 (14)	100	0[a]	1–2	~30	8
Timor Leste	*R. apiculata*	Wet	16 (6)	69 (7)	–	81 (13)	85	–	20	~30	9
Timor Leste	*R. mucronata*	Dry	28 (7)	37 (6)	–	61 (10)	100	–	44	~30	9

(continued)

Table 6.4 (continued)

Location	Dominant tree	Season	O_2 air	O_2 water	CO_2 air	CO_2 water	SRR	FeR	Salinity	Temp[b]	Ref
			mmol m^{-2} d^{-1}				%	%		°C	
Timor Leste	*S. alba*	Dry	19 (10)	15 (39)	–	224 (8)	97	–	35	~30	9
Hurun Bay, Indonesia	*R. stylosa*	Dry	29 (24)	38 (15)	23 (9)	34 (2)	66	0[a]	–	~30	10
Hurun Bay, Indonesia	*R. stylosa*	Wet	203 (19)	39 (3)	26 (6)	45 (6)	77	0[a]	–	~30	10
Matang, Malaysia	*R. apiculata*	Dry	15 (4)	–	17 (4)	–	95	–	24	28–31	11
Matang, Malaysia	*R. apiculata*	Wet	58 (34)	13 (6)	114 (44)	11 (15)	63	11	–	~30	12
Mekong Delta, Vietnam	*R. apiculata*	Dry	55 (18)	–	53 (25)	–	5	0[a]	14	29–33	13
Mekong Delta, Vietnam	*R. apiculata*	Wet	13 (2)	–	80 (16)	28 (8)	48	0[a]	14	29–33	13
Jiulongjiang, China	*K. candel*	Dry	83 (13)	–	40 (14)	–	100	0[a]	17	28	14
IWP: S Asia											
Indus delta, Pakistan	*A. marina*	Dry	45 (9)	17 (1)	–	50 (10)	64	–	30	30	15
Pichavaram, India	*A. marina*	Dry	–	80 (16)	–	74 (13)	85	1[a]	72	32	16
Pichavaram, India	*A. marina*	Wet	–	59 (4)	–	196 (38)	85	–	63	24	16
Pichavaram, India	*R. apiculata*	Dry	–	58 (3)	–	165 (11)	65	3[a]	45	32	16
Pichavaram, India	*R. apiculata*	Wet	–	66 (5)	–	214 (18)	65	–	39	24	16
IWP: East Africa											
Mtoni, Tanzania	*S. alba*	Dry	45 (13)	32 (5)	88 (3)	57 (6)	12	77	34	28	5
Ras Dege, Tanzania	*R. mucronata*	Dry	55 (9)	54 (12)	69 (7)	88 (8)	46	37	37	28	5
AEP: America											
Cananeia, Brazil	*L. rasemosa*	Dry	–	30 (5)	156 (17)	15 (6)	–	–	23	28–30	17

(1) Alongi et al. (1999); (2) Alongi et al. (2000a); (3) Kristensen and Alongi (2006); (4) Kristensen et al. (1991); (5) Kristensen et al. (1994); (6) Holmer et al. (1999); (7) Kristensen et al. (2000); (8) Alongi et al. (2001); (9) Alongi et al. (2012); (10) Alongi et al. (2008); (11) Alongi et al. (1998); (12) Alongi et al. (2004); (13) Alongi et al. (2000b); (14) Alongi et al. (2005a); (15) Kristensen et al. (1992); (16) Alongi et al. (2005b); (17) Kristensen et al. (2011); (18) Quintana et al. (unpublished)

[a]Iron reduction determined from accumulation of dissolved Fe^{2+}

[b]Water temperature for inundated and sediment temperature for air-exposed measurements

(e.g., salinity, hydroperiod), biota composition (e.g., plant and crab species), and biogeographical factors (e.g., temperature, precipitation) at the local scale. Aerobic degradation of labile materials near the mangrove sediment surface is usually so rapid that O_2 rarely penetrates more than 2 mm into the sediment (Kristensen et al. 1994). Most of the sediment, therefore, remains largely anoxic, except for translocation of oxygen deep into the sediment through a network of roots and infauna burrows (Kristensen and Alongi 2006).

Aerobic respiration and anaerobic sulfate reduction are usually considered the most important C oxidation pathways in mangrove sediments, with a typical share of 30–50% each (e.g. Alongi et al. 2000a; Kristensen et al. 2011). Other metabolic pathways such as denitrification, Mn, and Fe respiration have traditionally been considered unimportant for the C cycling of old-growth forests due to limited availability of the electron acceptors NO_3^-, Mn(IV), and Fe(III) (e.g., Rivera-Monroy and Twilley 1996; Alongi et al. 2000a; Kristensen et al. 2000). Yet, recent evidence suggests that the role of Fe respiration may be comparable to or higher than that of sulfate reduction in Fe-rich mangrove sediments (Fig. 6.4; Kristensen et al. 2000; Kristensen and Alongi 2006; Kristensen et al. 2011). A critical consideration when assessing the relative role of microbial Fe reduction is related to how this process is measured. Most studies use anaerobic sediment incubations to provide reliable measures of solid-phase Fe reduction (Kristensen and Alongi 2006). However, some studies have only included dissolved Fe^{2+} accumulation (e.g. Alongi et al. 2000a), which greatly underestimates actual rates of Fe reduction. The drawback of this approach is that most Fe^{2+} produced from Fe reduction will rapidly precipitate with sulfide, carbonates, and phosphates or be chelated into sheet silicates and organic matrices (Thamdrup 2000). Instead, it is recommended that reactive solid phase Fe(III) and Fe(II) are extracted over time with a sufficiently strong extractant (e.g., 0.5 M HCl, Lovley and Phillips 1987). Using this method, the decrease in extracted Fe(III) and corresponding increase in extracted Fe(II) provide a reliable measure of Fe reduction (e.g., Kristensen and Alongi 2006). This approach to measure Fe reduction has been applied in some mangrove settings (Table 6.4) and other coastal sediments (e.g., saltmarshes, Kostka et al. 2002; Gribsholt et al. 2003).

The rates of microbial C oxidation and partitioning of electron acceptors within mangrove sediments are also dependent on other factors. These include forest age, species diversity, forest density, root physiological activity, extent of water logging and flooding duration, and the intensity of faunal burrowing activities. For example, sulfate reduction accounts for less of the total sediment respiration in young (i.e., 20–30%) than old (i.e., >50%) *Avicennia marina* and *Rhizophora apiculata* forests (Alongi et al. 1998, 2000b). Similarly, aerobic respiration usually dominates in permeable sandy sediments under young *Rhizophora* stands characterized by low plant biomass and high exposure to tidal forcing. Conversely, in older forests where iron reduction and sulfate reduction are the dominant respiration processes, most of the oxygen uptake is driven by oxidation of reduced metabolites (e.g., HS^- and Fe^{2+}) diffusing from deeper sediment layers (Canfield et al. 2005; Kristensen et al. 2011). Furthermore, the impact of water logging is evident in regions with distinct dry and

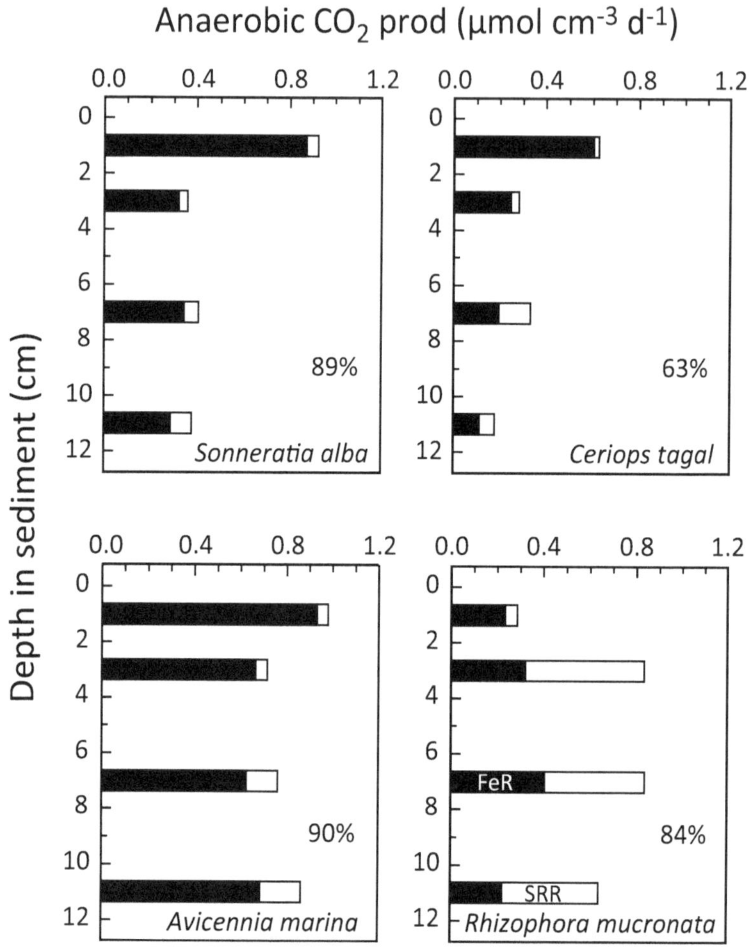

Fig. 6.4 Partitioning of anaerobic CO_2 production into Fe reduction (FeR) and sulfate reduction (SRR) in sediments vegetated by four mangrove species. Results are from two Tanzanian mangrove forests. The percentages given in the figure indicated how much FeR + SRR accounts for of the total C-oxidation in the sediments (Modified from Kristensen et al. 2011)

wet seasons. When the water table is low during the dry season, oxygen penetrates deep into the sediment through gas-filled crab burrows and cracks, which enhances the oxidizing effect and promotes oxidation of organic C via denitrification and Fe reduction. In contrast, prolonged flooding during the rainy season prevents transport of oxygen into the sediment and sulfate reduction becomes the dominant pathway (Clark et al. 1998; Marchand et al. 2004).

The least energy-yielding step in the sedimentary metabolic pathway of all aquatic environments is methanogenesis. This process is controlled by the organic content of the sediment, oxygen concentrations, temperature, rainfall, substrate acidity, and the presence of inhibitors like sulfide (Livesley and Andrusiak 2012;

Dutta et al. 2013; Konnerup et al. 2014). The formation of CH_4 usually occurs deep in sediments where sulfate supplies are exhausted. Methanogenesis has traditionally been considered negligible in mangrove sediments due to the competitive dominance of sulfate reduction (Livesley and Andrusiak 2012; Nóbrega et al. 2016). Yet, recent discoveries of high numbers of active methanogenic archaea in mangrove sediments have challenged this opinion (Lyimo et al. 2002, 2009). For example, Marinho et al. (2012) found four to five times higher methanogenesis in sediment vegetated by *Rhizophora mangle* than in adjacent seagrass beds and unvegetated subtidal marine sediments. In addition, it was demonstrated that otherwise competitive sulfate-reducing and methanogenic bacteria can coexist in mangrove sediments with ample organic matter supplies (Lyimo et al. 2009; Otero et al. 2014; Chauhan et al. 2015). High rates of methanogenesis can also occur in mangrove environments influenced by freshwater (e.g., rivers, groundwater and precipitation) when sulfate reduction is hampered due to sulfate dilution (Lu et al. 1999; Maher et al. 2015).

6.3.2 *The Importance of Nitrogen and Phosphorus Nutrients*

Nitrogen (N) and phosphorus (P) are critical nutrients that regulate the magnitude and spatial distribution of mangrove forest productivity and structural properties (Lovelock et al. 2009; Feller et al. 2010; Reef et al. 2010; Dangremond and Feller 2014). Although N transformations are generally slow in mangrove wetlands, the actual rates vary among mangrove ecotypes and depend strongly on local (e.g., nutrient gradients, salinity), regional (e.g., geomorphology), and anthropogenic impacts (Alongi 2009; Kristensen et al. 2000; Keuskamp et al. 2015). Generation of newly available N by N-fixation ($N_2 \rightarrow NH_3$) and loss of available N through denitrification ($NO_3^- \rightarrow N_2$) occur, but to a much lower extent than in other estuarine environments (Twilley and Rivera-Monroy 2009). Given the low and almost similar rates of denitrification and N-fixation (Table 6.5), recycling through mineralization is probably the source of most inorganic N for primary producers in mangrove forests (Feller et al. 2003; Alongi 2011), except when there are significant anthropogenic sources. In fact, eutrophication (effluents from, e.g., aquaculture and human developments) has in recent years significantly changed the nutrient balance and thus impacted biogeochemical cycles and productivity of many mangrove environments (Alongi 2009).

Denitrification in mangrove sediments is primarily controlled by the supply of the electron acceptor nitrate (NO_3^-) and electron donors in the form of labile organic matter, and a number of secondary factors including the presence of macrofauna, macrophytes, benthic microalgae, H_2S, and FeS as mentioned earlier (Sect. 6.3.1). Although the C and N cycles in this way are coupled in mangrove wetlands, NO_3^- removal via denitrification can limit the production of organic matter (Rivera-Monroy et al. 2010). This may under certain conditions be counteracted by dissimilatory nitrate reduction to ammonium (DNRA) that effectively conserves

Table 6.5 Mangrove sediment nitrogen fixation and denitrification at sites from the IWP and the AEP

Region	Site	Mangrove type	Vegetation	N fixation µmol m^{-2} h^{-1}	Denitrification µmol m^{-2} h^{-1}	Ref
IWP	Makham Bay (Thailand)	Mid-intertidal	*Rhizophora apiculata*	12	2	1
	Sawi Bay (Thailand)	Managed mid- and high-intertidal	*Avicennia alba Ceriops decandra Rhizophora apiculata*	0–24	0–160	2
	Mekong Delta (Vietnam)	Managed high-intertidal	*Rhizophora apiculata*	10–59	0–92	3
	Matang Reserve (Malaysia)	Managed mid-intertidal	*Avicennia marina Rhizophora apiculate*	0–125	16–458	4
	Jiulongjiang Estuary (China)	Managed low-, mid- and high-intertidal	*Kandelia candel*	0–2	46–158	5
AEP	Joyuda Lagoon (Puerto Rico)	Fringe	No specified	13–31	1–161	6
	Oyster Bay (Jamaica)	Fringe center, rear	*Avicennia germinans Rhizophora mangle*	0–100	0–83	7
	Twin Cays (Belize)	Fringe transition, dwarf	*Avicennia germinans Laguncularia racemosa Rhizophora mangle*	0–17	0–8	8

1) Kristensen et al. (1998); 2) Alongi et al. (2002); 3) Alongi et al. (2000b); 4) Alongi et al. (2004); 5) Alongi et al. (2005a); 6) Morell and Corredor (1993); 7) Nedwell et al. (1994); 8) Lee and Joye (2006)

and recirculates N. Thus, the occurrence of DNRA in mangrove forests has important implications for maintaining N levels and sustaining primary productivity (Fernandes et al. 2012a). However, only few studies have measured DNRA in mangrove wetlands, and it is not yet clear how this pathway contributes to the overall N budgets (Giblin et al. 2013; Molnar et al. 2013). Also, the anaerobic conversion of NO_2^- and NH_4^+ to N_2 (anammox), which in conjunction with denitrification represent a sink of N, requires further study. Estimates of anammox in mangrove sediments are scarce (Li and Gu 2013; Wang et al. 2013), but two experimental studies have revealed rates that account for <10% of total N_2 production (Meyer et al. 2005; Fernandes et al. 2012b). Work in other coastal ecosystems suggests that denitrification is also much higher than anammox, particularly in sediments receiving a high load of reactive organic matter (Dalsgaard et al. 2005; Fernandes et al. 2012b).

P availability within mangrove wetlands is, in contrast to N, strongly dependent on the dynamic interactions of P with Fe and S cycling (Nóbrega et al. 2014; Deborde et al. 2015). For example, phosphate (PO_4^{3-}) is readily adsorbed and retained by Fe(III) oxyhydroxides in near-surface sediments, around crab burrows and around rhizospheres, thus limiting plant production (Clark et al. 1998). However, the adsorbed PO_4^{3-} can be released back to dissolved form and be available again for primary producers when Fe(III) oxyhydroxides are reduced in anoxic sediment. This oxidation–reduction cycle depends on either transport of particles between oxic and anoxic zones or is due to temporal expansion and contraction of oxic zones. Crabs typically mediate the former mechanism when they rework surface and subsurface sediments, whereas the latter mechanism is primarily due to tidal and seasonal changes in redox conditions.

The use of fertilization experiments under field conditions has advanced our understanding of the complex interaction and relative role of N and P availability for mangrove structural development and productivity (Lovelock et al. 2006; Feller et al. 2007; Simpson et al. 2013). The response of ecological processes to nutrient enrichment depends on site characteristics, species composition and dominance, and the nature of nutrient limitation (Feller et al. 2010; Reef et al. 2010). For example, the resorption of P from senescent tissue by *R. mangle* is under P-limited conditions much higher (≈70%) than that for N (≈45%). N fertilization does not change this pattern, but P fertilization decreases P resorption (<50%), whereas N resorption (≈70%) increases (Feller et al. 1999). Scrub mangrove forests (e.g., *R. mangle* and *A. germinans*) growing in P limited carbonate sediments always respond to P fertilization, while surrounding fringing mangroves (e.g., *R. mangle*) respond mostly to N fertilization, and those exposed to intermediate tidal influence respond to both N and P fertilization (Lovelock et al. 2006) as hydroperiod interacts with nutrient availability (Twilley and Rivera-Monroy 2009). The response of mangrove wetlands to nutrient additions appears to be similar in both the IWP and AEP biogeographical region. Large-scale experimental work on N and P limitation was initially performed in the Caribbean region (Belize), Central America (Panama), and North America (Florida), but has been expanded to areas in Australia and New Zealand, particularly when assessing effects of nutrient availability on C sequestration (Alongi 2011). For example, large-scale comparisons (Caribbean, Australia, New Zealand) revealed that P is less limiting to plant metabolism at higher than lower latitudes (Lovelock et al. 2007). Although this and other large-scale latitudinal comparisons to evaluate differences in N and P responses between the IWP and AEP have been undertaken, there are still large gaps in the overall conceptual framework for mangrove wetlands.

Table 6.6 Burrow and disturbance depths by various invertebrate and vertebrate taxa in mangrove environments from different geographical regions

Geographical region	Taxon (genus)	Burrow/Disturbance Depth (cm)	Density[a] (m^{-2})	Ref
Indo-west-Pacific				
	Crab			
	Mictyris	~10–30	226	1–2
	Uca	~10–100	~100	3–4
	Ocypode	~16–40	~1	5–6
	Sesarma	~100–120	~12	7
	Helice	~35	36	8
	Neoepisesarma	~80	0.2	9–10
	Chiromantes	~10	18	11–12
	Amphipod			
	Victoriopisa	~10	3500	13
	Callianassid, penaeid, alpheid shrimp			
	Trypaea	~120	200	14
	Metapenaeus	~1		15–16
	Alpheus	~50	56	17
	Thalassinid lobster			
	Thalassina	~250	0.5	18
	Sipunculid worm			
	Siphonosoma	~50		19
	Bivalve			
	Geloina	Upper sediment	40[b]	20
	Teleost fish			
	Periophthalmus	>10		21
	Elasmobranch fish (Ray)			
	Himantura	~5		22
Atlantic-east-Pacific				
	Crab			
	Uca	~40	~70	23[c]
	Ucides	~200	~3	24
	Ocypode		~0.7	25 42–43
	Penaeid shrimp			
	Penaeus	~5		26
	Thalassinid lobster			
	Thalassina	~250	0.5	27[c]
	Sipunculid worm			
	Sipunculus	Upper sediment	~240	28
	Bivalve			
	Mytella	Upper sediment	>20[b]	29

(continued)

Table 6.6 (continued)

Geographical region	Taxon (genus)	Burrow/Disturbance Depth (cm)	Density[a] (m^{-2})	Ref
	Teleost fish			
	Lutjanus	Upper sediment		30
	Elasmobranch fish (Ray)			
	Dasyatis	~20	~0.6	31

[1] Rossi and Chapman (2003); [2] Shih (1995); [3] Gillikin (2000); [4] Qureshi and Saher (2012); [5] Chan et al. (2006); [6] Dubey et al. (2013); [7] Stieglitz et al. (2000); [8] Mchenga et al. (2007); [9] Thongtham and Kristensen (2005); [10] Kristensen (2008); [11] Gillikin and Kamanu (2005); [12] Xiong et al. 2010); [13] Dunn et al. (2009); [14] Kerr (2001); [15] Joshi et al. (1979); [16] Primavera and Lebata (1995); [17] Dworshak and Pervesler (2002); [18] Kartika and Patria (2012); [19] Zhou and Li (1990); [20] Morton (1976); [21] Clayton and Snowden (2000); [22] O'Shea et al. (2012); [23] Kristensen (2008); [24] Pülmanns et al. (2014); [25] da Silva Castiglioni and Negreiros-Fransozo (2005); [26] Fuss (1964); [27] Dworshak et al. (2012); [28] Rice et al. (1995); [29] Bacon (1975); [30] Vasleta et al. (2012); [31] Cross and Curran (2000)

[a]Example density of burrow/disturbance

[b]Represents individuals in lieu of burrow density. Upper sediments represent bioturbation predominantly in shallow surface sediments with no documented depth (cm) from literature

[c]Includes references therein

6.3.3 The Impact of Benthic Fauna

A broad diversity of benthic animals lives or feeds in and on mangrove sediments. Most of these are invertebrates, including crustaceans, polychaetes, sipunculids, and molluscs, whereas teleost fish and rays may be present occasionally (Table 6.6). Brachyuran crabs are particularly dominant because their hard and compact carapace provides good locomotion ability and protection in the harsh mangrove environment, offering evolutionary advantage over other invertebrates (Krobicki and Zatoń 2008). However, the diversity of brachyuran crabs associated with mangroves varies considerably; from over a hundred species in Southeast Asia to only a few dozen species in East African and the Americas (Gillikin and Schubart 2004; Lee 2008). Nevertheless, crabs (especially Grapsoidea and Ocypodoidea) usually dominate both in numbers and biomass in mangrove ecosystems all over the world (Jones 1984).

As expected from their prominent distribution and occurrence, these decapods are key species for regulating and controlling mangrove ecological and biogeochemical functioning. Crabs are known to reduce leaf litter export to adjacent open waters by burial and consumption of leaves (Table 6.7). Litter handled by crabs eventually enters the microbial food chain either in the form of uneaten remains and

Table 6.7 Examples from around the world of removal (litter taken into burrows) and consumption of leaf litter by leaf-eating mangrove crabs

Species	Site	Rate of consumption or removal (% of total litter fall)	Ref
Sesarmid crabs	North Queensland, Australia. High intertidal	Removal: *C. tagal* forest: 71%, *B. exaristata* forest: 79%, *A. marina* forest: 33%	1
Neoepisesarma versicolor	Bangrong, Phuket, Thailand Mid intertidal	*R. apiculata* forest: Removal: 87% Consumption: 65%	2
Sesarma meinerti	Mgazana river estuary South Africa High intertidal	Consumption: *A. marina* forest: 44%	3
Sesarma meinerti	South Africa. High intertidal	*Bruguiera gymnorrhiza* forest Removal: 99%; Consumption: 64%	4
Ucides cordatus	Pará, North Brazil. High intertidal	Consumption: *R. mangle* forest: 81%	5

(1) Robertson and Daniel (1989); (2) Thongtham et al. (2008); (3) Emmerson and McGwynne (1992); (4) Steinke et al. (1993); (5) Nordhaus et al. (2006)

fecal material buried in the sediment or as crab carcasses (Giddins et al. 1986; Robertson 1986; Lee 1997, Twilley et al. 1997) and therefore contributes to nutrient recycling within mangrove forests (Nordhaus et al. 2006). Maceration of plant material during ingestion and contact with digestive enzymes in the gut appears to facilitate microbial attack on insoluble carbohydrates. The rate of microbial decay of recalcitrant leaf litter in sediments is therefore facilitated after passage through crab guts (Kristensen and Pilgaard 2001). In contrast, the intact plant material that escapes crab handling is readily exported by tides and thus lost from the mangrove ecosystem.

A striking performance of bioturbating animals in mangrove ecosystems is the formation of deep and branched burrows and reworking of sediment particles. The animals involved are primarily crabs, but other crustaceans (amphipods, alpheid shrimp, and thalassinid lobsters) and sipunculids also form burrows, whereas mytilid bivalves and fish produce less structured disturbances of surface sediments (Table 6.6). Although the diversity of bioturbating animals in general is greater in IWP than AEP mangrove forests, the functional types of bioturbation are similar, as are the typical densities of burrows (Table 6.6).

As mentioned in Sect. 6.3.3., bioturbation by crabs has a profound effect on mangrove sediments by constantly disrupting the vertical redox zonation and increasing the complexity of the sediment system (Kristensen 2008). Functionally, bioturbation regulates and controls biogeochemical processes such as organic C oxidation and benthic nutrient fluxes (Table 6.8). Sesarmid and ocypodid crabs are the most prominent taxa affecting biogeochemical transformations in mangrove

Table 6.8 Effects of bioturbating crabs on biogeochemical parameters relevant to mangrove sediments (↑ represents increase and ↓ represents decrease of analyte concentration/process)

Geographical region[a]	Bioturbator	Analyte concentration						Process				Ref
		Chl-a	*Org C*[b]	NH_4^+	S^{2-}	*Eh*	*Salinity*	SO_4^{2-} *red*	CO_2 *efflux*	*Prod*	*Resp*	
Indo-West Pacific												
Australasia	*Mictyris longicarpus*	↓		↓					↑	↓	↓	1
	Uca vocans	↓	↓	↓	↓			↓↑	↓	↓	↓	2
	Sesarma messa						↓					3
	Sesarma messa		↑									4
	Sesarma messa *Sesarma semperi longicristatum*			↓	↓							5
Indo-Malesia	*Helice formosensis*					↑						6
	Helice formosensis			↓↑								7
	Uca spp.							↓			↑	8
	Neoepisesarma versicolor				↓							9
East Africa	*Uca* spp., *Chiromantes* spp., *Perisesarma* spp. *Neosarmatium africanum*		↑									10
	Uca annulipes, *Uca inversa*								↑			11
Atlantic East Pacific												
East America	*Uca marcoani*, *Ucides cordatus*					↑						12
	Uca spp.				↓							13

(continued)

Table 6.8 (continued)

Geographical region[a]	Bioturbator	Analyte concentration						Process				Ref
		Chl-a	*Org C*[b]	NH_4^+	S^{2-}	*Eh*	*Salinity*	SO_4^{2-} *red*	CO_2 *efflux*	*Prod*	*Resp*	
	Uca uruguayensis, Uca rapax		↓									14
	Ucides cordatus		↑									15

[1] Webb and Eyre (2004); [2] Kristensen and Alongi (2006); [3] Stieglitz et al. (2000); [4] Robertson (1986); [5] Smith et al. (1991); [6] Mchenga et al. (2007); [7] Mchenga and Tsuchiya (2008); [8] Nielsen et al. (2003); [9] Thongtham and Kristensen (2003); [10] Andreetta et al. (2014); [11] Penha-Lopes et al. (2010); [12] Araújo et al. (2012); [13] Ferreira et al. (2007a); [14] Sayão-Aguiar et al. (2012); [15] Nordhaus et al. (2006)

[a] Geographical regions zoned according to Duke et al. (1998)

[b] Reported as loss-on-ignition or total organic carbon

ecosystems. Regardless of the species involved and its behavior, bioturbation considerably enhances the transport of O_2 and other electron acceptors to deeper sediment layers and promotes the upward translocation of metabolites (e.g. CO_2) and reduced compounds (i.e. iron sulfides)(Fig. 6.5). Several studies have demonstrated effects of bioturbation on various sediment parameters, including microphytobenthic chlorophyll, C turnover, nutrient (N, P) cycling, and S distribution (Kristensen and Alongi 2006; Bartoli et al. 2009). Some biogeochemical processes are consistently increased or decreased among animal taxa, whereas the impact of others are more genus or species specific, i.e., by shifting the dominance of sulfate reduction to other respiration pathways or vice versa (Alongi et al. 2001; Kristensen 2008). The capacity to promote such biogeochemical shifts depends on crab behavior, such as burrowing and refuge strategies, as well as feeding and mating behavior (Kristensen 2008).

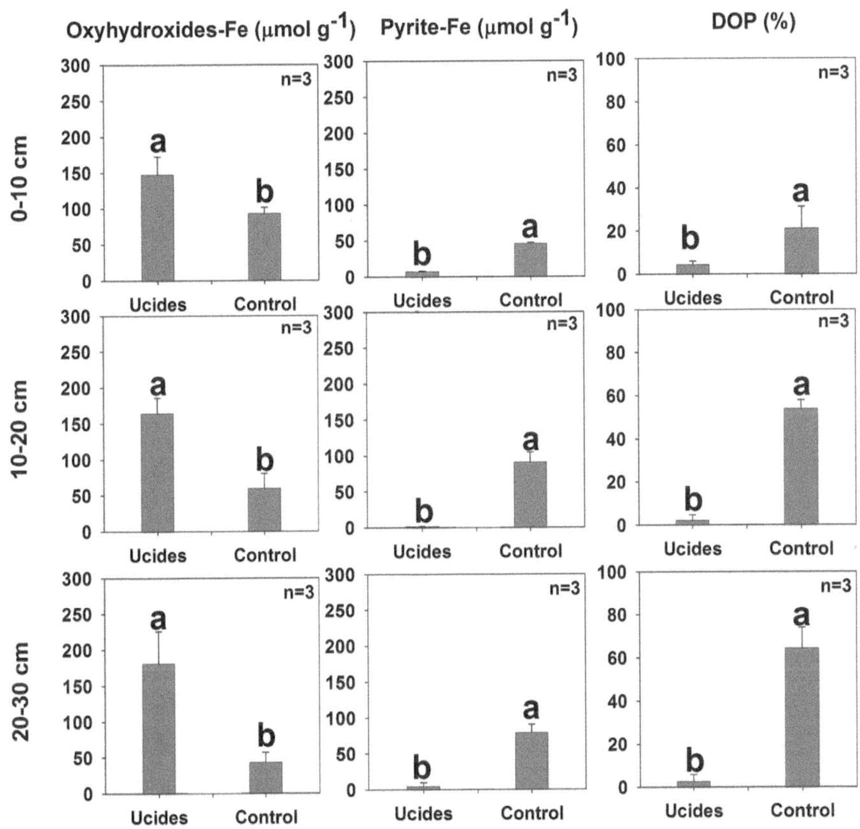

Fig. 6.5 Field exclusion experiment in mangrove forests of NE Brazil. The contribution of oxidized Fe, pyrite Fe, and the degree of iron pyritization (DOP) in different depth layers (mean ± SD) of sediment taken from plots with and without *Ucides cordatus* is shown. *Bars* with *different letters* indicate significant differences at $p < 0.05$ (Modified from Araújo et al. 2012)

Because C accumulation is a key process (and an important Ecosystem Service) in mangrove wetlands, O_2 translocation into deeper layers of the sediment by bioturbation can enhance organic C degradation processes and consequently increase CO_2 emissions to the atmosphere (see Sect. 6.4). Thus, crabs are not only capable of retaining C within mangrove ecosystems through burrowing and mixing of sediment particles, but are also important actors in the opposite process of exporting C by favoring more efficient decomposition pathways and thus boosting CO_2 emissions (Pülmanns et al. 2014). This is evident in Tanzanian mangrove forests where crab burrows are responsible for 36–62% of the total CO_2 emission from the sediments (Kristensen et al. 2008b).

6.3.4 The Importance of Hydroperiod and Hydrology

Hydrology is recognized by the hydroperiod, which is defined as the depth, duration, and frequency of inundation (Fig. 6.6) (Reddy and Delaune 2008; Twilley and Rivera-Monroy 2009). Mangrove wetland hydroperiod encompasses all aspects of water budgets (rainfall, evaporation, as well as subsurface and surface flow) regardless of the water source (Mazda and Wolanski 2009). In combination with resources (e.g., N, P, light, space) (Tilman 1982) and regulators (e.g., salinity, sulfide, pH, Eh), gradients in hydroperiod control mangrove structural and functional properties (Ellison and Farnsworth 1997; Twilley 1997; Twilley and Rivera-Monroy 2005; Berger et al. 2008). One of the major causes of mangrove mortality and loss at the global scale is the direct and indirect impact of human activities on hydrology, which is perhaps "[…] the single most important determinant of the establishment and maintenance of specific types of wetlands and wetlands processes" (Mitsch and Gosselink 2015). As hydroperiod controls plant growth, it also has a significant role for the availability and concentration of essential elements in mangrove sediments (Twilley and Day 2013; Mitsch and Gosselink 2015). Indeed, studies assessing the impact of hydroperiod include how mangrove species zonation (e.g., Crase et al. 2013), development of sediment physicochemical gradients (i.e., sediment and water quality) (Alongi 2009), and physiological traits of mangrove species (e.g., growth rate, photosynthesis performance, nutrient use efficiency, biomass allocation) and regulator gradients (i.e., salinity) respond to changes in flooding frequency and duration.

As mentioned earlier, there is uncertainty in the magnitude and spatio-temporal variability of biogeochemical transformations in mangrove forests, which in certain cases can be associated with the lack of data on hydrologic parameters (Mazda and Wolanski 2009). For example, mangrove studies assessing the impact of flooding on biogeochemical transformations use water level recorders positioned along tidal creeks at some distance (km) from the study site, whereas other studies rely on data from hydrographic stations installed in nearby ports and coastal cities. Hydrographs or tide tables based on astronomical calculations are then combined with elevation measurements in single points or along transects to produce inundation frequencies (e.g., Mendoza et al. 2012). Few studies have actually deployed water level

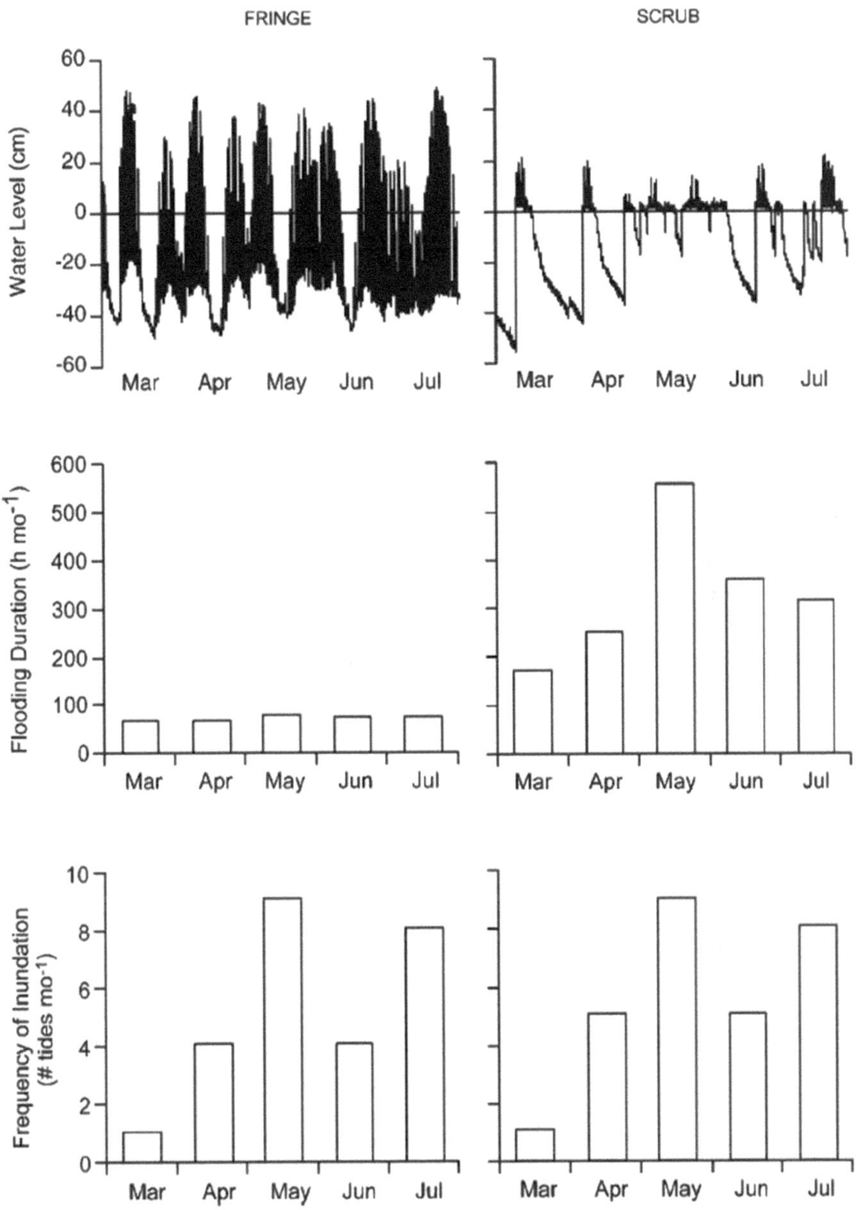

Fig. 6.6 Water level, flooding duration, and frequency of inundation in fringe and scrub mangrove zones of the San Bernardo Estuary, Gulf of Fonseca, Honduras. The zero mark in the upper panel is relative to the ground surface in this site (Modified from Castañeda-Moya et al. 2006)

recorders inside the forest to register long-term flooding frequency and depth of inundation. In fact, modeling of hydrological patterns in estuaries and inside mangrove areas shows distinct differences in tidal patterns and symmetry due to creek geomorphology, local weather, and forest structure (riverine, basin, fringe) (see Chap. 11) (Lugo and Snedaker 1978; Mazda et al. 1995), reinforcing the need to directly measured spatial and temporal variation of hydroperiod parameters at both local and regional scales.

Duration of inundation is critical because it directly controls sediment redox conditions, and thus microbial transformations and exchange processes within the sediment, whereas inundation depth determines the net material exchange (e.g. organic matter and nutrients) between mangrove forests and adjacent coastal waters (Twilley and Rivera-Monroy 2005; Adame and Lovelock 2011). The direct effect of hydroperiod on O_2 diffusion into mangrove substrata is well documented (e.g., Kristensen 2008; Twilley and Rivera-Monroy 2009), but there is still a lack of knowledge directly linking frequency and duration of inundation to biogeochemical transformations such as denitrification, iron reduction, sulfate reduction, and methanogenesis, or even CO_2 and CH_4 fluxes at sediment–air and sediment–water interfaces. These measurements are necessary to evaluate the uncertainty and applicability of fluxes using sediment under laboratory conditions to extrapolate values to larger scales (see Chap. 11).

Among the variables directly measured along with hydroperiod under experimental and field conditions is salinity of overlying water and sediment porewater. Salinity is an excellent proxy of physicochemical status because it integrates a number of factors controlling hydrology and biogeochemistry in coastal regions; from tidal inundation, evapotranspiration, and river discharge to partitioning between iron reduction, sulfate reduction, and methanogenesis in sediments. Salinity is an easy parameter to measure in hydrological studies as reflected by the number of studies relating hydroperiod and salinity patterns in mangrove zonation studies. For example, Crase et al. (2013) showed that the spatial partition of three mangrove species (*Sonneratia alba, Rhizophora stylosa, Ceriops tagal*) in northern Australia is significantly associated to hydroperiod and porewater salinity. Similarly, Castaneda-Moya et al. (2006) found clear distinctions in the spatial distribution of mangrove ecotypes (fringe vs scrub forests) and species (*Rhizophora mangle* and *Avicennia germinans*) as a result of the interaction between hydroperiod and salinity (Fig. 6.7). In general, porewater salinity is lower in the fringe zone dominated by *Rhizophora* spp. (<40) than in transition (60) and particularly scrub (>70) mangrove zones dominated by *Avicennia* spp. (Castaneda-Moya et al. 2006).

Although analysis of mangrove hydrology has improved (Mazda and Wolanski 2009), there is still a need to identify the mechanisms by which hydroperiod controls sediment biogeochemistry within different mangrove ecotypes (e.g., riverine, fringe, basin) (Lugo and Snedaker 1978; Woodroffe 1992, 2002). In situ and experimental work is needed to assess how horizontal and vertical hydrodynamics, modified by forest tree density or root structure (Mazda et al. 1995, 2005), enhances or limits the relative role of mangroves wetlands as sinks, sources, and/or transformers of biologically important elements (C, N, P, S) in coastal regions with different ecogeomorphic characteristics (Woodroffe 2002).

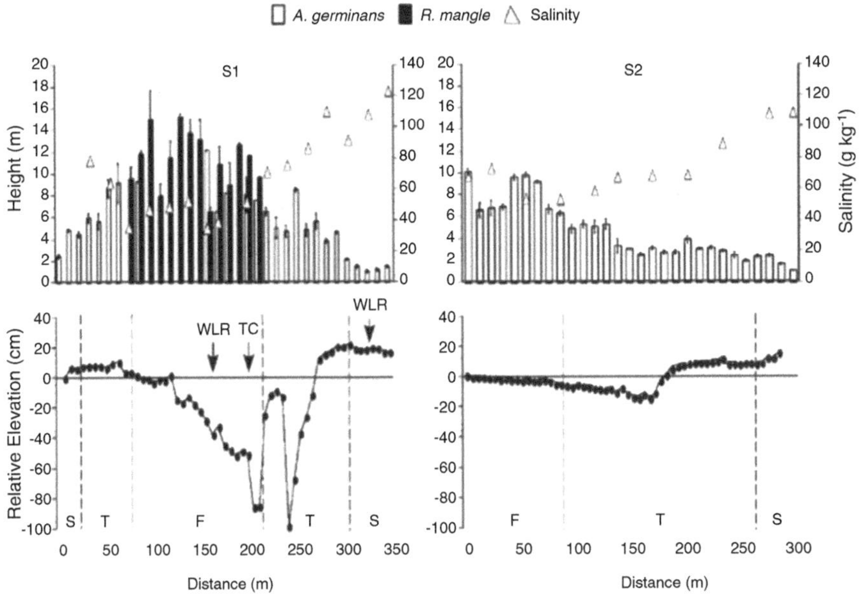

Fig. 6.7 The impact of sediment porewater salinity and relative elevation on the spatial distribution of *Avicennia germinans and Rhizophora mangle* at two locations (*S1* and *S2*) in the Gulf of Fonseca, Honduras, Central America. *Upper panel*: tree height and sediment porewater salinity. *Lower panel*: the relative elevation. *Dashed lines* indicate the extent of mangrove zones/ecotypes: (*F*) fringe, (*T*) transition, and (*S*) scrub. *TC* indicates the location of the tidal creek. *WLR* indicates location of water level recorders (Modified from Castañeda-Moya et al. 2006)

6.4 Greenhouse Gas (GHG) Balance of Mangrove Ecosystems

Mangrove wetlands have a potentially high impact on the global C budget because of their high net productivity (218 ± 72 Tg C y^{-1}, Bouillon et al. 2008) and C storage capacity (18–29 Tg C y^{-1}, Alongi 2012), despite their simple forest structure, low biodiversity, and limited areal cover (less than 150,000 km^2 worldwide) (Polidoro et al. 2010; Giri et al. 2011). This means that an equivalent of as much as ~10% of the organic C produced in mangrove wetlands escapes export to coastal waters by tidal flushing and pumping, or release to the atmosphere in the form of CO_2. The waterlogged and anoxic mangrove sediments with slow decomposition, therefore, allow a substantial long-term organic C accumulation (Fig. 6.8), and thus function as sinks for atmospheric CO_2 (Donato et al. 2011; Murdiyarso et al. 2015). Accordingly, CO_2 releases at the sediment–air interface in mangrove wetlands are lower than those measured in both temperate and tropical terrestrial environments (Leopold et al. 2015). This is not only true for CO_2 emissions, but also for other GHG such as methane (CH_4) and nitrous oxide (N_2O). However, recent studies report a high variability of GHG emission from mangrove areas depending on their productivity, position in the tidal range, anthropogenic impact and seasons (Livesley

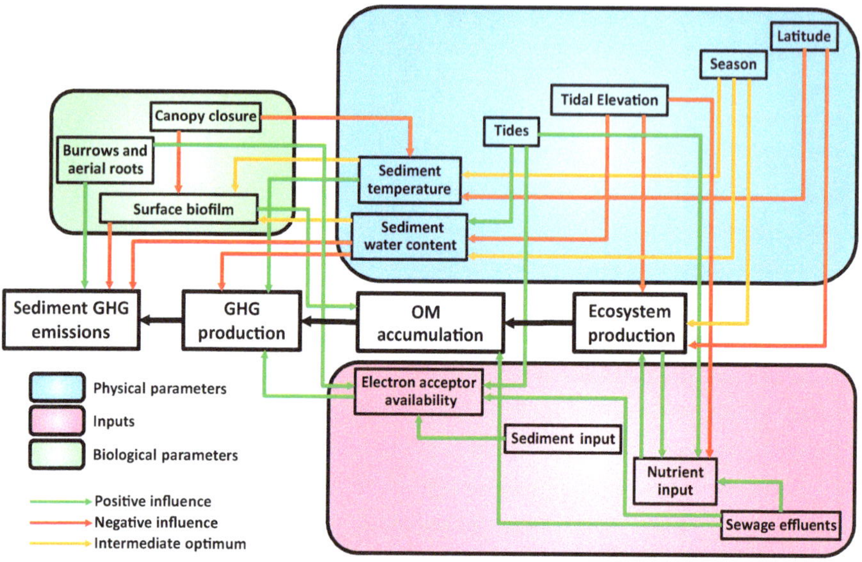

Fig. 6.8 Flow diagram of carbon from fixation of CO_2 by ecosystem production to organic carbon (OC) accumulation, greenhouse gas (GHG) production, and sediment GHG emission. Important controlling factors are indicated, and marked with *green arrows* if they have a positive effect on processes, *red arrows* if they have a negative effect, and *yellow arrows* if they have an intermediate optimum

and Andrusiak 2012; Leopold et al. 2013; Chen et al. 2014; Nóbrega et al. 2016). Thus, it seems, from the studies to date, that particularly mangrove eutrophication caused by discharge of untreated wastewater can substantially enhance GHG production (Fig. 6.8) (Konnerup et al. 2014).

Bouillon et al. (2008) pointed out that the current uncertainty of published mangrove C budgets at the global scale were largely due to lack of reliable sediment-air and water-air CO_2 flux measurements. Several studies have subsequently provided new information about key drivers influencing CO_2 emission at the sediment-air interface (Kristensen et al. 2008b; Chen et al. 2014; Lovelock et al. 2014; Leopold et al. 2015). These studies clearly demonstrate the importance of features such as C stocks, sediment water content, temperature, and crab burrows/pneumatophores density as regulators of CO_2 emission from mangrove sediments (Fig. 6.8). Thus, Lovelock et al. (2014) observed a direct relationship between sediment respiration and aboveground net primary production (a proxy of C accumulation). The role of sediment water content for CO_2 emission is related to the impact of water on transport processes and decomposition efficiency within the sediment. Molecular diffusion is faster in gas than water, and CO_2 fluxes may increase when sediments become partly dry during air exposure (see Sect. 6.3.1), while organic matter mineralization is slow in waterlogged sediment due to the dominance of anaerobic processes (Kristensen et al. 2008a). Consequently, sediment-air CO_2 emissions may decrease under low redox, as observed by the wet conditions during the monsoon season in

India (Chanda et al. 2013). Conversely, CO_2 emission also changes with temperature due to thermal sensitivity of microbial processes. Studies assessing the temperature dependence of CO_2 emission from mangrove sediments reported Q_{10} values between 2 and 3, which are comparable to other forested ecosystems (Lovelock 2008; Leopold et al. 2015) and marine sediments (Thamdrup et al. 1998). In addition, Leopold et al. (2015) showed that temperature was the main driving factor for seasonal variations of sediment-air CO_2 emissions in *Rhizophora* and *Avicennia* dominated stands in a semi-arid climate. They also suggest that the surface biofilm at sediment surfaces may limit CO_2 efflux towards the atmosphere by forming a protective barrier consisting of extracellular polymeric substances (EPS) and due to microbial assimilation (Fig. 6.8). As mentioned earlier, crab burrows and aerial roots (e.g., pneumatophores) act as efficient conduits that allow the ascent of excess deep CO_2 to the atmosphere (Kristensen et al. 2008b; Troxler et al. 2015). However, due to lack of comparable studies and data on how sediment-air CO_2 fluxes are controlled in various mangrove forest ecotypes, it is presently not possible to perform a thorough analysis of CO_2 emissions among biogeographical regions.

CO_2 release measured across the sediment-water/air interfaces does not represent the total mineralization of organic matter derived from mangrove forests. Part of their net production (e.g. litter) is exported into tidal creeks and the adjacent estuarine and coastal waters through tidal-driven export of particulate materials (Jennerjahn and Ittekkot 2002; Alongi 2014) and dissolved organic C from porewater seepage (Fig. 6.9) (Bouillon et al. 2007; Kristensen et al. 2008a, Maher et al. 2013; Stieglitz et al. 2013). These organic materials may be quickly mineralized in the water column and the generated CO_2 emitted to the atmosphere (Bouillon et al. 2003). As a result, waters surrounding mangrove forests are usually oversaturated in CO_2 (Kone and Borges 2008). This source of CO_2 in tidal creek and estuarine open waters must be considered when assessing C budgets in mangrove ecosystems. Further work is urgently needed to directly measure CO_2 emissions from mangrove waters to the atmosphere, including their seasonal variability from the upper watersheds to the coastal zone (Fig. 6.9).

The largest CO_2 exchange between mangrove forests and the atmosphere is through canopy uptake and release. This exchange has traditionally been derived by extrapolating leaf metabolism measurements and leaf-area index (LAI) data (Alongi 2009). However, the recent implementation of the eddy-covariance (EC) technique in mangrove forests has improved our understanding of the gas exchange between forest canopies and the atmosphere (Fig. 6.9) (Barr et al. 2010, Rivera-Monroy et al. 2013). For example, Barr et al. (2010) used the EC technique in a subtropical region of the AEP to evaluate the seasonality in riverine mangrove forest CO_2 assimilation. They showed that assimilation was highest during the winter dry season due to an increase in diffuse solar irradiance to the forest canopy. Furthermore, net ecosystem exchange (NEE) decreased during low tide under both night and daylight conditions. More recently, Troxler et al. (2015) integrated all components of CO_2 fluxes at the same location to the whole forest scale, and found a larger than expected contribution of the below-canopy respiration components to total forest ecosystem respiration (ER). This underscores the need to improve our understanding of

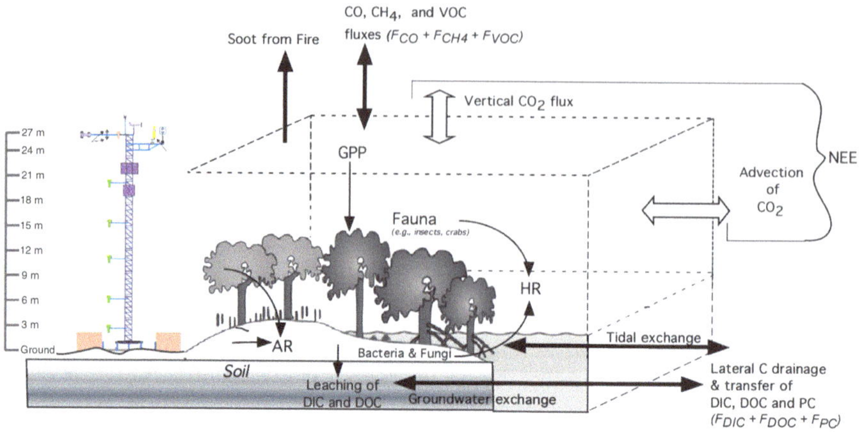

Fig. 6.9 Carbon fluxes identified as the main drivers for net ecosystem C balance (NECB) (solid *black arrows*) and net ecosystem CO_2 exchange (*NEE*) obtained with the eddy covariance method (*open arrows*). The *dotted box* represents the mangrove forest exchanging C with coastal waters and the atmosphere. The processes contributing to NECB are diffusive and advective exchange of CO_2 (i.e., NEE), CH_4, CO, volatile organic C (VOC) with the atmosphere (including soot emission during fires), lateral leaching and tidal exchange of dissolved organic C (*DOC*), dissolved inorganic C (*DIC*), and particulate C (*PC*) in mangrove waters. The biological processes regulating net ecosystem production are gross primary production (*GPP*), autotroph respiration (*AR*), and heterotroph respiration (*HR*) (Modified from Chapin et al. 2006; Rivera-Monroy et al. 2013)

below-canopy CO_2 fluxes to elucidate how mangrove C cycling respond to various disturbance events (e.g., hurricanes) and long-term changes (e.g., climate change and sea-level rise). Thus, global mangrove forest C budgets must include both canopy and below-canopy ER at larger geographical scales to fully clarify the role of these forests in mitigating climate change (Rivera-Monroy et al. 2013; Giri et al. 2011).

Methane (CH_4) emissions from mangrove sediments are usually undetectable (Alongi et al. 2000a; Kreuzwieser et al. 2003) or very low (<1% of the total CO_2 emissions; Kristensen et al. 2008b; Nóbrega et al. 2016), mainly because it is oxidized by methanotrophs before reaching the sediment surface (Canfield et al. 2005). Actual methanogenesis within mangrove sediments can be orders of magnitude higher than fluxes measured at the sediment surface (Sotomayor et al. 1994; Giani et al. 1996; Lyimo et al. 2002). Significant CH_4 emissions from mangrove sediments will therefore only occur when environmental conditions (e.g., sediment exposure during low tide) allow rapid transport to the atmosphere. As mentioned earlier in the case of CO_2 flux, crab burrows and aerial roots (e.g. pneumatophores) can also act as efficient conduits for CH_4 emissions. Furthermore, porewater seepage from creek banks during low tide generate a discharge of porewater supersaturated in CH_4, potentially becoming a source to the atmosphere (Call et al. 2015). Eutrophication can also enhance CH_4 emission as result of large labile OM inputs that increase sediment metabolic activity, leading to near-surface depletion of sulfate and a shift to methanogenesis (Sotomayor et al. 1994; Purvaja and Ramesh 2001).

Nitrous oxide (N_2O) is a potent GHG with a high radiant forcing potential (Lashof and Ahuja 1990). This gas can be produced either by denitrification under anoxic conditions or by nitrification under oxic conditions (Canfield et al. 2005). Mangrove sediments are generally characterized by low levels of porewater NH_4^+ and NO_3^- (Alongi et al. 1998; Deborde et al. 2015), and are considered small N_2O sources to the atmosphere (Chen et al. 2011; Livesley and Andrusiak 2012). N_2O fluxes measured at the sediment-air interface in mangrove environments are closely controlled by oxygen availability, organic C content, tidal exchange, Fe and SO_4^{2-} concentrations, and N availability. However, mangrove forests are areas of intense dissolved inorganic nitrogen (DIN) consumption since NH_4^+ and NO_3^- are assimilated by plants as soon as they are produced (Alongi et al. 2002). Consequently, plant uptake strongly limits N_2O emissions to the atmosphere, and high rates from mangrove sediments have only been reported under excessive anthropogenic DIN input (Konnerup et al. 2014; Chauhan et al. 2015).

6.5 Ecosystem Services: Local and Global Perspectives

Mangrove wetlands provide a number of ecosystem services (ES) through their biogeochemical functions. Using the Millennium Ecosystem Assessment classification (Carpenter et al. 2006, 2009), these services include climate change mitigation, flood regulation and water purification (see Chap. 8). A monetary value has been proposed for other mangrove ES (e.g., raw material and food, coastal protection, erosion control, maintenance of fisheries, tourism/recreation), but there are no comprehensive economic estimates for functions linked to biogeochemical transformations (Barbier et al. 2011). A possible exception is the role of C sequestration for climate change mitigation as recent studies have provided C storage estimates in some IWP and AEP locations. A major reason for the lack of economic information is the nonmarket benefits and social values of these biogeochemical functions in contrast to the more straightforward estimates for services like fish and fish habitats (Alongi 2011; Barbier et al. 2011). Another key service provided by mangroves is the export of organic material and nutrients that drives foodwebs in adjacent habitats such as coral reefs and seagrass meadows and support fisheries (Bouillon and Connolly 2009). The monetary values of this service have not been rated, but quantification (e.g. Davis et al. 2014) and incorporation into more comprehensive assessments of the ecological value of mangrove export has been initiated (Sheaves et al. 2015).

Furthermore, there is a need to assign the potential role of mangrove forests and their biogeochemical functions as sinks, sources, and transformers of C, nutrients, and heavy metals correctly in relation to the nature and location of the original inputs. Because different mangrove types (fringe, basin, riverine) provide different ES (Ewel et al. 1998), it is critical to determine the precise functional role and ES capacity of each mangrove type. Yet, given the diverse range of mangrove ecogeomorphic settings, further information is needed about, for example, the range of nutrient loadings and removal efficiencies along complex hydrological gradients

(Rivera-Monroy et al. 1999). Currently, this information is lacking for a wide range of coastal environments, including mangrove forests in both IWP and AEP regions (Barbier et al. 2011).

Discussions about the optimal use of mangrove wetlands for wastewater treatment have emerged regularly since the 1980s (Clough et al. 1983). However, there is lack of reliable field data to support the implementation of this management strategy at appropriate spatial scales (i.e., hectares) (Robertson and Phillips 1995; Rivera-Monroy et al. 1999), particularly in areas strongly impacted by aquaculture and agriculture practices and urban development (Lee et al. 2014). A potential opportunity to advance mangrove biogeochemical studies at local and regional scales is through climate-related and inspired restoration and rehabilitation projects coupled with C sequestration studies (Manez et al. 2014). Although the economic valuation of C sequestration in mangrove areas is still under discussion (Alongi 2012; Hemati et al. 2015; Huxham et al. 2015), the local incentive for such efforts can be stimulated by including nutrient removal and sewage treatments in the long-term estimates of the economic and social value of these presently nonmarket ES.

6.6 Conclusions and Research Directions

Mangrove forests are ecotones between marine and terrestrial environments. These productive wetlands possess attributes of both environments, but also have intrinsic ecological mechanisms and processes that clearly differentiate them from their immediate surroundings, defined by distinct hydrological and elevation gradients. Because of their transitional position in coastal regions, mangrove forests around the world are increasingly vulnerable to anthropogenic impacts (i.e. deforestation and urban settlement), and associated environmental alterations (e.g. erosion and fluvial contamination and alteration of marine currents). Biogeochemistry research has focused on C, N, P, and mineral cycling in mangrove ecosystems, but most surveys have been short-term (months or less) and lacking a comprehensive regional coverage to evaluate processes and transformations at the landscape level. Our current understanding of the complex interactions between multiple factors and processes that characterize mangrove biogeochemical transformations is limited and prevents extrapolation of information from one biogeographical region to another. More detailed studies must be performed at larger temporal (decadal) and spatial (watershed, regional) scales to characterize and compare environmental processes controlled by both natural and human disturbances.

Mangrove forests are particularly sensitive to rising sea level because of their direct hydrological interactions with coastal waters (Lovelock et al. 2015). The Intergovernmental Panel on Climate Change predicts that the sea level rise for the years 2081–2100 relative to 1986–2005 will range from 40 to 60 cm (IPCC 2013). Detailed studies of the response of mangrove environments to changes in sea level are, therefore, urgently required. Research initiatives should include an international network of permanent plots for long-term monitoring of biogeochemical processes, biodiversity, and plant structure among selected mangrove locations where mangrove distribution is rich and extensive in both the IWP and AEP regions. These proposed studies should include the establishment of eddy-covariance stations along biogeographical and anthropogenic gradients to improve our assessment of factors controlling mangrove net ecosystem productivity and GHG dynamics. Such EC systems will also facilitate our understanding of how recurrent natural disturbance (cyclones, tsunamis) and climate change, causing seasonal and interannual variability in precipitation, temperature, river discharge, will affect mangrove spatial distribution, productivity, and CO_2 exchange.

The interaction between mangrove forests and neighboring ecosystems with respect to exchange of materials and energy is another important aspect for consideration in future studies. For example, inland zones at higher relative elevation adjacent to mangrove forests (e.g. hypersaline sandy flats) are considered potential refuges for mangrove expansion as response to sea level rise. Although mangrove ecosystems are legally protected against deforestation or land use conversions in many countries, those inland areas are not included in conservation plans and are consistently developed for human use. Given the increasing demand for space as human populations expand into the coastal zone at an increasing rate, sometimes triggered by a high demand of mangrove ES (e.g., fisheries, recreation), there is an urgent need to develop conservation alternatives tailored to local and regional needs. Managers and decision makers must acknowledge this problem now to avoid further mangrove "squeeze" effects in the future that may cause a dramatic loss of mangrove area.

A clear example of the need to understand how mangrove biogeochemical cycles interact with pressing management priorities is the use of mangrove wetlands as tertiary treatment of waste water (e.g., aquaculture, urban). To evaluate the consequences of such disturbance, more knowledge is urgently needed about cascading effects of hydroperiod on sediment redox conditions and how this may alter, e.g., C, N, P or S cycling. The current lack of understanding could result in destructive effects on both forest productivity and survival and even conversion of mangrove forests from functional C sinks to sources of GHG. The main issue is not if this type of management (tertiary treatment) is feasible and effective because this approach has been used in other temperate wetlands, but rather to ensure correct implementation of water and nutrient management practices (aquaculture, agriculture, urban) at adequate spatial scales in combination with an understanding of the critical biogeochemical transformations.

We have undoubtedly advanced our understanding of the most critical biogeochemical transformations controlling mangrove productivity during the last 35 years, but the disproportion in the scope and quantity of mangrove biogeochemical studies across and within the IWP and AEP regions is troublesome. We advocate for collaboration and expansion of biogeochemical studies around the world. This is urgent because of the increasing rate of mangrove fragmentation and area loss at continental scales. The scientific community should soon translate current data and information about the complexity of mangrove biogeochemistry (i.e., supporting ES) into robust and applicable performance measures in management programs. It is our hope that such action may advance the conservation and protection of one of the most productive coastal ecosystems in the world.

Acknowledgements We thank all our colleagues for providing the content for this chapter. We could not have done this without their intensive work on mangrove biogeochemistry during the last decades. EK was supported by grant #12-132701 from The Danish Council for Strategic Research and grant #12-127012 from the Danish Council for Independent Research/Natural Sciences. RMC was supported by the CSIRO Coastal Carbon Biogeochemistry Cluster and thanks Ryan Dunn for advice on bioturbation. The Florida Coastal Everglades Long-Term Ecological Research program (grant nos. DBI-0620409 and DEB-1237517), NASA-JPL project "Vulnerability Assessment of Mangrove Forest Regions of the Americas" (LSU Subcontract no. 1452878), and the South Central–Climate Science Center (USA Department of Interior) provided partial funding for VHRM. XLO was supported by the Proyecto PROMETEO (SENESCYT-Ecuador).

References

Adame MF, Lovelock CE (2011) Carbon and nutrient exchange of mangrove forests with the coastal ocean. Hydrobiologia 663:23–50

Adame MF, Neil D, Wright SF, Lovelock CE (2010) Sedimentation within and among mangrove forests along a gradient of geomorphological settings. Estuar Coast Shelf Sci 86:21–30

Aller RC (2014) Sedimentary diagenesis, depositional environments, and benthic fluxes. In: Holland HD, Turekian KK (eds) Treatise on geochemistry, vol 8, 2nd edn. Elsevier, Oxford, pp 293–334

Alongi DM, Trott LA, Rachmansyah, Tirendi F, McKinnon AD, Undu MC (2008) Growth and development of mangrove forests overlying smothered coral reefs, Sulawesi and Sumatra, Indonesia. Mar Ecol Prog Ser 370:97–109

Alongi DM (2009) The energetics of mangrove forests. Springer, Breinigsville

Alongi DM (2011) Carbon payments for mangrove conservation: ecosystem constraints and uncertainties of sequestration potential. Environ Sci Pol 14:462–470

Alongi DM (2012) Carbon sequestration in mangrove forests. Carbon Manage 3:313–322

Alongi DM (2014) Carbon cycling and storage in mangrove forests. Annu Rev Mar Sci 6:195–219

Alongi DM, Dixon P (2000) Mangrove primary production and above and below-ground biomass in Sawi Bay, southern Thailand. Phuket Mar Biol Cent Spec Publ 22:31–38

Alongi DM, de Carvalho NA, Amaral AL, da Costa A, Trott L, Tirendi F (2012) Uncoupled surface and below-ground soil respiration in mangroves: implications for estimates of dissolved inorganic carbon export. Biogeochemistry 109:151–162

Alongi DM, Pfitzner J, Trott LA, Tirendi F, Dixon P, Klumpp DW (2005a) Rapid sediment accumulation and microbial mineralization in forests of the mangrove Kandelia candel in the Jiulongjiang Estuary, China. Estuar Coast Shelf Sci 63:605–618

Alongi DM, Ramanathan AL, Kannan L, Tirendi F, Trott LA, Bala Krishna Prasad M (2005b) Influence of human-induced disturbance on benthic microbial metabolism in the Pichavaram mangroves, Vellar-Coleroon complex, India. Mar Biol 147:1033–1044

Alongi DM, Sasekumar A, Chong VC, Pfitzner J, Trott LA, Tirendi F, Dixon P, Brunskill GJ (2004) Sediment accumulation and organic material flux in a managed mangrove ecosystem: estimates of land–ocean–atmosphere exchange in peninsular Malaysia. Mar Geol 208:383–402

Alongi DM, Sasekumar A, Tirendi F, Dixon P (1998) The influence of stand age on benthic decomposition and recycling of organic matter in managed mangrove forests of Malaysia. J Exp Mar Biol Ecol 225:197–218

Alongi DM, Tirendi F, Clough BF (2000a) Below-ground decomposition of organic matter in forests of the mangroves *Rhizophora stylosa* and *Avicennia marina* along the arid coast of Western Australia. Aquat Bot 68:97–122

Alongi DM, Tirendi F, Dixon P, Trott LA, Brunskill GJ (1999) Mineralization of organic matter in intertidal sediments of a tropical semi-enclosed delta. Estuar Coast Shelf Sci 48:451–467

Alongi DM, Tirendi F, Trott LA, Xuan TT (2000b) Benthic decomposition rates and pathways in plantations of the mangrove *Rhizophora apiculata* in the Mekong delta, Vietnam. Mar Ecol Prog Ser 194:87–101

Alongi DM, Trott LA, Wattayakorn G, Clough BF (2002) Below-ground nitrogen cycling in relation to net canopy production in mangrove forests of southern Thailand. Mar Biol 140:855–864

Alongi DM, Wattayakorn G, Pfitzner J, Tirendi F, Zagorskis I, Brunskill GJ, Davidson A, Clough BF (2001) Organic carbon accumulation and metabolic pathways in sediments of mangrove forests in southern Thailand. Mar Geol 179:85–103

Andreetta A, Fusi M, Cameldi I, Cimo F, Carnicelli S, Cannicci S (2014) Mangrove carbon sink. Do burrowing crabs contribute to sediment carbon storage? Evidence from a Kenyan mangrove system. J Sea Res 85:524–533

Araújo JMC, Otero XL, Marques AGB, Nóbrega GN, Silva JRF, Ferreira TO (2012) Geochemistry of iron in mangrove soils in a semiarid tropical climate: effects of the burrowing activity of the crabs *Ucides cordatus* and *Uca maracoani*. Geo-Mar Lett 32:289–300

Bacon PR (1975) Shell form, byssal development and habitat of *Mytella guyanensis* (Lamarck) and *M. falcata* (Orbigny) (Pelecypoda:Mytilidae) in Trinidad, West Indies. Proc Malacol Soc Lond 41:511–520

Barbier EB, Hacker SD, Kennedy C, Koch EW, Stier AC, Silliman BR (2011) The value of estuarine and coastal ecosystem services. Ecol Monogr 81:169–193

Barr JG, Engel V, Fuentes JD, Zieman JC, O'Halloran TL, Smith TJ, Anderson GH (2010) Controls on mangrove forest-atmosphere carbon dioxide exchanges in western Everglades National Park. J Geophys Res Biogeosci 115:G02020

Bartoli M, Longhi D, Nizzoli D, Como S, Magni P, Viaroli P (2009) Short term effects of hypoxia and bioturbation on solute fluxes, denitrification and buffering capacity in a shallow dystrophic pond. J Exp Mar Biol Ecol 381:105–113

Berger U, Rivera-Monroy VH, Doyle TW, Dahdouh-Guebas F, Duke NC, Fontalvo-Herazo ML, Hildenbrandt H, Koedam N, Mehlig U, Piou C, Twilley RR (2008) Advances and limitations of individual-based models to analyze and predict dynamics of mangrove forests: a review. Aquat Bot 89:260–274

Bouillon S (2011) Storage beneath mangroves. Nat Geosci 4:1–2

Bouillon S, Borges AV, Castañeda-Moya E, Diele K, Dittmar T, Duke NC, Kristensen E, Lee SY, Marchand C, Middelburg JJ, Rivera-Monroy VH, Smith TJ III, Twilley RR (2008) Mangrove production and carbon sinks: a revision of global budget estimates. Glob Biogeochem Cycles 22:GB2013. doi:10.1029/2007GB003052

Bouillon S, Connolly RM (2009) Carbon exchange among tropical coastal ecosystems. In: Nagelkerken I (ed) Ecological connectivity among tropical coastal ecosystems. Springer, Dordrecht, pp 45–70

Bouillon S, Dahdouh-Guebas F, Rao AVVS, Koedam N, Dehairs F (2003) Sources of organic carbon in mangrove sediments: variability and possible ecological implications. Hydrobiologia 495:33–39

Bouillon S, Middelburg JJ, Dehairs F, Borges AV, Abril G, Flindt M, Ulomi S, Kristensen E (2007) Importance of intertidal sediment processes and porewater exchange on the water column biogeochemistry in a pristine mangrove creek (Ras Dege, Tanzania). Biogeosciences 4:311–322

Bunt JS (1995) Continental scale patterns in mangrove litter fall. Hydrobiologia 295:135–140

Burdige DJ (2006) Geochemistry of marine sediments. Princeton University Press, Princeton, p 609

Butler IB, Rickard D (2000) Framboidal pyrite formation via the oxidation of iron (II) monosulfide by hydrogen sulphide. Geochim Cosmochim Acta 64:2665–2672

Call M, Maher DT, Santos IR, Ruiz-Halpern S, Mangion P, Sanders CJ, Erler DV, Oakes JM, Rosentreter J, Murray R, Eyre BD (2015) Spatial and temporal variability of carbon dioxide and methane fluxes over semi-diurnal and spring–neap–spring timescales in a mangrove creek. Geochim Cosmochim Acta 150:211–225

Canfield DE, Kristensen E, Thamdrup B (2005) Aquatic geomicrobiology. Academic, Amsterdam

Canfield DE, Raiswell R, Bottrell S (1992) The reactivity of sedimentary iron minerals toward sulfide. Am J Sci 292:659–683

Carpenter SR, DeFries R, Dietz T, Mooney HA, Polasky S, Reid WV, Scholes RJ (2006) Millennium ecosystem assessment: research needs. Science 314:257–258

Carpenter SR, Mooney HA, Agard J, Capistrano D, DeFries RS, Diaz S, Dietz T, Duraiappah AK, Oteng-Yeboah A, Pereira HM, Perrings C, Reid WV, Sarukhan J, Scholes RJ, Whyte A (2009) Science for managing ecosystem services: beyond the millennium ecosystem assessment. Proc Natl Acad Sci U S A 106:1305–1312

Castañeda-Moya E, Rivera-Monroy VH, Twilley RR (2006) Mangrove zonation in the dry life zone of the Gulf of Fonseca, Honduras. Estuar Coasts 29:751–764

Castaneda-Moya E, Twilley RR, Rivera-Monroy VH (2013) Allocation of biomass and net primary productivity of mangrove forests along environmental gradients in the Florida Coastal Everglades, USA. For Ecol Manag 307:226–241

da Silva Castiglioni D, Negreiros-Fransozo ML (2005) Comparative population biology of *Uca rapax* (smith, 1870) (Brachyura, Ocypodidae) from two subtropical mangrove habitats on the Brazilian coast. J Nat Hist 39:1627–1640

Chan BKK, Chan KKY, Leung PCM (2006) Burrow architecture of the ghost crab *Ocypode ceratophthalma* on a sandy shore in Hong Kong. Hydrobiologia 560:43–49

Chanda A, Akhand A, Manna S, Dutta S, Hazra S, Das I, Dadhwal VK (2013) Characterizing spatial and seasonal variability of carbon dioxide and water vapour fluxes above a tropical mixed mangrove forest canopy, India. J Earth Syst Sci 122:503–513

Chapin FS, Woodwell GM, Randerson J, Lovett G, Baldocchi D, Clark DA, Harmon ME, Rastetter E, Schimel D, Valentini R, Wirth C, Cole J, Goulden M, Harden J, Heimann M, Howarth R, Matson P, McGuire AD, Melillo J, Mooney H, Neff J, Houghton R, Pace M, Ryan MG, Running S, Sala O, Schlesinger W, Schulze D (2006) Reconciling carbon cycle concepts, terminology, and methods. Ecosystems 9:1041–1050

Chauhan R, Datta A, Ramanathan AL, Adhya TK (2015) Factors influencing spatio-temporal variation of methane and nitrous oxide emission from a tropical mangrove of eastern coast of India. Atmos Environ 107:95–106

Chen GC, Tam NFY, Wong YS, Ye Y (2011) Effect of wastewater discharge on greenhouse gas fluxes from mangrove soils. Atmos Environ 45:1110–1115

Chen GC, Ulumuddin YI, Pramudji S, Chen SY, Chen B, Ye Y, Ou DY, Ma ZY, Huang H, Wang JK (2014) Rich soil carbon and nitrogen but low atmospheric greenhouse gas fluxes from North Sulawesi mangrove swamps in Indonesia. Sci Total Environ 487:91–96

Chmura GL, Anisfeld SC, Cahoon DR, Lynch JC (2003) Global carbon sequestration in tidal, saline wetland soils. Glob Biogeochem Cycles 17:1111

Clark MW, McConchie D, Lewis DW, Saenger P (1998) Redox stratification and heavy metal partitioning in *Avicennia*-dominated mangrove sediments: a geochemical model. Chem Geol 149:147–171

Clayton DA, Snowden R (2000) Surface activity in the mudskipper, *Periophthalmus waltoni* Koumans 1941 in relation to prey activity and environmental factors. Trop Zool 13:239–249

Clough BF, Boto KG, Attiwill PM (1983) Mangroves and sewage: a re-evaluation. In: Teas HJ (ed) Tasks for vegetation science. W. Junk, The Hague, pp 151–161

Cornel RM, Schwertmann U (1996) The iron oxide. VCH, New York, p 377

Coupland GT, Paling EI, McGuinness KA (2005) Vegetative and reproductive phenologies of four mangrove species from northern Australia. Aust J Bot 53:109–117

Crase B, Liedloff A, Vesk PA, Burgman MA, Wintle BA (2013) Hydroperiod is the main driver of the spatial pattern of dominance in mangrove communities. Glob Ecol Biogeogr 22:06–817

Cross RE, Curran MC (2000) Effects of feeding pit formation by rays on an intertidal meiobenthic community. Estuar Coast Shelf Sci 51:293–298

Dalsgaard T, Thamdrup B, Canfield DE (2005) Anaerobic ammonium oxidation (anammox) in the marine environment. Res Microbiol 156:457–464

Dangremond EM, Feller IC (2014) Functional traits and nutrient limitation in the rare mangrove *Pelliciera rhizophorae*. Aquat Bot 116:1–7

Davis JP, Pitt KA, Fry B, Olds AD, Connolly RM (2014) Seascape-scale trophic links for fish on inshore coral reefs. Coral Reefs 33:897–907

Day JW, Coronado-Molina C, Vera-Herrera FR, Twilley R, Rivera-Monroy VH, Alvarez-Guillen H, Day R, Conner W (1996) A 7 year record of above-ground net primary production in a southeastern Mexican mangrove forest. Aquat Bot 55:39–60

Deborde J, Marchand C, Molnar N, Patrona LD, Meziane T (2015) Concentrations and fractionation of carbon, iron, sulfur, nitrogen and phosphorus in mangrove sediments along an intertidal gradient (semi-arid climate, New Caledonia). J Mar Sci Eng 3:52–72

Demas GP, Rabenhorst MC (1999) Subaqueous soils: pedogenesis in a submersed environment. Soil Sci Soc Am J 63:1250–1257

Demas GP, Rabenhorst MC, Stevenson JC (1996) A pedological approach to the study of shallow-water habitats. Estuaries 19:229–237

Donato DC, Kauffman JB, Murdiyarso D, Kurnianto S, Stidham M, Kanninen M (2011) Mangroves among the most carbon-rich forests in the tropics. Nat Geosci 4:293–297

Dubey SK, Chakraborty DC, Chakraborty S, Choudhuyry A (2013) Burrow architecture of red ghost crab *Ocypode macrocera* (H. Milne-Edwards, 1852): a case study in Indian Sundarbans. Explor Anim Med Res 3:136–144

Duke NC, Ball MC, Ellison J (1998) Factors influencing biodiversity and distributional gradients in mangroves. Glob Ecol Biogeogr Lett 7:27–47

Dunn RJK, Welsh DT, Jordan MA, Teasdale PR, Lemckert CJ (2009) Influence of natural amphipod (Victoriopisa australiensis) (Chilton, 1923) population densities on benthic metabolism, nutrient fluxes, denitrification and DNRA in sub-tropical estuarine sediment. Hydrobiologia 628:95–109

Dutta MK, Chowdhury C, Jana TK, Mukhopadhyay SK (2013) Dynamics and exchange fluxes of methane in the estuarine mangrove environment of the Sundarbans, NE coast of India. Atmos Environ 77:631–639

Dworshak PC, Felder DL, Tudge C (2012) Infraorders Axiidea de Saint Laurent, 1979 and Gebiidea de Saint Laurent, 1979 (formerly known collectively as Thalassinidea). In: Schram FR, von Vaupel Klein JC, Charmantier-Daures M, Frost J (eds) Treatise on zoology – anatomy, taxonomy, biology – the Crustacea, Decapoda, volume 9 part B Decapoda: Astacidea P.P. (Enoplometopoidea, Nephropoidea), Glypheidea, Axiidea, Gebiidea, and Anomura. Koninklijke Brill NV, Leiden

Dworshak PC, Pervesler P (2002) *Alpheus migrans* Lewinsohn & Holthuis, 1978 (Decapoda, Caridea, Alpheidae): burrow morphology and first record from the Red Sea. Crustaceana 75:351–357

Ellison AM, Farnsworth EJ (1997) Simulated sea level change alters anatomy, physiology, growth, and reproduction of red mangrove (*Rhizophora mangle* L.) Oecologia 112:435–446

Emmerson WD, McGwynne LE (1992) Feeding and assimilation of mangrove leaves by the crab *Sesarma meinerti* de Man in relation to leaf-litter production in Mgazana, a warm-temperate southern African mangrove swamp. J Exp Mar Biol Ecol 157:41–53

Ewel K, Twilley R, Ong J (1998) Different kinds of mangrove forests provide different goods and services. Glob Ecol Biogeogr 7:83–94

Feller IC, Lovelock CE, Berger U, McKee KL, Joye SB, Ball MC (2010) Biocomplexity in mangrove ecosystems. Annu Rev Mar Sci 2:395–417

Feller IC, Lovelock CE, McKee KL (2007) Nutrient addition differentially affects ecological processes of *Avicennia germinans* in nitrogen versus phosphorus limited mangrove ecosystems. Ecosystems 10:347–359

Feller IC, Whigham DF, McKee KL, Lovelock CE (2003) Nitrogen limitation of growth and nutrient dynamics in a disturbed mangrove forest, Indian River Lagoon, Florida. Oecologia 134:405–414

Feller IC, Whigham DF, O'Neill JP, McKee KL (1999) Effects of nutrient enrichment on within-stand cycling in a mangrove forest. Ecology 80:2193–2205

Fernandes SO, Bonin PC, Michotey VD, Garcia N, Bharathi PAL (2012a) Nitrogen-limited mangrove ecosystems conserve N through dissimilatory nitrate reduction to ammonium. Sci Rep 2. doi:10.1038/srep00419

Fernandes SO, Michotey VD, Guasco S, Bonin PC, Bharathi PAL (2012b) Denitrification prevails over anammox in tropical mangrove sediments (Goa, India). Mar Environ Res 74:9–19
Ferreira TO, Otero XL, de Souza VS, Vidal-Torrado P, Macias F, Firme LP (2010) Spatial patterns of soil attributes and components in a mangrove system in Southeast Brazil (Sao Paulo). J Soils Sediments 10:995–1006
Ferreira TO, Otero XL, Vidal-Torrado P, Macías F (2007b) Effects of bioturbation by root and crab activity on iron and sulfur biogeochemistry in mangrove substrate. Geoderma 142:36–46
Ferreira TO, Otero XL, Vidal-Torrado P, Macías M (2007c) Redox processes in mangrove soils under Rhizophora mangle in relation to different environmental conditions. Soil Sci Soc Am J 71:484–491
Ferreira TO, Vidal-Torrado P, Otero XL, Macías F (2007a) Are mangrove forest substrates sediments or soils? A case study in southeastern Brazil. Catena 70:79–91
Fiala K, Hernandez L (1993) Root biomass of a mangrove forest in southwestern Cuba (Majana). Ekologia-Bratislava 12:15–30
Fischer WR, Pfanneberg T (1984) An improved method for testing the rate of iron(III) oxide reduction by bacteria. Zentralblatt für Mikrobiologie 139:163–166
Furukawa K, Wolanski E, Mueller H (1997) Currents and sediment transport in mangrove forests. Estuar Coast Shelf Sci 44:301–310
Fuss CM (1964) Observations on burrowing behavior of the pink shrimp, *Penaeus duorarum* Burkenroad. Bull Mar Sci Gulf Caribb 14:62–73
Giani L, Bashan Y, Holguin G, Strangmann A (1996) Characteristics and methanogenesis of the Balandra lagoon mangrove soils, Baja California Sur, Mexico. Geoderma 72:149–160
Giblin AE, Tobias CR, Song B, Weston N, Banta GT, Rivera-Monroy VH (2013) The importance of dissimilatory nitrate reduction to ammonium (DNRA) in the nitrogen cycle of coastal ecosystems. Oceanography 26:124–131
Giddins RL, Lucas JS, Neilson MJ, Richards GN (1986) Feeding ecology of the mangrove crab *Neosarmatium smithi* (Crustacea, Decapoda, Sesarmidae). Mar Ecol Prog Ser 33:147–155
Gillikin DP (2000) Factors controlling the distribution of Kenyan brachyuran mangrove crabs: salinity tolerance and ecophysiology of two Kenyan *Neosarmatium* species. M.Sc. Thesis, Free University of Brussels, Brussels, Belgium
Gillikin DP, Kamanu CP (2005) Burrowing in the east African mangrove crab, *Chiromantes ortmanni* (crosnier, 1965) (Decapoda, Brachyura, Sesarmidae). Crustaceana 78:1273–1275
Gillikin DP, Schubart CD (2004) Ecology and systematics of mangrove crabs of the genus *Perisesarma* (Crustacea: Brachyura: Sesarmidae) from East Africa. Zool J Linnean Soc 141:435–445
Giri C, Ochieng E, Tieszen LL, Zhu Z, Singh A, Loveland T, Masek J, Duke N (2011) Status and distribution of mangrove forests of the world using earth observation satellite data. Glob Ecol Biogeogr 20:154–159
Gonneea ME, Charette MA, Liu Q, Herrera-Silveira JA, Morales-Ojeda SM (2014) Trace element geochemistry of groundwater in a karst subterranean estuary (Yucatan peninsula, Mexico). Geochim Cosmochim Acta 132:31–49
Gribsholt B, Kostka JE, Kristensen E (2003) Impact of fiddler crabs and plant roots on sediment biogeochemistry in a Georgia salt marsh. Mar Ecol Prog Ser 259:237–251
Hemati Z, Hossain M, Emenike CU, Rozainah MZ (2015) Rate of carbon storage in soil of natural and degraded mangrove forest in peninsular Malaysia. Clean-Soil Air Water 43:14–619
Hernes PJ, Benner R, Cowie GL, Goni MA, Bergamaschi BA, Hedges JI (2001) Tannin diagenesis in mangrove leaves from a tropical estuary: a novel molecular approach. Geochim Cosmochim Acta 65:3109–3122
Holmer M, Andersen FØ, Kristensen E, Thongtham N (1999) Transformation and exchange processes in the Bangrong mangrove forest – seagrass bed system, Thailand. Seasonal and spatial variations in benthic metabolism and sulfur biogeochemistry. Aquat Microb Ecol 20:203–212
Holmer M, Kristensen E, Banta GT, Hansen K, Jensen MH, Bussarawit N (1994) Biogeochemical cycling of sulfur and iron in sediments of a south-east Asian mangrove, Phuket Island, Thailand. Biogeochemistry 26:145–161

Huxham M, Emerton L, Kairo J, Munyi F, Abdirizak H, Muriuki T, Nunan F, Briers RA (2015) Applying climate compatible development and economic valuation to coastal management: a case study of Kenya's mangrove forests. J Environ Manag 157:68–181

Imgraben S, Dittmann S (2008) Leaf litter dynamics and litter consumption in two temperate south Australian mangrove forests. J Sea Res 59:83–93

IPCC (2013) In: Stocker TF, Qin D, Plattner G-K, Tignor M, Allen SK, Boschung J, Nauels A, Xia Y, Bex V, Midgley PM (eds) Climate change 2013: the physical science basis. Contribution of working group I to the fifth assessment report of the intergovernmental panel on climate change. Cambridge University Press, Cambridge, UK/New York, p 1535

Jennerjahn TC, Ittekkot V (2002) Relevance of mangroves for the production and deposition of organic matter along tropical continental margins. Naturwissenschaften 89:23–30

Jones DA (1984) Crabs of the mangal ecosystem. In: Por FD, Dor I (eds) Hydrobiology of the mangal. W. Junk, The Hague, pp 89–109

Joshi PK, Kulkarni GK, Nagabhushanam R (1979) Studies on the behaviour and substratum preference in juveniles of the marine prawn, *Metapenaeus monoceros* (Fabricius). Hydrobiologia 65:195–198

Kartika WD, Patria MP (2012) Nest characteristics of mud lobster *Thalassina anomala* (Herbst 1804) in Tanjung Jabung Barat, Jambi, Sumatra, Indonesia. Int J Basic Appl Sci 12:245–248

Kerr G (2001) Ecological aspects of *Trypaea australiensis*. Ph.D. Thesis, Southern Cross University, Lismore, New South Wales

Keuskamp JA, Hefting MM, Dingemans BJJ, Verhoven JTA, Feller I (2015) Effects of nutrient enrichment on mangrove leaf litter decomposition. Sci Total Environ 508:402–410

Kone YJ-M, Borges AV (2008) Dissolved inorganic carbon dynamics in the waters surrounding forested mangroves of the Ca Mau Province (Vietnam). Estuar Coast Shelf Sci 77:409–421

Konnerup D, Betancourt-Portela JM, Villamil C, Parra JP (2014) Nitrous oxide and methane emissions from the restored mangrove ecosystem of the Cienaga Grande de Santa Marta, Colombia. Estuar Coast Shelf Sci 140:43–51

Kostka JE, Gribsholt B, Petrie E, Dalton D, Skelton H, Kristensen E (2002) The rates and pathways of carbon oxidation in bioturbated saltmarsh sediments. Limnol Oceanogr 47:230–240

Kreuzwieser J, Buchholz J, Rennenberg H (2003) Emisson of methane and nitrous oxide by Australian mangrove ecosystems. Plant Biol 5:423–431

Kristensen E (2008) Mangrove crabs as ecosystem engineers, with emphasis on sediment processes. J Sea Res 59:30–43

Kristensen E, Alongi DM (2006) Control by fiddler crabs (*Uca vocans*) and plant roots (*Avicennia marina*) on carbon, iron and sulfur biogeochemistry in mangrove sediment. Limnol Oceanogr 51:1557–1571

Kristensen E, Andersen FØ, Holmboe N, Holmer M, Thongtham N (2000) Carbon and nitrogen mineralization in sediment of the Bangrong mangrove area, Phuket, Thailand. Aquat Microb Ecol 22:199–213

Kristensen E, Bouillon S, Dittmar T, Marchand C (2008a) Organic carbon dynamics in mangrove ecosystems. Aquat Bot 89:201–219

Kristensen E, Devol AH, Ahmed SI, Saleem M (1992) Preliminary study of benthic metabolism and sulfate reduction in a mangrove swamp of the Indus Delta, Pakistan. Mar Ecol Prog Ser 90:287–297

Kristensen E, Flindt MR, Borges AV, Bouillon S (2008b) Emission of CO_2 and CH_4 to the atmosphere by sediments and open waters in two Tanzanian mangrove forests. Mar Ecol Prog Ser 370:53–67

Kristensen E, Holmer M, Bussarawit N (1991) Benthic metabolism and sulfate reduction in a south-east Asian mangrove swamp. Mar Ecol Prog Ser 73:93–103

Kristensen E, Jensen MH, Banta GT, Hansen K, Holmer M, King GM (1998) Transformation and transport of inorganic nitrogen in sediment of a southeast Asian mangrove forest. Aquat Microb Ecol 15:165–175

Kristensen E, King GM, Banta GT, Holmer M, Jensen MH, Hansen K, Bussarawit N (1994) Sulfate reduction, acetate turnover and carbon metabolism in sediments of the Ao Nam Bor mangrove, Phuket, Thailand. Mar Ecol Prog Ser 109:245–255

Kristensen E, Mangion P, Tang M, Flindt MR, Ulomi S (2011) Benthic metabolism and partitioning of electron acceptors for microbial carbon oxidation in sediments of two Tanzanian mangrove forests. Biogeochemistry 103:143–158

Kristensen E, Pilgaard R (2001) The role of fecal pellet deposition by leaf-eating sesarmid crabs on mineralization processes in a mangrove sediment (Phuket, Thailand). In: Aller JY, Woodin SA, Aller RC (eds) Organism-sediment interactions. University South Carolina Press, Columbia, pp 369–384

Kristensen E, Rabenhorst MC (2015) Do marine rooted plants grow in sediment or soil? Earth-Sci Rev 145:1–8

Krobicki M, Zatoń M (2008) Middle and Late Jurassic roots of brachyuran crabs: palaeoenvironmental distribution during their early evolution. Palaeogeogr Palaeoclimatol Palaeoecol 263:30–43

Lacerda LD, Ittekkot V, Patchineelam SR (1995) Biogeochemistry of mangrove soil organic matter: a comparison between *Rhizophora* and *Avicennia* soils in south-eastern Brazil. Estuar Coast Shelf Sci 40:713–720

Lashof DA, Ahuja DR (1990) Relative contributions of greenhouse gas emissions to global warming. Nature 344:529–531

Lee RY, Joye SB (2006) Seasonal patterns of nitrogen fixation and denitrification in oceanic mangrove habitats. Mar Ecol Prog Ser 307:127–141

Lee SY (1997) Potential trophic importance of the faecal material of the mangrove sesarmine crab *Sesarma messa*. Mar Ecol Prog Ser 159:275–284

Lee SY (2008) Mangrove macrobenthos: assemblages, services, and linkages. J Sea Res 59:16–29

Lee SY (2016) From blue to black: anthropogenic forcing of carbon and nitrogen influx to mangrove-lined estuaries in the South China Sea. Mar Pollut Bull 109:682–690

Lee SY, Primavera JH, Dahdouh-Guebas F, McKee K, Bosire JO, Cannicci S, Diele K, Fromard F, Koedam N, Marchand C, Mendelssohn I, Mukherjee N, Record S (2014) Ecological role and services of tropical mangrove ecosystems: a reassessment. Glob Ecol Biogeogr 23:726–743

Leopold A, Marchand C, Deborde J, Allenbach M (2015) Temporal variability of CO_2 fluxes at the sediment-air interface in mangroves (New Caledonia). Sci Total Environ 502:617–626

Leopold A, Marchand C, Deborde J, Chaduteau C, Allenbach M (2013) Influence of mangrove zonation on CO_2 fluxes at sediment-air interface (New-Caledonia). Geoderma 202–203:62–70

Li M, Gu J-D (2013) Community structure and transcript responses of anammox bacteria, AOA, and AOB in mangrove sediment microcosms amended with ammonium and nitrite. Appl Microbiol Biotechnol 97:9859–9874

Livesley SJ, Andrusiak SM (2012) Temperate mangrove and salt marsh sediments are a small methane and nitrous oxide source but important carbon store. Estuar Coast Shelf Sci 97:19–27

Lovley DR, Phillips EJP (1987) Rapid assay for microbially reducible ferric iron in aquatic sediments. Appl Environ Microbiol 53:1536–1540

Lovelock CE (2008) Soil respiration and belowground carbon allocation in mangrove forests. Ecosystems 11:342–354

Lovelock CE, Ball MC, Martin KC, Feller IC (2009) Nutrient enrichment increases mortality of mangroves. PLoS One 4. doi:10.1371/journal.pone.0005600

Lovelock CE, Feller IC, Ball MC, Ellis J, Sorrell B (2007) Testing the growth rate vs. geochemical hypothesis for latitudinal variation in plant nutrients. Ecol Lett 10:1154–1163

Lovelock CE, Feller IC, Ball MC, Engelbrecht BMJ, Ewe ML (2006) Differences in plant function in phosphorus- and nitrogen-limited mangrove ecosystems. New Phytol 172:514–522

Lovelock CE, Feller IC, Reef R, Ruess RW (2014) Variable effects of nutrient enrichment on soil respiration in mangrove forests. Plant Soil 379:135–148

Lovelock CE, Cahoon DR, Friess DA, Guntenspergen GR, Krauss KW, Reef R, Rogers K, Saunders ML, Sidik F, Swales A, Saintilan N, Thuyen LX, Triet T (2015) The vulnerability of indo-Pacific mangrove forests to sea-level rise. Nature 526:559–563

Lu CY, Wong YS, Tam NFY, Ye Y, Lin P (1999) Methane flux and production from sediments of a mangrove wetland on Hainan Island, China. Mangrove Salt Marshes 3:41–49

Lugo AE, Snedaker SC (1978) The ecology of mangroves. Annu Rev Ecol Syst 5:39–64

Lyimo TJ, Pol A, Jetten MSM, Op den Camp HJM (2009) Diversity of methanogenic archaea in a mangrove sediment and isolation of a new Methanococcoides strain. FEMS Microbiol Lett 291:247–253

Lyimo TJ, Pol A, Op den Camp HJM (2002) Sulfate reduction and methanogenesis in sediments of Mtoni mangrove forest, Tanzania. Ambio 31:614–616

Maher DT, Cowley K, Santos IR, Macklin P, Eyre BD (2015) Methane and carbon dioxide dynamics in a subtropical estuary over a diel cycle: insights from automated in situ radioactive and stable isotope measurements. Mar Chem 168:69–79

Maher DT, Santos IR, Golsby-Smith L, Gleeson J, Eyre BD (2013) Groundwater-derived dissolved inorganic and organic carbon exports from a mangrove tidal creek: the missing mangrove carbon sink? Limnol Oceanogr 58:475–488

Manez KS, Krause G, Ring I, Glaser M (2014) The Gordian knot of mangrove conservation: disentangling the role of scale, services and benefits. Glob Environ Chang-Hum Policy Dimens 28:120–128

Marchand C, Albéric P, Lallier-Vergès E, Baltzer F (2006) Distribution and characteristics of dissolved organic matter in mangrove sediments pore waters along the coastline of French Guiana. Biogeochemistry 81:59–75

Marchand C, Allenbach M, Lallier-Vergès E (2011) Relationships between heavy metals distribution and organic matter cycling in mangrove sediments of Conception Bay, new-Caledonia. Geoderma 160:444–456

Marchand C, Baltzer F, Lallier-Verges E, Alberic P (2004) Pore-water chemistry in mangrove sediments: relationship with species composition and developmental stages (French Guiana). Mar Geol 208:361–381

Marchand C, Disnar J-R, Lallier-Vergès E, Lottier N (2005) Early diagenesis of carbohydrates and lignin in mangrove sediments submitted to variable redox conditions (French Guiana). Geochim Cosmochim Acta 69:131–142

Marinho CC, Campos EA, Guimaraes JRD, Esteves FA (2012) Effect of sediment composition on methane concentration and production in the transition zone of a mangrove (Sepetiba Bay, Rio de Janeiro, Brazil). Braz J Biol 72:429–436

Matsui N (1998) Estimated stocks of organic carbon in mangrove roots and sediments in Hinchinbrook Channel, Australia. Mangrove Salt Marshes 2:199–204

Mazda Y, Kanazawa N, Wolanski E (1995) Tidal asymmetry in mangrove creeks. Hydrobiologia 295:51–58

Mazda Y, Kobashi D, Okada K (2005) Tidal scale hydrodynamics within mangrove swaps. Wetl Ecol Manag 13:647–655

Mazda Y, Wolanski E (2009) Hydrodynamics and modeling of water flow in mangrove areas. In: Perillo GME, Wolanski E, Cahoon D, Brinson M (eds) Coastal wetlands. Elsevier, Amsterdam, pp 69–88

Mchenga ISS, Mfilinge PL, Tsuchiya M (2007) Bioturbation activity by the grapsid crab *Helice formosensis* and its effects on mangrove sedimentary organic matter. Estuar Coast Shelf Sci 73:316–324

Mchenga ISS, Tsuchiya M (2008) Nutrient dynamics in mangrove crab burrow sediments subjected to anthropogenic input. J Sea Res 59:103–113

Mendoza UN, da Cruz CC, Menezes MP, Lara RJ (2012) Flooding effects on phosphorus dynamics in an Amazonian mangrove forest, Northern Brazil. Plant Soil 353:107–121

Meyer RL, Risgaard-Petersen N, Allen DE (2005) Correlation between anammox activity and microscale distribution of nitrite in a subtropical mangrove sediment. Appl Environ Microbiol 71:6142–6149

Middleton BA, McKee KL (2001) Degradation of mangrove tissues and implications for peat formation in Belizean island forests. J Ecol 89:818–828

Mitsch WJ, Gosselink JG (2015) Wetlands, Third edn. Wiley, New York

Molnar N, Welsh DT, Marchand C, Deborde J, Meziane T (2013) Impacts of shrimp farm effluents on water quality, benthic metabolism and N-dynamics in a mangrove forest (New Caledonia). Estuar Coast Shelf Sci 117:12–21

Morell JM, Corredor JE (1993) Sediment nitrogen trapping in a mangrove lagoon. Estuar Coast Shelf Sci 37:203–212

Morton B (1976) The biology and functional morphology of the Southeast Asian mangrove bivalve, *Polymesoda* (*Geloina*) erosa (Solander, 1786) (Bivalvia: Corbiculidae). Can J Zool 54:482–500

Murdiyarso D, Purbopuspito J, Kauffman JB, Warren MW, Sasmito SD, Donato DC, Manuri S, Krisnawati H, Taberima S, Kurnianto S (2015) The potential of Indonesian mangrove forests for global climate change mitigation. Nat Clim Chang 5:1089–1092

Nedwell DB, Blackburn TH, Wiebe WJ (1994) Dynamic nature of the turnover of organic carbon, nitrogen and sulphur in the sediments of a Jamaican mangrove forest. Mar Ecol Prog Ser 110:223–231

Nielsen OI, Macintosh DJ, Kristensen E (2003) Impact of fiddler crabs (*Uca* spp.) on rates and pathways of benthic mineralization in deposited shrimp pond waste. J Environ Mar Biol Ecol 289:59–81

Nóbrega GN, Ferreira TO, Neto MS, Queiroz HM, Artur AG, Mendonça EDS, Silva EO, Otero XL (2016) Edaphic factors controlling summer (rainy season) greenhouse gas emissions (CO_2 and CH_4) from semiarid mangrove soils (NE-Brazil). Sci Total Environ 542:685–693

Nóbrega GN, Otero XL, Macías F, Ferreira TO (2014) Phosphorus geochemistry in a Brazilian semi-arid mangrove soil affected by shrimp farm effluents. Environ Monit Assess 186:5749–5762

Noel V, Marchand C, Juillot F, Ona-Nguema G, Viollier E, Marakovic G, Olivi L, Delbes L, Gelebart F, Morin G (2014) EXAFS analysis of iron cycling in mangrove sediments downstream a lateritized ultramafic watershed (Vavouto Bay, New Caledonia). Geochim Cosmochim Acta 136:211–228

Nordhaus I, Wolff M, Diele K (2006) Litter processing and population food intake of the mangrove crab *Ucides cordatus* in a high intertidal forest in northern Brazil. Estuar Coast Shelf Sci 67:239–250

Opsahl S, Benner R (1999) Characterization of carbohydrates during early diagenesis of five vascular plant tissues. Org Geochem 30:83–94

O'Shea OR, Thums M, van Keulen M, Meekan M (2012) Bioturbation by stingrays at Ningaloo Reef, Western Australia. Mar Freshw Res 63:189–197

Otero XL, Ferreira TO, Huerta-Diaz MA, Partiti CSM, Souza V, Vidal-Torrado P, Macias F (2009) Geochemistry of iron and manganese in soils and sediments of a mangrove system, Island of Pai Matos (Cananeia — SP, Brazil). Geoderma 148:318–335

Otero XL, Ferreira TO, Vidal-Torrado P, Macias F (2006) Spatial variation in pore water geochemistry in a mangrove system (Pai Matos Island, Cananeia-Brazil). Appl Geochem 21:2171–2186

Otero XL, Lucheta AR, Ferreira TO, Huerta-Díaz MA, Lambais MR (2014) Archaeal diversity and the extent of iron and manganese pyritization in sediments from a tropical mangrove creek (Cardoso Island, Brazil). Estuar Coast Shelf Sci 146:1–13

Pendleton L, Donato DC, Murray BC, Crooks S, Jenkins WA, Sifleet S, Craft C, Fourqurean JW, Kauffman JB, Marba N, Megonigal P, Pidgeon E, Herr D, Gordon D, Baldera A (2012) Estimating global "blue carbon" emissions from conversion and degradation of vegetated coastal ecosystems. PLoS One 7:e43542

Penha-Lopes G, Kristensen E, Flindt M, Mangion P, Bouillon S, Paula J (2010) The role of biogenic structures on the biogeochemical functioning of mangrove constructed wetlands sediments – a mesocosm approach. Mar Pollut Bull 60:560–572

Polidoro BA, Carpenter KE, Collins L, Duke NC, Ellison AE, Ellison JC, Farnsworth EJ, Fernando ES, Kathiresan K, Koedam NE, Livingstone SR, Miyagi T, Moore GE, Nam VN, Ong JE, Primavera JH, Salmo SG, Sanciangco JC, Sukardjo S, Wang Y, Yong JWH (2010) The loss of species: mangrove extinction risk and geographic areas of global concern. PLoS One 5:e10095

Poulton SW, Krom MD, Raiswell R (2004) A revised scheme for the reactivity of iron (oxyhydr) oxide minerals towards dissolved sulfide. Geochim Cosmochim Acta 68:3703–3715

Primavera JH, Lebata J (1995) Diel activity patterns in *Metapenaeus* and *Penaeus* juveniles. Hydrobiologia 295:295–302
Pülmanns N, Diele K, Mehlig U, Nordhaus I (2014) Burrows of the semi-terrestrial crab *Ucides cordatus* enhance CO_2 release in a north Brazilian mangrove forest. PLoS One 9:e109532
Purvaja R, Ramesh R (2001) Natural and anthropogenic methane emission from wetlands of south India. Environ Manag 27:547–557
Qureshi NA, Saher NU (2012) Burrow morphology of three species of fiddler crab (*Uca*) along the coast of Pakistan. Belg J Zool 142:114–126
Reddy KR, Delaune DR (2008) Biogeochemisty of wetlands: science and applications. CRC Press, Boca Raton
Reef R, Feller IC, Lovelock CE (2010) Nutrition of mangroves. Tree Physiol 30:1148–1160
Rice ME, Piraino J, Reichardt HF (1995) A survey of the Sipuncula of the Indian River lagoon. Bull Mar Sci 57:128–135
Rivera-Monroy VH, Castañeda-Moya E, Barr JG, Engel V, Fuentes JD, Troxler TG, Twilley RR, Bouillon S, Smith TJ, O'Halloran TL (2013) Current methods to evaluate net primary production and carbon budgets in mangrove forests. In: DeLaune RD, Reddy KR, Richardson CJ, Megonigal JP (eds) Methods in biogeochemistry of wetlands, Soil science society of America book series 10. Soil Science Society of America, Madison, pp 243–288
Rivera-Monroy VH, Lenaker P, Twilley RR, Delaune RD, Lindau CW, Nuttle W, Habib E, Fulweiler RW, Castaneda-Moya E (2010) Denitrification in coastal Louisiana: a spatial assessment and research needs. J Sea Res 63:157–172
Rivera-Monroy VH, Torres LA, Nixon B, Newmark F, Twilley RR (1999) The potential use of mangrove forests as nitrogen sinks of shrimp aquaculture pond effluents: the role of denitrification. J World Aquacult Soc 30:12–25
Rivera-Monroy VH, Twilley RR (1996) The relative role of denitrification and immobilization in the fate of inorganic nitrogen in mangrove sediments. Limnol Oceanogr 41:284–296
Robertson AI (1986) Leaf-burying crabs: their influence on energy flow and export from mixed mangrove forests (*Rhizophora* spp.) in north eastern Australia. J Exp Mar Biol Ecol 102:237–248
Robertson AI, Daniel PA (1989) The influence of crabs on litter processing in high intertidal mangrove forests in tropical Australia. Oecologia 78:191–198
Robertson AI, Phillips MJ (1995) Mangroves as filters of shrimp pond effluent: predictions and biogeochemical research needs. Hydrobiologia 295:311–321
Roden EE, Zachara JM (1996) Microbial reduction of crystalline iron(III) oxides: Influence of oxide surface area and potential for cell growth. Environ Sci Technol 30:1618–1628
Rossi F, Chapman MG (2003) Influence of sediment on burrowing by the soldier crab *Mictyris longicarpus* Latreille. J Exp Mar Biol Ecol 289:181–195
Sakho I, Mesnage V, Copard Y, Deloffre J, Faye G, Lafite R, Niang I (2015) A cross-section analysis of sedimentary organic matter in a mangrove ecosystem under dry climate conditions: the Somone estuary, Senegal. J Afr Earth Sci 101:220–231
Saenger P, Snedaker SC (1993) Pantropieal trends in mangrove above-ground biomass and annual litterfall. Oecologia 96:293–299
Sanders CJ, Santos IR, Barcellos R, Silva Filho EV (2012) Elevated concentrations of dissolved Ba, Fe and Mn in a mangrove subterranean estuary: consequence of sea level rise? Cont Shelf Res 43:86–94
Sanders CJ, Smoak JM, Naidu AS, Sanders LM, Patchineelam SR (2010) Organic carbon burial in a mangrove forest, margin and intertidal mud flat. Estuar Coast Shelf Sci 90:168–172
Sayão-Aguiar B, Pinheiro MAA, Colpo KD (2012) Sediment bioturbation potential of *Uca rapax* and *Uca uruguayensis* as a result of their feeding activity. J Crustac Biol 32:223–229
Seiter K, Hensen C, Schröter J, Zabel M (2004) Organic carbon content in surface sediments—defining regional provinces. Deep-Sea Res Part I 51:2001–2026
Sheaves M, Baker R, Nagelkerken I, Connolly RM (2015) True value of estuarine and coastal nurseries for fish: incorporating complexity and dynamics. Estuar Coasts 38:401–414

Shih J-T (1995) Population densities and annual activities of *Mictyris brevidactylus* (Stimpson, 1858) in the Tanshui mangrove swamp of northern Taiwan. Zool Stud 34:96–105
Silva CAR, Lacerda LD, Ovalle AR, Rezende CE (1998) The dynamics of heavy metals through litterfall and decomposition in a red mangrove forest. Mangrove Salt Marshes 2:149–157
Simonson RW (1959) Outline of a generalized theory of soil genesis. Soil Sci Soc Am J 23:152–156
Simpson LT, Feller IC, Chapman SK (2013) Effects of competition and nutrient enrichment on *Avicennia germinans* in the salt marsh-mangrove ecotone. Aquat Bot 104:55–59
Smith TJ III, Boto KG, Frusher SD, Giddins RL (1991) Keystone species and mangrove forest dynamics: the influence of burrowing by crabs on soil nutrient status and forest productivity. Estuar Coast Shelf Sci 33:419–432
Soil Survey Staff (1999) Soil taxonomy. Agricultural Handbook AH-436, 2nd ed US Department of Agriculture - Natural Resources Conservation Service, Washington, DC
Sotomayor D, Corredor JE, Morell JM (1994) Methane flux from mangrove sediments along the southwestern coast of Puerto Rico. Estuaries 17:140–147
Sousa WP, Dangremond EM (2011) Trophic interactions in coastal and estuarine mangrove forest ecosystems. Treatise Estuar Coast Sci 6:43–93
Souza-Júnior VS, Vidal-Torrado P, González Tessler M, Ruiz Pessenda LC, Ferreira TO, Otero XL, Macías F (2007) Evolução quaternária, distribuição de partículas nos solos e ambientes de sedimentaçao em manguezais do estado de São Paulo. Braz J Soil Sci 31:753–769
Stieglitz TC, Clark JF, Hancock GJ (2013) The mangrove pump: the tidal flushing of animal burrows in a tropical mangrove forest determined from radionuclide budgets. Geochim Cosmochim Acta 102:12–22
Stieglitz T, Ridd P, Muller P (2000) Passive irrigation and functional morphology of crustacean burrows in a tropical mangrove swamp. Hydrobiologia 421:69–76
Steinke TD, Rajh A, Holland AJ (1993) The feeding behavior of the red mangrove crab *Sesarma meinerti* De Man, 1887 (Crustacea, Decapoda, Grapsidae) and its effect on the degradation of mangrove leaf litter. S Afr J Mar Sci 13:151–160
Thamdrup B (2000) Bacterial manganese and iron reduction in aquatic sediments. In: Schink B (ed) Advances in microbial ecology. Kluwer, New York, pp 41–84
Thamdrup B, Hansen JW, Jørgensen BB (1998) Temperature dependence of aerobic respiration in a coastal sediment. FEMS Microbiol Ecol 25:189–200
Thongtham N, Kristensen E (2003) Physical and chemical characteristics of mangrove crab (*Neoepisesarma versicolor*) burrows in the Bangrong mangrove forest, Phuket, Thailand; with emphasis on behavioural response to changing environmental conditions. Vie et Milieu 53:141–151
Thongtham N, Kristensen E (2005) Carbon and nitrogen balance of leaf-eating sesarmid crabs (*Neoepisesarma versicolor*) offered different food sources. Estuar Coast Shelf Sci 65:213–222
Thongtham N, Kristensen E, Phungprasan S-Y (2008) Leaf removal by leaf-eating sesarmid crabs in the Bangrong mangrove forest, Phuket, Thailand; with emphasis on the feeding ecology of *Neoepisesarma versicolor*. Estuar Coast Shelf Sci 80:583–590
Tilman D (1982) Resource competition. Princeton University Press, Princeton
Tomlinson PB (1994) The botany of mangroves. Cambridge Tropical Biology Series. Cambridge University Press, p 419
Tremblay L, Benner R (2006) Microbial contributions to N-immobilization and organic matter preservation in decaying plant detritus. Geochim Cosmochim Acta 70:133–146
Troxler TG, Barr JG, Fuentes JD, Engel V, Anderson G, Sanchez C, Lagomasino D, Price R, Davis SE (2015) Component-specific dynamics of riverine mangrove CO_2 efflux in the Florida coastal Everglades. Agric For Meteorol 213:273–282
Twilley RR (1997) Mangrove wetlands. In: Messana G, Connor W (eds) Southern forested wetlands: ecology and management. CRC Press, Boca Raton, pp 445–473
Twilley RR, Chen RH, Hargis T (1992) Carbon sinks in mangrove forests and their implications to the carbon budget of tropical coastal ecosystems. Water Air Soil Pollut 64:265–288
Twilley RR, Day JW (2013) Mangrove wetlands. Estuarine ecology. Wiley, Hoboken, pp 165–202

Twilley RR, Pozo M, Garcia VH, Rivera-Monroy VH, Zambrano R, Bodero A (1997) Litter dynamics in riverine mangrove forests in the Guayas River estuary, Ecuador. Oecologia 111: 109–122

Twilley RR, Rivera-Monroy VH (2005) Developing performance measures of mangrove wetlands using simulation models of hydrology, nutrient biogeochemistry, and community dynamics. J Coast Res SI40:79–93

Twilley RR, Rivera-Monroy VH (2009) Ecogeomorphic models of nutrient biogeochemistry for mangrove wetlands. In: Perillo GME, Wolanski E, Cahoon DR, Brinson MM (eds) Coastal wetlands: an integrated ecosystem approach. Elsevier, Amsterdam, p 641

Twilley RR, Rivera-Monroy VH, Chen R, Botero L (1998) Adapting an ecological mangrove model to simulate trajectories in restoration ecology. Mar Pollut Bull 37:404–419

Valdemarsen T, Kristensen E (2010) Degradation of dissolved organic monomers and short-chain fatty acids in sandy marine sediment by fermentation and sulfate reduction. Geochim Cosmochim Acta 74:1593–1605

Vasleta A, Phillips DL, France C, Feller IC, Baldwin CC (2012) The relative importance of mangroves and seagrass beds as feeding areas for resident and transient fishes among different mangrove habitats in Florida and Belize: evidence from dietary and stable-isotope analyses. J Exp Mar Biol Ecol 434–435:81–93

Wang Y-F, Feng Y-Y, Ma X, Gu J-D (2013) Seasonal dynamics of ammonia/ammonium-oxidizing prokaryotes in oxic and anoxic wetland sediments of subtropical coastal mangrove. Appl Microbiol Biotechnol 97:7919–7934

Webb AP, Eyre BD (2004) The effect of natural populations of the burrowing and grazing soldier crab (*Mictyris longicarpus*) on sediment irrigation, benthic metabolism and nitrogen fluxes. J Exp Mar Biol Ecol 309:1–19

Weng BS, Xie XY, Yang JJ, Liu JC, Lu HL, Yan CL (2013) Research on the nitrogen cycle in rhizosphere of Kandelia obovata under ammonium and nitrate addition. Mar Pollut Bull 76:227–240

Wolanski E, Mazda Y, King B, Gay S (1990) Dynamics, flushing and trapping in Hinchinbrook channel, a giant mangrove swamp, Australia. Estuar Coast Shelf Sci 31:555–579

Woodroffe C (1992) Mangrove sediments and geomorphology. In: Robertson AI, Alongi DM (eds) Tropical mangrove ecosystems. American Geophysical Union, Washington, DC, pp 7–42

Woodroffe C (2002) Coasts: form, process and evolution. Cambridge Unversity Press, Cambridge

Xiong L-H, Wang A, Lu J-J (2010) The validity of burrow counting in estimating the densities of *Chiromantes dehaani* and *Ilyoplax deschampsi* in tidal flat at Yangtze Estuary. Acta Ecol Sin 30:40–44

Zhang XF, Chua VP, Cheong HF (2015) Hydrodynamics in mangrove prop roots and their physical properties. J Hydro Environ Res 9:281–294

Zhou H, Li FL (1990) Sipunculans from the South China Sea. In: Morton B (ed) The marine biology of the South China Sea. Hong Kong University Press, Hong Kong, pp 129–142

Chapter 7
Mangrove Ecosystems under Climate Change

T.C. Jennerjahn, E. Gilman, K.W. Krauss, L.D. Lacerda, I. Nordhaus, and E. Wolanski

7.1 Introduction

From fossil records, evidence suggests that mangroves first started colonizing the intertidal zone around the Tethys Sea in the Upper Cretaceous (Ellison et al. 1999). Since then, they have formed an interface between land and sea in most tropical and sub-tropical regions worldwide, and are now expanding into temperate regions of multiple continents (Saintilan et al. 2014). Environmental impacts on mangrove plants and sediments make them susceptible to climate change outcomes, including sea level rise (SLR), temperature shifts, and alterations of atmospheric gas composition and moisture over long temporal scales. All of these changes, as well as anthropogenic impacts from land use and water resources in the watershed are likely to significantly affect mangrove distribution and performance directly during this century. The consequences for mangrove ecosystems, and the ecosystem services that they provide, are likely to vary on a local to regional scale (Lee et al. 2014). Effective

T.C. Jennerjahn (✉) • I. Nordhaus
Leibniz Centre for Tropical Marine Research (ZMT), Bremen, Germany
e-mail: tim.jennerjahn@leibniz-zmt.de; inga.nordhaus@leibniz-zmt.de

E. Gilman
College of Natural Sciences, Hawaii Pacific University, Honolulu, HI, USA
e-mail: EGilman@fisheriesresearchgroup.org

K.W. Krauss
U.S. Geological Survey, Wetland and Aquatic Research Center, Lafayette, LA, USA
e-mail: kkrauss@usgs.gov

L.D. Lacerda
Labomar, Universidade Federal do Ceara, Fortaleza, Brazil
e-mail: ldrude@pq.cnpq.br

E. Wolanski
TropWATER, James Cook University, Townsville, Queensland, Australia
e-mail: eric.wolanski@jcu.edu.au

V.H. Rivera-Monroy et al. (eds.), *Mangrove Ecosystems: A Global Biogeographic Perspective*, https://doi.org/10.1007/978-3-319-62206-4_7

conservation and management therefore requires a local to regional scale assessment and understanding of mangrove responses to climate change.

Hazards to mangroves not only arise from climate change, but also from human interventions like, for example, land use change, urbanization, alterations to river catchment hydrology, overexploitation of natural resources, and coastal construction. Moreover, since the fourth IPCC (Intergovernmental Panel on Climate Change) report (Solomon et al. 2007), "Anthropocene" climate change and warming of the earth, i.e., during the past approximately 250 years and in particular, since the 1950s, is suspected to be largely related to human interventions, i.e., the burning of fossil fuels, land use change, and other human-induced changes in the atmosphere's composition. Mangroves have an adaptive capability as demonstrated by the fact that they have survived harsh environmental conditions and climate and sea level changes for millions of years, but at the same time, they now face rates of environmental change that are unprecedented in their history.

In recent decades, it has become clear that mangroves and their connectivity to adjacent terrestrial and marine ecosystems provide important ecological functions and ecosystem services (e.g., Ewel et al. 1998; Mumby et al. 2004; Wells et al. 2006). Moreover, in recent years, political and societal awareness of the importance of ecosystem services has increased, mainly due to the activities of global institutions such as the IPCC and the Millennium Ecosystem Assessment (2005). Political decisions in an economy-driven world hardly consider the value of a resource or service that is not directly marketable. Therefore, attempts have been made to put a value on ecosystem services (ES) from coastal wetlands. For example, one such study resulted in a value of 10,000 USD ha^{-1} $year^{-1}$ for mangroves (Costanza et al. 1997). However, giving these ES a direct economic value is difficult and remains debatable.

This chapter will first identify the possible outcomes of climate change that are likely to affect mangrove ecosystems, then report how mangrove forests respond to these outcomes, and from that define the most vulnerable regions. The chapter will also discuss the interaction of climate change with human interventions, the effects on ecosystem services, and adaptation and management options. It will close with an outlook on knowledge gaps and priority research needed to fill these gaps.

7.2 Climate Change Effects

There are well-known direct effects of climate change, which have the potential to alter the structure, functions, and ES of mangroves. However, these effects additionally alter the physical, chemical, biological, and geomorphological setting of the mangrove environment. These, in turn, can indirectly alter the structure, functions, and ES of mangroves through an altered exchange with neighboring ecosystems, including effects of multiple climate change stressors and other anthropogenic and natural stressors on mangrove systems (Table 7.1). There are also potential

Table 7.1 Direct and indirect effects of climate change and direct human effects on mangroves

Direct effects of climate change	Indirect effects of climate change	Direct human impacts
• Sea level rise • Warming of surface waters • Warming of atmosphere • Changing atmospheric moisture transport and precipitation • Changing atmospheric gas composition (higher CO_2)	• Changing surface ocean circulation affecting tidal exchange and geospatial dispersal of mangrove propagules • Changing salinity gradients affecting tidal exchange • Surface water acidification • Changing freshwater inflow • Changing allochthonous sediment input • Changes in extreme weather events • Increased frequency of extreme high water events • Changes in seasonality • Degradation of ecosystems that are functionally linked to mangroves	• Clearing of mangroves • Changing hydrology and tidal flushing by roads and levees, and dredging navigation channels • Changing freshwater inflows by river damming and diversion • Changing riverine sediment inflows by increased erosion and river damming • Pollution (e.g., nutrients, garbage, sewage, dredge spoil, oil spills) • Overharvesting of forest and fisheries • Subsidence due to extraction of water, petroleum, and gas

feedback mechanisms of climate change outcomes that may affect mangrove resistance and resilience (e.g., Cahoon et al. 2006; Krauss et al. 2014a).

7.2.1 *Direct Effects*

Sea level rise is a prominent direct effect of climatic warming on mangrove ecosystems. However, what is important is not the absolute sea level change, but rather the relative sea level change that mangroves experience locally. The concept of relative SLR has been described in various ways by different scientific disciplines (see Cahoon 2015), but for mangrove vulnerability assessment, relative SLR must account for a number of regional and local factors (e.g., impoundment, harvesting effects on subsidence) not all of which are related directly to climate change (Webb et al. 2013; Lang'at et al. 2014; Wolanski and Elliott 2015; Kiwango et al. 2015). For example, in some regions and settings, subsurface processes, such as mangrove root production and decomposition, can be a primary control of mangrove sediment surface elevation and hence, local-scale trends in relative sea level, within an individual mangrove site (Cahoon et al. 2006; McKee et al. 2007; Krauss et al. 2014a). One significant external factor of regional significance is the glacial rebound that presently affects mangroves differently in various areas of the world. Morphology and relief (i.e., a mangrove site's physiographic setting) certainly play an important role for the survival of mangroves facing relative SLR (Krauss et al. 2014a). Another significant factor is the tidal regime deformation in shallow waters facing large river deltas such as that of the Mekong, which can increase the tidal range and hence tidal flooding (Nhan 2016).

The global mean sea level rose by 3.2 ± 0.4 mm $year^{-1}$ in the past two decades, but with large regional variations (Wolanski and Elliott 2015). For example, sea level fell by 1–2 mm $year^{-1}$ along the western coast of North America and northern coast of South America, but it rose by 5–20 mm $year^{-1}$ in the Southeast Asia/Western Pacific region (Nicholls and Cazenave 2010). A recent IPCC report projects sea level to rise 26–55 cm (average 40 cm) by 2081–2100 for the "best case" scenario RCP2.6 (RCP = representative concentration pathways) and between 45 and 82 cm (average 63 cm) for the "worst case" scenario RCP8.5 (relative to 1986–2005). This rise will not be uniform across regions and approximately 70% of the global coastline will experience a change of ±20% of the global mean (IPCC 2014).

Another important factor directly affecting mangrove wetlands is the rise in air temperature and sea surface temperature. The projected increase of both can vary largely according to the four IPCC scenarios (RCP2.6, RCP4.5, RCP6.0, and RCP8.5) and it differs between land and ocean. The increase in sea surface temperature (SST) will likely be lowest as the ratio of land to ocean warming is 1.4–1.7 (Collins et al. 2013). Moreover, the temperature increase is not spatially uniform (Fig. 7.1); thus the impact on mangroves worldwide will vary according to location. The degree to which changing SSTs affect extreme winter air temperatures may be most important for determining whether mangroves will continue to expand into temperate zones, and supplant salt marsh. In the second half of the twentieth century, the SST increase was generally more pronounced in higher latitudes and in the northern hemisphere. For surface air temperature, the general trend was similar, but with a few low latitude regions exhibiting temperature increases in the same magnitude as in polar regions. In particular, arid to semiarid areas in eastern South America and western North Africa experienced an increase of 1.5–2.5 °C (Hartmann et al. 2013). According to IPCC projections, the increase in SST will be larger in the low latitudes in the coming two decades before it becomes larger in the high latitudes at the end of the twenty-first century. Surface air temperature will generally increase more rapidly and the increase will be larger in high latitude regions, in particular in the northern hemisphere (Collins et al. 2013; Kirtman et al. 2013).

Changes will also occur in atmospheric moisture content and transport and hence in precipitation. Again, this change will not be uniform across the globe. With respect to mangrove distribution, the most relevant projected changes are increases in precipitation in the equatorial Pacific, East Africa and the Middle East, and India. While eastern South America and western Africa will probably become drier, changes in Australasia and Southeast Asia are not that clear (Christensen et al. 2013; Hartmann et al. 2013). As a result of changes in precipitation and evapotranspiration, the runoff will change. The largest increase in runoff is projected for the Southeast Asia/West Pacific region while the strongest decrease is to be expected for Central America and eastern South America (Collins et al. 2013; Fig. 7.2).

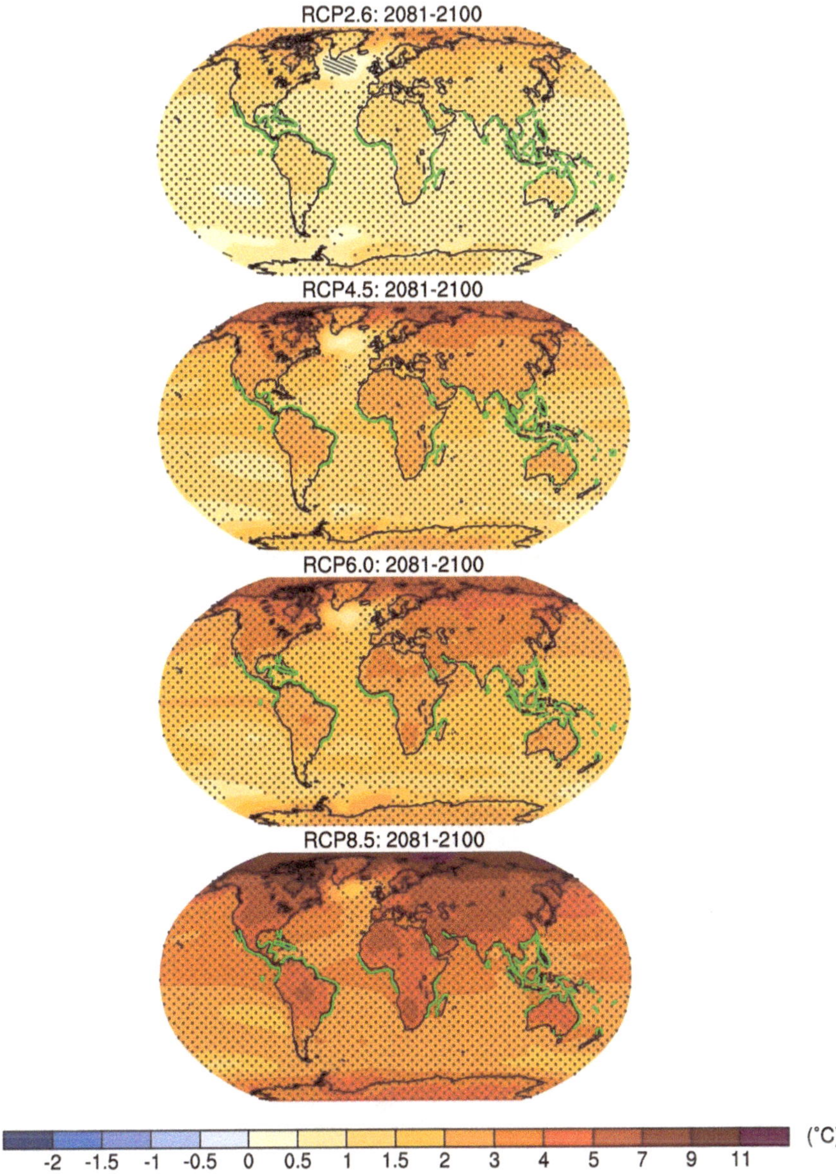

Fig. 7.1 IPCC projections for 2081–2100 of surface temperature changes for four emission scenarios (*RCP* representative concentration pathways; RCP2.6, RCP4.5, RCP6.0, RCP8.5) and the present-day mangrove distribution (*green lines*) according to Giri et al. (2011). Figure modified from Collins et al. (2013)

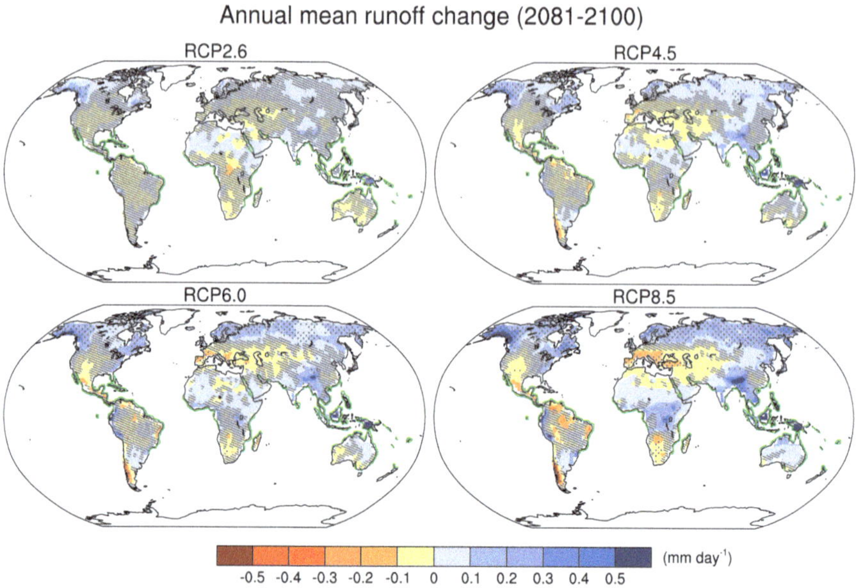

Fig. 7.2 IPCC projections for 2081–2100 of annual mean runoff changes for four emission scenarios (RCP2.6, RCP4.5, RCP6.0, RCP8.5) and the present-day mangrove distribution (*green lines*) according to Giri et al. (2011). Figure modified from Collins et al. (2013)

Last, the atmospheric concentration of greenhouse gases will increase and the increasing CO_2 can directly affect mangrove productivity (Ball et al. 1997; McKee and Rooth 2008). The IPCC projections foresee an increase of atmospheric CO_2 concentration up to 450 ppm in the "best case" scenario (RCP2.6) and of >1000 ppm in the "worst case" (RCP8.5) until 2100 (IPCC 2014).

7.2.2 Indirect Effects

Climate change will also indirectly affect mangrove wetlands in several ways (Table 7.1). Changes in the global heat budget and the hydrological cycle can alter the circulation in the coastal ocean, which may alter mangrove propagule dispersal (Duke et al. 1998; Chap. 2) and impact the flushing rate of estuaries and their fringing mangroves. For instance, warming and changes in rainfall and evaporation possibly increase the formation of an estuarine and coastal salinity barrier that prevents the exchange of mangrove water with the coastal ocean. This may result in stagnation that can last several months and formation of hypersaline waters that are stressful to mangroves (Wolanski et al. 1992; Andutta et al. 2011). In places where rainfall increases, mangroves may expand, such as in Moreton Bay, Australia (Dale et al. 2013). In general, changes in precipitation and evaporation will affect the water budget of estuaries and thus, the salinity of their fringing wetlands, including mangroves (Robins et al. 2014). Such changes in salinity are expected to modify the ecology and mangrove species distribution (Osland et al. 2014b).

The expected changes in storminess and coastal hydrodynamics will be crucial for mangroves. An increase in frequency and intensity of tropical storms has been observed in some areas (Knutson et al. 2010), such as the tropical North Atlantic for the period 1983–2005 (Kossin et al. 2007). Identifying general trends is hampered by the scarcity and inconsistency of data. Despite the uncertainties in projections of future changes until 2100, it is likely that the intensity of tropical storms will increase in the North Atlantic and western North Pacific, which already are the regions with highest tropical storm activity (Christensen et al. 2013).

Increasing extreme weather events will lead to physical damage that will be exacerbated if the mangroves are weakened by the formation of hypersaline waters or pollution. These weather events may defoliate the trees, make them more susceptible to attack by wood borers, and erode the soils. Although younger trees may be able to refoliate more rapidly, the process may still take several decades (Salmo et al. 2013). In addition, mangrove sediment elevation can be altered through soil erosion, soil deposition, peat collapse, and soil compression (e.g., Cahoon 2006; Piou et al. 2006; Aung et al. 2013). The frequency of extreme high water events is projected to increase over coming decades. It may affect the position and health of mangroves and associated coastal ecosystems by altering salinity, recruitment, and inundation, in addition to changing the wetland sediment budget. This is due to the same atmospheric and oceanic factors causing global sea level to rise, and possibly additional issues such as variations in regional climate and change in storminess and resulting storm surges (Woodworth and Blackman 2004).

Ocean acidification, a prominent outcome of climate change (e.g., Hoegh-Guldberg and Bruno 2010), will also reach mangrove ecosystems through estuarine flushing and will most likely affect their flora and fauna as well as biogeochemical cycles, although there are no published studies yet to quantify this impact. Mangrove forests are functionally linked to neighboring coastal ecosystems, including seagrass beds, coral reefs, and terrestrial coastal habitats, but the functional links are not fully understood (e.g., Mumby et al. 2004). Coral reefs, seagrass beds, estuaries, beaches, and coastal upland ecosystems may experience reduced area and health from climate change outcomes (e.g., Hoegh-Guldberg and Bruno 2010). For example, one indirect impact of increased temperature and CO_2 on mangroves is the degradation of adjacent coral reefs caused by mass bleaching and impaired growth (Hoegh-Guldberg 1999). The damage to coral reefs reduces their capability to protect mangroves from wave action which, in turn, makes them prone to erosion. On the other hand, mangrove forests may provide a non-reef coral refuge from thermal stress and ocean acidification. On St. John, Virgin Islands, shaded scleractinian corals that grow attached to and under mangrove prop roots do not show signs of bleaching in contrast to unshaded colonies. In addition, it was concluded that the combination of substrate and habitat heterogeneity, hydrographic conditions, and chemical water conditions acts as a buffer against ocean acidification (Yates et al. 2014).

7.3 Response to Climate Change

The response of mangroves to climate change effects will be manifold (e.g., Alongi 2015). The most important expected responses are shifts in (i) distribution (latitudinal and keeping pace with SLR), diversity, and community composition, (ii) physiological processes of flora and fauna, (iii) water budget, (iv) productivity and respiration, (v) carbon storage in biomass and sediments, and (vi) filter function for elements beneficial or harmful to life. These responses will likely vary from region to region, at least between the two large biogeographical regions, the Atlantic East Pacific (AEP) and the Indo West Pacific (IWP), which display profound differences in the diversity of mangrove trees and key faunal species (Tomlinson 1986; Lee 2008).

7.3.1 Distribution, Diversity, and Community Composition

7.3.1.1 Geographic Distribution and Shoreline Position

Global warming and SLR will directly lead to shifts in mangrove distribution. Using the Tropics of Capricorn and Cancer at 23.5 degrees northern and southern latitude is a common way to delimit the tropical regions. However, there are other climatological and geographic features, mainly temperature, precipitation patterns and atmospheric circulation that allow for other classifications and define slightly different geographical limits. Taking these into account, climate observations from 1979 to 2004 indicate a poleward shift of the tropical belt by 2–4.5 degrees latitude. For comparison, model simulations for the same period found an expansion of the tropical belt by about 2 degrees latitude (Seidel et al. 2008 and references therein). Changes in surface temperatures have been observed to be affecting the latitudinal limits of mangrove distribution at least in the North Atlantic and North Pacific, where range is limited by temperature only (e.g., Saintilan et al. 2014).

Osland et al. (2013) discovered that a 0–2 °C increase in mean annual minimum temperatures would facilitate mangrove expansion onto salt marshes on an additional 740 km^2 of coastline in southeastern United States, while a 2–4 °C increase could give rise to an additional 9860 km^2 of mangrove vegetation. Such temperature-mediated shifts in mangrove expansion are becoming widely documented (Cavanaugh et al. 2014; Saintilan et al. 2014), and a strong global effort is underway to understand the impact of temperate-zone marsh-to-mangrove habitat shifts on carbon storage, nutrient processing, surface elevation change, and structural provisioning for wildlife (e.g., Perry and Mendelssohn 2009; Osland et al. 2014a). New efforts have even started to identify the influences of rainfall regime in areas having active temperature-mediated mangrove encroachment (Osland et al. 2014b).

A global-scale examination of temperature variation and latitudinal limits of the two mangrove genera *Avicennia* and *Rhizophora* did not find common isotherms characterizing the poleward limit of their distribution. *Rhizophora* did not expand towards the colder *Avicennia* limit along the same coast although all other environmental conditions were similar (Quisthoudt et al. 2012). Nevertheless, a 28-year time-series of mangrove distribution at the northeast coast of Florida revealed a large increase in mangrove area at its northern latitudinal distribution limit related to a decreased frequency of extreme cold events (days colder than −4 °C; Cavanaugh et al. 2014).

Mangrove poleward proliferation corresponds well with the extension of temperature thresholds observed on five continents in the past 50 years (Saintilan et al. 2014). However, the response is not evenly distributed; a detailed study of the southern limit of mangroves in the western South Atlantic showed no expansion to the south (Soares et al. 2012). Similarly, at the southern limit of mangroves along the western Pacific Ocean in Peru, the cold and arid conditions hinder mangrove expansion south of a small stand of mangrove vegetation in the Piura River Estuary (Clüsener and Breckle 1987). Obviously, species-specific responses to warming in combination with other abiotic and biotic factors play a role in the poleward expansion of mangroves. Although temperature is a major control of mangrove distribution, it appears that there is no consistent response of poleward mangrove expansion to temperature increase (e.g., Bianchi et al. 2013).

The estuarine and coastal ocean circulation may also be altered, in a manner not necessarily reflecting the atmospheric change. This could have an impact on the variability of flushing and salinity of estuaries and their fringing mangrove vegetation because of changes in rainfall and evaporation. Mangroves trap sediment at rising tides, especially so during river floods (Victor et al. 2004); thus a change in the river hydrology will affect the sedimentation rate in mangroves and their ability to keep up with SLR.

The first and simplest response of mangrove distribution to SLR would be to keep up with the sea level and/or migrate landward where not obstructed. Mangroves keep pace with changing sea level when the accretion of the mangrove sediment surface is at least similar to the rate of change in relative sea level. The change in mangrove sediment surface elevation is controlled by both surface and subsurface processes (Krauss et al. 2014a). Whether mangroves can keep up with the sea level depends, in part, on the net sediment budget, which includes both the organic matter build-up through the growth of mangrove roots and deposition of litter, as well as the inorganic sediment inflow from the estuary and coastal waters (Lovelock et al. 2015a). These processes depend strongly on human activities in the catchment (Table 7.1) and on the local sea level change generating a prograding or an eroding coast. However, the "true" landward migration and possible change in area covered also depends on a number of other factors determining the environmental setting, including rainfall (Woodroffe 1992; Eslami-Andargoli et al. 2009; Woodroffe et al.2016). The relatively few data available show that sediment accretion rates in healthy mangrove forests are often higher than current rates of SLR (Alongi 2008; Lovelock et al. 2015a), but this assessment can be highly variable among forests. In those environmental settings that have sufficient allochthonous sediment input and/or production and accumulation of organic matter and a suitable gradient of land surface elevation, SLR should not impose a risk on mangrove distribution. This generally holds true for environmental settings that are dominated by rivers and tides with abundant sediment supply on prograding coasts only (Woodroffe 1992; Woodroffe et al. 2016).

However, mangrove surface elevation is not exclusively determined by sediment accretion, but also by (i) land uplift or subsidence, (ii) groundwater influx, (iii) vegetation and soil processes, and (iv) whether the coast is prograding or eroding (Krauss et al. 2003, 2014a; McKee et al. 2007; Cherry et al. 2009; Lovelock et al. 2015a). Taking these factors into account, a review of the relatively sparse global database concluded that surface elevations in many mangrove forests will not be able to keep pace with SLR for very long under the projected highest IPCC scenario RCP8.5 until the end of 2100. This holds true for mangrove areas in small island locations with little allochthonous sediment input, e.g., in the Caribbean, East Africa and parts of the Indo-Pacific region, even for the lowest scenario RCP2.6 (Sasmito et al. 2015); although there are noteworthy exceptions (McKee et al. 2007). Using surface elevation change records at 27 sites, a recent study assessed that the elevation gain is significantly correlated with sediment accretion in Indo-Pacific mangrove forests to SLR (Lovelock et al. 2015b). The surface elevation gain in 69% of the records was lower than the long-term rate of SLR. Model simulations based on these data sets and a moderate SLR (0.48 m by 2100, RCP6 scenario) indicate submergence of mangrove forests by 2070 in the Gulf of Thailand, the southeast coast of Sumatra, the northeast coasts of Java, Papua New Guinea and the Solomon Islands. In general, mangrove forests may be submerged by 2080 in regions of the Indo-Pacific with low tidal range and low allochthonous sediment supply under the moderate IPCC RCP6 scenario, while mangroves receiving higher sediment loads and larger tidal ranges are more buffered from submergence (Lovelock et al. 2015b). Thus, in their current shoreline positions, some mangroves will be converted to deeper water habitats, and where not obstructed, they may expand landward of their current positions as relative sea level rises.

The landward migration response is determined by a complex setting of local and regional environmental aspects modulating the effects of climate change. For example, mangrove forest area along the coastline southeast of the Amazon mouth expanded by 718 km^2 over a 12-year period (1996–2008), which seems to be typical for this region (Nascimento Jr. et al. 2013). Also, along the semiarid coast of northeast Brazil, mangroves are expanding rapidly in response to reduced rainfall due to climate change and land use alteration in the local watersheds, in particular damming. For example, in the estuary of the Jaguaribe River, the largest river in the state of Ceará, the mangrove-covered area increased by 24 ha between the years 1992 and 2003. A combination of land use-induced increased sedimentation and a decrease in rainfall over the river basin allowed mangroves to rapidly colonize newly formed islands in the estuary (Godoy and Lacerda 2014).

The complexity of factors controlling mangrove distribution makes it difficult to assess the potential net loss in areas related to climate change in this century. For example, based on the IPCC maximum SLR scenario, Gilman et al. (2006) predicted an up to 13% loss of Pacific island mangrove until the year 2100. Similarly,

Alongi (2008) concluded that a global loss rate of mangroves related to climate change on the order of 10–15% over a long time frame is a realistic projection.

7.3.1.2 Diversity and Community Composition

The future distribution of 12 common mangrove trees under climate change were projected by applying species and community distribution models (Record et al. 2013). A range of SLR projections and a set of climatic, hydrological, and geomorphological variables associated with mangrove distribution patterns were used in these models. For half of the modeled species (*Avicennia germinans*, *Laguncularia racemosa*, *Rhizophora mangle*, *R. racemosa*, *R. mucronata*, *Lumnitzera littorea*), a poleward shift of 2 degrees of latitude or more was projected. In addition, losses in the total area of suitable coastal habitat available were predicted (Record et al. 2013). Four species (*A. marina*, *Ceriops tagal*, *Lumnitzera racemosa*, *R. apiculata*) will gain suitable coastal habitat with mean latitudinal gains of less than 2 degrees regardless of the amount of predicted SLR. *Sonneratia alba* and *R. stylosa* were forecasted to gain coastal habitat and experience a decrease in absolute mean latitude. Regarding species richness, a gain is predicted across much of southeastern Asia, southern Brazil, northern Chile, eastern Australia, southeastern Africa, parts of northern Africa, and parts of northwestern Mexico. By contrast, a loss in species richness is projected for the Caribbean islands, parts of Central America, and parts of northern Australia (Record et al. 2013).

For tropical and subtropical regions, species-specific changes in growth, productivity, and competitive ability will probably lead to changes in tree species composition (Pernetta 1993). It has been suggested that mangrove crabs and molluscs expand poleward and that the length of inactive periods of native species in certain locations may be reduced in response to a shorter cold season (Nordhaus 2008).

7.3.2 Physiology of Flora and Fauna

Global warming, the altered hydrological cycle, and the increase of atmospheric CO_2 will directly affect the physiology of mangrove trees and animals which, in turn, will entail changes in productivity and respiration, community composition, distribution, and biogeochemical transformations.

7.3.2.1 Flora

Mangrove trees are very plastic in their response to shifts in salinity and inundation, as they are able to adjust photosynthetic water use efficiency and leaf morphological characteristics. As the environment changes over time, there may be a shift in species composition to more stress-tolerant trees within a forest (Reef and Lovelock

2015; Lovelock et al. 2016). While the majority of studies investigating adjustments in water use efficiency in mangroves along environmental gradients focus on seedlings, recent evidence suggests that mangroves can also adjust their water use efficiencies and CO_2 uptake capabilities at the stand level (Barr et al. 2013). This ability of mangrove stands to take up CO_2 and facilitate growth and maintenance despite reductions in rainfall or increases in salinity positions mangrove stands for resiliency.

The few sap flow studies that are available from mangrove trees, generally confirm their water use efficiency values documented from the seedling leaf–gas exchange literature (Lovelock et al. 2016): Water use values for individual trees are typically <31 L H_2O d^{-1}, which is approximately one-third the water use of non-mangrove trees of similar size (Lovelock et al. 2016). Thus, water use by most mangrove stands is rather conservative relative to evapotranspiration (Krauss et al. 2015), and can rival the water use reductions documented by C_4 salt marsh grasses in some settings (Krauss et al. 2014b). The ability to conserve water would buffer mangroves against harmful rainfall variability and fluctuations in salinity associated with both climate and land use alterations. How water is partitioned is also strongly dependent upon forest structural characteristics. For example, stand water use of trees ≥5 cm in diameter from two mangrove forests occurring within 2 km of each other in Rookery Bay, Florida, ranged from 373–481 mm $year^{-1}$ over a 2-year period (or 36–47% of evapotranspiration), with the higher water use being related to larger individual trees of one species (*Laguncularia racemosa*) (Fig. 7.3; Krauss et al. 2015). Thus, any environmental, land use, or climate factor that alters forest struc-

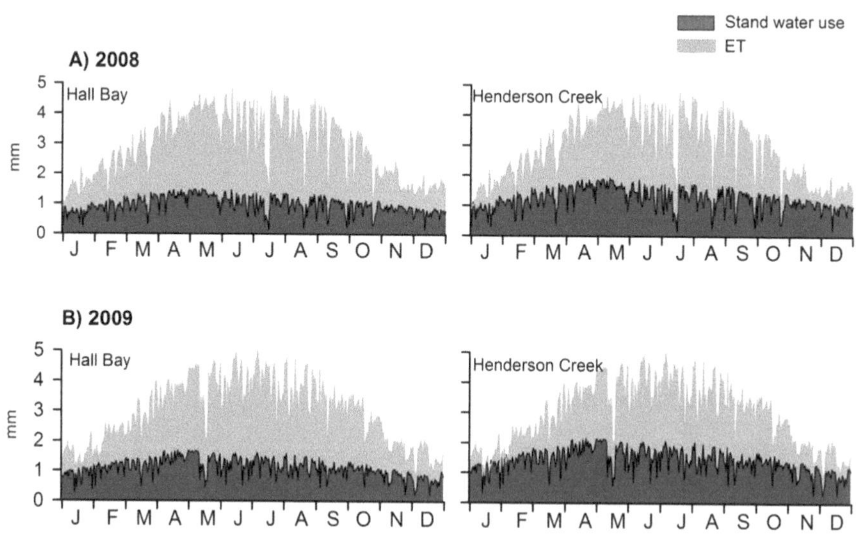

Fig. 7.3 Stand water use by dominant mangrove forest vegetation versus regional evapotranspiration (ET) for two forests (Hall Bay, Henderson Creek) differing slightly in forest structural attributes in Rookery Bay, Florida for the modeled year (**a**) 2008 and (**b**) 2009 (Data from Krauss et al. 2015)

ture also has the potential to alter water use requirements of mangroves significantly. In fact, while mangrove forest vegetation seemed to restrict losses to evapotranspiration from those two locations in Rookery Bay, water use by mangrove forests with larger trees in a different location was 872 mm year^{-1}, and accounted for a higher percentage of regional evapotranspiration (63–66%; Krauss et al. 2015).

7.3.2.2 Fauna

Benthic invertebrates in mangrove forests experience harsh environmental conditions during their life phases. Juveniles and adults of crustaceans and mollusks have to cope with daily fluctuations of temperature, salinity, and pH associated with the tidal cycle. However, most have planktonic larvae that are less exposed to these fluctuations. This is beneficial because larvae can have increased susceptibility to environmental stress (Pechenik 1999) and are often regarded as the bottleneck for success of marine species with regard to ocean warming and acidification (e.g., Byrne and Przeslawski 2013).

Considering distribution patterns, thermotolerant species with a wide latitudinal range of distribution are often robust to warming (Byrne and Przeslawski 2013). By contrast, many tropical species are considered to live close to their upper thermal tolerance (McMahon 2001; Nilsson et al. 2009) and may thus respond with reduced respiratory scope to even small temperature increases as reported for coral reef fishes (e.g., Nilsson et al. 2009). The heat coma temperature of intertidal gastropods from mangroves, rocky shores, and salt marshes displayed an overall negative quadratic relationship with latitude. While there was no trend with latitude in tropical regions, heat coma temperature decreased strongly with increasing latitude in extratropical regions. The highest thermal tolerance was found for gastropods of the group Littorinoidea from mangroves and salt marshes (McMahon 2001).

Two crab species occurring in tropical and subtropical mangrove forests demonstrated species- and population-specific responses to temperature increases. *Uca urvillei* is a thermal generalist and tolerates a broad range of temperatures across its geographical distribution. By contrast, the thermal tolerance of *Perisesarma guttatum* is much lower, but populations showed adaptations to local conditions. Tropical populations of *P. guttatum* had higher tolerances to acute heat stress than subtropical populations and are likely less vulnerable to global warming (Fusi et al. 2015). Another example is the mangrove oyster *Crassostrea rhizophorae*, which survived aerial exposure at 45, 42, and 35 °C for 2, 5, and 24 hours, respectively. Thus, adaptation to high temperatures typically occurs in mangrove forests (Littlewood 1989).

So far, the combined effects of warming and acidification on mangrove invertebrates have not been investigated, but information is available for other marine and estuarine species. A recent review on multistressor impacts of ocean warming and acidification on marine invertebrates found effects to vary across life stages and species (Byrne and Przeslawski 2013). Pre-larval stages are particularly sensitive to slight warming (+2 °C), whereas larvae of some species may tolerate near-future warming and acidification (+2 °C/pH 7.8). By contrast, deleterious effects on larvae, e.g., reduced sizes and survival rates, were recorded in experiments with higher deviance from present conditions (ca. ≥4 °C/pH <7.6). Calcifying organisms, e.g., corals, mollusks, and the larvae of echinoderms will be most negatively impacted by ocean acidification (Kroeker et al. 2013). Calcifying larvae need more energy for calcification as a result of decreased saturation of carbonate minerals (Byrne and Przeslawski 2013). For instance, the exposure to near-future elevations of pCO_2 and temperature had deleterious effects on the oysters *Magallana gigas* (formerly *Crassostrea gigas*) and *Saccostrea glomerata*, including a reduction in the fertilization success of gametes, a reduction in the development of embryos, size of larvae and spat, and an increase in abnormal morphology of larvae (Parker et al. 2010). More active organisms, such as mobile crustaceans and fish, may be less sensitive to acidification (Kroeker et al. 2013). Crustacean species that inhabit fluctuating environments are considered to be the most tolerant to ocean acidification as they are capable of compensating for acid–base disturbances via ion exchange mechanisms (Whiteley 2011). However, experiments with coastal prawns (*Palaemon pacificus*) from Japan revealed effects of ocean acidification on their molting frequency and growth (Kurihara et al. 2008).

A serious shortcoming of these laboratory experiments is that adaptations to the more gradual change of temperature and pH in the ocean over several decades cannot be evaluated. Species with short generation times may be able to tolerate near-future oceanic change through acclimatization and/or adaptation (Byrne and Przeslawski 2013). However, considering the data available, a clear signal of acclimatization is lacking (Kroeker et al. 2013). In addition, most experiments were conducted with single species only. Responses were more pronounced when species were exposed to acidification in multispecies assemblages, which led to the conclusion that indirect effects also have to be considered (Kroeker et al. 2013).

Based on the few studies available, a high thermal tolerance can be expected of the adults of a number of benthic species in mangrove forests. However, not only species-specific but also population-specific responses and adaptations (Fusi et al. 2015), varying responses of the different life history stages (Byrne and Przeslawski 2013) and species interactions (Kroeker et al. 2013) have to be considered to predict the consequences of climate change. The negative effects of warming and acidification on many planktonic larvae may result in a decrease of benthic populations and shifts in community composition if species are not able to adapt accordingly.

7.3.3 Water Budget

The water budget and associated variations in salinity are also important controls of mangrove distribution. The demise of the large mangrove area in the arid Indus Delta region illustrates this well (Kidwai et al. 2016). Damming of the Indus in the middle of the last century and the use of water for numerous purposes reduced the freshwater flow to about one-fifth of its original flow of 180 km^3 $year^{-1}$ and the sediment load from 400 to 100 million tons $year^{-1}$. In combination with the arid climate, the salinity in the Indus Delta region increased to 50 and as a consequence, the number of mangrove species decreased from 8 to 3; the community is now dominated by the highly salt-tolerant and dwarfed *Avicennia marina* (Amjad et al. 2007). Although caused by human interventions, this example demonstrates how the combination of reduced freshwater flow and increasing temperature as a consequence of climate change can endanger the survival of mangroves. A similar case was observed in the Tanzanian Wami River. There, a decrease in precipitation during the second half of the last century in combination with increasing human uses of river water for agriculture, irrigation, industry, and drinking water led to increasing dryness in the estuary during dry seasons (Kiwango et al. 2015). Even with no anthropogenic changes in the hydrology of rivers and estuaries, the combination of increasing temperature and changing atmospheric moisture from climate change will affect the precipitation/evaporation ratio and the freshwater flux and salinity of mangrove environments. The largest threat to mangroves in terms of the water budget is the combination of increasing temperature with decreasing rainfall.

Higher temperatures and reduced rainfall may lead to a reduction in mangrove area because upper tidal zones are converted to hypersaline flats (Gilman et al. 2008). Despite the water conservation strategies of mangroves discussed previously (Sect. 7.3.2.1), increasing pore water salinity and decreasing water availability will reduce productivity, growth, and seedling survival (Field 1995; Ellison 2000) and the reduced inflow of nutrient-laden freshwater will reduce overall ecosystem productivity. In areas of low rainfall, mangrove forests are characterized by dwarfed trees, narrower margins, and interspersed salt flats mainly because of salt stress (Ellison 2000). Species-specific responses will also entail changes in community composition and most probably, a reduction in biodiversity in these areas. In contrast, mangroves in regions with increasing rainfall will benefit as the increased water availability will reduce salinity and physiological stress of plants and increase sediment and nutrient inflow. Usually, mangroves are more productive, taller, and more diverse in regions with high rainfall compared to those with lower rainfall (Field 1995; Duke et al. 1998). Moreover, enhanced rainfall may lower salinity and allow mangroves to outcompete salt marshes and hence, increase mangrove area (Rogers et al. 2005).

7.3.4 Productivity and Remineralization

The rise in atmospheric CO_2, air temperature, and water temperature will likely increase the productivity and respiration of mangroves (e.g., Field 1995; Ball et al. 1997). A globally uniform trend, however, is unlikely because of the complex interplay of the aforementioned factors with other regional conditions and environmental settings. In general, mangrove productivity will probably not change much in the inner tropics, while the increase will be largest near the poleward temperature limits (Osland et al. 2013; Cavanaugh et al. 2014). However, in arid and semiarid regions where temperature may exceed a threshold of 33 °C, the photosynthetic rate of mangrove plants declines (Cheeseman 1994; see also Osland et al. 2014b). This will mainly concern the Red Sea and Persian Gulf regions.

A linear response of mangrove productivity to CO_2 increase is not likely because of the interplay of photosynthesis with other factors, namely salinity, water use efficiency, and nutrient availability (Lovelock et al. 2016). In an experiment on Caribbean mangroves *Rhizophora mangle*, *Avicennia germinans*, *Conocarpus erectus*, and *Laguncularia racemosa* under exposure to instantaneous pulses of increased CO_2 (361–485 ppm), trees displayed decrease in stomatal conductance and transpiration and increase in transpiration efficiency. However, net primary productivity decreased in *L. racemosa* and did not change in the other species. Therefore, it was suggested that a global rise in atmospheric CO_2 may result in a competitive disadvantage to *L. racemosa* in mixed mangrove communities and may lead to alteration of the community composition in these locations (Snedaker and Araujo 1998).

Rhizophora mangle trees displayed enhanced growth rates and biomass in a one-year experiment under doubled CO_2 and became reproductive much earlier than usual in mangrove forests of Belize, Central America (Farnsworth et al. 1996). Growth of *Rhizophora apiculata* and *Rhizophora stylosa* did not increase under doubled CO_2 over 14 weeks in high salinity areas, but increased in low salinity areas during an experiment in Australia (Ball et al. 1997). In general, exposure to elevated CO_2 enhanced mangrove seedling growth by 12–47% (up to a maximum 71%) and enhanced water use efficiency by 2–218% across a range of studies (Krauss et al. 2008; Lovelock et al. 2016). It was concluded that an increase of CO_2 may enhance mangrove productivity when carbon uptake is limited by the evaporative demand of the leaves, but not when it is limited by salinity at the roots (Ball et al. 1997; Gilman et al. 2008). The partly contradicting results of these experiments suggest species-specific responses to elevated CO_2 and the need for further research.

7.3.5 Carbon Storage in Biomass and Sediments

The well-known high carbon storage potential of mangrove ecosystems (Twilley et al. 1992; Jennerjahn and Ittekkot 2002; Bouillon et al. 2008; Alongi 2014; Chaps. 5, 6) gained scientific and public attention in past years when the climate change debate started focusing on the identification and conservation of natural carbon sinks in the ocean, the so-called "blue carbon" concept (Nellemann et al. 2009). Mangroves are part of the vegetated coastal habitats with very high relative carbon storage rates per unit area, much higher than on shelves or in the open ocean (Alongi 2014). Being among the most productive ecosystems on earth, mangroves store a larger amount of carbon in their above- and belowground biomass than terrestrial forests. In addition, they accumulate carbon-rich sediments composed of dead plant material and, depending on the environmental setting, large amounts of allochthonous mineral sediments and carbon (e.g., Jennerjahn and Ittekkot 2002; Donato et al. 2011).

Climate change can affect the carbon storage of mangrove biomass and sediments in manifold ways, first and foremost through changes in area, second through related changes in gain and loss terms of autochthonous (photosynthesis) and allochthonous (exchange with land, sea, atmosphere) carbon. Mangrove ecosystems gain carbon mainly through photosynthesis of (i) mangrove plants and (ii) macro- and microalgae colonizing roots and the sediment surface and (iii) through import from adjacent terrestrial and marine systems. Major carbon losses occur through respiration and tidal export, while secondary production and respiration by mangrove macrofauna are of minor importancc (Alongi 2014).

Starting off with the first global mangrove carbon storage estimate by Twilley et al. (1992), efforts to quantify global mangrove carbon storage have increased in the past decade. Interestingly, when normalized to an area of 138,000 km^2 (Giri et al. 2011), the annual total global mangrove carbon storage estimates increased from 14–16 Tg $year^{-1}$ (Twilley et al. 1992; Jennerjahn and Ittekkot 2002) to 22–24 Tg $year^{-1}$ (Alongi 2012; Breithaupt et al. 2012) over one decade. In the light of the ongoing mangrove area loss per year, such an increase is surprising and the possible reasons are manifold. First, despite the advance in identifying and quantifying gain and loss terms of carbon, there are still large uncertainties, for example, in the loss through dissolved inorganic carbon (e.g., Bouillon et al. 2008). Second, mangrove carbon accumulation rates can vary by one to two orders of magnitude both among and within sites (e.g., Breithaupt et al. 2012). Global-scale extrapolation from the relatively small database therefore can have a quite large degree of uncertainty. Third, it is possible that the increasing atmospheric CO_2 and eutrophication of coastal waters enhance mangrove productivity and burial of carbon in sediments. Several studies have shown that nutrient enrichment can enhance mangrove growth (e.g., Feller et al. 2002, 2003; Lovelock et al. 2007) and possibly also increase carbon burial in sediments. The increase in global estimates by a factor of two over relatively short time is probably a result of all of these factors.

The average of all available budgets results in an annual global mangrove carbon storage of 22 ± 6 Tg $year^{-1}$. Approximately 75% of this carbon is stored in sediments, the rest is above- and belowground biomass (Donato et al. 2011; Alongi 2014). Interestingly, there are differences between the two major biogeographic regions. Based on the global set of data used by Breithaupt et al. (2012), the average carbon accumulation rate is 265 ± 161 g C m^{-2} $year^{-1}$ in the IWP and 194 ± 155 g C m^{-2} $year^{-1}$ in the AEP. Despite the large variability in numbers, this difference is consistent with the finding of a significantly higher aboveground biomass in the IWP than in the AEP mangrove forests at the same tree height. It potentially reflects a higher primary productivity in the IWP region, the reasons of which can be manifold, e.g., differences in climate, tree density, stand age, species architecture, and anthropogenic disturbances (Lee 2008 and references therein). The low latitudes of the Asian/western Pacific region receive the highest fluxes of dissolved nutrients and suspended particulate matter worldwide and are also strongly affected by human interventions (Smith et al. 2003; Syvitski et al. 2005; Milliman and Farnsworth 2011; Jennerjahn 2012). Accordingly, the high nutrient input may promote high productivity and, in combination with the higher allochthonous organic matter input, also a higher carbon accumulation in IWP than in AEP mangroves.

Overall, the total annual average mangrove carbon sequestration of 22 ± 6 Tg year^{-1} accounts for approximately 0.2% of the 9200 Tg year^{-1} (estimate for the period 2002–2011) of anthropogenic carbon emissions (Ciais et al. 2013). Assuming that climate change related loss of carbon is equal to the estimated 10–15% loss of mangrove area by the year 2100 (Alongi 2008), the annual carbon storage will be reduced to 19 Tg year^{-1}. In this context, mangroves appear to be a negligible sink for anthropogenic CO_2. However, the carbon that is released from the destruction of mangroves is quantitatively more important. A recent estimate of the total mangrove carbon stock worldwide, accounting for the uncertainties of available data, arrived at the broad range of 4000–20,000 Tg (Donato et al. 2011), which adds significantly to the total tropical wetland forest carbon storage of 82,000–92,000 Tg (Page et al. 2011). Climate change-related 10–15% mangrove loss until 2100 would mean an annual release of 4.7–35.3 Tg C year^{-1} and sum up to a total of 400–3000 Tg C. Accordingly, a complete loss, though unlikely, would mean an annual release of 47–235 Tg C year^{-1}. Compared to the annual anthropogenic carbon emissions, it remains a small, but at 2.5% not a negligible amount (Fig. 7.4). Because of the large uncertainties in fluxes, it remains unclear whether mangroves in the year 2100 will

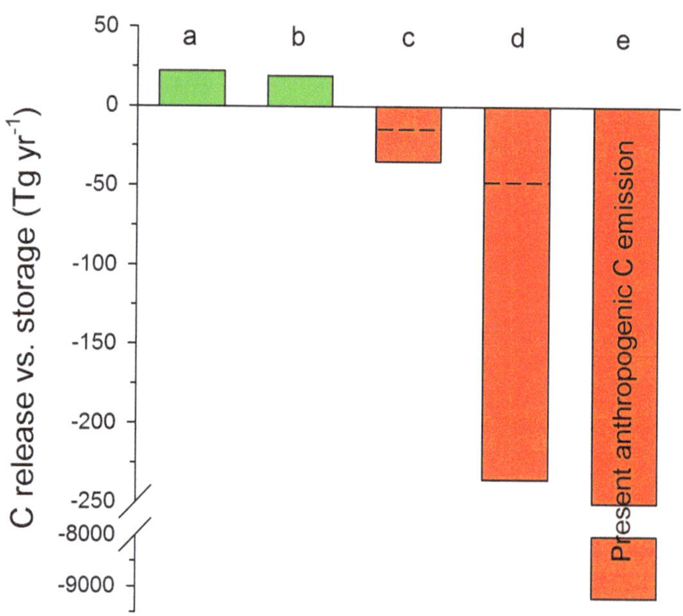

Fig. 7.4 Annual mangrove carbon storage (*green*) and release (*red*) today (*a*) and under 10–15% loss (*b* and *c*) and total loss scenarios (*d*) until the year 2100 compared to the present-day (i.e., 2012) anthropogenic carbon emissions (*e*). *Dashed lines* denote the lower limit of carbon release from mangroves as reported in the text (Data sources: Ciais et al. (2013), Donato et al. (2011) and data sources in the text. Note the break in the *Y*-axis)

be carbon sinks or sources. However, from these calculations, it is conceivable that the consequences of climate change alone can turn mangroves from sinks into sources of carbon by the end of the century.

7.3.6 *Filter Function for Elements Beneficial or Harmful to Life*

The many characteristics making mangroves such a unique ecosystem in the land–ocean transition also make them an important filter for land-derived substances that are beneficial or harmful to life. They trap mineral sediments, organic matter, and nutrients. Climate change related SLR, temperature increase, altered hydrology, and ocean acidification will alter these properties. As a consequence, land-derived substances retained and processed in mangroves will decrease if the mangrove area decreases 10–15% by 2100 (Alongi 2008).

For example, eutrophication, i.e., the exposure of coastal waters to excess nutrients, is a major man-made phenomenon. In a global-scale analysis of estuarine systems, Valiela and Cole (2002) showed that salt marshes and mangroves can protect seagrass meadows from land-derived nitrogen loads. Fringing wetlands are capable of absorbing up to 20 kg N ha^{-1} $year^{-1}$. Nitrogen loads between 20 and 100 kg N ha^{-1} $year^{-1}$ are known as a critical range for seagrass meadows (Valiela and Cole 2002). Climate-related loss of mangroves will therefore reduce the retention capacity and increase the exposure of downstream ecosystems to land-derived nutrients.

Tidal exchange of suspended matter (TSS), as well as total (TP), particulate (Part-P), and soluble reactive phosphorus (SRP) in northeast Brazil showed that the retention of phosphorus in mangroves varies with anthropogenic inputs. A mangrove site receiving effluents containing 1.2–5.2 kg h^{-1} of total phosphorus from nearly 3000 ha of shrimp ponds could trap about 40%, 45%, 47%, and 70% for TSS, TP, SRP, and Part-P, respectively, of the incoming phosphorus flux. However, a non-impacted mangrove site receiving phosphorus from only 10 ha of ponds (0.2 kg h^{-1}) tidal balances retained 92% of the total input of TSS and 100% of all other P fractions. This suggests that mangrove phosphorus retention capacity decreases with increasing nutrient input and limits the potential of mangroves as a natural nutrient barrier (Fonseca et al. 2014).

Similarly, mangroves display an elevated capacity for storing calcophylic metals as sulfides precipitates. Metal sulfides formed as a consequence of bacterial dissimilatory sulfate reduction is efficiently buried in anoxic mangrove sediments due to high accretion and waterlogged conditions. Mangrove sediments immobilize 40.6 g m^{-2} of Zn, 3.3 g m^{-2} of Cu and 43 mg m^{-2} of Hg in the highly polluted Guanabara Bay, Rio de Janeiro, southeastern Brazil (Machado et al. 2002). Some metals, such as Hg, may suffer a different fate in mangrove environments due to their high affinity to organic complexes. Inorganic Hg forms the major pool of Hg entering mangroves through tides or river transport. It will efficiently accumulate in sediments, but a significant fraction, rather than precipitating as sulfides will be sequestered by the large amount of dissolved organic complexes (DOC) present in pore waters due to the incomplete oxidation of organic matter. Formation of complexes with DOC then will increase the amount of highly bioavailable organic Hg complexes being exported by mangroves relative to the incoming fluxes of these chemical species. A 5-year study in the Jaguaribe estuary showed an increasing export of dissolved Hg to adjacent waters over time. It was enhanced in dry periods, which are more frequent because of an overall decreasing annual rainfall, and because of damming in the river basin (Lacerda et al. 2013).

The impact of global climate change on the aforementioned processes will depend on local geomorphological dynamics. In areas with expected mangrove expansion, such as those in northeastern Brazil and in the southeastern USA, accumulation and storage of carbon and nutrients as well as of pollutants may be increased accordingly. However, along constrained coastlines where mangroves cannot expand landward or on low-lying islands, erosion of the deposited sediments may result in the release of nutrients and pollutants to the adjacent waters.

7.4 Vulnerability of Regions

The four most important factors determining the vulnerability of mangrove forests to climate change are SLR and associated increase in frequency/intensity of storms, temperature, and aridity. In this context, mangrove forests are best protected in river- and tide-dominated settings where allochthonous inputs from land and ocean can help keep pace with the rising ocean, i.e., along macrotidal coasts, in river estuaries and in wet coastal areas (e.g., Woodroffe 1992; Woodroffe et al. 2016). Applying these criteria allows for an identification of the most vulnerable mangrove regions to climate change worldwide as depicted in Fig. 7.5. While low-lying islands in the equatorial Pacific are mainly threatened by SLR, areas further north and south are possibly additionally threatened by an increased intensity of storms. The same holds true for the Caribbean. In the already arid regions of the Red Sea and the Persian Gulf, IPCC projections foresee increasing temperature and aridity (Christensen et al. 2013; Collins et al. 2013), which may go beyond ecological thresholds and therefore pose an additional risk to the existence of mangroves. Other vulnerable areas include mangrove forests along coasts where narrow coastal plains are restricted by mountains, such as in southeastern Brazil and on mountain islands. It appears, though, that more mangrove systems are vulnerable to the adverse effects of climate change in the IWP than in the AEP region (Fig. 7.5). A recent assessment of climate change impacts on mangrove forests, based on IPCC projected regional changes in salinity, precipitation, and SLR (Alongi 2015), came to a similar conclusion.

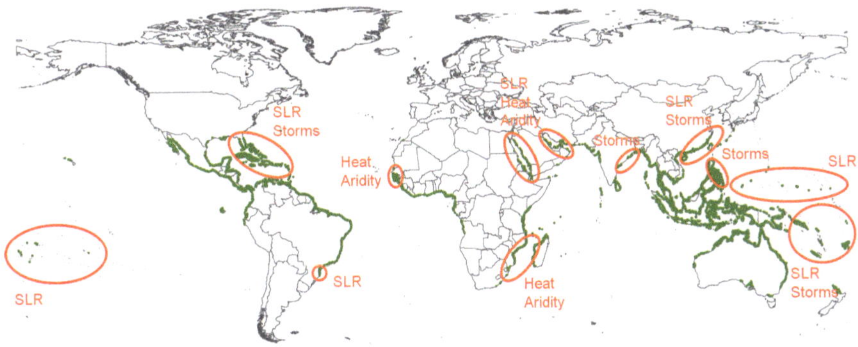

Fig. 7.5 Mangrove regions most vulnerable to consequences of climate change (sea level rise, increased frequency/intensity of storms, increasing temperature and aridity) are marked in *red*. Regions were chosen according to most severe changes projected by the IPCC AR5. Mangrove distribution (*green lines*) (According to Giri et al. (2011); data set taken from UNEP World Conservation Monitoring Centre, URL: data.unep-wcmc.org/datasets/21. World borders data set taken from Thematic Mapping Engine, URL: thematicmapping.org)

7.5 Interaction with Human Interventions

Human activities play an important, if not the dominant, role in the decline of mangrove forests (Valiela et al. 2001), although mangrove loss has slowed over the last decades (Spalding et al. 2010). It is still alarmingly high. Assuming that deforestation continues at the usually cited loss rates of 1–2% per year (Alongi 2002), mangroves will be lost before the end of the century, rendering climate-related changes almost negligible (Alongi 2008). However, recent studies indicate that the average loss rate was an order of magnitude lower since the year 2000. Between 2000 to 2012 global mangrove deforestation varied between 0.16% and 0.39% per year (Hamilton and Casey 2016) and was on average 0.18% per year in Southeast Asia (Richards and Friess 2016), the region which contains the greatest diversity of mangrove species and roughly one third of the global mangrove area. Taking this into account mangrove vulnerability to climate change threats may be larger than previously thought.

The major factor determining mangrove resilience to climate change related SLR is landward migration as shorelines retreat. The survival of the ecosystem as a whole is then mainly determined by hinterland topography and/or sediment accretion allowing the system as such to migrate landward and to maintain a suitable surface elevation. Simulations of landward migration in Gazi Bay, Kenya, until the year 2100, have shown that mangroves under low to medium SLR scenarios can manage without significant losses. However, under a maximum SLR scenario, further landward migration is obstructed by a strong increase of the topographical gradient (Di Nitto et al. 2014). What becomes even more important in this context is coastal development, i.e., settlements and hard engineering coastal protection (e.g., seawalls, revetments), that hinders landward migration of mangroves. Coasts suffering from mangrove clearing can become destabilized even when coastal engineering measures are taken. In Guyana, for example, the coastal dikes constructed after mangrove clearing were not as efficient in stabilizing the muddy coast as the mangroves were (Anthony and Gratiot 2012). However, with respect to SLR and an increasing storminess, coastal engineering measures and mangroves may even complement each other as shown by an example from the Mekong Delta. There, keeping or restoring mangroves helps to keep the necessary dikes lower and hence less expensive than without a seaward mangrove belt (Albers and Schmitt 2015).

Other important human factors are the regulation of hydrology and land use change, which affect fluxes of sediments and all other dissolved and particulate constituents from terrestrial runoff. Land use change, mainly deforestation, settlements and infrastructure, leads to an increase of erosion and higher river fluxes of suspended sediments, nutrients, and other substances. The opposite happens in the case of river damming and other hydrological regulations (e.g., Vörösmarty et al. 2003; Walling and Fang 2003; Jennerjahn 2012). A global analysis of river fluxes showed that human activities have increased the suspended sediment transport to the ocean by 2.3×10^9 t year^{-1} through soil erosion, and simultaneously decreased the transport by 3.7×10^9 t year^{-1} through trapping in reservoirs behind dams, resulting in a net loss of 1.4×10^9 t year^{-1} suspended sediment during the Anthropocene (Syvitski et al. 2005). However, large regional variations in the net effect of both processes and increases in the sediment load during the Anthropocene are almost exclusively observed in the tropical belt. While the coasts of Africa, large parts of South America, and West Asia suffer from reduced sediment input, sediment fluxes increased in parts of South and Southeast Asia (mainly Indonesia and the Philippines) and in Central America and the Amazon region (Syvitski et al. 2005). In the latter regions, the increased sediment input can probably to some extent offset the adverse effects of SLR and may even promote an increase in mangrove area. In the former coastal regions, sediment starvation may aggravate the loss of mangrove areas to SLR.

Eutrophication is another anthropogenic threat to coastal ecosystems. Human discharges of nitrogen and phosphorus into the ocean have increased by a factor of three from the 1970s to the 1990s (Smith et al. 2003). The Mississippi–Gulf of Mexico and the Danube–NW Black Sea are two prominent large-scale examples of how human interventions drastically change the amount and composition of nutrients with associated effects on biogeochemical cycles and food webs (e.g., Humborg et al. 1997; Rabalais et al. 2000). However, the inputs and ecological consequences can be very different in tropical regions (e.g., Jennerjahn et al. 2008; Jennerjahn 2012). Flooding with nutrient enriched coastal waters and increased input of fluvial nutrients may enhance productivity of mangrove forests, but may also cause changes in community composition of flora and fauna. However, nutrient enrichment may also increase mangrove mortality (Reef et al. 2010). Since nutrient enrichment favors the growth of shoots over roots, the higher water demand of the shoots has to be met by the roots. This, in turn, increases the vulnerability of mangrove trees to environmental stress such as high salinity, low humidity, and low rainfall (Lovelock et al. 2009). Seaward fringing forests are less vulnerable than landward scrub forests due to differences in inundation frequency and these differences become larger along gradients in aridity. The higher ratio of above- to belowground biomass could also make mangrove forests more vulnerable to windthrow and waves (Alongi 2008; Lovelock et al. 2009). It is therefore conceivable that eutrophication will exacerbate climate change–related mangrove dieback particularly in areas exposed to tropical storms and where temperature and aridity increase.

The increasing settlement and use of coastal zones also changes the physical conditions. Large deltas of major world rivers have always been preferred locations for human settlement. Tropical deltas also host the most luxurious mangrove forests because of the high sediment and nutrient supply, and wide intertidal zones. The subsidence of deltas has increased dramatically in the past decades, mainly due to the reduced sediment supply caused by river damming, extraction of oil, gas, and water from delta sediments, and floodplain engineering (Syvitski et al. 2009). Subsidence due to natural compaction is generally $\leq$3 mm year^{-1} (Syvitski 2008). Accelerated compaction due to human activities can be orders of magnitude higher as, for example, in the Chao Phraya, where it ranges between 50 and 150 mm year^{-1} (Saito et al. 2007). An analysis of 33 representative world deltas revealed that many of them are sinking relative to the global sea level mainly because of human activities making them particularly vulnerable to flooding and storm surges. Many of the major world river deltas in the tropics are at great risk, particularly in Asia. Relative rates of SLR in those regions are generally >2 mm year^{-1} and can be as high as 150 mm year^{-1}. Moreover, the northern hemisphere Asian deltas are exposed to the highest frequency and intensity of tropical storms (Syvitski et al. 2009). Taken together, it appears that mangrove areas that were supposed to be least vulnerable to SLR and other consequences of climate change because of high allochthonous inputs of sediment and nutrients, may be more vulnerable than previously thought due to the human-induced sinking of deltas (Lovelock et al. 2015b).

7.6 Effects on Ecosystem Services

The ecosystem services (ES) of mangroves will change during this century because of climate change and associated SLR, but probably much more by other human interventions. In particular, the "provisioning" and the "cultural" ES, which are directed towards sustaining livelihoods will be much more affected by increases in exploitation. The climate change related overall loss of mangrove areas will also lead to an overall loss of ES, but apart from that the consequences of climate change will be more relevant for the "regulating" and the "supporting" ES and probably display large regional variations (Table 7.2).

At the lower end of the temperature range, global warming may positively affect almost all ES. The "provisioning" ES may benefit from the temperature-induced increased productivity, i.e., at the northern and southern latitudinal limits of mangrove distribution, it could lead to an enhanced supply of fuelwood and building materials. It could also enhance carbon sequestration ("regulating") and even tourism and recreation ("cultural") could benefit from more comfortable temperatures. Such a proliferation at latitudinal limits will often occur at the expense of salt marshes. As both these ecosystems have almost similar ES, also in monetary values (Barbier et al. 2011), this ecosystem change may not change the overall ES of such a coastal zone. However, there are clear exceptions, such as Hawaii, where mangroves are non-native introduced species. Changes resulting from mangroves replacing other coastal habitats have been interpreted as reducing ES, including reduced stream flow with concomitant flooding risk in adjacent inhabited areas, eliminating open water habitats of native waterbird species, and obstructing culturally valuable fish ponds (Allen 1998; Chimner et al. 2006).

At the upper end of the temperature range, further temperature increase can exceed species-specific thresholds and result in less productivity, hence reducing "provisioning" ES. This is of particular importance in areas that may suffer from reduced freshwater supply in the future, e.g., in the Persian Gulf and the Red Sea. In general, increasing aridity and decreasing freshwater inflow probably impair the "regulating" ES such that, for example, the filter function for nutrients, sediments, and other elements will be reduced. In areas that are projected to expect an increasing freshwater and allochthonous sediment input, the "regulating" and "supporting" ES may benefit. According to the IPCC AR5, this is mainly the case for the Southeast Asia/West Pacific region and the west coast of South America (Fig. 7.2). However, in the former region, the large number of dams and other regulations of hydrology will counteract such enhanced freshwater and sediment inflow into the coastal zone while increased erosion due to deforestation may support it. Species-specific responses of organisms to warming and changing water availability may lead to changes in the community composition, which can have consequences for ES. For example, changes in tree community composition can directly affect sediment trapping, water quality maintenance, and nutrient cycling functions, while a change in benthic fauna through altered burrowing and feeding activities may affect nutrient cycling and carbon sequestration (Table 7.2).

Rapid SLR will mainly impair the "regulating" ES by increasing beach and soil erosion and, in turn, also lower the coastline protection. This is of particular importance in regions with low freshwater and allochthonous sediment input and where increasing heat and aridity are projected to be most vigorous like, in the Red Sea and the Persian Gulf. Increases in the frequency and intensity of storms will probably impair all the "regulating" ES and tourism may also suffer strongly in affected regions (Fig. 7.5). During floods, storms, and extreme high water events exacerbated by climate change, the increased turbidity and direct hydrodynamic damage will change the "supporting" ES in a way that will alter nutrient cycling, probably destroy nursery habitats and change biodiversity (Table 7.2). The most deleterious effects on ecosystem services can be expected in areas that are most vulnerable to the combined effect of SLR and storms like, for example, in the Caribbean, on Pacific islands and along the Chinese and Japanese coasts (Fig. 7.5).

Table 7.2 Climate change-related effects on mangrove ecosystem services: Ecosystem services of mangroves defined by the UNEP World Conservation Monitoring Center (Wells et al. 2006; second column) are grouped in the four categories defined by the Millennium Ecosystem Assessment (2005; first column)

Category	Mangrove ecosystem service	Response to climate change
Provisioning	(a) Subsistence and commercial fisheries (b) Aquaculture (c) Honey (d) Fuelwood (e) Building materials (f) Traditional medicines	• Area loss decreases all ES • Shift in community composition may lead to change of particular ES value (c) to (f) (one species may have a higher value than another) • Increasing temperature and reduced freshwater inflow may impede (a) and (b) • Increasing productivity (because of increasing temperature and CO_2) increases (d) to (f)
Regulating	(a) Protection of beaches and coastlines from storm surges, waves, and floods (b) Reduction of beach and soil erosion (c) Stabilization of land by trapping sediments (d) Water quality maintenance (nitrogen and pollutant filter) (e) Climate regulation (carbon sequestration)	• Area loss decreases all ES • Rapid SLR decreases (a) to (d) • Increasing intensity of storms decreases (a) to (e) • Reduced freshwater inflow decreases (d), increased freshwater inflow may increase (d) (if freshwater not laden with anthropogenic nutrients and pollutants) • Increased/reduced allochthonous sediment input related to increased/reduced freshwater input may increase/decrease (a) to (e) • Species-specific responses and changing community composition may alter (c) to (e)
Cultural	(a) Tourism and recreation (b) Spiritual—Sacred sites	• Area loss decreases all ES • Changes in freshwater and sediment inflow may impair (a) • Increases in storm intensity decrease (a)
Supporting	(a) Cycling of nutrients (b) Nursery habitats (c) Biodiversity	• Area loss decreases all ES • Reduced/increased freshwater inflow decreases/increases (a) • Increases in storm intensity decrease (a) and (b) • Species-specific response will alter, probably reduce (c)

The degree of resource exploitation is another critical factor that may accelerate the climate change related loss of mangroves. Natural resources are a major supplier of mangrove ecosystem services, i.e., the "provisioning" ES (Table 7.2), directly sustaining livelihoods of people, but at the same time also a major threat to mangrove survival, if not controlled in a sustainable way. The "regulating" ES of mangroves are to a large extent determined by the physical setting and its interaction with the flora and fauna. The vegetation, in particular the roots, largely determines the dissipation of tidal and wave energy in mangrove forests, which promotes the settling of particles and hence, the formation of sediments. It stabilizes the whole system and contributes to an increase in area (e.g., Mazda et al. 2007). As such, the mangrove vegetation provides protection from waves and storms and supplies the substrate and preconditions on which the "provisioning" ES build. Similarly, ecosystem engineers such as crabs also have important functions like, for example, retaining nutrients in the system (e.g., Robertson 1986; Robertson and Daniel 1989; Nordhaus et al. 2006). Overuse of the "provisioning" ES by, for example, extraction of too much mangrove wood, probably leads to a destabilization of the substratum and eventually to the loss of sediments and nutrients, i.e., a weakening of the "regulating" and "supporting" ES. The overexploitation-induced destabilization can therefore accelerate the loss of mangrove areas related to SLR and increased intensity of tropical storms.

7.7 Adaptation and Management Options

Being located at the interface between land and ocean in the high temperature range of the globe and exposed to large gradients in physicochemical properties of water, mangrove forests as a whole and their assemblage of organisms are by nature successful "adaptors." They have developed strategies that allow them to survive in such harsh environments and even to move in response to sea level oscillations as long as topographic gradients and surface elevations allow. Large-scale oscillations of sea level on glacial–interglacial timescales led to major disruptions of mangrove distribution during the Pleistocene. However, since around 7000 years ago, when the speed of SLR slowed down and sea level almost reached its present position, mangroves have colonized a large part of the tropical intertidal zone (Woodroffe et al. 1985; Woodroffe 1992).

Considering SLR, it is obvious that systems with little or no allochthonous sediment input are generally the most vulnerable. From Holocene records, it was concluded that mangrove ecosystems can keep pace with rising sea level on the order of 8–9 cm in 100 years, are under stress at rates of 9–12 cm in 100 years, and may collapse at higher rates (Ellison and Stoddart 1991). This view has been challenged by examples where mangrove swamps kept pace with higher rates of SLR (e.g., Snedaker et al. 1994 and references therein). Nevertheless, the aforementioned rates appear to be valid for carbonate settings and/or low-lying islands where mangrove growth almost exclusively builds on the autochthonous production. The various scenarios of the recent IPCC report suggest an average global SLR of 40–63 cm until the year 2100 (Church et al. 2013), clearly higher than the threshold value for mangrove settings without allochthonous sediment input. Increasing temperature, atmospheric CO_2, and nutrient input may increase mangrove productivity and growth in carbonate settings and/or low-lying islands and hence, raise the aforementioned thresholds. However, such an increase will not at all match even the lowest projected rate of SLR. As a consequence, many of the mangroves mainly in the western Pacific and the Caribbean have little chance to survive long term.

In the numerous regions where human exploitation dominates, direct management options mitigating potential adverse consequences of climate change are limited.

An overview of adaptation options to augment mangrove resistance and resilience to climate change (Gilman et al. 2008) suggests that it is rather the human-induced degradation that can be readily mitigated. For example, reducing or eliminating the main non-climate related stressors on mangroves and functionally-linked adjacent coastal ecosystems would contribute to reducing mangrove susceptibility to climate change outcomes. Managing coastal land uses to facilitate gradual landward mangrove migration at sites where this is predicted to occur in response to relative SLR is an additional adaptation method.

In combination with these and other options for augmenting mangrove resistance and resilience to climate change, there is a need for (i) regional monitoring networks to provide a stronger basis for inferring whether global versus local stressors are causing observed changes in mangrove position, structure, and processes and (ii) outreach and education activities to increase awareness of the value of mangrove ecosystem services among the public and decision makers. The future of mangroves therefore to a large extent depends on the degree of human interventions and their interactions with climate-related changes.

7.8 Knowledge Gaps and Future Directions

- The physiological response of mangrove plants and animals as well as the response of communities to climate change are not well understood and require further study, in particular on the response to interacting multistressors.
- The carbon storage potential has large uncertainties and needs to be improved.
- There are large knowledge gaps on how the interaction of human interventions and climate change will affect mangrove ecosystem structure, functions, and services; multi- and interdisciplinary studies on these are required.
- Further efforts to increase the awareness among the public and decision makers of the value of mangrove ecosystem services will contribute to decreasing the risk of mangrove loss related to climate change.

References

Albers T, Schmitt K (2015) Dyke design, floodplain restoration and mangrove co-management as parts of an area coastal protection strategy for the mud coasts of the Mekong Delta, Vietnam. Wetl Ecol Manag 23:991–1004

Allen JA (1998) Mangroves as alien species: the case of Hawaii. Glob Ecol Biogeogr 7:61–71

Alongi DM (2002) Present state and future of the world's mangrove forests. Environ Conserv 29:331–349

Alongi DM (2008) Mangrove forests: resilience, protection from tsunamis, and responses to global climate change. Estuar Coast Shelf Sci 76:1–13

Alongi DM (2012) Carbon sequestration in mangrove forests. Carbon Manag 3:313–322

Alongi DM (2014) Carbon cycling and storage in mangrove forests. Annu Rev Mar Sci 6:195–219

Alongi DM (2015) The impact of climate change on mangrove forests. Curr Clim Change Rep 1:30–39

Amjad AS, Kasawani I, Kamaruzaman J (2007) Degradation of Indus Delta mangroves in Pakistan. Int J Geol 1:27–34

Andutta F, Ridd PV, Wolanski E (2011) Dynamics of hypersaline coastal waters in the Great Barrier Reef. Estuar Coast Shelf Sci 94:299–305

Anthony EJ, Gratiot N (2012) Coastal engineering and large-scale mangrove destruction in Guyana, South America: averting an environmental catastrophe in the making. Ecol Eng 47:268–273

Aung TT, Mochida Y, Than MM (2013) Prediction of recovery pathways of cyclone-disturbed mangroves in the mega delta of Myanmar. Forest Ecol Manag 293:103–113

Ball MC, Cochrane MJ, Rawason HM (1997) Growth and water use of the mangroves *Rhizophora apiculata* and *R. stylosa* in response to salinity and humidity under ambient and elevated concentration of atmospheric CO_2. Plant Cell Environ 20:1158–1166

Barbier ED, Hacker SD, Kennedy C, Koch EW, Stier AC, Silliman BR (2011) The value of estuarine and coastal ecosystem services. Ecol Monogr 81:69–193

Barr JG, Engel V, Fuentes JD, Fuller DO, Kwon H (2013) Modeling light use efficiency in a subtropical mangrove forest equipped with CO_2 eddy covariance. Biogeosciences 10:2145–2158

Bianchi TS, Allison MA, Zhao J, Li X, Comeaux RS, Feagin RA, Kulawardhana RW (2013) Historical reconstruction of mangrove expansion in the Gulf of Mexico: linking climate change with carbon sequestration in coastal wetlands. Estuar Coast Shelf Sci 119:7–16

Bouillon S, Borges AV, Castaneda-Moya E, Diele K, Dittmar T, Duke NC, Kristensen E, Lee SY, Marchand C, Middelburg JJ, Rivera-Monroy VH, Smith TH III, Twilley RR (2008) Mangrove production and carbon sinks: a revision of global budget estimates. Glob Biogeochem Cycles 22:GB2013. https://doi.org/10.1029/2007GB003052

Breithaupt JL, Smoak JM, Smith TJ III, Sanders CJ, Hoare A (2012) Organic carbon burial rates in mangrove sediments: strengthening the global budget. Glob Biogeochem Cycles 26:GB3011

Byrne M, Przeslawski R (2013) Multistressor impacts of warming and acidification of the ocean on marine invertebrates´ life histories. Integr Comp Biol 53:582–596

Cahoon DR (2006) A review of major storm impacts on coastal wetland elevations. Estuar Coasts 29:889–898

Cahoon DR (2015) Estimating relative sea-level rise and submergence potential at a coastal wetland. Estuar Coasts 38:1077–1084

Cahoon DR, Hensel PF, Spencer T, Reed DJ, McKee KL, Saintilan N (2006) Coastal wetland vulnerability to relative sea-level rise: wetland elevation trends and process controls. In: JTA V, Beltman B, Bobbink R, Whigham D (eds) Wetlands and natural resource management. Ecological studies, vol 190. Springer, Berlin/Heidelberg, pp 271–292

Cavanaugh KC, Kellner JR, Fordem AJ, Gruner DS, Parker JD, Rodriguez W, Feller IC (2014) Poleward expansion of mangroves is a threshold response to decreased frequency of extreme cold events. P Natl Acad Sci USA 111:723–727

Cheeseman J (1994) Depressions of photosynthesis in mangrove canopies. In: Baker NR, Bowyer JR (eds) From molecular mechanisms to the field. Bios Scientific Publishers, Oxford, pp 377–389

Cherry JA, McKee KL, Grace JB (2009) Elevated CO_2 enhances biological contributions to elevation change in coastal wetlands by offsetting stressors associated with sea-level rise. J Ecol 97:67–77

Chimner RA, Fry B, Kaneshiro MY, Cormier N (2006) Current extent and historical expansion of introduced mangroves on O'ahu, Hawai'i. Pac Sci 60:377–383

Christensen JH, Krishna Kumar K, Aldrian E, An S-I, Cavalcanti IFA, de Castro M, Dong W, Goswami P, Hall A, Kanyanga JK, Kitoh A, Kossin J, Lau N-C, Renwick J, Stephenson DB, Xie SP, Zhou T (2013) Climate phenomena and their relevance for future regional climate change. In: Stocker TF, Qin D, Plattner G-K, Tignor M, Allen SK, Boschung J, Nauels A, Xia Y, Bex V, Midgley PM (eds) Climate change 2013: The Physical Science Basis. Contribution of Working Group I to the Fifth Assessment Report of the Intergovernmental Panel on Climate Change. Cambridge University Press, Cambridge/New York, pp 1217–1308

Church JA, Clark PU, Cazenave A, Gregory JM, Jevrejeva S, Levermann A, Merrifield MA, Milne GA, Nerem RS, Nunn PD, Payne AJ, Pfeffer WT, Stammer D, Unnikrishnan AS (2013) Sea level change. In: Stocker TF, Qin D, Plattner G-K, Tignor M, Allen SK, Boschung J, Nauels A, Xia Y, Bex V, Midgley PM (eds) Climate change 2013: The Physical science basis. Contribution of Working Group I to the Fifth Assessment Report of the Intergovernmental Panel on Climate Change. Cambridge University Press, Cambridge/New York, pp 1137–1216

Ciais P, Sabine C, Bala G, Bopp L, Brovkin V, Canadell J, Chhabra A, DeFries R, Galloway J, Heimann M, Jones C, Le Quéré C, Myneni RB, Piao S, Thornton P (2013) Carbon and other biogeochemical cycles. In: Stocker TF, Qin D, Plattner G-K, Tignor M, Allen SK, Boschung J, Nauels A, Xia Y, Bex V, Midgley PM (eds) Climate change 2013: The physical science basis. Contribution of Working Group I to the Fifth Assessment Report of the Intergovernmental Panel on Climate Change. Cambridge University Press, Cambridge/New York, pp 465–570

Clüsener M, Breckle SW (1987) Reasons for the limitation of mangrove along the west coast of northern Peru. Vegetatio 68:173–177

Collins M, Knutti R, Arblaster J, Dufresne J-L, Fichefet T, Friedlingstein P, Gao X, Gutowski WJ, Johns T, Krinner G, Shongwe M, Tebaldi C, Weaver AJ, Wehner M (2013) Long-term climate change: projections, commitments and irreversibility. In: Stocker TF, Qin D, Plattner G-K, Tignor M, Allen SK, Boschung J, Nauels A, Xia Y, Bex V, Midgley PM (eds) Climate change 2013: The physical science basis. Contribution of Working Group I to the Fifth Assessment Report of the Intergovernmental Panel on Climate Change. Cambridge University Press, Cambridge/New York, pp 1029–1136

Costanza R, d'Arge R, de Groot R, Farber S, Grasso M, Hannon B, Limburg K, Naeem S, O'Neill RV, Paruelo J, Raskin RG, Sutton P, van den Belt M (1997) The value of the world's ecosystem services and natural capital. Nature 387:253–260

Dale P, Eslami-Andargoli L, Knight J (2013) The impact of encroachment of mangroves into saltmarshes on saltwater mosquito habitats. J Vector Ecol 38:330–338

Di Nitto D, Neukermans G, Koedam N, Defever H, Pattyn F, Kairo JG, Dahdouh-Guebas F (2014) Mangroves facing climate change: landward migration potential in response to projected scenarios of sea level rise. Biogeosciences 11:857–871

Donato DC, Kauffman JB, Murdiyarso D, Kurnianto S, Stidham M, Kanninen M (2011) Mangroves among the most carbon-rich forests in the tropics. Nat Geosci 4:293–297

Duke NC, Ball MC, Ellison JC (1998) Factors influencing biodiversity and distributional gradients in mangroves. Glob Ecol Biogeogr 7:27–47

Ellison J (2000) How South Pacific mangroves may respond to predicted climate change and sea level rise. Chapter 15. In: Gillespie A, Burns W (eds) Climate change in the South Pacific: impacts and responses in Australia, New Zealand, and Small Islands States. Kluwer Academic Publishers, Dordrecht, pp 289–301

Ellison JC, Stoddart DR (1991) Mangrove ecosystem collapse during predicted sea-level rise: Holocene analogues and implications. J Coast Res 7:151–165

Ellison AM, Farnsworth EJ, Merkt RE (1999) Origins of mangrove ecosystems and the mangrove biodiversity anomaly. Glob Ecol Biogeogr 8:95–115

Eslami-Andargoli L, Dale P, Sipe N, Chaseling J (2009) Mangrove expansion and rainfall patterns in Moreton Bay, Southeast Queensland, Australia. Estuar Coast Shelf Sci 85:292–298

Ewel KC, Twilley RR, Ong JE (1998) Different kinds of mangrove forests provide different goods and services. Glob Ecol Biogeogr 7:83–94

Farnsworth EJ, Ellison AM, Gong WK (1996) Elevated CO_2 alters anatomy, physiology, growth, and reproduction of red mangrove (*Rhizophora mangle* L.) Oecologia 108:599–609

Feller IC, Whigham DF, McKee KL, O'Neill JP (2002) Nitrogen vs. phosphorus limitation across an ecotonal gradient in a mangrove forest. Biogeochem 62:145–175

Feller IC, Whigham DF, McKee KL, Lovelock CE (2003) Nitrogen limitation of growth and nutrient dynamics in a mangrove forest, Indian River Lagoon, Florida. Oecologia 134:405–414

Field CD (1995b) Impact of expected climate change on mangroves. Hydrobiologia 295:75–81

Fonseca IV, Marins RV, Lacerda LD (2014) Phosphorus and suspended mater balance in a mangrove forest receiving shrimp farms effluents in the Jaguaribe River, NE Brazil. Braz J Aquat Sci Technol 18:53–60
Fusi M, Giomi F, Babbini S, Daffonchio D, McQuaid CD, Porri F, Cannicci C (2015) Thermal specialization across large geographical scales predicts the resilience of mangrove crab populations to global warming. Oikos 124:784–795
Gilman E, Van Lavieren H, Ellison J, Jungblut V, Wilson L, Areki F, Brighouse G, Bungitak J, Dus E, Henry M, Sauni I, Kilman M, Matthews E, Teariki-Ruatu N, Tukia S, Yuknavage K (2006) Pacific Island mangroves in a changing climate and rising sea. UNEP Regional Seas Reports and Studies No. 179. UNEP, Nairobi
Gilman EL, Ellison J, Duke NC, Field C (2008) Threats to mangroves from climate change and adaptation options: a review. Aquat Bot 89:237–250
Giri C, Ochieng E, Tieszen LL, Zhu Z, Singh A, Loveland T, Masek J, Duke N (2011) Status and distribution of mangrove forests of the world using earth observation satellite data. Glob Ecol Biogeogr 20:154–159
Godoy MDP, Lacerda LD (2014) River-island response to land-use change within the Jaguaribe River, Brazil. J Coast Res 30:399–410
Hamilton SE, Casey D (2016) Creation of a high spatio-temporal resolution global database of continuous mangrove forest cover for the 21st century (CGMFC-21). Glob Ecol Biogeogr 25:729–738
Hartmann DL, Klein Tank AMG, Rusticucci M, Alexander LV, Brönnimann S, Charabi Y, Dentener FJ, Dlugokencky EJ, Easterling DR, Kaplan A, Soden BJ, Thorne PW, Wild M, Zhai PM (2013) Observations: atmosphere and surface. In: Stocker TF, Qin D, Plattner G-K, Tignor M, Allen SK, Boschung J, Nauels A, Xia Y, Bex V, Midgley PM (eds) Climate change 2013: the physical science basis. Contribution of Working Group I to the Fifth Assessment Report of the Intergovernmental Panel on Climate Change. Cambridge University Press, Cambridge/New York, pp 159–254
Hoegh-Guldberg O (1999) Climate change, coral bleaching and the future of the world's coral reefs. Mar Freshw Res 50:839–866
Hoegh-Guldberg O, Bruno JF (2010) The impact of climate change on the world's marine ecosystems. Science 318:1523–1528
Humborg C, Ittekkot V, von Cociasu A, Bodungen B (1997) Effect of Danube River dam on Black Sea biogeochemistry and ecosystem structure. Nature 386:385–388
IPCC (2014) In: Pachauri RK, Meyer LA (eds) Climate Change 2014: Synthesis Report. Contribution of Working Groups I, II and III to the Fifth Assessment Report of the Intergovernmental Panel on Climate Change [Core Writing Team]. IPCC, Geneva, p 151
Jennerjahn TC (2012) Biogeochemical response of tropical coastal systems to present and past environmental change. Earth Sci Rev 114:19–41
Jennerjahn TC, Ittekkot V (2002) Relevance of mangroves for the production and deposition of organic matter along tropical continental margins. Naturwissenschaften 89:23–30
Jennerjahn TC, Soman K, Ittekkot V, Nordhaus I, Sooraj S, Priya RS, Lahajnar N (2008) Effect of land use on the biogeochemistry of dissolved nutrients and suspended and sedimentary organic matter in the tropical Kallada River and Ashtamudi estuary, Kerala, India. Biogeochemistry 90:29–47
Kidwai, FP, Ahmed W, Tabrez M, Zhang J, Khan MW (2016) Practicality of marine protected areas - Can there be solutions for the River Indus delta? Estuar Coast Shelf Sci 183:349–359
Kirtman B, Power SB, Adedoyin JA, Boer GJ, Bojariu R, Camilloni I, Doblas-Reyes FJ, Fiore AM, Kimoto M, Meehl GA, Prather M, Sarr A, Schär C, Sutton R, van Oldenborgh GJ, Vecchi G, Wang HJ (2013) Near-term climate change: projections and predictability. In: Stocker TF, Qin D, Plattner G-K, Tignor M, Allen SK, Boschung J, Nauels A, Xia Y, Bex V, Midgley PM (eds) Climate change 2013: the physical science basis. Contribution of Working Group I to the Fifth Assessment Report of the Intergovernmental Panel on Climate Change. Cambridge University Press, Cambridge/New York, pp 953–1028

Kiwango H, Njau KN, Wolanski E (2015) The need to enforce minimum environmental flow requirements in Tanzania to preserve estuaries: case study of mangrove-fringed Wami River estuary. Ecohydrol Hydrobiol. https://doi.org/10.1016/j.ecohyd.2015.09.002

Knutson TR, McBride JL, Chan J, Emmanuel K, Holland G, Landsea C, Held I, Kossin JP, Srivastava AK, Sugi M (2010) Tropical cyclones and climate change. Nat Geosci 3:157–163

Kossin JP, Knapp KR, Vimont DJ, Murnane RJ, Harper BA (2007) A globally consistent reanalysis of hurricane variability and trends. Geophys Res Lett 34:L04815. https://doi.org/10.1029/2006GL028836

Krauss KW, Allen JA, Cahoon DR (2003) Differential rates of vertical accretion and elevation change among aerial root types in Micronesian mangrove forests. Estuar Coast Shelf Sci 56:251–259

Krauss KW, Lovelock CE, McKee KL, López-Hollman L, Ewe SML, Sousa WP (2008) Environmental drivers in mangrove establishment and early development: a review. Aquat Bot 89:105–127

Krauss KW, McKee KL, Hester MW (2014a) Water use characteristics of black mangrove (*Avicennia germinans*) communities along an ecotone with marsh at a northern geographical limit. Ecohydrology 7:354–365

Krauss KW, McKee KL, Lovelock CE, Cahoon R, Saintilan N, Reef R, Chen L (2014b) How mangrove forests adjust to rising sea level. New Phytol 202:19–34

Krauss KW, Barr JG, Engel V, Fuentes JD, Wang H (2015) Approximations of stand water use versus evapotranspiration from three mangrove forests in southwest Florida, USA. Agri Forest Meteorol 213:291–303. https://doi.org/10.1016/j.agrformet.2014.11.014

Kroeker KJ, Kordas RL, Crim R, Hendriks IE, Ramajo L, Singh GS, Duarte CM, Gattuso JP (2013) Impact of ocean acidification on marine organisms: quantifying sensitivities and interaction with warming. Glob Chang Biol 19:1884–1896

Kurihara H, Matsui M, Furukawa H, Hayashi M, Ishimatsu A (2008) Long-term effects of predicted future seawater CO_2 conditions on the survival and growth of the marine shrimp *Palaemon pacificus*. J Exp Mar Biol Ecol 367:41–46

Lacerda LD, Dias FJS, Marins RV, Soares TM, Godoy JM, Godoy MLDP (2013) Pluriannual watershed discharges of Hg into a tropical semi-arid estuary of the Jaguaribe River, NE Brazil. J Braz Chem Soc 24:1719–1731

Lang'at JKS, Kairo JG, Mencuccini M, Bouillon S, Skov M, Waldron S, Huxham M (2014) Rapid losses of surface elevation following tree girdling and cutting in tropical mangroves. PlosOne 9:e107868

Lee SY (2008) Mangrove macrobenthos: assemblages, services, and linkages. J Sea Res 59:16–29

Lee SY, Primavera JH, Dahdouh-Guebas F, McKee K, Bosire JO, Cannicci S, Diele K, Fromard F, Koedam N, Marchand C, Mendelssohn I, Mukherjee N, Record S (2014) Ecological role and services of tropical mangrove ecosystems: a reassessment. Glob Ecol Biogeogr 23:726–743

Littlewood DT (1989) Thermal tolerance and the effects of temperature on air-gaping in the mangrove oyster, *Crassostrea rhizophorae*. Comp Biochem Physiol 93A:395–397

Lovelock CE, Feller IC, Ellis J, Hancock N, Schwarz AM, Hancock N, Nichols P, Sorrell B (2007) Mangrove growth in New Zealand estuaries: the role of nutrient enrichment at sites with contrasting rates of sedimentation. Oecologia 153:633–641

Lovelock CE, Ball MC, Martin KC, Feller IC (2009) Nutrient enrichment increases mortality of mangroves. PlosOne 4. https://doi.org/10.1371/journal.pone.0005600

Lovelock CE, Adame MF, Bennion V, Hayes M, Reef R, Santini M, Cahoon DR (2015a) Sea level and turbidity controls on mangrove soil surface elevation change. Estuar Coast Shelf Sci 153:1–9

Lovelock CE, Cahoon DR, Friess DA, Guntenspergen GR, Krauss KW, Reef R, Rogers K, Saunders ML, Sidik F, Swales A, Saintilan N, Thuyen LX, Triet T (2015b) The vulnerability of Indo-Pacific mangrove forests to sea-level rise. Nature 526:559–563

Lovelock CE, Krauss KW, Osland MJ, Reef R, Ball MC (2016) The physiology of mangrove trees with changing climate. In: Goldstein G, Santiago L (eds) Tropical tree physiology: adaptations and responses in a changing environment. Springer, New York, pp 149–179

Machado W, Moscatelli M, Rezende LG, Lacerda LD (2002) Mercury, zinc and copper accumulation in mangrove sediments affected by landfill wastewater. Environ Pollut 120:455–461

Mazda Y, Wolanski E, Ridd PV (2007) Part I: outline of the physical processes within mangrove systems. In: Mazda Y, Wolanski E, Ridd PV (eds) The role of physical processes in mangrove environments: manual for the preservation and utilization of mangrove ecosystems. Terrapub Publishers, Tokyo, pp 3–64

McKee KL, Rooth JE (2008) Where temperate meets tropical: multi-factorial effects of elevated CO_2, nitrogen enrichment, and competition on a mangrove-salt marsh community. Glob Chang Biol 14:971–984

McKee KL, Cahoon DR, Feller IC (2007) Caribbean mangroves adjust to rising sea level through biotic controls on change in soil elevation. Glob Ecol Biogeogr 16:545–556

McMahon RF (2001) Acute thermal tolerance in intertidal gastropods relative to latitude, superfamily, zonation and habitat with special emphasis on the Littorinoidea. J Shellfish Res 20:459–467

Millennium Ecosystem Assessment (2005) Ecosystems and human well-being: synthesis. Island Press, Washington, DC. www.unep.org/maweb

Milliman JD, Farnsworth KL (2011) River discharge to the coastal ocean – a global synthesis. Cambridge University Press, Cambridge

Mumby P, Edwards A, Arlas-Gonzalez J, Lindeman K, Blackwell P, Gall A, Gorczynska M, Harbone A, Pescod C, Renken H, Wabnitz C, Llewellyn G (2004) Mangroves enhance the biomass of coral reef fish communities in the Caribbean. Nature 427:533–536

Nascimento WR Jr, Souza-Filho PW, Proisy C, Lucas RM, Rosenqvist A (2013) Mapping changes in the largest continuous Amazonian mangrove belt using object-based classification of multisensor satellite imagery. Estuar Coast Shelf Sci 117:83–93

Nellemann C, Corcoran E, Duarte CM, Valdés L, De Young C, Fonseca L, Grimsditch G (eds) (2009) Blue Carbon. A Rapid Response Assessment. United Nations Environment Programme, GRID-Arendal, www.grida.no

Nicholls RJ, Cazenave A (2010) Sea-level rise and its impact on coastal zones. Science 328:1517–1520

Nilsson GE, Crawley N, Lunde IG, Munday PL (2009) Elevated temperature reduces the respiratory scope of coral reef fishes. Glob Chang Biol 15:1405–1412

Nhan NN (2016) Tidal regime deformation by sea level rise along the coast of the Mekong Delta. Estuar Coast Shelf Sci 183: 382–391

Nordhaus I (2008) Global climate and regional environmental change affecting coastal ecosystems: 2. Ecology and Resources Of Mangrove Forests. In: Datta RK (ed) Coastal ecosystems – hazards, management and rehabilitation. Centre for Science and Technology of the Non-Aligned and Other Developing Countries (NAM S&T Centre), Daya Publishing House, Delhi, pp 78–99

Nordhaus I, Wolff M, Diele K (2006) Litter processing and population food intake of the mangrove crab *Ucides cordatus* in a high intertidal forest in northern Brazil. Estuar Coast Shelf Sci 67:239–250

Osland MJ, Enwright N, Day RH, Doyle TW (2013) Winter climate change and coastal wetland foundation species: salt marshes vs. mangrove forests in the southeastern United States. Glob Chang Biol 19:1482–1494

Osland MJ, Day RH, Larriviere JC, From AS (2014a) Aboveground allometric models for freeze-affected black mangroves (*Avicennia germinans*): equations for a climate sensitive mangrove-marsh ecotone. PlosOne 9:e99604

Osland MJ, Enwright N, Stagg CL (2014b) Freshwater availability and coastal wetland foundation species: ecological transitions along a rainfall gradient. Ecology 95:2789–2802

Page SE, Rieley JO, Banks CJ (2011) Global and regional importance of the tropical peatland carbon pool. Glob Chang Biol 17:798–818

Parker LM, Ross PM, O'Connor WA (2010) Comparing the effect of elevated pCO2 and temperature on the fertilization and early development of two species of oysters. Mar Biol 157:2435–2452

Pechenik JA (1999) On the advantages and disadvantages of larval stages in benthic marine invertebrate life cycles. Mar Ecol Prog Ser 177:269–297

Pernetta JC (1993) Mangrove forests, climate change and sea level rise: hydrobiological influences on community structure and survival, with examples from the Indo-West Pacific. A marine conservation and development report. IUCN, Gland

Perry CL, Mendelssohn IA (2009) Ecosystem effects of expanding populations of *Avicennia germinans* in a Louisiana salt marsh. Wetlands 29:396–406

Piou C, Feller IC, Berger U, Chi F (2006) Zonation patterns of Belizean offshore mangrove forests 41 years after a catastrophic hurricane. Biotropica 38:365–372

Quisthoudt K, Schmitz N, Randin CF, Dahdouh-Guebas F, Robert EMR, Koedam N (2012) Temperature variation among mangrove latitudinal range limits worldwide. Trees 26:1919–1931

Rabalais NN, Turner RE, Justic D, Dortch Q, Wiseman WJ Jr, Sen Gupta BK (2000) Gulf of Mexico biological system responses to nutrient changes in the Mississippi River. In: Hobbie J (ed) Estuarine science: a synthetic approach to research and practice. Island Press, Washington, DC, pp 241–268

Record S, Charney ND, Zakaria RM, Ellison AM (2013) Projecting global mangrove species and community distributions under climate change. Ecosphere 4:1–23

Reef R, Lovelock CE (2015) Regulation of water balance in mangroves. Ann Bot-London 115:385–395

Reef R, Feller IC, Lovelock CE (2010) Nutrition of mangroves. Tree Physiol 30:1148–1160

Richards DR, Friess DA (2016) Rates and drivers of mangrove deforestation in Southeast Asia, 2000–2012. Proc Nat Acad Sci 113:344–349

Robertson AI (1986) Leaf-burying crabs: their influence on energy flow and export from mixed mangrove forests (*Rhizophora* spp.) in northeastern Australia. J Exp Mar Biol Ecol 102:237–248

Robertson AI, Daniel PA (1989) The influence of crabs on litter processing in high intertidal mangrove forests in tropical Australia. Oecologia 78:191–198

Robins PE, Lewis MJ, Simpson JH, Howlett ER, Malham SK (2014) Future variability of solute transport in a macrotidal estuary. Estuar Coast Shelf Sci 151:88–99

Rogers K, Saintilan N, Heijns H (2005) Mangrove encroachment of salt marsh in Western Port Bay, Victoria: the role of sedimentation, subsidence and sea level rise. Estuaries 28:551–559

Saintilan N, Wilson NC, Rogers K, Rajkaran A, Krauss KW (2014) Mangrove expansion and salt marsh decline at mangrove poleward limits. Glob Chang Biol 20:147–157

Saito Y, Chaimanee N, Jarupongsakul T, Syvitski JPM (2007) Shrinking megadeltas in Asia: sea-level rise and sediment reduction impacts from case study of the Chao Phraya delta. Inprint Newsletter of the IGBP/IHDP Land Ocean Interaction in the Coastal Zone 2:3–9

Salmo SG, Lovelock CE, Duke NC (2013) Assessment of vegetation and soil conditions in restored mangroves interrupted by severe tropical typhoon 'Chan-hom' in the Philippines. Hydrobiologia 733:85–102

Sasmito SD, Murdiyarso D, Friess DA, Kurnianto S (2015) Can mangroves keep pace with contemporary sea level rise? A global data review. Wetl Ecol Manag. https://doi.org/10.1007/s11273-015-9466-7

Seidel DJ, Fu Q, Randel WJ, Reichler TJ (2008) Widening of the tropical belt in a changing climate. Nat Geosci 1:21–24

Smith SV, Swaney DP, Talaue-McManus L, Bartley JD, Sandhei PT, McLaughlin CJ, Dupra VC, Crossland CJ, Buddemeier RW, Maxwell BA, Wulff F (2003) Humans, hydrology, and the distribution of inorganic nitrogen loading to the ocean. Bioscience 53:235–245

Snedaker SC, Araujo RJ (1998) Stomatal conductance and gas exchange in four species of Caribbean mangroves exposed to ambient and increased CO_2. Mar Freshw Res 49:325–327

Snedaker SC, Meeder JF, Ross MS, Ford RG (1994) Discussion of "Ellison, Joanna C. and Stoddart, David R., 1991. Ellison, J.C., Stoddart, D.R., 1991. Mangrove ecosystem collapse during predicted sea-level rise: holocene analogues and implications. J Coastal Res 7:151-165". J Coast Res 10:497–498

Soares MLG, Estrada GCD, Fernandez V, Tognella MMP (2012) Southern limit of the Western South Atlantic mangroves: assessment of the potential effects of global warming from a biogeographical perspective. Estuar Coast Shelf Sci 101:44–53

Solomon S, Qin D, Manning M, Alley RB, Berntsen T, Bindoff NL, Chen Z, Chidthaisong A, Gregory JM, Hegerl GC, Heimann M, Hewitson B, Hoskins BJ, Joos F, Jouzel J, Kattsov V, Lohmann U, Matsuno T, Molina M, Nicholls N, Overpeck J, Raga G, Ramaswamy V, Ren J, Rusticucci M, Somerville R, Stocker TF, Whetton P, Wood RA, Wratt D (2007) Technical summary. In: Solomon S, Qin D, Manning M, Chen Z, Marquis M, Averyt KB, Tignor M, Miller HL (eds) Climate change 2007: the physical science basis. Contribution of Working Group I to the Fourth Assessment Report of the Intergovernmental Panel on Climate Change. Cambridge University Press, Cambridge/New York

Spalding M, Kainuma M, Collins L (2010) World atlas of mangroves. Earthscan, Washingon, DC

Syvitski JPM (2008) Deltas at risk. Sustain Sci 3:23–32

Syvitski JPM, Vörösmarty CJ, Kettner AJ, Green P (2005) Impact of humans on the flux of terrestrial sediment to the global coastal ocean. Science 308:376–380

Syvitski JPM, Kettner AJ, Overeem I, Hutton EWH, Hannon MT, Brakenridge GR, Day J, Vörösmarty CJ, Saito Y, Giosan L, Nicholls RJ (2009) Sinking deltas due to human activities. Nat Geosci 2:681–686

Tomlinson PB (1986) The botany of mangroves. Cambridge University Press, Cambridge

Twilley RR, Chen RH, Hargis T (1992) Carbon sinks in mangroves and their implications to carbon budget of tropical coastal ecosystems. Water Air Soil Pollut 64:265–288

Valiela I, Cole ML (2002) Comparative evidence that salt marshes and mangroves may protect seagrass meadows from land-derived nitrogen loads. Ecosystems 5:92–102

Valiela I, Bowen JL, York JK (2001) Mangrove forests: one of the world's threatened major tropical environments. Bioscience 51:807–815

Victor S, Golbuu Y, Wolanski E, Richmond R (2004) Fine sediment trapping in two mangrove-fringed estuaries exposed to contrasting land-use intensity, Palau, Micronesia. Wetl Ecol Manag 12:277–283

Vörösmarty CJ, Meybeck M, Fekete B, Sharmad K, Green P, Syvitski JPM (2003) Anthropogenic sediment retention: major global impact from registered river impoundments. Glob Planet Chang 39:169–190

Walling DE, Fang D (2003) Recent trends in the suspended sediment loads of the world's rivers. Glob Planet Chang 39:111–126

Webb EL, Friess DA, Krauss KW, Cahoon DR, Guntenspergen GR, Phelps J (2013) A global standard for monitoring coastal wetland vulnerability to accelerated sea-level rise. Nat Clim Chang 3:458–465

Wells S, Ravilous C, Corcoran E (2006) In the front line: shoreline protection and other ecosystem services from mangroves and coral reefs. Cambridge, United Nations Environment Programme World Conservation Monitoring Centre

Whiteley NM (2011) Physiological and ecological responses of crustaceans to ocean acidification. Mar Ecol Prog Ser 430:257–271

Wolanski E, Elliott M (2015) Estuarine ecohydrology. An introduction. Elsevier, Amsterdam

Wolanski E, Mazda Y, Ridd P (1992) Mangrove hydrodynamics. In: Robertson AI, Alongi DM (eds) Tropical mangrove ecosystems, AGU, Washington, Coastal and Estuarine Studies, vol 41, pp 43–62

Woodroffe CD (1992) Mangrove sediments and geomorphology. In: Robertson AI, Alongi DM (eds) Tropical mangrove ecosystems, AGU, Washington, Coastal and Estuarine Studies, vol 41, pp 7–41

Woodroffe CD, Thom BG, Chappell J (1985) Development of widespread mangrove swamps in mid-Holocene times in northern Australia. Nature 317:711–713

Woodroffe, CD, Rogers, K, McKee, KL, Lovelock, CE, Mendelssohn, IA, Saintilan, N (2016) Mangrove Sedimentation and Response to Relative Sea-Level Rise. Annual Review of Marine Science 8:243–266

Woodworth PL, Blackman DL (2004) Evidence for systematic changes in extreme high waters since the mid-1970s. J Clim 17:1190–1197

Yates KK, Rogers CS, Herlan JJ, Brooks GR, Smiley NA, Larson RA (2014) Diverse coral communities in mangrove habitats suggest a novel refuge from climate change. Biogeosciences 11:4321–4337

Chapter 8
Mangroves and People: Local Ecosystem Services in a Changing Climate

Mark Huxham, Amrit Dencer-Brown, Karen Diele, Kandasamy Kathiresan, Ivan Nagelkerken, and Caroline Wanjiru

There is no wealth but life
– John Ruskin

8.1 Introduction

8.1.1 The Value of Local Ecosystem Services

Mangrove forests are exceptionally useful ecosystems, but understanding the ways in which they improve human welfare and communicating this so that it makes a difference is difficult. A popular approach involves economic valuation. A range of economic tools is used to capture and summarise the value of different services in monetary terms. There are good arguments for this approach; 'money talks' so nature can be heard. However, it risks encouraging partial or distorted views – often to the frustration of economists themselves who do not conflate 'the price' with the value.

M. Huxham (✉) • K. Diele
School of Applied Sciences, Edinburgh Napier University, Edinburgh EH11 4BN, UK
e-mail: m.huxham@napier.ac.uk; k.diele@napier.ac.uk

A. Dencer-Brown
Institute for Applied Ecology, Auckland University of Technology,
Auckland 1142, New Zealand
e-mail: amritrules@hotmail.com

K. Kathiresan
Centre of Advanced Study in Marine Biology, Marine Sciences, Annamalai University,
Parangipettai 608 502, TN, India
e-mail: kathirsum@rediffmail.com

I. Nagelkerken
Southern Seas Ecology Laboratories, School of Biological Sciences, The University of
Adelaide, Darling Building, DX 650 418, Adelaide, SA 5005, Australia
e-mail: ivan.nagelkerken@adelaide.edu.au

C. Wanjiru
Kenya Marine and Fisheries Research Institute, P.O. Box 81651, Mombasa, Kenya
e-mail: carolwanjiru80@gmail.com

V.H. Rivera-Monroy et al. (eds.), *Mangrove Ecosystems: A Global Biogeographic Perspective*, https://doi.org/10.1007/978-3-319-62206-4_8

It is particularly so for ecosystem services of most immediate benefit to local (and often poor) people, the focus for this chapter. An example from Gazi Bay in Kenya helps illustrate the point. Here the value of wood to be used as fuel for cooking is only around 5% of the total economic value of the forest, whilst carbon sequestration makes up 38% (Huxham et al. 2015). This follows the typical pattern in which regulating services such as coastal protection and carbon capture and storage dominate the estimated values of mangrove forests when using economic methods. However, 71% of households in the village rely on mangrove firewood and 96% of the individuals responsible for collecting this wood are women (who as a group are poorer than men). Throughout the tropics, it is common for women to carry the burden of fuel wood collection (e.g., in Brazil; Glaser 2003); hence, under-valuation of the ecosystem service of firewood production may marginalise women as well as the poor. In Kenya, firewood is used daily in the preparation of food, whilst the value of carbon sequestration to local people is a theoretical one at most sites (although not entirely at Gazi; Plan Vivo 2015). Hence, firewood assumes an immediate and pressing importance in the lives of some of the poorest people associated with mangrove forests, an importance that may be obscured if we only rely on global economic valuations. From the local perspective, the value of mangrove forests in national or global policy may be irrelevant and, hence, do little to help encourage local action for sustainable management. Social equity and practical conservation demand that we remember the importance of the local.

8.1.2 *Defining Local Services*

Ecosystem services are contributions to human welfare made by the natural world. As such, all human beings are stakeholders in ecosystem services if we follow Hein et al. (2006) in defining stakeholders in this context as 'any group or individual who can affect or is affected by the ecosystem's services'; ultimately, all of us depend on ecosystems. However, different services are generated and used at different spatial scales. In defining here what we mean by 'local' ecosystem services, we focus on the social rather than the ecological characteristics, since these two elements might differ. For example, services might be generated at very small ecological scales (such as truffle fungi growing in a few patches of a forest) but enjoyed over large or distant scales (in exclusive restaurants gracing distant capitals). In economic language, the 'distance decay' of a service might be fast (the shade provided by a tree?) or non-existent (the value of Antarctic wilderness, as perceived by Europeans). Hein et al. (2006) use international, national, state/provincial, municipal, family and individual levels in their analysis of the institutional scales pertinent to ecosystem services. Our definition of 'local' incorporates their individual to municipal levels. Figure 8.1 summarises the spatial extent of key ecosystem services provided by mangrove forests. It is based on the usual and maximum distances recorded in the literature over which the benefits of a service from a specific mangrove site were

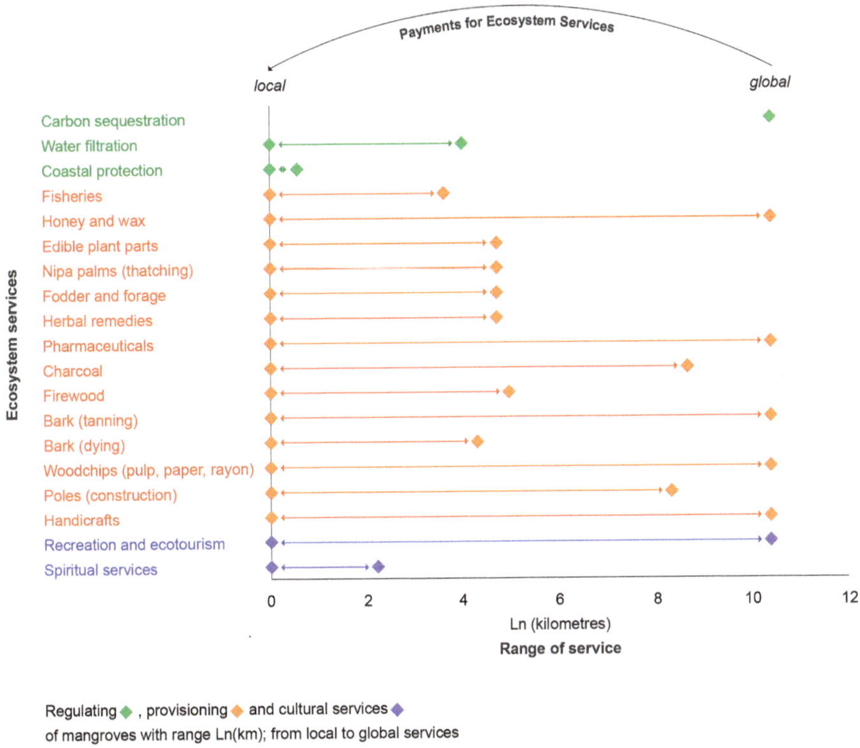

Fig. 8.1 Mangrove ecosystem services by category range. Range refers to the maximum distance over which the benefits of the service are spread, as recorded in a review of relevant literature. For example honey and wax are utilised and sold immediately adjacent to forests and in global markets hundreds of kilometres away. In contrast, commercial pharmaceuticals serve only global or distant markets. Our definition of 'local services' are those of most benefit to people in close proximity to mangrove environments. Whilst payments for ecosystem services schemes can make distant services of local benefit we exclude them from this review. Full details of the methodology underpinning Figs. 8.1, 8.2 and 8.5 are provided in supplementary materials

spread; for example, the distance to the market of a product such as charcoal or the area of coastal land protected from storm surges. Hence, the ecosystem services considered in this chapter are primarily those of most benefit to people living in close proximity (from zero to tens of kilometres) of mangrove forests. This definition excludes services such as carbon sequestration and storage that benefit all humanity, since these benefits accrue more internationally than locally. In an increasingly globalised and interconnected world, such distinctions between 'local' and 'global' scales of impact and benefit has become harder. A striking example is the inequality in the distribution of costs borne and benefits realised from intensive shrimp aquaculture in mangrove swamps. High income groups (the global rich) are responsible for 44% of the estimated costs of this destructive practice but suffer <0.5% of the damages (mainly because they usually live at a great distance from the affected sites); the equivalent data for low income groups are 29% and 54% (using 'equity weighted' sums that correct for poverty; Srinivasan et al. 2008). One complication here is the existence of a range of market and policy mechanisms that may help transfer these global benefits (or compensation for costs) back to local levels, creating real income for local people. Examples include payment for ecosystem services (PES) schemes in which local stakeholders are paid for protecting and enhancing services enjoyed by others (Locatelli et al. 2014; Plan Vivo 2015). Whilst acknowledging their potential for helping to 'make the global local', such schemes are still rare in mangrove areas and are beyond the scope of this chapter.

The local mangrove services we consider here include provisioning (fuel wood and charcoal, fodder, timber, crabs and fin-fish) and regulating (coastal protection and fisheries nursery functions) services. We chose these because the literature emphasises their local importance and because there is information to explore general trends; other services, such as water filtration, may be vital locally but are poorly studied. The forests that deliver these services range in size from a few hundred trees to thousands of hectares, may consist of just one or dozens of tree species and may grow in vast deltas or on tiny over-wash islands; but the diversity of our subject matter is not limited to the biological and geographical features. The types of human communities contiguous to these forests range from megacities to tiny villages and consist of some of the poorest and the wealthiest people on the planet. Any general patterns that emerge in the face of such variety are likely to be imprecise and to ignore multiple exceptions. This does not mean that attempts to classify mangrove forests and their ecosystem services at the global scale are doomed to lack utility. For example, a scheme based on only three variables – soil organic matter, suspended particles and tidal range – provides a useful indication of the likely best approaches to mangrove restoration across the tropics (Thorsten and Friess 2016). Two guiding and interacting themes inform the approach taken in this chapter: the roles of biological diversity and of poverty. For each of our local services, we consider whether patterns of usage – involving, for example, particular species or families of mangrove trees – are consistent between biogeographical regions. We examine the role of biological diversity in determining local use and value of mangrove forests, asking whether people exploit a wider range of species, which provide a wider range of services, in forests with higher levels of floral diversity. We also explore the social factors, and in particular poverty, that may determine forest use. Finally, we conclude with thoughts on the future. Mangrove ecosystems face numerous challenges, not least from climate change; are there biological or social factors that are likely to make the provision of local mangrove services more or less vulnerable in the face of these stresses?

8.2 Fuel Wood and Charcoal

Mangrove trees are used for fuel wood and charcoal production throughout the tropics, making this ecosystem service one of the most widespread (Walters et al. 2008). Many studies report that mangrove wood is highly prized as fuel. In Pakistan, about

0.1 million people use 18,000 tonnes of mangrove firewood each year (Vannucci 2002). In the Philippines, 'mangrove wood is unanimously viewed as superior for firewood to non-mangrove wood' (Walters 2005). This preference may reflect the ease of access to the wood, or the fact that it is not owned or actively managed at some sites, i.e., that it is an open access resource in contrast to wood from trees deliberately cultivated. Thus, forests that are adjacent to settlements and are easy to reach are likely to experience higher rates of exploitation. This can sometimes reflect seasonal differences in accessibility; for example, flooding permits easier access to mangrove firewood by canoe in Cameroon (Munji et al. 2014). However, the superior qualities of the wood, such as its ability to burn for a longer time and at high temperatures combined with the production of particular flavours of wood smoke may also be important. Walters (2005) describes how the use of mangroves as firewood increases dramatically during fiestas in the Philippines, when even households that do not routinely use it, prefer it for roasting pig. Because species in the family Rhizophoracea (such as *Rhizophora* spp. and *Ceriops* spp.) produce particularly dense wood that sustains hot fires, some literature identifies them as preferred sources of fuel wood. Less desired species, such as *Avicennia germinans*, may be left untouched as part of a management strategy (for example in the Caribbean; Smith and Berkes 1993), since maintaining canopy cover can help prevent deterioration of sediment quality and help ensure natural regeneration (Huxham et al. 2010). More commonly, however, other species are also exploited if the preferred sources are depleted. The tendency for communities to prefer Rhizophoraceae (and specifically *R. mangle)* is most pronounced in the Atlantic-East-Pacific region (Table 8.1), which is also the region with the lowest overall species richness of mangroves. With increasing availability of different species (or perhaps decreasing dominance of one or two key species), there is less evidence of a clear preference. Hence, communities in Asia are recorded as utilising a wide range of different species with no clear preferences (Table 8.1). For fuel wood, this broad comparison suggests that increasing biological diversity (that is, increasing numbers of mangrove species) generally correlates with increasing diversity of human use and perhaps, greater flexibility in provision of this ecosystem service.

Wood is the main source of domestic fuel throughout the tropics, although rates of use vary widely between countries. Much of this variation is driven by climate, with desert and dryland nations having much lower wood resources than those in wetter areas. For example, people in Chad use an estimated 0.3 kg/cap/day compared with 1.5 kg/cap/day in wetter Nigeria (Yevich and Logan 2003). However, social factors exert at least as strong an influence as climate. Africa has the lowest global per capita use of fossil fuels because of its relative poverty. Hence, most people, and especially those in rural areas, rely heavily on biofuels, principally wood and charcoal. For example, rural households in Kenya consume an average of 2.14 kg/cap/day of fuel wood, more than 15 times the average used in urban areas (Kituyi et al. 2001). Urban dwellers have a higher reliance on charcoal, at 0.37 kg/cap/day, than rural users (0.26 kg/cap/day), which partly compensates for this lower fuel wood use (Kituyi et al. 2001). Rates of fuel wood use are even higher in other

Table 8.1 Mangrove species used for fuel wood and timber in the Atlantic-East-Pacific (AEP) and Indo-West- Pacific (IWP) regions. Species in bold are those highlighted in studies that report more than one species being used but identify those species as of particular importance. Information here and in subsequent tables was taken following a literature search for detailed case studies of particular sites as well as more general overviews of countries and regions. Full references and more methodological information for Tables 8.1, 8.2, 8.3, and 8.4 are available as supplementary material

AEP region			IWP region		
Americas	Caribbean	W Africa	E Africa	Asia	Oceania
Fuel wood					
R. mangle *A. germinans*	*R. mangle*	*A. germinans* *C. erectus* *L. racemosa* *R. racemosa* ***R. mangle***	*A. marina* *B. gymnorrhiza* *C. tagal* *R. mucronata* *S. alba*	*A.officinalis* *A. corniculatum* *A. marina* *B.cylindrical* *B. gymnorrhiza* *B. parviflora* *C.decandra* *E. agallocha* *H.fomes* *L. racemosa* *R. apiculata* *R. mucronata* *S. apetala* *S. caseolaris* *X. granatum*	*B. gymnorhiza* *R. apiculata*
Timber					
A. germinans *L. racemosa* ***R. mangle***		*R. harisonii* *A. germinans* *R. mangle*	*B. gymnorrhiza* *C. tagal* ***R. mucronata***	*A. marina* *A. officinalis* *B. cylindrical* *B. parviflora* *E. agallocha* *H. fomes* *R. apiculata* *R. mucronata* *S. alba* *S. apetala* *X. granatum* *B. gymnorrhiza*	*B. gymnorhiza* ***R. apiculata***

African countries where relatively abundant supplies of wood combine with poverty and lack of access to fossil fuels; thus, Zambia tops the per capita consumption rates at 3.24 kg/cap/day (Yevich and Logan 2003). Mangrove forests provide an interesting case study of how poverty and social factors influence the use of natural wood resources; most users of mangrove fuel wood collect under de facto open access conditions from unplanted forests growing close to their homes. Reviewing the literature for case studies that identify proportional reliance (the percentage of households using mangrove wood for fuel) in communities living adjacent to mangrove forests shows a strong influence of average income (as reflected by GDP per capita

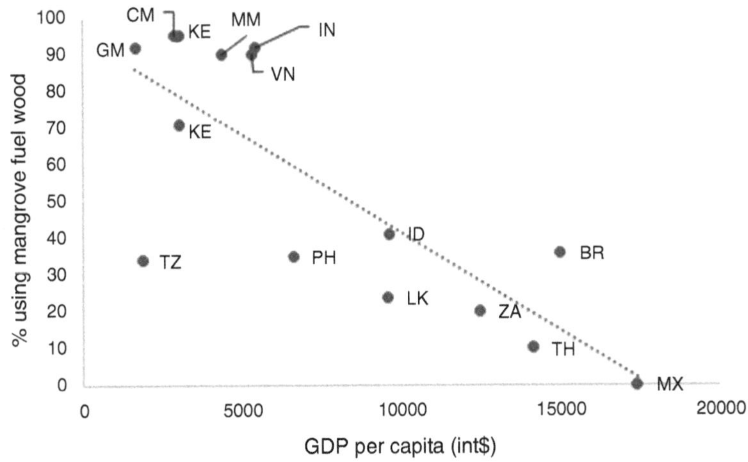

Fig. 8.2 Local dependence on mangroves fuel wood versus national average per capita GDP. Dependence data come from case studies of communities living adjacent to forests, and show the percentage of households reported using mangrove fuel wood. GDP per capita are in international dollars (adjusted for spending parity) taken from IMF 2014. Linear regression $R^2 = 0.63$, $P < 0.001$. Countries are *BR* Brazil, *CM* Cameroon, *GM* Gambia, *ID* Indonesia, *IN* India, *KE* Kenya, *LK* Sri Lanka, *MM* Myanmar, *MX* Mexico, *PH* Philippines, *TH* Thailand, *TZ* Tanzania, *VN* Vietnam and *ZA* South Africa

for the country) on the proportional use, with wealthier countries showing less reliance (Fig. 8.2). However, the considerable scatter in this relationship also reflects the importance of local conditions including governance and access regimes; for example, the data from Tanzania represent an outlier, with unexpectedly low percentage use, because of the effective regulation in force in the area of the case study (McNally et al. 2011).

Just as mangrove wood is considered a superior fuel, charcoal derived from mangroves is also highly prized as long burning, with a low ash and moisture content and resistance to spitting. Whilst exploitation for fuel wood and charcoal often co-occur and share some of the same driving factors, there are significant differences in the socio-economic characteristics of these two related services. In particular, charcoal is often produced for sale to markets that may be at a great distance (more than 2000 km) from their source (Fig. 8.1). Hence, its definition as a local ecosystem service is less consistent than for firewood. Although there are cases of large-scale cutting of mangrove forests to supply commercial markets with fuel wood (for example, in Cameroon, where mangrove wood is also used for smoking fish; Atheull et al. 2009), most exploitation for fuel wood is artisanal, unmanaged and outside formal markets. In contrast, some mangrove forests are planted and managed specifically for commercial charcoal production. For example, plantations of *R. apiculata* have been managed for charcoal in Thailand for 50 years (Kridiborworn et al.

2012), whilst the Matang mangrove forest reserve in Malaysia has been a model of mangrove silviculture, producing commercial charcoal and poles for timber, for more than a century (Goessens et al. 2014).

8.3 Timber, Thatch and Fodder

Mangrove wood is used for a range of timber products, from small household items, to fish traps through to large beams for construction (Walters et al. 2008). Harvesting for timber is often species and size specific. For example, in Kenya, poles with diameters of 8–13 cm are preferred for construction and *R. mucronata* has been the main species targeted for these. This has in some areas led to shifts in forest composition towards less valuable species such as *Ceriops tagal* (Kairo et al. 2002). In Colombia, *R. mangle* is targeted for poles, whilst *A. germinans* is preferred for planks; this selective extraction has led to shifts towards *Laguncularia racemosa* dominance (Blanco et al. 2012). Whilst a general preference for Rhizophoraceae is recorded in the literature, and is reflected in the deliberate cultivation in managed forest (Goessens et al. 2014), it is not universal (Table 8.1). As with firewood, the largest number of exploited species is recorded from Asia where the largest number of available species grows.

The mangrove palm *Nypa* is often highly prized. The fronds of this plant are used as thatch for houses and outer boundary walls. One hectare of *Nypa* plantation provides about 15,300 palm leaves each year. In addition, mats, baskets, hats and rain caps are also woven from leaf fibres. Young leaves are used as wrappers for food, while the ribs are used as fuel. The sap of young inflorescences is tapped for sugar production, alcohol distillation and vinegar production. The soft endosperm of the fruit is edible and highly esteemed in Thailand, Indonesia and Philippines. The hard shells of the ripe fruits are used to make buttons.

Some mangrove species, particularly *Avicennia* spp., provide cheap and nutritive feed for buffaloes, sheep, goats and camels; this use is common in arid areas of India, Pakistan and the Persian Gulf region. It is believed that cattle feeding on mangroves yield highly nutritious milk. Camel herding is one of the activities practiced by the pastoral communities known as 'Maldharis' in Gujarat, India. The Maldharis are in the habit of shifting along with their livestock to distant areas in search of fodder for their cattle, and degradation and restriction of access to mangrove forests have critically impacted their livelihoods (Kathiresan 2015).

8.4 Mangrove Crab Fisheries

An abundant and diverse fish, mollusc and crustacean fauna inhabit mangrove forests and estuaries. Many species are exploited by small-scale and artisanal fisheries (sensu FAO http://www.fao.org/fishery/topic/14753/en) for subsistence and income.

Fig. 8.3 1000 to 3000-year old clay artefact of the mangrove crab *Ucides cordatus* (Museo Forte do Castelo, Belém, Pará). Excavated at Marajó Island near the mouth of the Amazon river, Brazil (Museo Forte do Castelo, Belém, Pará, personal communication) (Photo credit: Karen Diele)

These fisheries deliver the major value of marketed mangrove resources (Walters et al. 2008). We focus on the fishery of crabs here, a key faunal component of mangrove ecosystems around the globe, playing important functional roles whilst also delivering significant provisioning services and being culturally important. In northern Brazil, for example, the obligate mangrove forest dweller *Ucides cordatus* is the most frequently used mangrove resource and supports a valuable market-driven yield of up to 7 tonnes per km^2 *Rhizophora mangle* forest yr.$^{-1}$ (Diele et al. 2010). About 60% of interviewed households indicated its use for subsistence and 40% for marketing (Glaser 2003). In Asia, mud crabs (i.e., *Scylla* spp.) reached a production value of US$ 252 million in 2004 (FAO 2006; cited after Ellison 2008). Mangrove crabs were probably already harvested in pre-historic times in the Americas, as suggested by a richly ornamented clay artefact of *U. cordatus* (Fig. 8.3), dating back to the Marajó culture (approximately AD 400–1350) (Diele et al. 2010). Coastal villages in northern Brazil have retained a strong cultural relation to the fisheries resources provided by adjacent mangrove ecosystems, exemplified par excellence during annual thanksgiving festivals when dancers dress in beautiful 'natural costumes', such as those made entirely from crab shells (Fig. 8.4).

Table 8.2 summarizes a literature search for mangrove and mangrove associated crabs fished for subsistence and/or commercial purposes. The harvested crabs include supratidal species of the family Gecarcinidae, intertidal forest-dwelling grapsid, ocypodid, sesarmid and ucidid crabs foraging at low tide, as well as mostly subtidal/mangrove–estuarine families (Calappidae, Matutidae, Menippidae, Oziidae, Panopeidae, Portunidae and Varunidae). The list is not exclusive, particularly for species captured for subsistence that are unlikely to be adequately documented in the scientific literature. They are also not included in fisheries statistics

Fig. 8.4 Girl dancing at thanks giving festival in Acarajó, north Brazil. Her dress is made entirely out of the shell from *Ucides cordatus* legs (Photo credit: Karen Diele)

that typically focus on marketed species only, despite the fact that local communities may harvest significant amounts of these non-marketed fisheries resources. In Micronesia, for example, the proportion of non-marketed catch to total catch was estimated at 90% (Naylor and Drew 1998). Clearly, to obtain a realistic view of livelihood dependencies on mangrove forests, of fishery impact on the full range of harvested species (including the delivery of functional roles of the targeted species) and of economic valuation of mangrove ecosystems, these non-marketed resources need consideration. The poorest and less educated parts of coastal populations often harvest supratidal and forest dwelling intertidal crabs, since their capture does not require costly equipment. These species are mostly burrowing and can be caught with relative ease (yet requiring professional experience) during low tide, e.g., by digging them out by hand, grabbing them carefully with a hooked stick or with simple baited traps (e.g., Brown 1993; Rodriguez-Fourquet and Sabat 2009; Diele et al. 2005). The relative 'affordability' of this fishery permits self-employment, so harvesting supra- and intertidal mangrove crabs holds an important poverty alleviation function for those with few or no other income options. Harvesting subtidal mangrove-associated species such as portunid crabs, e.g., *Callinectes* in the Neotropics and *Charybdis* in the Paleotropics, generally requires more financial input such as nets, and often involves trawling from smaller boats, resulting in a lower number of self-employed fishers. Crabs are also targeted by recreational fisheries in some countries, e.g., *Scylla serrata* in Australia.

Two main patterns emerge from the literature review regarding the identity and diversity of the targeted species (Table 8.2). First, targeted species are mostly mid (4–5 cm carapace width) to large- sized (> > 7 cm carapace width). Eating crabs is a laborious task due to their hard carapace and small specimens may not contain enough meat to make the effort worthwhile. This also explains why the number of supra- and intertidal species that are fished is only a small proportion of the much larger number of crab species associated with these habitats globally (about 300 mangrove species of brachyuran crabs reported; Ellison 2008), since few of these species grow large, e.g., crabs of the genera *Cardisoma* and *Ucides*. The size selectivity of crab fisheries further explains the globally much higher number of targeted subtidal crabs compared to the intertidal ones. The portunid swimming crabs, for example, include the large and fast growing species of the genera *Callinectes*, *Charybdis* and *Scylla*, representing over half of the species listed in Table 8.2. The size selectivity of mangrove crab fisheries, however, only accounts for the yield directed to live-crab and meat-processing markets. In contrast, large numbers of megalopae and small juveniles ('crablets') of mud crabs, *Scylla* spp., are collected as seed stock to supply crab farms in many Asian countries. The continuous capture of these early life stages due to expanding export markets is threatening wild populations (Quinitio et al. 2001).

The second pattern emerging from the literature review is that the number of recorded species that are harvested is by far largest in Asia. This probably reflects the overall higher faunal (and floral) diversity in this region (e.g., Ellison 2008), rather than simply being a function of a possibly higher number of fishers. For example, the number of medium-sized "fishable" sesarmid species is much higher in this region than in the Americas (Lee 1998). Overall, the literature review yielded 27 exploited mangrove crab species in the AEP and 40 in the IWP. This divergence matches the global pattern observed for the use of mangrove fuel wood, charcoal and timber.

8.5 Mangrove Fin-Fisheries

Fish may utilise different habitats at different life stages (Kimirei et al. 2013) and mangrove forests often act as important nursery habitats for marine fish (e.g., Nagelkerken 2009). Local households often depend directly on mangrove forests for fish, deriving their income and subsistence from fishing practices within and around forests (Barbier 2006). There can be serious social and economic consequences for local people if this fisheries function is impaired. For example, mangrove-rich areas in India provide up to 70 times more catch and income than similar mangrove-poor areas (Kathiresan and Rajendran 2002). This service is often well understood at a local level. Mangrove forests are described as 'the roots of the

Table 8.2 Mangrove crabs harvested for subsistence or commercial use in the AEP and IWP regions. The list is not exclusive, particularly for species harvested for subsistence, which is often not reported in the scientific literature

AEP region			IWP region		
Americas[a]	Caribbean	W Africa	E Africa	Asia	Oceania
Supra/intertidal (most active in air)					
Gecarcinidae *Cardisoma guanhumi*[WA] *C. crassum*[EP] Grapsidae *Goniopsis cruentata*[WA] Ucididae *Ucides cordatus*[WA] *U. occidentalis*[EP]	Gecarcinidae *Cardisoma guanhumi* Ucididae *Ucides cordatus*	Gecarcinidae *Cardisoma armatum* Grapsidae *Goniopsis pelii* Sesarmidae *Sesarma angolense* Ocypodidae *Uca tangeri*	Gecarcinidae *Cardisoma carnifex* Sesarmidae *Neosarmatium meinerti*	Gecarcinidae *Cardisoma carnifex* Sesarmidae *Episesarma chentongense* *E. mederi* *E. palawanense* *E. singaporense* *E. versicolor*	
Subtidal					
Portunidae *Callinectes bocourti*[WA] *C. danae*[WA] *C. exasperatus*[WA] *C. marginatus*[WA] *C. sapidus*[WA] *C. arcuatus*[EP] *C. bellicosus*[EP] *C. toxotes*[EP]	Menippidae *Menippe mercenaria* Portunidae *Callinectes bocourti* *C. danae* *C. exasperatus* *C. marginatus* *C. rathbunae* *C. sapidus*	Menippidae *Menippe nodifrons* Panopeidae *Panopeus africanus* Portunidae *Callinectes amnicola* *C. marginatus* *C. pallidus* *Cronius ruber* *Portunus hastatus* *Sanquerus validus* *Thalamita sp.*	Matudidae *Ashtoret lunaris* Portunidae *Charybdis feriata* *C. natator* *Podophthalmus vigil* *Portunus pelagicus* *P.sanguinolentus* *Scylla serrata*[a] *Thalamita crenata* Varunidae *Varuna litterata*	Calappidae *Calappa lophos* *C. pustulosa* Matudidae *Ashtoret lunaris* *Matuta planipes* Menippidae *Myomenippe fornasinii* *M. hardwickii* Oziidae *Baptozius vinosus* *Epixanthus dentatus*	Portunidae *Charybdis natator* *Portunus pelagicus* *Scylla serrata*[a]

				Ozius guttatus *O. tuberculosus* Portunidae *Charybdis affinis* *C. feriata* *C. granulata* *C. helleri* *C. lucífera* *C. miles* *C. natator* *C. orientalis* *C. rostrata* *C. truncata* *C. variegata* *Portunus pelagicus* *P. sanguinolentus* *Scylla olivácea*[a] *S. paramamosain*[a] *S. serrata*[a] *S. tranquebarica* *Thalamita crenata* *Thalamita danae* Varunidae *Varuna litterata* *V. yui*	
Total: 13 species	*Total: 9 species* (of which 7 also occur in continental Americas)	*Total: 13 species* (one occurring in Americas/ Caribbean)	*Total: 11 species* (8 of them occurring in Asia also)	*Total: 37 species* (8 of them occurring in E-Africa also)	*Total: 3 species* (all fished in Asia also)

[a]Also occurring in the lower intertidal, but most active in water
WA Western Atlantic, *EP* Eastern Pacific

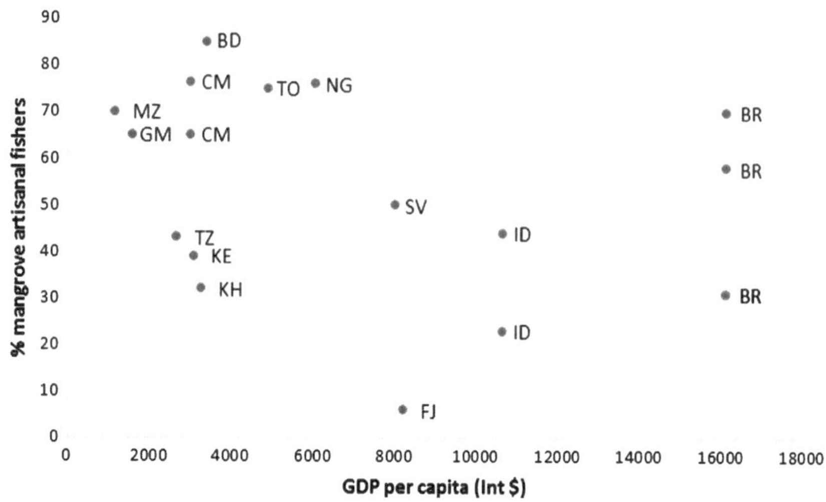

Fig. 8.5 Local dependence on mangrove fish species versus national average per capita GDP. Dependence data come from case studies of communities living adjacent to forests, and show the percentage of households reported as being reliant on fishing in and around mangrove waters as their main household income. GDP per capita are in international dollars taken from IMF 2014. Countries are *BD* Bangladesh, *BR* Brazil, *CM* Cameroon, *KH* Cambodia, *SV* El Salvador, *FJ* Fiji, *GM* Gambia, *ID* Indonesia, *KE* Kenya, *MZ* Mozambique, *NG* Nigeria, *TO* Tonga and *TZ* Tanzania

sea' in Asia; their loss would leave no fish and the sea would behave like a tree without roots.

Whilst fin-fish can be vitally important for individual households and whole communities adjacent to mangrove forests, there is wide variation that cannot simply be explained by crude measures of poverty. In contrast with fuel wood, there is no clear relationship between the GDP of the country and reported local reliance on mangrove forests for fish (Fig. 8.5). Fishing is often an occupation of the poor, but it can also secure high and stable incomes and may be central to cultural identity. Furthermore, recreational fisheries are important in many countries with a high GDP. Hence, in contrast to fuel wood, this local service may not decline in importance with increasing wealth.

Many of the world's commercial fish species rely on mangrove areas during their life cycle (FAO 2006). Table 8.3 shows the main mangrove fish taxa caught by artisanal fishers in the AEP and IWP, identified in a literature search of studies looking at artisanal mangrove fisheries; 'key species' (or taxa) were defined as those cited from more than one study in any one region. Fish from the families Lutjanidae and Gerridae are the most prevalent species and both commercial and artisanal fishers target these.

Table 8.3 Key fish species and families utilising mangrove estuaries caught by artisanal fisheries in the AEP and IWP regions

AEP region			IWP region		
Americas	Caribbean	W Africa	E Africa	Asia	Oceania
Anchoa spp.	*H. flavolineatum H.*	*Ethmalosa spp.*	*Chanos chanos*	*Chelon spp.*	*Acanthopagrus spp.*
Centropomus	*sciurus*	*E. fimbriata*	*Gerres filamentosus*	*Gerres spp.*	*Liza spp.*
undecimalis	*H. parra*	*Ilisha Africana*	*G. oyena*	*G. oyena*	
Eugerres spp.	*L. apodus*	*Liza grandisquamis*	*Lethrinus spp.*	*Liza spp.*	
Diapterus spp.	*L. griseus*	*Lutjanus spp.*	*L. harak*	*Scolopsis spp.*	
Lutjanus griseus	*L. analis*	*Pseudotolithus spp.*	*L. fulviflamma Lutjanus spp.*	*Sillago sihama*	
Mugil spp.	*Negaprion brevirostris*	*P. elongates*	*Mugil spp.*	*Lates calcarifer*	
Cynoscion spp.	*Gerres cinereus*	*Elops lacerta*	*M. cephalus*	*Tenualosa ilisha*	
Stellifer spp.	*Sphyraena barracuda*	*Tilapia spp.*	*Siganus spp.*	*Sardinella spp.*	
Bairdiella spp.	*Scaridae spp.*		*S. canaliculatus*	*Megalops*	
	Epinephelus itajara		*Sphyraena spp.*	*cyprinoides*	
			Tylosurus spp.	*Rastrelliger spp.*	
			Valamugil seheli		
			Lutjanus argentimaculatus		

Table 8.4 Species of mangroves recorded as planted or managed specifically or primarily for the purposes of coastal and storm protection. This includes 'storm holes' for sheltering boats from cyclones in Australia and the Caribbean

AEP region			IWP region		
Americas	Caribbean	W Africa	E Africa	Asia	Oceania
Avicennia germinans *Laguncularia racemosa* *Rhizophora mangle*	All species available	*Rhizophora sp.*	*Sonneratia alba*	*Avicennia spp.* *A. marina* *A. officinalis* *Kandelia spp.* *K. candel* *Sonneratia spp.* *S. apetala* *S. caseolaris* *Rhizophora spp.* *R. apiculata* *R. stylosa*	All species available

8.6 Coastal Protection

The idea that mangrove forests can act to protect the shoreline is old. Saenger and Bellan (1995) describe misguided advice to stabilise the banks of the Suez Canal by planting mangroves and report official calls in 1911 to protect the coastline of Cameroon using mangroves. Some local communities have also long recognised this function; for example, Fijians traditionally maintained mangrove forests for coastal protection, and areas that have continued this tradition suffer less erosion now (Mimura and Nunn 1998). Both the profile and the understanding of this service have developed rapidly in the past two decades, driven by dramatic events (such as the Asian tsunami) and impending sea level rise. McIvor et al. (2012) summarise the evidence for protection against major events (storm surges) whilst Thampanya et al. (2006) present a clear demonstration of how mangrove forests control erosion; forested sites across Thailand suffered, over the past 30 years, significantly less (or no) loss of shoreline compared to sites where mangrove trees had been cleared.

The shoreline protection service often features prominently in official government and NGO campaigns and in estimates of the total economic value of mangrove forests. It may, however, be of less immediate concern to local people than provisioning services, since its benefits can be subtle (gradual erosion) or demonstrated during relatively rare events (such as storm surges), and the poor may not have the luxury of privileging such longer-term considerations over more immediate subsistence. It is certainly harder to find examples in the literature where this service is the main incentive for local communities in their use, management and restoration of mangrove sites than for other services (Table 8.4). However, knowledge of the ability of mangrove forests to protect the coast is widespread and is often reported as an additional benefit in projects and surveys focused on other services such as fish and timber. This is particularly true in Asia, where some of the worst tropical storms

Fig. 8.6 A jetty in south east India destroyed by the 2004 tsunami, in contrast to the resilient adjacent mangroves (Photo credit: K. Kathiresan)

occur and where recovery from the 2004 tsunami continues. Barbier (2006) reports how communities in Thailand that are most dependent on mangrove fish are most likely to invest time in restoration, whilst also being aware of the protective function of the forests. Dangerous storms provide tragic demonstrations for local people to see for themselves the protective functions of mangrove vegetation, and local anecdotes are supported by wider surveys and theory. For example, in 2008, cyclone 'Nargis' caused the deaths of over 30,000 people in Myanmar, with mortality worst in areas with no or degraded mangrove forests (Kathiresan 2015). Mangrove vegetation is often surprisingly resilient in the face of extreme events, in sharp and visible contrast to human infrastructure and artificial coastal protection (Fig. 8.6)

Theoretical and empirical studies have identified some of the mangrove characteristics most likely to reduce the depth of storm surges, the heights of waves and to ensure surface elevation in the face of sea level rise. The clearest (and most obvious) is the size of the forest. Wave heights during a typhoon were reduced by 50% after passage through ~380 m of *K. candel* forest, and by ~90% after 1 km (Barbier et al. 2008). The density of forest is also important (Kumara et al. 2010; Thampanya et al. 2006). Some species are more effective than others in reducing wave energy, in particular those with denser aerial roots (hence, *S. caseolaris* is three times better than *K. candel*; Barbier et al. 2008), and some (such as *R. mangle* compared with *L. racemosa)* are better at recovering after storm damage (McIvor et al. 2012 and refs therein). Experimental work suggests that mixing species with different root profiles may boost forest productivity and therefore resilience (Lang'at et al. 2013). However, where coastal protection or storm shelter are the key objectives of restoration or management, the species of tree used are normally not selected specifically or exclusively for their ability to stabilise and protect shorelines. The planning and implementation of such work is rarely sophisticated enough to utilise such knowledge, but must focus instead on considerations of ensuring successful planting and growth and appropriate local tenure and governance (Primavera and Esteban 2008). As with other services, Asia supports more species providing coastal protection than in other regions (Table 8.4). This probably reflects the greater range of species available at individual sites rather than any deliberate attempt to utilise different ecological or hydrological properties.

8.7 The Vulnerability of Local Mangrove Services to Climate Change

Mangrove forests face a wide range of stressors and threats. Any of these – such as aquaculture, coastal development, diversion of freshwater and silt by dams and agricultural conversion – can undermine the provision of their local services and require urgent amelioration (Van Lavieren et al. 2012). Here, we focus on the possible impacts of climate change for two reasons. First, this growing global threat is relevant to all mangrove forests regardless of their location. Second, the factors that make local mangrove services at any given site particularly vulnerable to anticipated climate change tend also to reduce their resilience to other threats. Hence, considering climate change vulnerability helps clarify general points about the resilience and fragility of local mangrove service provision and can thus improve the management. The combination of potential climate change impacts and vulnerability of individual species will help guide the type of management needed. For example, when both variables are high, a more active and directed management approach (e.g., habitat re-creation, species translocations) may be required, while passive management (e.g., monitoring, ecosystem-based management) would suffice when both are low (Koehn et al. 2011).

8.7.1 Mangrove Forests and Sea Level Rise

The positioning of mangrove forests at the interface of land and sea makes them directly susceptible to sea level rise. Higher sea levels result in extended exposure to seawater reducing growth, survival and reproduction (Krauss et al. 2008). The degree to which any given forest is affected by sea level rise largely depends on local physical processes, coastal geomorphology, interactions with other environmental factors and ecosystem interdependencies (Mcleod et al. 2010; Alongi 2015). The balance between sediment accretion rate and sea level rise is critical in determining whether mangroves drown, persist or expand at their seaward and landward edges. Mangrove forests will be more resilient where sediment sources are plentiful and unobstructed (Field 1995). Rivers are important sources of freshwater and sediment for riverine mangrove forests and their natural setting within the seascape is therefore important. However, human activities modify freshwater and sediment loads of rivers systems in many ways (Davis et al. 2015). For example, alterations to river flows due to construction of dams and channels, and extraction of freshwater, have consequences for the influx of freshwater and sediment into tropical river deltas. The degree of rainfall also regulates the magnitude of freshwater and sediment flow, and climate change is predicted to alter regional precipitation patterns, most likely leading to less rainfall in dry regions and more rainfall in wet ones (Alongi 2015).

Mangrove forests may adapt to sea level rise by extending landwards (Di Nitto et al. 2014), but the degree and type of coastal development at the landward fringes will determine how likely this is. Where mangrove forests occur next to coastal developments, their extension to higher elevation is prevented and forests are squeezed in the coastal zone. Topography also plays a major role as steep slopes prevent horizontal extension, while mangrove forests on flat, low-lying islands quickly run out of space at higher elevations. For example, shoreline retreat in Gambia has been predicted to be 6.8 m in cliffy areas and ~880 m in flat, sandy areas, and most areas lost to inundation will be associated with wetland and mangrove ecosystems (Jallow et al. 1996). Ecosystem interdependencies may further alter the impacts of sea level rise. In cases where coral reefs or seagrass beds occur close to mangrove forests, the wave energy is usually reduced by these more seaward located ecosystem structures (Gillis et al. 2014). However, coral reefs are unlikely to keep up with the rate of sea level rise, leading to more wave energy moving into back reef areas (Saunders et al. 2014). This effect will be further exacerbated by direct destructive impacts of local communities on reefs and reduced reef calcification due to ocean warming and acidification. Whilst elevated CO_2 can lead to enhanced mangrove primary productivity, this may only occur at low salinity and high humidity (Ball et al. 1997). Hence, the combination of higher aridity (from global warming) and salinity (from sea level rise) might counteract any positive effects of elevated CO_2 on mangrove growth. Thus, climate change will most likely cause multiple stressors to interact (see Chap. 7).

The degree to which local mangrove ecosystem services are affected by climate change will depend on local environmental and geomorphological conditions and whether humans exacerbate or mitigate climate change effects. For example, the way in which river flows are modified can alter the responses to sea level rise. Some mangrove forests will increase in surface area in response to sea level inundation (Traill et al. 2011) and may provide more opportunities for harvesting wood and forest products by local communities, whereas in other parts of the world, mangrove areas might quickly decline (Saleem Khan et al. 2012) leading to lower primary and secondary productivity. Higher sea levels can also have positive effects such as better access by canoe to forest areas otherwise inaccessible to harvest mangrove wood (Munji et al. 2014). Altered inundation patterns of mangrove wetlands may also change their nursery function and alter fish community structures (Igulu et al. 2014; Hylkema et al. 2015) and may affect local harvests. Not all ecosystem services will be equally affected by climate change. In a case study in India, agriculture, aquaculture and mangrove forests were the three natural resources most at risk of inundation from sea level rise (Saleem Khan et al. 2012). Aquaculture in this area is based on brackish water farming of prawns, leaving this industry and the communities that depend on this prone to salinisation from seawater intrusion.

8.7.2 Range Extensions of Mangrove Forests

Mangroves are limited at their high-latitude ranges by low winter temperatures. Climate change is facilitating range extensions of various mangrove species to higher latitudes (Osland et al. 2013; Saintilan et al. 2014). Modelling studies suggest that Central America and the Caribbean will lose relatively more mangrove cover than elsewhere in the world due to regional decreases in rainfall (Record et al. 2013). Because saltmarshes occupy a similar ecological niche to mangrove forests, this often leads to invasion of salt marshes by mangroves. In cases where they already co-occur, mangrove forests have been observed to extend their ranges landwards and invade salt marshes (Saintilan and Williams 1999).

Whilst mangrove forests and salt marshes are both highly productive and support fisheries, they harbour very different biological communities and therefore changes in coastal vegetation due to climate change may have large consequences for local marine fauna and for the people that depend on it. We know very little about the response of associated marine species to mangrove range extensions, or range contractions of salt marshes. As a potential analogue to range extensions, one study found that introduced mangrove forests in Hawaii did not have a negative effect on local fish communities and may act as nurseries for local as well as exotic fish species (MacKenzie and Kryss 2013). Although the ultimate effects are still difficult to predict, range extensions of mangrove forests will lead to losses as well as gains for local human communities.

8.7.3 Range Extensions of Fisheries Species

As the oceans warm and foundation species like mangroves extend their ranges, so will those species that depend on them. Many of these mangrove-associated species are of high value to local fishers. Various marine species have already extended their ranges to higher latitudes (Poloczanska et al. 2013); this is an on-going process driven by ocean warming. Most emphasis on marine range extensions has been on ecosystems and species from coastlines or the open ocean, with barely anything known about inshore ecosystems or mangrove-associated fauna like fishes, crabs, shrimp and bivalves. It is very likely that these taxa will also extend their ranges, but we know little of the rate at which this might occur and the species that will respond fastest. Mangrove-dependent species might be limited by the rate at which mangrove habitats move, while others may outpace mangrove movement and utilise novel habitats they encounter (Riley et al. 2014). Because many species that live in mangrove forests and estuaries are more tolerant to fluctuating environmental conditions than oceanic species (Gillanders et al. 2011), their responses might differ from the latter. Life history strategy will also determine how fisheries species are impacted, because some species use mangrove estuaries during their entire life, some for part of their life, and some as transient areas to move between the ocean and freshwater systems (Crook et al. 2015). Because the rates of community response to climate change will vary with habitat type, species with the most complex life history dependencies might be particularly vulnerable (Nagelkerken et al. 2015). These include species that require mangroves, seagrasses and corals in close proximity, such as various species of Haemulidae, Lutjanidae and Scaridae (Nagelkerken et al. 2001).

The impact of species range extensions and changes in species assemblages due to climate change on local communities will largely depend on the specific species that are targeted by artisanal mangrove and estuarine fisheries. In addition, local environmental factors and geomorphology will play a role, with (partially) enclosed mangrove estuaries perhaps being less prone to range-extending species than ocean-facing mangrove forests and mangrove forests along continental coastlines being more likely to experience range extensions by post-settlement movement of marine species than oceanic islands surrounded by deep waters.

8.7.4 *Effects of Ocean Acidification, Warming, Salinity and Hypoxia on Fisheries Species*

Elevated CO_2 and temperature can have positive effects on primary producers, thereby enhancing benthic and pelagic marine primary production and increasing food availability for consumers (Roessig et al. 2004), although this is more likely to be the case for temperate rather than tropical species. Higher temperatures elevate metabolism usually leading to a greater demand for food (Roessig et al. 2004). In cases where food is not limiting, growth rates of some mangrove and estuarine fauna may therefore increase with temperature leading to higher reproduction and secondary productivity (Hare et al. 2010). However, global meta-analyses suggest that although primary production might increase in some regions, this does not translate to higher (fisheries) productivity of most consumer species (Nagelkerken and Connell 2015). Moreover, elevated CO_2 can have detrimental effects on species survival rates through altered animal behaviour (Nagelkerken and Munday 2016) and by reducing the growth of calcifying organisms (Fabry et al. 2008). Mangrove forests harbour a range of calcifying species such as oysters, mussels and various species of crustaceans, and ocean acidification is likely to have large impacts on such seafood species (Branch et al. 2013).

The degree to which mangrove fisheries species are affected on a local scale will depend not only on species identity, but also on multi-stressor effects. For example, many shallow coastal areas and estuaries are increasingly turning into hypoxic 'dead zones' due to human eutrophication, exacerbated by ocean warming (Altieri and Gedan 2015), whereas artificially drained mangrove wetlands can be sources of highly acidic sulphur soils (Sammut et al. 1996). Furthermore, because climate change is predicted to either increase or decrease rainfall (and therefore river flow and salinity of estuaries) depending on region, and because many species have life cycles driven by temperature and salinity gradients, climate change will have consequences for many species using estuaries (Gillanders et al. 2011; Igulu et al. 2014). This will be particularly evident for fish species from freshwater and marine environments that temporarily move into and utilise (mangrove) estuaries as spawning grounds or nurseries (Boucek and Rehage 2014), as well as in offshore fisheries that are often positively correlated with freshwater runoff and rainfall (Meynecke et al. 2006). How such local and global stressors interact at levels relevant to local communities, in combination with local hydrology and geomorphology, will determine the impact on local fisheries. Little is known about multi-stressor effects on mangrove fisheries species specifically, but due to different species sensitivities to such stressors (or their interactive effects), there is the possibility that local fisheries might need to adjust the species they target in the near future.

8.7.5 Socio-economic Implications and Climate Adaptation Options

The impacts of climate change on mangrove fisheries will vary among different types of fishers (Roessig et al. 2004). Commercial fishers can often adapt to changes in the range and season of fish stocks since their large boats can stay at sea for long periods. They also have access to storage facilities and more distant markets. Artisanal fishers are usually restricted to areas close to their homes and have fewer financial resources for adaptation. Changes in fish species, abundances, migrations and body sizes may therefore have a stronger impact on local communities, and artisanal fishers may need to adjust the species they target and the gear they use. For example, modelling studies have forecasted smaller fish body sizes and lower catchability in the tropics (Rashid Sumaila et al. 2011). Recreational fishers are also likely to be affected at local scales. A narrow range of fish species form the basis of the recreational fishing industry in various countries. Some of these are associated with mangrove estuaries, e.g., tarpon (*Megalops atlanticus*), bonefish (*Albula vulpes*), grey snapper (*Lutjanus griseus*), and barracuda (*Sphyraena barracuda*) in the Caribbean, and barramundi (*Lates calcarifer*) in Australia. Recreational fishing, especially in westernised countries, can be a large industry with significant financial flow-on effects to support businesses (e.g., hotels, restaurants, car rentals, boat charters, sale of fishing gear and bait). For example, the annual recreational fishing industry in Australia is worth $1.8 billion based on 3.36 million fishers (Stephan and Hobsbawn 2014). In Queensland alone, the recreational fishery is worth ~ $320–400 million annually, based on 700,000 fishers. About 43% of this state's recreational catch is from mangrove estuaries, with the main target species being yellowfin bream (*Acanthopagrus australis*), sand whiting (*Sillago ciliata*), trumpeter whiting (*Sillago maculata*), dusky flathead (*Platycephalus fuscus*), pikey bream (*Acanthopagrus pacificus*) and barramundi (Taylor et al. 2012). Range extensions of highly esteemed recreational species will probably have strongest effect on local businesses that cannot move as easily as the moving species, whereas the fishers themselves could more easily target the species in their new ranges. The aquaculture sector will also feel the impacts of climate change. Sea level rise and increased storm frequency can lead to salinisation of inshore areas, having a negative effect on local communities that culture fish in freshwater ponds close to shore, but potentially having a positive effect on shrimp farming (Ahammad et al. 2013).

Local communities most at risk from sea level rise are those living in low-lying flat coastal areas where risk and extent of inundation will be largest. Sea level rise and any increase in the intensity of storms following climate change will exacerbate the risks of coastal erosion and storm damage. Hence, the value of coastal protection provided by mangrove forests is likely to increase, provided the forests can adapt to the new conditions. Healthy forests with adequate sediment supply should be able to track sea level rise, and management that encourages dense growth may help (Kumara et al. 2010). Modelling saltmarsh carbon burial suggests that initial responses to increased CO_2 may be an increase in organic matter accumulation (Kirwan and Mudd 2012); if mangrove forests respond similarly, then there may be initial positive impacts for this service. Also at risk are communities that rely heavily on local resources (e.g., recreational fishing tourism, specific local fisheries species, nearby fishing grounds) and cannot adequately adapt, i.e., diversify livelihoods, migrate, change to alternative fishing areas, change to alternative forms of protein, have insufficient capital to switch gear (Ruckelshaus et al. 2013). Reliance on mangrove fisheries varies considerably (see Fig. 8.5) with economic fisheries values in the range 0.2–12,305 US$/ha/yr. (Hutchison et al. 2014), showing that different local communities will not be affected to the same extent. Fishers that depend on traditional knowledge (rather than technology), using weather and tidal patterns to predict abundance and catch potential of fisheries species, are at high risk (Marschke et al. 2014).

Options for climate adaptation by local communities include switching gear and fisheries target species, mangrove reforestation, fish species and habitat protection, and greater reliance on ecotourism (Roessig et al. 2004). Ecotourism can provide a buffer against further decline from exploitation and create income for local communities (Marschke et al. 2014). Likewise, mangrove habitat protection and reforestation can buffer mangrove loss due to climate change and provide people with access to a more diversified range of products making them more resilient to climate change (Pramova et al. 2012). Another option for climate adaptation involves changing from wild-caught fisheries to aquaculture, because the latter has higher control over water quality (Richards et al. 2015). The early life stages of fish are particularly sensitive to global change stressors, and in aquaculture, these early stages are often kept in culture environments where water quality can be monitored and adjusted. While global stressors like warming and acidification are difficult to halt, appropriate actions toward local stressors, such as eutrophication, acid soils and river flow alteration, can mitigate cumulative stressor effects (Gilman et al. 2008) and provide opportunities for species to acclimate. Climate change presents a major threat to many oceanic habitats (and their services), but mangrove forests are by nature amongst the most resilient to the anticipated impacts. For example, whilst risk of impacts on warm water corals of an increase in sea surface temperature of 1.5–2 °C is predicted to be high or very high, the direct risk to mangroves of a similar change is undetectable or moderate (Gattuso et al. 2015). Hence, there is hope that careful management of local stressors could allow mangrove forests to flourish under the less extreme climate predictions. Severe concurrent impacts on more sensitive ecosystems such as corals, and the predicted loss of their services, may make those provided by mangrove forests even more important to many local communities.

8.8 Conclusions

We should do what we can to conserve and restore the world's mangrove forests; that is a general message that ecologists, economists, political scientists and mangrove-dependent communities agree on (see Chap. 10). But the urgency of this message, the chances of its success and the winners and losers from sustainable mangrove management vary greatly from one place to another. As this chapter shows, biological and socio-economic factors work together to determine the degree of reliance of people on particular local mangrove services and species, and their options in the face of decline or change in these services. In general, higher biological diversity translates to a wider range of exploited species, and therefore greater possible redundancy (or 'insurance') in the event of the loss or decline of any particular species. Similarly, greater wealth brings more options for fulfilling some needs – fossil fuels and electricity, protein from farms instead of the sea, land that is less vulnerable to salinisation and flooding. Similar conclusions are likely to apply to climate change impacts. Hence, forests with many tree species may have the biological resources to allow shifts in species composition in response to changes in rainfall and salinity, whilst those with fewer species may not achieve this. Wealthy communities can also invest in methods of fishing and storing fish that allow adaptation to changing fishing locations, species and times. Put crudely, vulnerability to current and predicted future stresses increases as we move from diverse and lush mangrove locations to biologically depauperate and economically poor ones. Of course, such a simplification ignores important caveats and may mislead thinking, particularly where there are complex dependencies between these variables. For example, mangrove forests that are biologically productive and diverse may be so because of their place in a complex connected sea-scape of other forests and related habitat (Brander et al. 2012). Their higher diversity may thus be a sign of a broader vulnerability to environmental damage in fragile neighbouring habitats such as coral reefs.

There is a general trend of greater variety of exploited fuel wood, timber, crab species and coastal protection in the IWP (and particularly in Asia) than in the AEP. This probably reflects the higher floral and faunal diversity of mangrove forests in this region presenting a greater range of options for harvesting and planting. Alternative explanations are possible; for example, sites with higher population densities and greater anthropogenic pressure may see people forced into the use of a wider range of less preferable species. However, this is unlikely to explain the broad pattern, since many sites in the AEP region report intense use of all the available, but limited, pool of species (for example for fuel wood in West Africa). Finfish are an exception to this pattern and provide support for this opportunistic explanation, since we are not aware of higher diversity of mangrove-related fish species in Asia compared with other regions.

Income is a key predictor of the dependence on some local mangrove services. Firewood is the best example here, with people changing to alternative fuels when they can afford them. The use of less desirable (smaller) crab species is also probably closely linked to wealth. But this pattern, of 'liberation' from use of the local forest, is not seen for other ecological services such as fisheries and coastal protection. Whilst wealth may bring a broadening of options, these often include an increased expenditure on desirable fish protein or investment in infrastructure that is sheltered by mangrove forests. Hence, economic development may increase local resilience to environmental change, but does not imply a reduction in the value, economic or ecological, of mangrove forests. It will rather result in a shift in importance, often from provisioning towards regulating services and from less preferred to higher valued products. Whilst the world as a whole benefits from mangrove forests, it is local people, rich or poor, who are the key beneficiaries and who can best act as their champions and protectors.

Acknowledgements Thanks to Sarah Murray for help with Fig. 8.1 and to Victor Rivera-Monroy for his help with editorial work. Essam Mohammed and an anonymous reviewer provided helpful comments on an earlier draft.

References

Ahammad R, Nandy P, Husnain P (2013) Unlocking ecosystem based adaptation opportunities in coastal Bangladesh. J Coast Conserv 17:833–840

Alongi DM (2015) The impact of climate change on mangrove. Curr Clim Chang Rep 1:30–39

Altieri AH, Gedan KB (2015) Climate change and dead zones. Glob Chang Biol 21:1395–1406

Atheull AN, Din N, Longonje SN, Koedam N, Dahdouh-Guebas F (2009) Commercial activities and subsistence utilization of mangrove forests around the Wouri estuary and the Douala-Edea reserve (Cameroon). J Ethnobiol Ethnomed 5:35

Ball MC, Cochrane MJ, Rawson HM (1997) Growth and water use of the mangroves *Rhizophora apiculata* and *R-stylosa* in response to salinity and humidity under ambient and elevated concentrations of atmospheric CO_2. Plant Cell Environ 20:1158–1166

Barbier EB (2006) Natural barriers to natural disasters: replanting mangroves after the tsunami. Front Ecol Environ 6:124–131

Barbier EB, Koch EW, Silliman BR, Hacker SD, Wolanski E, Primavera J, Granek EF, Polasky S, Aswani S, Cramer LA, Stoms DM, Kennedy CJ, Bael D, Kappel CV, Perillo GME, Reed DJ (2008) Coastal ecosystem-based management with nonlinear ecological functions and values. Science 319:321–323

Blanco JF, Estrada EA, Ortiz LF, Urrego LE (2012) Ecosystem-wide impacts of deforestation in mangroves: the Urabá Gulf (Colombian Caribbean) case study. ISRN EcolArticle ID 958709, 14 pages. doi:https://doi.org/10.5402/2012/958709

Boucek RE, Rehage JS (2014) Climate extremes drive changes in functional community structure. Glob Chang Biol 20:1821–1831

Branch TA, DeJoseph BM, Ray LJ, Wagner CA (2013) Impacts of ocean acidification on marine seafood. Trends Ecol Evol 28:178–186

Brander L, Wagtendonk A, Hussain S, Mcvittie A, Verburg PH, De Groot RS, Van Der Ploeg S (2012) Ecosystem service values for mangroves in Southeast Asia: a meta-analysis and value transfer application. Ecosyst Serv 1(1):62–69

Briffa M, de la Haye K, Munday PL (2012) High CO_2 and marine animal behaviour: potential mechanisms and ecological consequences. Mar Pollut Bull 64:1519–1528

Brown IW (1993) Mangrove crabs. In: Wright A, Hill L (eds) Nearshore marine resources of the South Pacific: information for fisheries management and development. Fisheries Agency, Honiara, pp 611–642

Costanza R, Arge R, De Groot R, Farberk S, Grasso M, Hannon B, Limburg K, Naeem S, Neill RVO, Paruelo J, Raskin RG, Suttonkk P, Farber S, Sutton P (1997) The value of the world's ecosystem services and natural capital. Nature 387:253–260

Crook DA, Lowe WH, Allendorf FW, Eros T, Finn DS, Gillanders BM, Hadwen WL, Harrod C, Hermoso V, Jennings S, Kilada RW, Nagelkerken I, Hansen MM, Page TJ, Riginos C, Fry B, Hughes JM (2015) Human effects on ecological connectivity in aquatic ecosystems: integrating scientific approaches to support management and mitigation. Sci Total Environ 534:52–64

Davis J, O'Grady AP, Dale A, Arthington AH, Gell PA, Driver PD, Bond N, Casanova M, Finlayson M, Watts R, Capon SJ, Nagelkerken I, Tingley R, Fry B, Page TJ, Specht A (2015) When trends intersect: the challenge of protecting freshwater ecosystems under multiple land use and hydrological intensification scenarios. Sci Total Environ 534:65–78

Di Nitto D, Neukermans G, Koedam N, Defever H, Pattyn F, Kairo JG, Dahdouh-Guebas F (2014) Mangroves facing climate change: landward migration potential in response to projected scenarios of sea level rise. Biogeosciences 11:857–871

Diele K, Araújo AR, Glaser M, Salzmann U (2010) Fishery of the mangrove crab *Ucides cordatus* in N-Brazil: first steps to successful co-management. In: Saint-Paul U, Schneider H (eds) Mangrove dynamics and management in North Brazil, Ecol Stud, vol 211. Springer, Heidelberg, pp 287–297

Diele K, Koch V, Saint-Paul U (2005) Population structure and catch composition of the exploited mangrove crab *Ucides cordatus* in the Caeté estuary, North Brazil: indications for overfishing? Aquat Living Resour 18:169–178

Ellison AM (2008) Managing mangroves with benthic biodiversity in mind: moving beyond roving banditry. J Sea Res 59:2–15

Fabry VJ, Seibel BA, Feely RA, Orr JC (2008) Impacts of ocean acidification on marine fauna and ecosystem processes. ICES J Mar Sci 65:414–432

FAO (2006) FISHSTAT plus: universal software for fishery statistical time series, version 2.3.1. FAO Fisheries Department, Fishery

Field CD (1995) Impact of expected climate-change on mangroves. Hydrobiologia 295:75–81

Gattuso JP, Magnan A, Bille R, Cheung WWL, Howes EL, Joos F, Allemand D, Bopp L, Cooley SR, Eakin CM, Hoegh-Guldberg O, Kelly RP, Portner HO, Rogers AD, Baxter JM, Laffoley D, Osborn D, Rankovic A, Rochette J, Sumaila UR, Treyer S, Turley C (2015) Contrasting futures for ocean and society from different anthropogenic CO_2 emissions scenarios. Science 349(6243):4722-1–4722-10

Gillanders BM, Elsdon TS, Halliday IA, Jenkins GP, Robins JB, Valesini FJ (2011) Potential effects of climate change on Australian estuaries and fish utilising estuaries: a review. Mar Freshw Res 62:1115–1131

Gillis LG, Bouma TJ, Jones CG, van Katwijk MM, Nagelkerken I, Jeuken CJL, Herman PMJ, Ziegler AD (2014) Potential for landscape-scale positive interactions among tropical marine ecosystems. Mar Ecol Prog Ser 503:289–303

Gilman EL, Ellison J, Duke NC, Field C (2008) Threats to mangroves from climate change and adaptation options: a review. Aqua Bot 89:237–250

Glaser M (2003) Interrelations between mangrove ecosystem, local economy and social sustainability in Caeté Estuary, North Brazil. Wetl Ecol Manag 11:265–272

Goessens A, Satyanarayana B, Van der Stocken T, Quispe Zuniga M, Mohd-Lokman H, Sulong I, Dahdouh-Guebas F (2014) Is Matang mangrove forest in Malaysia sustainably rejuvenating after more than a century of conservation and harvesting management? PLoS One 9(8):e105069. https://doi.org/10.1371/journal.pone.0105069

Hare JA, Alexander MA, Fogarty MJ, Williams EH, Scott JD (2010) Forecasting the dynamics of a coastal fishery species using a coupled climate-population model. Ecol Appl 20:452–464

Hein L, van Koppen K, de Groot RS, van Ierland EC (2006) Spatial scales, stakeholders and the valuation of ecosystem services. Ecol Econ 57:209–228

Hutchison J, Spalding MZU, Ermgassen P (2014) The role of mangroves in fisheries enhancement. The Nature Conservancy and Wetlands International, Cambridge, UK, p 54

Huxham M, Emerton L, Kairo J, Munyi F, Abdirizak H, Muriuki T, Nunan F, Briers RA (2015) Applying Climate Compatible Development and economic valuation to coastal management: a case study of Kenya's mangrove forests. J Environ Manag 157:168–181

Huxham M, Kumara MP, Jayatissa LP, Krauss KW, Kairo J, Langat J, Mencuccini M, Skov MW, Kirui B (2010) Intra- and interspecific facilitation in mangroves may increase resilience to climate change threats. Philos Trans R Soc Lond B Biol Sci 365:2127–2135

Hylkema A, Vogelaar W, Meesters HWG, Nagelkerken I, Debrot AO (2015) Fish species utilization of contrasting sub-habitats distributed along an ocean-to-land environmental gradient in a tropical mangrove and seagrass lagoon. Estuar Coasts 38:1448–1465

Igulu MM, Nagelkerken I, Dorenbosch M, Grol MGG, Harborne AR, Kimirei IA, Mumby PJ, Olds AD, Mgaya YD (2014) Mangrove habitat use by juvenile reef fish: meta-analysis reveals that tidal regime matters more than biogeographic region. PLoS One 9(12):e114715. https://doi.org/10.1371/journal.pone.0114715

Jallow BP, Barrow MKA, Leatherman SP (1996) Vulnerability of the coastal zone of The Gambia to sea level rise and development of response strategies and adaptation options. Clim Res 6:165–177

Kairo JG, Dahdouh-Guebas F, Gwada PO, Ochieng C, Koedam N (2002) Regeneration status of mangrove forests in Mida creek, Kenya: a compromised or secured future? Amb 31:562–568

Kathiresan K (2015) Ecology of mangroves. In: Ocean coast ecological. Scientific Publishers, Jodhpur, pp 169–207

Kathiresan K, Rajendran N (2002) Fishery resources and economic gain in three mangrove areas on the southeast coast of India. Fish Manag Ecol 49:277–283

Khan AS, Ramachandran A, Usha N, Punitha S, Selvam V (2012) Predicted impact of the sea-level rise at Vellar-Coleroon estuarine region of Tamil Nadu coast in India: mainstreaming adaptation as a coastal zone management option. J Ocean Coast Manag 69:327–339

Kimirei IA, Nagelkerken I, Trommelen M, Blankers P, van Hoytema N, Hoeijmakers D, Huijbers CM, Mgaya YD, Rypel AL (2013) What drives ontogenetic niche shifts of fishes in coral reef ecosystems? Ecosystems 16:783–796

Kirwan ML, Mudd SM (2012) Response of salt-marsh carbon accumulation to climate change. Nat 489:550–553

Kituyi E, Marufu L, Huber BO, Wandiga SO, Jumba IO, Andreae M, Helas G (2001) Biofuel consumption rates and patterns in Kenya. Biomass Bioenergy 20:83–99

Koehn JD, Hobday AJ, Pratchett MS, Gillanders BM (2011) Climate change and Australian marine and freshwater environments, fishes and fisheries: synthesis and options for adaptation. Mar Freshw Res 62:1148–1164

Krauss KW, Lovelock CE, McKee KL, Lopez-Hoffman L, Ewe SML, Sousa WP (2008) Environmental drivers in mangrove establishment and early development: a review. Aquat Bot 89:105–127

Kridiborworn P, Chidthaisong A, Yuttitham M, Tripetchkul S (2012) Carbon sequestration by mangrove forest planted specifically for charcoal production In: Yeesarn, SamutSongkram. J Sustain Energy Environ 3:87–92

Kumara MP, Jayatissa LP, Krauss KW, Phillips DH, Huxham M (2010) High mangrove density enhances surface accretion, surface elevation change, and tree survival in coastal areas susceptible to sea-level rise. Oecologia 164:545–553

Lang'at JKS, Kirui BKY, Skov MW, Kairo JG, Mencuccini M, Huxham M (2013) Species mixing boosts root yield in mangrove trees. Oecologia 172:271–278

Van Lavieren H, Spalding M, Alongi DM, Kainuma M, Clüsener-godt M, Adeel Z (2012) Securing the future of mangroves. A Policy Brief. doi: http://www.inweh.unu.edu

Lee SY (1998) Ecological role of grapsid crabs in mangrove ecosystems: a review. Mar Freshw Res 49:335–343

Locatelli T, Binet T, Kairo JG, King L, Madden S, Patenaude G, Upton C, Huxham M (2014) Turning the tide: how blue carbon and Payments for Ecosystem Services (PES) might help save mangrove forests. Amb 43:981–995

MacKenzie RA, Kryss CL (2013) Impacts of exotic mangroves and chemical eradication of mangroves on tide pool fish assemblages. Mar Ecol Prog Ser 472:219–237

Marschke M, Lykhim O, Kim N (2014) Can local institutions help sustain livelihoods in an era of fish declines and persistent environmental change? A Cambodian case study. Sustainability 6:2490–2505

McIvor A, Spencer T, Möller I (2012) Storm surge reduction by mangroves. Natural Coastal Protection Series. Report 2, Cambridge

McLeod E, Hinkel J, Vafeidis AT, Nicholls RJ, Harvey N, Salm R (2010) Sea-level rise vulnerability in the countries of the Coral Triangle. Sustain Sci 5:207–222

McNally CG, Uchida E, Gold AJ (2011) The effect of a protected area on the tradeoffs between short-run and long-run benefits from mangrove ecosystems. Proc Natl Acad Sci U S A 108:13945–13950

Meynecke JO, Lee SY, Duke NC, Warnken J (2006) Effect of rainfall as a component of climate change on estuarine fish production in Queensland, Australia. Estuar Coast Shelf Sci 69:491–504

Mimura N, Nunn P (1998) trends of beach erosion and shoreline protection in rural Fiji. J Coast Res 14:37–46

Munji CA, Bele MY, Idinoba ME, Sonwa DJ (2014) Floods and mangrove forests, friends or foes? Perceptions of relationships and risks in Cameroon coastal mangroves. Estuar Coast Shelf Sci 140:67–75

Nagelkerken I, Munday PL (2016) Animal behaviour shapes the ecological effects of ocean acidification and warming: moving from individual to community-level responses. Glob Chang Biol 22:974–989

Nagelkerken I, Connell SD (2015) Global alteration of ocean ecosystem functioning due to increasing human CO_2 emissions. Proc Natl Acad Sci U S A 112(43):13272–13277

Nagelkerken I, Sheaves M, Baker R, Connolly RM (2015) The seascape nursery: a novel spatial approach to identify and manage nurseries for coastal marine fauna. Fish Fisher 16:362–371

Nagelkerken I (2009) Evaluation of nursery function of mangroves and seagrass beds for tropical decapods and reef fishes: patterns and underlying mechanisms. In: Nagelkerken I (ed) Ecological connectivity among tropical coastal ecosystems. Springer Science and Business Media, Dordrecht, pp 357–399

Nagelkerken I, Kleijnen S, Klop T, van den Brand R, de la Moriniere EC, van der Velde G (2001) Dependence of Caribbean reef fishes on mangroves and seagrass beds as nursery habitats: a comparison of fish faunas between bays with and without mangroves/seagrass beds. Mar Ecol Prog Ser 214:225–235

Naylor RL, Drew M (1998) Valuing mangrove resources in Kosrae, Micronesia. Eviron Dev Econ 3:471–490

Osland MJ, Enwright N, Day RH, Doyle TW (2013) Winter climate change and coastal wetland foundation species: salt marshes vs. mangrove forests in the southeastern United States. Glob Chang Biol 19:1482–1494

Plan Vivo, MikokoPamoja. http://www.planvivo.org/projects/registeredprojects/mikoko-pamoja-kenya/. Accessed 10 May 2015

Poloczanska ES, Brown CJ, Sydeman WJ, Kiessling W, Schoeman DS, Moore PJ, Brander K, Bruno JF, Buckley LB, Burrows MT, Duarte CM, Halpern BS, Holding J, Kappel CV, O'Connor MI, Pandolfi JM, Parmesan C, Schwing F, Thompson SA, Richardson AJ (2013) Global imprint of climate change on marine life. Nat Clim Chang 3:919–925

Pramova E, Locatelli B, Djoudi H, Somorin OA (2012) Forests and trees for social adaptation to climate variability and change. Wiley Interdisciplinary Reviews. Clim Chang 3:581–596

Primavera JH, Esteban JMA (2008) A review of mangrove rehabilitation in the Philippines: successes, failures and future prospects. Wetl Ecol Manag 16:345–358

Quinitio ET, Parado-Etepa FD, Millamena OM, Rodriguez EM (2001) seed production of mud crab *Scylla serrata* juveniles. Asian Fish Sci 14:161–174

Rashid Sumaila U, Cheung WWL, Lam VWY, Pauly D, Herrick S (2011) Climate change impacts on the biophysics and economics of world fisheries. Nat Clim Chang 1:449–456

Record S, Charney ND, Zakaria RM, Ellison AM (2013) Projecting global mangrove species and community distributions under climate change. Ecosphere 4:34

Richards RG, Davidson AT, Meynecke JO, Beattie K, Hernaman V, Lynam T, van Putten IE (2015) Effects and mitigations of ocean acidification on wild and aquaculture scallop and prawn fisheries in Queensland, Australia. Fish Res 161:42–56

Riley ME, Johnston CA, Feller IC, Griffen BD (2014) Range Expansion of *Aratus pisonii* (Mangrove Tree Crab) into novel vegetative habitats. Southeast Nat 13:43–48

Rodriguez-Fourquet C, Sabat AM (2009) Effect of harvesting, vegetation structure and composition on the abundance and demography of the land crab *Cardisoma guanhumi* in Puerto Rico. Wetl Ecol Manag 17:627–640

Roessig JM, Woodley CM, Cech JJ, Hansen LJ (2004) Effects of global climate change on marine and estuarine fishes and fisheries. Rev Fish Biol Fish 14:251–275

Ruckelshaus M, Doney SC, Galindo HM, Barry JP, Chan F, Duffy JE, English CA, Gaines SD, Grebmeier JM, Hollowed AB, Knowlton N, Polovina J, Rabalais NN, Sydeman WJ, Talley LD (2013) Securing ocean benefits for society in the face of climate change. Mar Pol 40:154–159

Saenger P, Bellan MF (1995) The mangrove vegetation of the Atlantic Coast of Africa: a review. Université de Toulouse, Toulouse

Saintilan N, Williams RJ (1999) Mangrove transgression into saltmarsh environments in south-east Australia. Glob Ecol Biogeogr 8:117–124

Saintilan N, Wilson NC, Rogers K, Rajkaran A, Krauss KW (2014) Mangrove expansion and salt marsh decline at mangrove poleward limits. Glob Chang Biol 20:147–157

Sammut J, White I, Melville MD (1996) Acidification of an estuarine tributary in eastern Australia due to drainage of acid sulfate soils. Mar Freshw Res 47:669–684

Saunders MI, Leon JX, Callaghan DP, Roelfsema CM, Hamylton S, Brown CJ, Baldock T, Golshani A, Phinn SR, Lovelock CE, Hoegh-Guldberg O, Woodroffe CD, Mumby PJ (2014) Interdependency of tropical marine ecosystems in response to climate change. Nat Clim Chang 4:724–729

Smith AH, Berkes F (1993) Community-based use of mangrove resources in St. Lucia. Int J Environ Stud 43:123–131

Srinivasan UT, Carey SP, Hallstein E, Higgins PAT, Kerr AC, Koteen LE, Smith AB, Watson R, Harte J, Norgaard RB (2008) The debt of nations and the distribution of ecological impacts from human activities. Proc Natl Acad Sci U S A 105:1768–1773

Stephan M, Hobsbawn P (2014) Australian fisheries and aquaculture statistics 2013, Fisheries Research and Development Corporation project 2010/208. ABARES, Canberra, November. CC BY 3.0

Taylor S, Webley J, McInnes K (2012) 2010 Statewide Recreational Fishing Survey. State of Queensland. Department of Agriculture, Fisheries and Forestry

Thampanya U, Vermaat JE, Sinsakul S, Panapitukkul N (2006) Coastal erosion and mangrove progradation of Southern Thailand. Estuar Coast Shelf Sci 68:75–85

Thorsten B, Friess D (2016) Geomorphic knowledge for mangrove restoration: a pan-tropical categorization. Earth Surf Processes Landforms 41:231–239

Traill LW, Perhans K, Lovelock CE, Prohaska A, McFallan S, Rhodes JR, Wilson KA (2011) Managing for change: wetland transitions under sea-level rise and outcomes for threatened species. Divers Distrib 17:1225–1233

Vannucci M (2002) Indo-west pacific mangroves. In: Lacerda LD (ed) Mangrove ecosystems. Springer, Verlag, pp 123–215

Walters BB (2005) Patterns of local wood use and Cutting of Philippine mangrove forests. Econ Bot 59:66–76

Walters BB, Rönnbäck P, Kovacs JM, Crona B, Hussain SA, Badola R, Primavera JH, Barbier E, Dahdouh-Guebas F (2008) Ethnobiology, socio-economics and management of mangrove forests: a review. Aquat Bot 89:220–236

Yevich R, Logan J (2003) An assessment of biofuel use and burning of agricultural waste in the developing world. Glob Biogeochem Cycles 17:1–21

Chapter 9
Anthropogenic Drivers of Mangrove Loss: Geographic Patterns and Implications for Livelihoods

Rinku Roy Chowdhury, Emi Uchida, Luzhen Chen, Victor Osorio, and Landon Yoder

9.1 Introduction

National and global monitoring efforts of losses in mangrove areas are important to highlight the seriousness of threats facing mangroves worldwide (Valiela et al. 2001), as well as the outcomes of recovery efforts. They provide an invaluable evidence of changes to this dynamic and valuable ecosystem. However, multiple forces drive global mangrove losses, whose nature and dynamics can vary dramatically across regional and local contexts. If we are to truly understand the human dimensions of these systems, we need to understand the fundamental drivers of mangrove losses and how they interact at local levels. The fate of interdisciplinary mangrove science as well as current and future mangrove conservation efforts rests on this knowledge.

R.R. Chowdhury (✉)
Clark University, Worcester, MA, USA
e-mail: rroychowdhury@clarku.edu

E. Uchida
University of Rhode Island, Kingston, RI, USA
e-mail: euchida@uri.edu

L. Chen
Xiamen University, Xiamen, China
e-mail: luzhenchen@xmu.edu.cn

V. Osorio
Escuela Superior Politécnica y del Litoral, Guayaquil, Ecuador

Universidad de Almeria, Almeria, Spain
e-mail: vicosorio@gmail.com

L. Yoder
Indiana University, Bloomington, IN, USA
e-mail: yoderl@indiana.edu

V.H. Rivera-Monroy et al. (eds.), *Mangrove Ecosystems: A Global Biogeographic Perspective*, https://doi.org/10.1007/978-3-319-62206-4_9

In this chapter, we present a systematic, synoptic review of case studies involving mangrove ecosystems from Africa, Asia, and Latin America to compare the fundamental drivers of mangrove losses at a global scale. We then delve further into two significant drivers of mangrove losses: (a) mangrove-dependent subsistence economies and related poverty traps and (b) the global trade in shrimp. Next, we focus on Southeast Asia/China and Ecuador, representing two distinct geographic regions that have experienced rapid mangrove losses in the recent few decades, and examine specific drivers in those regional contexts. We discuss the implications of our findings for mangrove vulnerability and prospects for resilience. We argue for the benefits of a coupled system (specifically, a coupled socio-ecological system) approach to understand the bidirectional linkages between mangrove ecological dynamics on one hand, and the constellation of anthropogenic drivers of mangrove change, on the other.

9.2 Proximate Sources and Underlying Driving Forces of Mangrove Change: A Synoptic Approach

In this section, we draw from a representative cross section of subnational case studies to understand which factors are most prominent in driving mangrove loss. Following Geist and Lambin (2002), we identify proximate sources of mangrove loss (i.e., the immediate type of land conversion that directly provokes declines in mangrove area), as well as the underlying, deeper driving forces behind such losses (i.e., social and political processes behind the proximate land conversions, which are the indirect causal drivers of mangrove loss), and how multiple drivers interact to cause mangrove declines.

Specifically, we identified 43 case studies discussed in 38 published papers (33 journal articles, 2 book chapters, 1 dissertation, 1 conference paper, and 1 NGO report) for our analysis of both the proximate sources and underlying drivers of mangrove losses across Africa, Asia, and Latin America. Study periods ranged from single-year analyses to decadal reviews, dating back as far as 1903 to as recent as 2010. Criteria for including cases in the analysis were: (a) quantified estimates of mangrove loss during the study period—primarily as areal losses of mangrove cover, but sometimes verified through the reduction in mangrove tree stem diameter linked to unsustainable harvesting rates; (b) a detailed familiarity with the study area—as demonstrated via field-based interviews, surveys, documentation of social contexts or detailed case histories using a combination of primary sources and extensive literature reviews; and (c) a connection between specific proximate sources and underlying drivers identified by each author for the examined study area.

The overall patterns that emerged from this review provide evidence of distinct regional differences, but grounded on similar dynamics of change. The dominant proximate sources of change included land conversions to agriculture and aquaculture, harvesting of wood products (e.g., to produce charcoal for cooking, household and commercial construction), and infrastructure development (establishment of urban and rural settlements, roads, and dams). Underlying driving forces encompass demographic, economic, institutional, technological, and cultural factors. Our findings indicate that a driving force of mangrove loss from a given category (e.g., population increase as a demographic driver) did not occur in isolation, but rather along with drivers in other categories (e.g., aquaculture-market expansion as an economic driver).

Narrative descriptions of case studies suggest, moreover, that such coinciding drivers could also have strong interactive effects. Most notably, economic factors and policy/institutional factors have jointly driven mangrove loss across all regions. It indicates that our results herein are descriptive in nature, indicating common configurations (coinciding occurrences) of specific driving factors in cited cases of mangrove loss. They should not be interpreted as results indicating statistical significance or correlations among pairs of drivers in an analytical model, such as a regression. The main elements shaping these dynamics are the livelihood needs of people who directly utilize mangrove products, the opportunity to profit from aquaculture production, and government policies, property rights, and related institutional regimes that promote short-term uses over long-term mangrove protection.

9.2.1 *Proximate Sources of Mangrove Loss*

Three main proximate sources dominated the case studies on mangrove loss: infrastructure development, land use conversions linked to agriculture or aquaculture, and wood extraction. Among these, conversions for agriculture/aquaculture and harvesting of wood for both household and commercial use were the most commonly cited proximate sources of mangrove loss. Table 9.1 provides a breakdown of the main proximate sources, and Table 9.2 further details the particular types of mangrove conversion within these three main categories.

Agriculture and wood extraction are implicated in 29 (or 67%) of the 43 cases (Table 9.1). Infrastructure development is also an important factor, occurring in 20 of the cases (47%). However, it is clear that the three main proximate sources of mangrove losses most frequently co-occur, rather than occurring exclusively. For instance, none of the case studies cites infrastructure as the exclusive proximate source of mangrove loss. Rather, Table 9.1 indicates that infrastructure development acts in tandem with land use or with wood extraction in a two-factor causal chain driving mangrove loss (10 out of 43 cases), or with both land use and wood extraction in a three-factor causation (4 out of 43 cases). Of the limited number (16) of single-factor proximate sources, wood extraction is indicated for 6 African cases and 1 Latin American case, while land use conversions appear to be more frequent in Latin America or Asia (4 cases each).

We find additional regional-continental differences when we examine the particular types of proximate sources of mangrove loss (Table 9.2). Conversion of mangroves for aquaculture occurred in 26 of the cases (60%), overwhelmingly to create shrimp ponds. Conversion to agricultural uses, mostly for rice cultivation and pasture, occurred in nine cases (21%). Aquaculture occurs in 17 of 19 Asian cases,

Table 9.1 The three main proximate sources of mangrove loss include the development of infrastructure, land use conversions from forest to agriculture or aquaculture, and direct timber harvesting. The table shows the interactions between these different sources in driving mangrove losses

	All cases (n = 43)			Africa (n = 10)		Asia (n = 19)		Latin America (n = 14)	
	Abs.	Rel. (%)	Cum. (%)	Abs.	Rel. (%)	Abs.	Rel. (%)	Abs.	Rel. (%)
Single-factor causation (only one category or factor cited as the exclusive proximate source of mangrove loss for the case)									
Infrastructure	–	–	–	-	–	–	–	–	–
Land use (LU)	8	19	19	–	–	4	9	4	9
Wood extraction	7	16	35	6	14	–	–	1	2
Other	1	2	37	–	–	1	2	–	–
Two-factor causation									
Infra-LU	5	12	49	–	–	2	5	3	7
Infra-Wood	5	12	61	2	5	1	2	2	2
Infra-Other	–	–	–	–	–	–	–	–	–
LU-Wood	7	16	77	2	5	4	9	1	2
LU-Other	–	–	–	–	–	–	–	–	–
Wood-Other	–	–	–	–	–	–	–	–	–
Three-factor causation									
Infra-LU-Wood	4	9	86	–	–	3	7	1	2
Infra-LU-Other	–	–	–	–	–	–	–	–	–
Infra-Wood-Other	1	2	88	–	–	–	–	1	2
LU-Wood-Other	1	2	90	–	–	–	–	1	2
Four-factor causation									
All four factors	4	9	99	–	–	4	9	–	–
Total	**43**	**100**	**99**	**10**	**23**	**19**	**44**	**14**	**33**

Notes: Percentages rounded and may not total 100
abs absolute number, *rel* relative percentages, cum cumulative percentages

but is also a dominant force in Latin America (10 of 14 cases). Wood extraction is evenly split between harvests for household use for cooking or construction materials, and commercial harvesting (19 cases, 44% for each). Direct utilization of African mangrove forests is almost exclusively predicated on wood extraction, which is present in all 10 cases from the continent. In nearly all cases, household wood use is critical, typically reflecting a lack of affordable or accessible substitutes for cooking fuel. Moreover, in multiple African studies, mangrove harvesting is linked to fish smoking, an economically important industry requiring a significant amount of mangrove wood (Feka et al. 2009). Infrastructure development is another major proximate cause of losses in mangrove areas, as documented in 20 cases (47%). Of these, the establishment of rural or urban settlements is the leading causes of deforestation (11 cases, 26%); the construction of roads, ports/piers, and dams/dikes are collectively cited in 12 cases (28%).

Table 9.2 The types of activities identified in cases within the main proximate sources of mangrove loss

	All cases (n = 43)		Africa (n = 10)		Asia (n = 19)		Latin America (n = 14)	
	abs.	rel. %	abs.	rel. %	abs.	rel. %	abs.	rel. %
Agri/aquaculture conversion/expansion	29	67	2	5	17	40	10	23
Aquaculture conversion	26	60	–	–	17	40	9	21
Shrimp	23	53	–	–	14	33	9	21
Fish	6	14	–	–	5	12	1	2
Agriculture conversion	9	21	2	5	5	12	2	5
Rice	4	9	1	2	3	7	–	–
Grazing	3	7	1	2	2	5	–	–
Wood extraction	29	67	10	23	12	28	7	16
Household use (cooking, construction)	19	44	9	21	6	14	4	9
Commercial use	19	44	5	12	9	21	5	12
Infrastructure development	20	47	2	5	11	26	7	16
Urban, industrial, rural development	11	26	1	2	7	16	3	7
Dams and dikes	3	7	1	2	1	2	1	2
Roads	5	12	–	–	2	5	3	7
Port/pier construction	4	9	–	–	3	7	1	2
Boat traffic (wake erosion)	3	7	–	–	3	7	–	–
Use as landfills	3	7	–	–	1	2	2	5
Other	5	12	–	–	4	9	1	2
Mining	1	2	–	–	1	2	–	–
Pollution	3	7	–	–	2	5	1	2
War/conflict	1	2	–	–	1	2	–	–

Notes: Percentages rounded and may not total 100
abs absolute number, *rel* relative percentages, cum cumulative percentages

9.2.2 *Underlying Drivers of Mangrove Loss*

Underpinning these proximate mangrove conversion activities are fundamental, deeper or indirect anthropogenic drivers. Geist and Lambin (2002) identify five main types of driving forces of tropical deforestation: demographic, economic, institutional, technological and cultural. Table 9.3 provides a breakdown of the main categories of underlying drivers and their interactions, while Table 9.4 provides further details on these five categories. We have chosen to follow Geist and Lambin (2002:145, Table 9.1 in original article) to clarify single or multi-factor "causation" in our Table 9.3. Our empirical summary in these tables more closely presents the configurational conditions (e.g., collections of drivers that co-occur for a given case) and likely causal effects of the documented driving forces, rather than outlining causal mechanisms or causality per se (Meyfroidt 2015).

Table 9.3 The five main underlying drivers of mangrove loss include demographic, economic, political/institutional, cultural, and technological drivers. The table shows the interactions of these different drivers in causing mangroves losses. We follow Geist and Lambin (2002) in our usage of "causation" below. Our empirical summary presents the configurational conditions and implied causal effects behind mangrove loss, rather than outlining detailed causal mechanisms or causality per se (Meyfroidt 2015)

	All cases (n = 43)			Africa (n = 10)		Asia (n = 19)		Latin America (n = 14)	
	Abs.	Rel. (%)	Cum. (%)	Abs.	Rel. (%)	Abs.	Rel. (%)	Abs.	Rel. (%)
Single-factor causation									
Demographic	1	2	2	–	–	–	–	1	2
Economic	2	5	7	–	–	–	–	2	5
Political/Institutional	5	12	19	–	–	4	9	1	2
Two-factor causation									
Dem-Econ	2	5	24	–	–	1	2	1	2
Dem-Pol/Inst.	1	2	26	–	–	1	2	–	–
Econ-Pol/Inst.	8	19	45	2	5	3	7	3	7
Econ-Cultural	1	2	47	–	–	1	2	–	–
Three-factor causation									
Dem-Econ-Pol	3	7	54	2	5	–	–	1	2
Dem-Econ-Cult	1	2	56	1	2	–	–	–	–
Dem-Pol-Cult	1	2	58	–	–	1	2	–	–
Econ-Pol-Cult	3	7	65	–	–	1	2	2	5
Four-factor causation									
Dem-Econ-Pol-Cult	8	19	84	3	7	3	7	2	5
Dem-Econ-Pol-Tech	2	5	89	1	2	–	–	1	2
Dem-Econ-Cul-Tech	2	5	94	1	2	1	2	–	–
Five-factor causation									
All five factors	3	7	101	–	–	3	7	–	–
Total	**43**	**100**	**101**	**10**	**23**	**19**	**44**	**14**	**33**

Notes: Percentages rounded and may not total 100. Table omits factors and combinations of factors that were not present in any of the studies
abs absolute number, *rel* relative percentages, cum cumulative percentages

Our review indicates that the underlying driving forces of mangrove losses are overwhelmingly economic (cited in 34 cases, 79%) and policy/institutional factors (33 cases, 77%). In addition to these, demographic changes played an important role, featuring in 23 cases (50%), and cultural explanations also accounted as a driver in 20 cases (47%). Technological driving forces feature less prominently (7 cases, 16%). The leading factor among economic drivers causing mangrove decline are market and income opportunities (15 cases), increased local or tourist demand (9 cases), or export-oriented development (6 cases), totaling 70% of cases. Household subsistence needs and a lack of alternatives to mangrove-derived incomes also shaped economic drivers of mangrove loss (13 cases, 30%). Policy/

Table 9.4 The types of activities identified in cases within the major drivers of mangrove loss across continents

	All cases (n = 43)		Africa (n = 10)		Asia (n = 19)		Latin America (n = 14)	
	abs.	rel. %	abs.	rel. %	abs.	rel. %	abs.	rel. %
Economic Factors	34	79	10	23	13	30	11	26
Livelihood dependence/subsistence	13	30	8	19	2	5	3	7
Increased consumer demand	7	16	2	5	3	7	2	5
Market/income opportunity	15	35	2	5	7	16	6	14
Export opportunity	6	14	–	–	2	5	4	9
Policy and Institutional Factors	33	77	7	16	16	37	10	23
Formal policies facilitating loss	22	51	5	12	10	23	7	16
Government incentives	9	21	–	–	4	9	5	12
Weak enforcement/mismanagement	15	35	6	14	5	12	4	9
Foreign development	4	9	–	–	3	7	1	2
Demographic Factors	23	53	8	19	10	23	5	12
In-migration	8	19	2	5	3	7	3	7
Growing population density	6	14	2	5	4	9	–	–
Population pressure (proximity for access)	9	21	2	5	5	12	2	5
Resettlements (leads to clearing land)	3	7	2	5	1	2	–	–
Cultural Factors	20	47	6	14	10	23	4	9
Preference	10	23	4	9	5	12	1	2
Cooking flavor	7	16	3	7	3	7	1	2
Construction material	3	7	2	5	1	2	–	–
Values	7	16	2	5	5	12	–	–
Knowledge	3	7	–	–	–	–	3	7
Technological Factors	7	16	2	5	4	9	1	2
Alternative technology unused	2	5	1	2	1	2	–	–
Waste in harvesting/processing	1	2	1	2	–	–	–	–
Poor site selection/management	4	9	–	–	3	7	1	2

Notes: Percentages rounded and may not total 100. Table omits factors and combinations of factors that were not present in any of the studies
abs absolute number, *rel* relative percentages, cum cumulative percentages

institutional drivers consisted of two domains: formal institutions/policies that facilitated deforestation, and insecure land tenure. Formal institutions included policies and efforts to promote land development, most frequently shrimp aquaculture, or facilitate deforestation through government incentives or foreign development projects (13 cases, 30%). Weak enforcement or mismanagement of conservation or insecure land/forest rights is identified as a driver in 15 cases (35%). Land tenure conflicts are specifically cited in 14 cases (33%). For instance, conflicts may arise when collective usufruct (rights over mangrove resources held in common) of local communities can be undermined by other stakeholders that either treat mangroves as open access or attempt to privatize them for the development of aquaculture projects.

Demographic factors may also drive mangrove losses. For instance, increasing population density (6 cases, 14%), sometimes through immigration (8 cases, 19%), can lead to increasing demand for mangrove wood or the need to clear mangroves for settlements. Population pressure in proximity to mangroves leads to increased demand for mangrove products, harvesting rates, and urban pollution and waste that in turn impact mangroves (9 cases, 21%). Cultural drivers of forest loss relate to attitudes, beliefs, values, and preferences. The most frequently cited cultural driver of mangrove losses was a specific preference amongst mangrove wood users for the flavor that mangrove wood provides to food when used for cooking, as well as its burning efficiency. Technological drivers of mangrove losses were less frequently cited, indicative of broader coverage gaps among these factors behind global forest loss in the driving forces literature.

Nevertheless, poor site selection and mismanagement of aquaculture, especially intensive shrimp farming, has been an important driver of mangrove loss. In part, stocking densities of shrimp larvae and site selection can harm shrimp production through the increased risk of disease and water that is too acidic for shrimp production (Dewalt et al. 1996). Similarly, stocking densities can generate effluents that may cause human health risks, eutrophication, and harm shrimp production when there is insufficient mangrove area to absorb those pollutants (Dahdouh-Guebas et al. 2002; Mialhe et al. 2013). In some cases where shrimp production fails, especially due to disease outbreaks, farmers choose to abandon their ponds and clear more mangroves to establish new ponds (Thu and Populus 2007). Such problems affect the management, intensity, and sustainability of shrimp production, and are major causes to mangrove losses, since aquaculture is a leading proximate source of global mangrove conversions. While these issues were discussed in only a few case studies, they were raised at a general level by the selected papers. While a more focused review of mangrove losses specifically related to aquaculture is outside the scope of our general synopsis presented here, such an in-depth study would highlight specific management issues related to mangrove losses in regions of shrimp aquaculture expansion or intensification.

9.2.3 Interaction Effects of Multiple Causes

Echoing Geist and Lambin's (2002) findings for tropical deforestation, the interaction of multiple proximate sources and underlying drivers provides a better explanation of mangrove loss than single factors alone. Among proximate sources, multiple-factor land use impacts occurred in 27 cases (62%), while single-factor explanations account for 16 cases (37%). A single underlying driver explains only one-fifth of cases, while combinations of economic and policy/institutional drivers are present in 24 of the cases (57%). Multiple interacting drivers, most frequently economic factors in combination with policy/institutional and/or demographic ones, best explain why mangroves become deforested. A common scenario features government or market incentives to exploit mangrove habitat for shrimp aquaculture, prompting investment in aquaculture and the development of shrimp ponds (e.g., the case of Ecuador is detailed below in Sect. 9.4.2). This can occur in combination with unclear property rights that facilitate open-access conditions in coastal mangrove habitats. The long-term ecosystem services provided by mangroves, including protection of important subsistence economies such as artisanal coastal fisheries, is typically undervalued. This is further exacerbated by weak government protection for mangroves and insecure land tenure conditions (Binh et al. 2005; Hamilton 2011). Another common syndrome of mangrove degradation occurs when the proximity of a coastal population to mangroves facilitates primary extraction that becomes unsustainable as the population increases over time. Such increases can be facilitated by immigration following the building of roads, and can in turn lead to direct mangrove losses due to clearing for settlements, or indirect losses related to amplifying consumer demand for mangrove wood and other resources (Walters 2003).

It is also clear from these case studies, and from the broader literature on mangrove conservation, that when mangrove protection is poorly enforced or nonexistent, there are multiple and often devastating impacts on local communities. In the worst scenarios, conversion to intense shrimp aquaculture clears an area of mangrove trees, but then fails due to poor management that brings diseases to the shrimp ponds. The pond is then abandoned and the owners simply clear new mangrove areas and begin the process again (Thu and Populus 2007). In more promising circumstances, local communities may respond to the scarcity of mangrove wood by replanting mangroves and subsequently harvesting in a more sustainable manner (Walters 2003). In both of these scenarios, it is clear that economic and policy/institutional dynamics can combine to amplify either negative or positive consequences for mangrove conservation.

9.3 The Dynamics of Mangrove Dependence, Poverty, and Livelihoods

In this section, we delve further into poverty and livelihoods of rural coastal communities as one of the key underlying drivers of mangrove losses. Extreme poverty and natural resource degradation are correlated geographically and concentrated in rural areas where livelihoods depend disproportionately on natural resources (Barrett et al. 2011). Mangrove forests and the ecosystem services they provide are no exception. Mangroves mostly exist in rural coastal areas in tropical and subtropical countries, many in areas characterized by high poverty rates. Among the 118 countries and territories identified to have some mangrove forests in 2000 (Giri et al. 2011), the largest extent of mangroves is found in Asia (42%), followed by

Table 9.5 Most common direct and indirect use of mangroves by coastal communities

Direct use		Indirect use	
Forestry	Timber for building houses and boats	Shoreline protection	Blocking flood
	Fuel, e.g., firewood and charcoal		Slowing wind
	Food, e.g., seeds and honey		Preventing shoreline erosion
	Medical use	Agriculture and aquaculture	Water purification
Fishery	Inshore fishery		Nutrient retention
	Offshore fishery		
Recreational	Tourism		
	Education		

Africa (20%), North and Central America (15%), Oceania (12%) and South America (11%). Approximately 75% of global mangroves are concentrated in just 15 countries: Indonesia, Australia, Brazil, Mexico, Nigeria, Malaysia, Myanmar, Papua New Guinea, Bangladesh, Cuba, India, Guinea Bissau, Mozambique, Madagascar and the Philippines. All of these countries, except Australia, are categorized as low or middle income–based on GDP per capita. Moreover, within these countries, rural coastal areas with higher poverty rates tend to be where large areas of mangroves are located. This geographical overlap between mangrove forests and high prevalence of poverty sets the stage for a disproportionate reliance by the poor communities on mangrove-derived natural resources.

Mangrove forests and their ecosystem services support the livelihood of rural coastal poor communities. Mangrove forests are one of the most productive ecosystems in the world due to their special biological characteristics and adaptation to the coastal environment. They can benefit coastal communities in direct and indirect ways, including the provisioning of extractive goods (raw materials and energy) and nonextractive ecosystem services such as functioning as nurseries for fish and shrimp, shoreline protection from saltwater intrusion, erosion and floods (e.g., Bann1998; Gordon et al. 2009; Kuenzer and Tuan 2013, Table 9.5). Household and village-level surveys have documented the importance of these direct and indirect benefits of mangroves to the livelihoods of the rural poor, for example, in Malaysia (Bann 1999), India (Hussain and Badola 2010), Kenya (Dahdouh-Guebas et al. 2000; UNEP 2011), Tanzania (McNally et al. 2011), and Thailand (e.g., many works by Barbier, including Barbier 2010).

An extensive literature on valuation of mangrove forests demonstrates that their direct and indirect benefits are large. The majority of ecological-economic valuation studies are conducted in the context of rural developing countries, mostly in Asia, where communities are dependent on mangrove forests. Values have been estimated for fuelwood, fisheries and coastal protection. Estimates vary widely, especially for the functions of mangrove forests in protecting fisheries habitat. For example, Rönnbäck (1999) estimated that the value of mangrove forests for fisheries range from $16 per ha per year for penaeid shrimp in Indonesia to $113 per ha per year

(1999 prices) for sergestid shrimp in the Philippines. UNEP (2011) estimated that the mangrove forests in Kenya support local fisheries worth $44 per ha per year, while Islam and Ikejima (2009) found that the value was US$52–105 per ha per year in Thailand. Mangroves also support biodiversity by providing nutrients. Hussain and Badola (2008) estimated that each hectare of mangroves contains additional nutrients worth US$232 compared to non-mangrove areas, which translated to US$3.37 million worth of nutrients in 145 km^2 of mangrove forests.

Mangroves also play a key role in protecting shorelines by mitigating intensive waves, slowing down winds, and improving land productivity by accumulating nutrients from terrestrial runoff. Studies have found that mangrove ecosystems provided protection to life and property during the 2004 tsunami in Southeast Asia (Barbier 2006; Dahdouh-Guebas et al. 2005; Danielsen et al. 2005; Kathiresan and Rajendran 2005; UNEP 2005), although some have argued against this claim (Kerr and Baird 2007; Vermaat and Thampanya 2006). Studies in India showed that villages located behind mangrove forests suffered less damage than those directly exposed to the coast (Kathiresan and Rajendran 2005). In addition, Kathiresan (2012) suggested that the destructive power of the storm surge was exacerbated during Cyclone Nargis in 2008 by recent loss of mangroves in Myanmar, although no primary evidence to support these statements was presented. Some researchers who are skeptical about the ability of mangroves to protect against tsunamis have noted that mangroves might be more capable of protecting against tropical storm surges. Storm surges differ from tsunamis in having shorter wavelengths and relatively more of their energy near the water surface. The current consensus is seemingly that mangroves provide protection to a certain extent, but proving it scientifically is a challenge.

Damages from these natural hazards may be a source of poverty traps for households. In a comparative study in India, Badola and Hussain (2005) found that storm-related losses incurred per household were the greatest (US$154) in a village not sheltered by mangroves but had an embankment, followed by a village that was neither buffered by mangroves nor an embankment (US$44), and the village that was protected by mangrove forests suffered the lowest losses (US$ 33). In another study in India, mangroves were found to have reduced house damages by approximately US$177 per ha at 1999 prices (Das and Crépin 2013). Although these monetary values may be highly site-specific and some methodologies used in nonmarket valuation remain controversial (e.g., Kling et al. 2012), it is clear that mangrove forests and their ecosystem services benefit the coastal poor communities.

9.3.1 Linkages Between Dynamics of Mangrove Resources and Local Livelihoods

Certain socioeconomic contexts further heighten the importance of mangrove forests for the welfare of poor populations. First, poor households often have a severely limited capacity to purchase substitutable goods and services (World Bank 2008). Economic livelihoods of households from mangrove-rich villages have depended traditionally on the surrounding mangrove ecosystems with few other income sources (e.g., Aksornkoae et al. 2004). Additionally, poor households often have limited access to credit and insurance markets for risk management. Second, forest resources and their ecosystem services can also act as safety nets for the poor communities, for example, by providing additional income sources in times of need (Ewel et al. 1998; Barbier 2007b). However, the dependence of rural poor communities on natural resources can also drive mangrove degradation. Thus, the higher the demand for mangrove resources, the higher the pressure is on mangrove forests. This feedback mechanism necessitates a coupled systems framework to understand the impact of mangrove dynamics on local livelihoods and vice versa.

There are several potential mechanisms that link mangrove resources and local livelihoods. The most fundamental mechanism stems from the rural poor's dependence on extractive uses of mangroves for their livelihoods, which potentially leads to a poverty–environment trap—a vicious cycle between poverty and environment (e.g., WCED 1987; Vosti and Reardon 1997). Extraction of mangroves (i.e., use of their "provisioning" ecosystem services) at an unsustainable rate not only reduces future availability of mangrove ecosystem services but also generates contemporaneous losses of non-extractive (e.g., regulating) ecosystem services. When an ecosystem service diminishes, the productivity of labor can fall as well, as household members must expend more time and labor in meeting basic needs, for example, in securing fuelwood, drinking water, and fish and shrimp. In addition, such households may face a greater risk of asset damages from floods, coastal erosion, and other impacts of reduced coastal protection. Both types of vulnerability can lead to a poverty trap. Other mechanisms can exacerbate poverty and damage to mangrove ecosystem services. For example, institutions (e.g., property rights and cultural practices) may be poorly coordinated and thus fail to facilitate effective mangrove management or conservation. Lack of informational feedback can leave local communities ill-equipped to make decisions that can sustain ecosystem services.

Despite the importance of these mechanisms, much work remains to be done to adequately theorize and empirically document them. One exception is Barbier (2007b), who found that decrease in mangrove area has led to more participation of adults in off-farm employment for lower wages among coastal households in Thailand. The study showed that this effect is plausibly due to the fact that labor returns from mangrove-related livelihood activities diminished after mangroves forests were depleted. Other mechanisms that link mangrove valuation with local livelihood dynamics need to be examined conceptually and empirically. Such efforts will be essential for designing policies to reduce mangrove degradation and poverty.

9.4 Case Studies of Mangrove Loss

In this section, we highlight in greater depth the dynamics of mangrove loss using two regional case studies from Asia and Latin America, where much of the world's mangrove decline has been concentrated. The examples of China and Ecuador underscore how these two countries, characterized by important contrasts in political-economic conditions, have nevertheless seen similar trajectories of mangrove ecosystem change. One proximate source of mangrove loss that is common to both case study regions is the expansion of shrimp aquaculture. Shrimp farming originated in the middle of the twentieth century along tropical and subtropical coasts, particularly in Asia (e.g., China and Thailand), where three-quarters of the world's farmed shrimp is produced. Global production of shrimp increased from 1325 metric tons (MT) in 1950 to one million MT by 1982, 3.5 million MT in 2009, and peaking at 3.9 million MT in 2011. Production subsequently declined to 3.8 million MT in 2012 and 3.2 million MT in 2013, mainly due to diseases such as early mortality syndrome, which reduced shrimp production in China and South and Southeast Asia by 35% in 2015 (FAO 2016). The rapid expansion of shrimp production has spurred an intense debate regarding its attendant economic, social, and, environmental impacts. Particular concerns center on wetland losses, increased organic loading in coastal waters, introduction of exotic species, and the dispersal of harmful diseases (Boyd and Clay 1998; Primavera 2006). The sections below illustrate the role of shrimp farming in mangrove losses in the two case study areas, along with other drivers of mangrove loss. We reflect on the value of such case studies to highlight the causal dynamics of mangrove loss, and lessons for the future of mangrove conservation.

9.4.1 *Mainland China in the East/Southeast Asian Context*

In order to understand the trajectories of China's mangrove forests in the past decades, it is useful to relate it to the broader Asian and Southeast Asian context. Asia houses more than a third of the world's mangrove forests, and yet, has also lost more than one-third of its mangrove area since the 1980s. Aquaculture was the main proximate source of mangrove deforestations in Asia before year 2000, with land use conversions and extraction related to agriculture, while forestry and urbanization were of secondary importance in the region during that period (Valiela et al. 2001). However, agriculture and urbanization began to gain importance locally, and have been increasingly implicated in mangrove deforestation since 2000 (Lai et al. 2015; Richards and Friess 2016). Notably, Southeast Asia houses the greatest mangrove species diversity, and more than one-third of the world's mangrove area. A recent study quantified the proximate sources of mangrove deforestation across Southeast Asia between 2000 and 2012 (Richards and Friess 2016). Aquaculture and agriculture were found to be the primary sources of mangrove deforestation (Table 9.6 and Fig. 9.1).[1] Table 9.6 indicates, for instance, that during 2000–2012, Indonesia lost 1.72% of its net mangrove area. For this country, 48.6% of mangrove change during 2000–2012 reflected conversions to aquaculture, 0.1% to rice

[1] Table 9.6 and Fig. 9.1 reedited from Richards and Friess (2016). The original figure was cited with the permission of the journal.

Table 9.6 Percentage of mangrove areal loss in the period 2000–2012 in Southeast Asia, and the use/conversion categories

Country	Percent mangrove loss	Area (%) converted to distinct land use/covers					
		Aquaculture	Rice	Oil palm	Mangrove regrowth	Urban	Other category
Indonesia	1.72	48.6	0.1	15.7	22.6	1.9	11.2
Myanmar	5.53	1.6	87.6	1.1	0.5	1.6	7.6
Malaysia	2.83	14.7	0.1	38.2	17.6	12.8	16.7
Thailand	1.36	10.8	5.6	40.0	5.1	14.4	24.1
Philippines	0.50	36.7	0.9	11.1	7.3	2.7	41.3
Cambodia	2.28	27.7	1.5	8.9	9.8	4.6	47.6
Vietnam	0.25	21.0	10.4	0.5	0.6	62.5	4.9
Brunei	0.37	29.2	0	27.7	12.5	15.9	14.8
Timor-Leste	0.19	0	26.1	0	0	0	73.9[a]
Singapore	0	0	0	0	0	0	0
Total		29.9	21.7	16.3	15.4	4.2	12.3

Adapted from Richards and Friess (2016)
Countries are ordered by absolute areas of mangroves lost. Percentages might not sum to 100 owing to rounding
[a]The small amount of mangrove deforestation in Timor-Leste is due mainly to shoreline erosion

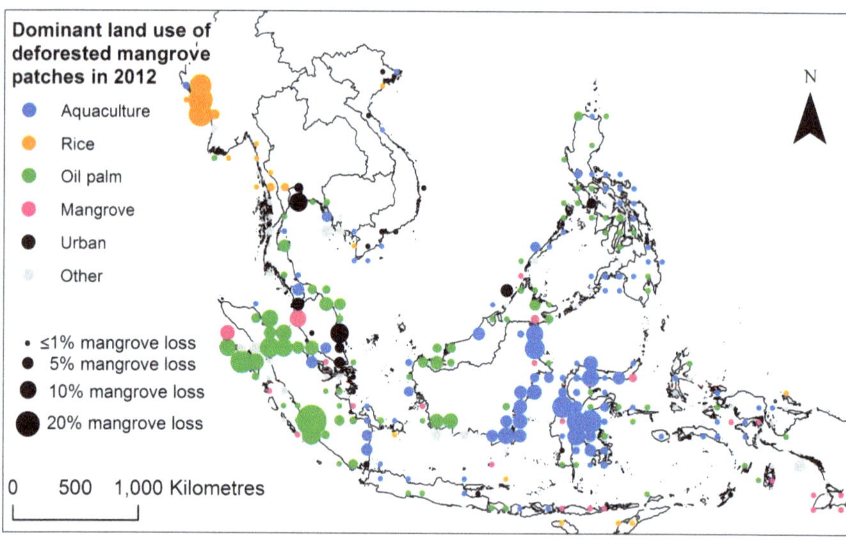

Fig. 9.1 Percentage of mangrove deforestation in Southeast Asia between 2000 and 2012, and dominant land uses of deforested areas in 2012. Land uses are summarized as the converted land use with the greatest area within each 1 decimal degree grid square. Circles are located in the center of each grid square; each circle size represents the percentage of the mangrove area in 2000 that has been lost (Adapted from Richards and Friess 2016)

farming, 15.7% to the expansion of oil palm plantations, 1.9% to urban expansion, 11.2% to other causes. Along with localized patterns of mangrove regrowth (22.6% for Indonesia), the rows (relative areal proportions under the listed land cover/use conversions) sum to approximately 100%. While aquaculture is a significant source of mangrove loss, rice agriculture and oil palm plantations were also identified as posing critical yet under-recognized threats to mangroves, especially in Myanmar, Indonesia, Thailand and Malaysia. Urbanization had small regional but locally important impacts in the Bangkok region of Thailand, southern Malaysia and Vietnam.

In China, almost all mangrove forests are located in the subtropical coastal regions, which are the northern limits for many mangrove species. Before their important ecological and economic values were publicly recognized in the early 1990s, mainland China's mangroves were greatly reduced from about 50,000 ha in the 1950s to 22,700 ha in 2001 (Chen et al. 2009). The main sources of mangrove deforestation were linked to China's growing economy since the 1950s, and differed from the case of Southeast Asia in terms of the proximate sources. From the 1950s to the 1990s, timber and firewood/charcoal used were the main threats to mangroves. As rice agriculture expanded, and seawalls were constructed along the coastline of China, mangroves underwent serious declines in the 1960s and 1970s. During these two decades, many seawalls built in the mid-intertidal zones resulted in the loss of mangrove forests in high-intertidal zones. Since 1980, large areas of mangroves continued to be reclaimed for the establishment of fish ponds. A total of 12,923 ha mangrove forests were lost and 96.7% of them converted into fish or shrimp ponds all over mainland China in the 1980s and 1990s (Wang and Wang 2007).

Urbanization was a source of major pressure on Chinese mangroves after the 1990s, even if it did not lead to areal declines immediately. National economic reforms begun in 1979 brought rapid economic growth in coastal areas in mainland China. Since then, national GDP growth has been closely and positively correlated with environmental degradation of coastal regions (He et al. 2014). Mangrove areas decreased significantly from the 1950s until reforestation programs were initiated after 1990 (Table 9.7; Liao and Zhang 2014).

Guangdong Province, which is among the most intensively developed regions in China following the economic reforms, saw extensive mangrove conversions to mariculture during the 1980s and subsequently to infrastructure expansion (Table 9.7). In 1979, Shenzhen city, located in the mangrove-rich Pearl River Estuary, was designated a special economic zone in Guangdong Province. Over the past three decades, Shenzhen has become a massive urban center. A national mangrove reserve was established in 1988, and has now become an iconic urban mangrove forest in China. A remote sensing study revealed mangrove patterns and spatial extents pre- and post-urbanization (Table 9.8, from Chen et al. 2012), demonstrating mangrove area decreased steadily from 1979 to 1998, but increased after 1998 due to mangrove conservation within and around the reserve.

Table 9.7 Historical mangrove extent (hectares) change in mainland China coastal provinces since the 1950s

Province	Year				
	1950s	1986	1995	2001[a]	2013
Hainan	9992	4667	4836	3930	4891
Guangxi	10,000	8000	5654	8375	6595
Guangdong	21,289	4000	3526	9084	12,131
Fujian	720	368	360	615	942
Zhejiang	0	0	21	20	0
Total	42,001	17,035	14,397	22,024	24,559

Adapted from Liao and Zhang (2014)

[a]Reforestation activities began in 1990s, which increased the area of mangroves

Table 9.8 Mangrove forests, fish ponds, and urban construction area (hectares) in Shenzhen city from 1979 to 2009

Year	Mangrove forests	Fish ponds	Urban built-up area
1979	79.56	227.25	508.95
1989	58.59	431.28	708.48
1998	52.65	100.26	1194.57
2003	72.99	60.39	1895.85
2009	81.00	30.15	2072.50

Reedited from Chen et al. (2012)

This local mangrove recovery was also replicated in other coastal regions after the 1990s. Total mangrove extents increased along the coastline of China, attributed to the establishment of mangrove reserves and reforestation programs. Now, almost all of the remaining natural mangroves in China are protected nature reserves (Chen et al. 2009). These efforts can help protect mangrove areas, and yet, forest degradation remains a major threat (Peng et al. 2016). Urbanization increases the proximity of human settlements to mangroves, and intensifies their direct and indirect impacts on them. In cities such as Shenzhen, mangrove reserves are surrounding by skyscrapers and many mangrove reserves became tourist sites (Fig. 9.2). In mangrove afforestation programs, monocultures and nonnative species of mangroves were intensively used to target rapid and visible forest regrowth, but these practices also degraded the long-term ecological diversity and health of mangroves.

The Chinese case study indicates the land use changes that have driven mangrove loss in the past decades are linked to significant economic growth and reforms. From primary extractive uses to recreational use, threats to mangrove have also changed, but driven by GDP growth. Over the course of massive urbanization, there was also a public recognition of the ecosystem services provided by mangrove ecosystems, and urban populations have come to rely much more on the tourism and educational potential of mangrove ecosystems. Such cultural and policy changes may help stem future areal losses and perhaps even lead to mangrove expansion. However, degradation caused by pollution, noise and human disturbance can become much more serious threats to mangroves in the coming years.

Fig. 9.2 Urbanization has become a new challenge to mangrove conservation. (**a**) A national mangrove nature reserve in Shenzhen City of Guangdong Province, China. (**b**) Tourism in a national mangrove nature reserve in surrounding region of Haikou City of Hainan Province, China

9.4.2 Ecuador

This section describes changes in the coverage of mangroves and salt flats along the Ecuadorian coast from 1991 to 2006, and discusses the development of shrimp aquaculture as the major driver of these changes. We use information from studies developed by CLIRSEN, the Ecuadorian authority for the monitoring of mangroves, during 1991, 1995, 1999, and 2006.

Mangroves, shrimp farms, and salt flats are found within 15 estuaries in Ecuador. Data from the CLIRSEN's 2006 survey show that 99.8% (146,971 ha) of mangroves and 97.1% (170,648 ha) of shrimp farms were concentrated in the following six estuaries (Fig. 9.3): (1) Cayapas Mataje, (2) Muisne River, (3) Cojimíes River, (4) Chone River, (5) Gulf of Guayaquil, and (6) the Jambelí Archipelago. The variation of mangroves, salt flats, and growing shrimp aquaculture in these estuaries are depicted in Figs. 9.4, 9.5, and 9.6, respectively.

Mangroves are mainly concentrated in the Gulf of Guayaquil (71.4%), Cayapas Mataje Estuary (14.5%) and Archipelago of Jambelí (10.3%). The most degraded estuaries are the estuaries of Muisne, Chone, and Cojimíes Rivers, with mangrove losses of 52.5%, 76.5%, and 79.1%, respectively. The least degraded estuary is the Cayapas Mataje, at 9.6% mangrove loss. Between 1969 and 2006, mangrove area declined from 203,625 to 147,229 ha and salt flats from 51,496 to 3705 ha. During the same period, the area devoted to shrimp aquaculture grew from zero to an impressive 1,706,489 ha.

The Gulf of Guayaquil and the Archipelago of Jambelí experienced the greatest loss of mangrove forest (68.6%) during 1991–1995, the former location experiencing an average deforestation rate of 1875 ha per year. Approximately 3000 ha of mangroves were lost in Cayapas Mataje from 1970–2008 (Hamilton 2012). In the Cojimies Estuary, 10,000 ha (66%) of mangrove forests were lost from 1971 to 2006. These losses were primarily due to shrimp aquaculture development (Hamilton and Collins 2013). The spatial extent of shrimp farms has surpassed that of man-

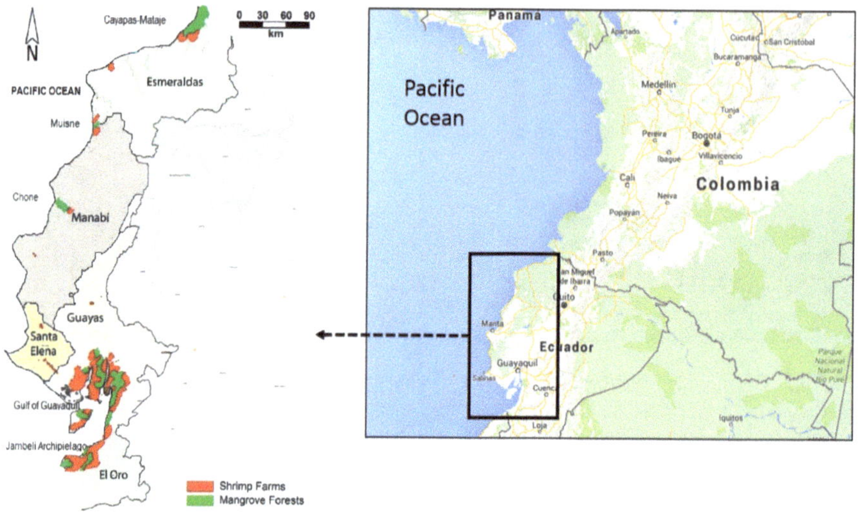

Fig. 9.3 Mangrove (*green*) and shrimp farm (*red*) locations on the coast of Ecuador

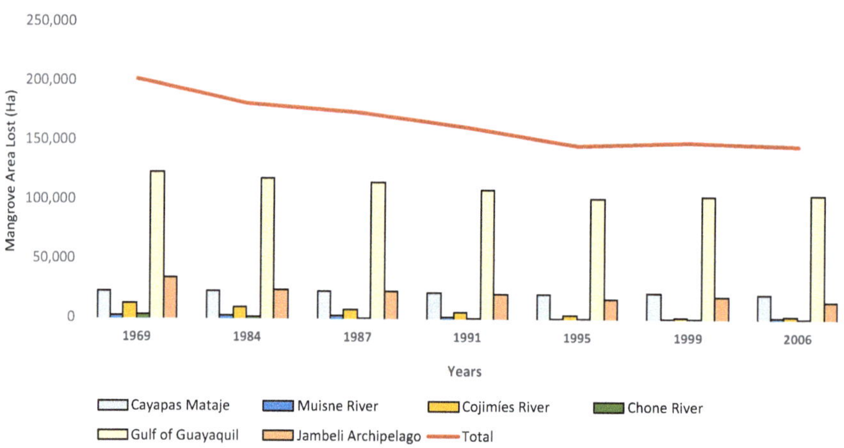

Fig. 9.4 Mangrove cover changes within the six estuaries containing 99.8% of the total mangrove area along the coast of Ecuador

groves in some estuaries. For instance, the Cojimíes River estuary has 4.8 ha of shrimp farms for each hectare of mangroves, and the Chones River estuary has 5.8 (Bravo 2010). Shrimp farms in Ecuador are located principally in the Gulf of Guayaquil (59%) and in the Archipelago of Jambelí (24%), distributed across 55,831 ha of mangroves (33%), 61,523 ha of uplands (36%), and 53,295 hectare of salt flats (31%) (Fig. 9.7). The expansion of shrimp farms reached its peak in the Gulf of Guayaquil during the period 1984–1987, at an average rate of increase of 7277 ha per year.

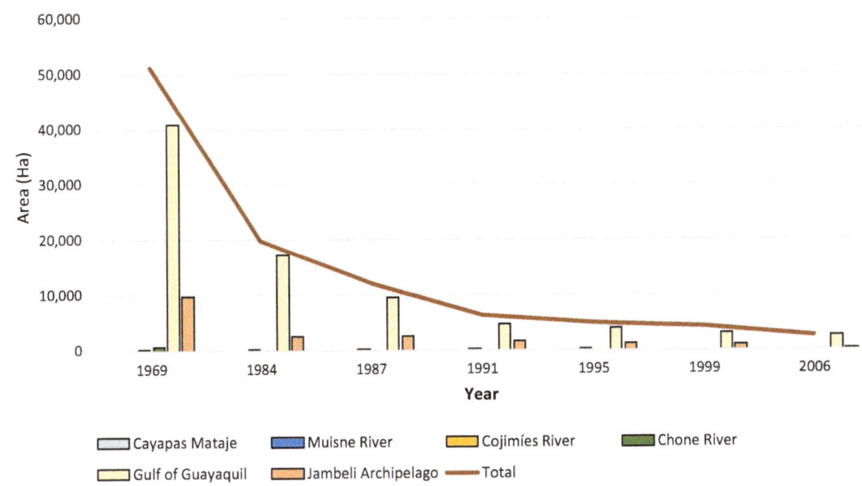

Fig. 9.5 Salt flat areal variation within the six estuaries containing 99.8% of the total mangroves of Ecuador

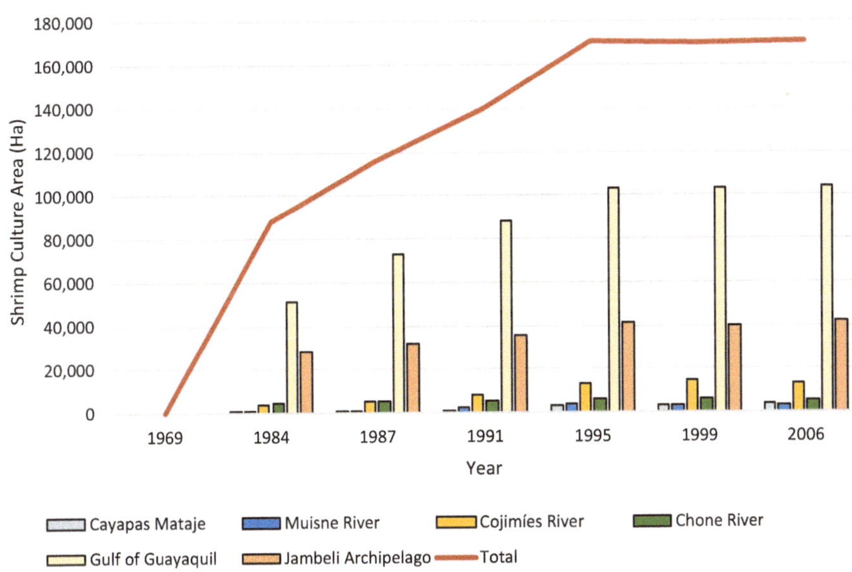

Fig. 9.6 Shrimp aquaculture areas within the six estuaries containing 99.8% of the total mangroves of Ecuador

The shrimp aquaculture industry was initiated in the late 1960s, and is based on the local species *Litopenaeus vannamei*. The growth of aquaculture has been steady, with periods of accelerated growth coinciding with El Niño cycles and low growth or declines during outbreaks of shrimp viral diseases such as white spot (Fig. 9.8). In response, the industry embraced extensive to semi-intensive stocking densities

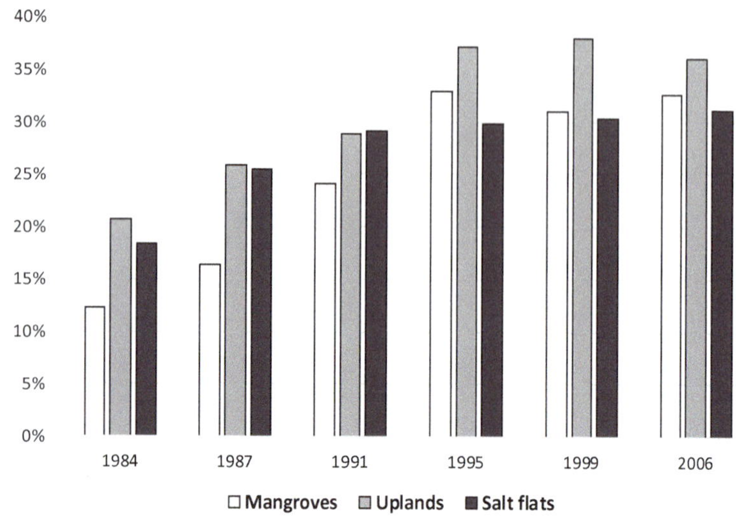

Fig. 9.7 Type of land converted to shrimp farms in Ecuador from 1984 to 2006

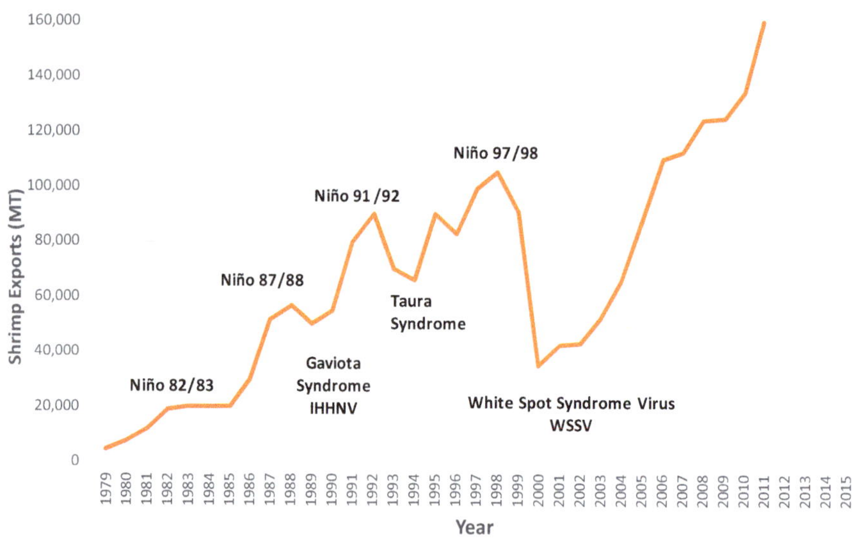

Fig. 9.8 Total shrimp exports (MT) by year for Ecuador

(7–12 post larvae/m^2) in order to continue production while lowering production risks and costs. Production reached historical highs following technical innovations to address white spot virus, including the development of white spot–resistant shrimp strains, application of biosecurity protocols during all phases of culture, better training of personnel, and other techno-managerial changes. Intensification of shrimp farming reached its peak in 2015 with a production of 297,000 MT without increasing the area of production (Ecuadorian National Aquaculture Chamber,2016).

Ecuadorian farms use locally developed technology at the pond-raising phase, and farms dating to the earliest phases of development of the industry continue to be active in shrimp production. State-of-the-art technology is used at shrimp breeding and seed-production phases, because the capture of wild seed and wild broodstock has been prohibited by law (Ministerial resolution 106, RO N° 685 of Oct. 17, 2002) since the occurrence of the white spot syndrome (WSSV). Broodstock is currently pond raised and disease-resistant strains are being developed. This combination of low- to medium-intensity production systems causes moderate environmental impacts associated with waste, feed, and the use of chemicals (Thompson 2014). Disease outbreaks on shrimp farms have been the driving force behind the shift of Ecuador's shrimp aquaculture from extensive to semi-intensive practices. Better aquaculture practices and the use of probiotics are the preferred forms of health management, but some use of controlled antibiotics (oxytetracycline or florfenicol) exists. Both oxytetracycline and florfenicol are listed by the World Health Organization (WHO) as "Highly Important" for human health (Briggs 2006a, b; Lightner 2011).

The majority of mangrove losses in Ecuador occurred before the late 1990s. Since then, a number of factors have greatly reduced mangrove losses due to shrimp farming, including: (a) regulatory protection and prohibitions on mangrove felling for any use; (b) the ability of the species (*Litopenaeus vannamei)* to grow in a broad range of salinity conditions (0 to 50+); and (c) the realization that former mangrove soils negatively affect pond water quality and shrimp coloration. Over a decade ago, Ecuador had a poor record of environmental protection due to widespread conversion of ecologically important habitats in the country. However, current national regulations seek to return mangrove areal extents to the official 1999 levels (Presidential Decree 1391, 2008) and prevent further mangrove and wetland loss during farm construction by instituting environmental impact assessments (EIA) (Thompson 2014).

Mandatory annual inspections carried out by the National Institute of Fisheries target the compliance of farms with existing environmental regulations. In comparison to intensive shrimp farming in Asia (stocking density of 150 post larvae per square meter), extensive shrimp culture in Ecuador requires the application of fertilizers to the ponds to stimulate natural feed production, while the supplemental use of feed in semi-intensive shrimp culture results in a relatively low (1–1.5) food conversion ratio (FCR), and therefore relatively low waste production from the shrimp. The relationship of mangrove loss and gain to the intensity of shrimp culture cannot be overemphasized. It stands to reason, and the Ecuadorean case illustrates, that an intensification of aquaculture along with attendant shifts in techno-managerial strategies can release pressure on mangrove forests. However, simply intensifying aquaculture in the absence of adequate regulatory protections against mangrove felling and water quality/toxicity problems can exacerbate threats to coastal ecosystems, local livelihoods and human health.

The current Ecuadorian Forestry Law stipulates that mangroves are public goods of the State, "are out of trade, not subject to possession or any other kind of appropriation, and can only be exploited by way of authorized concession granted under this law and its reclamation."[2] Since 1985, cutting, burning, and exploitation of mangroves have been prohibited (Bravo 2007). By the year 2000, the Ecuadorian State began to recognize the "ancestral" rights of "traditional user groups" to mangrove resources, paving the way for *custodias,* ten-year community-managed concessions. As of July 2011, a total of 37,818 ha have been granted to 41 different communities as part of the national strategy toward community-based conservation and management of mangrove resources (Rosero Moya and Santillan Salas 2011).

9.5 Conclusion: Mangroves as Critical Socio-Ecological Systems

Globally, mangroves constitute highly significant natural resources and ecosystems. They are globally significant land–ocean carbon exchange sites (France-Lanord and Derry 1997; see also Chaps. 5 and 6), house important hotspots of global biodiversity (Chap. 2), offer protection against extreme climate events such as cyclones and related storm surge (Chap. 7), and, in many parts of the world, have been utilized by local communities for generations. The provisioning, regulating, and cultural services provided by these important coastal ecosystems are extensive. It is, therefore, imperative to better understand the geographic patterns and processes that are driving continued losses and in some cases, gains in these vital, coupled socio-ecological systems.

Our synoptic review of the research literature clearly reveals that mangrove losses are spatially differentiated, time-dependent, and linked to varying sets of proximate activities and causal drivers. The expansion of agriculture and aquaculture is a major proximate source of mangrove losses worldwide. Extractive activities such as harvesting of timber and non-timber resources from mangroves are also linked to serious degradation of local mangrove resources, as is the significant increase in infrastructure development. These immediate sources of mangrove conversions are ultimately linked to deeper demographic, economic, institutional, technological and cultural driving forces. In particular, our review of a representative literature on mangrove loss and recovery indicates that such drivers rarely operate in isolation; we find, instead, that most studies reveal a complex set of interacting causal drivers of mangrove loss. Further, the combination of economic drivers (such as global shrimp commodity markets) with policy-institutional change (such as government policies and subsidies, property rights regimes, and conservation policies) has been the most powerful process shaping mangrove change globally.

[2](Author's translation). The original text of the Forestry Law *Titulo 1, Capitulo 1, Art. 1* published *in Registro Oficial* No. 418 (September 10, 2004) states: "Los manglares, aun aquellos existentes en propiedades particulares, se consideran bienes de Estado y están fuera del comercio, no son susceptibles de posesión o cualquier otro medio de apropiación y solamente podrán ser explotados mediante concesión otorgada, de conformidad con esta Ley y su reglamento."

The case studies from China and Ecuador exemplify these broader findings. In the Chinese context, national economic development and specific economic reforms led to important intensification of pressure on select geographic and coastal regions, impacting mangrove spatial extent and ecological health. The establishment of Special Economic Zones, and rampant urbanization therein served as the immediate, proximate source of pressure on mangrove resources, further threatened by the establishment of shrimp farming, rice agriculture, and associated changes in hydrological regimes. In the Ecuadorean case, shrimp aquaculture underwent phenomenal growth during the past decades, but especially during the mid-1980s. The expansion declined subsequently with a shift in the technological approach to shrimp aquaculture, de-emphasizing extensive shrimp farms and prioritizing intensive aquaculture practices instead.

In both China and Ecuador, increasing recognition of the ecological, socioeconomic and cultural value of mangroves has led to policy and regulatory shifts in favor of ecosystem protection and community-based conservation. Such institutional changes hold important implications for the sustainability of mangrove utilization as well as ecosystem recovery. These cases also reflect findings in the broader literature that economic factors and policy/institutional factors have jointly and overwhelmingly driven mangrove loss and gain across various geographic regions. These critical ecosystems are intimately tied to local livelihoods on one hand, and regional, national, and international economic and conservation priorities on the other. Future trends in mangrove degradation or recovery will ultimately depend on the design, implementation, and efficacy of policies that balance short-term priorities against long-term social and ecological goals of mangrove protection, and on how such goals fully and fairly engage the communities that depend the most on these important resources for their livelihoods and survival.

Acknowledgments The authors would like to thank the two anonymous reviewers who provided helpful comments on this manuscript. Uchida acknowledges that this material is based upon work supported by the National Science Foundation under Grant Number #1518471.

References

Aksornkoae S, Sugunnasil W, Sathirathai S (2004) Analytical background of the case studies and research sites: Ecological, historical and social perspectives. In: Barbier EB, Sathirathai S (eds) Shrimp farming and mangrove loss in Thailand. Edward Elgar, London, pp 73–95

Badola R, Hussain SA (2005) Valuing ecosystem functions: An empirical study on the storm protection function of Bhitarkanika mangrove ecosystem, India. Environmental Conservation 32(1), 85–92.

Bann C (1998) The Economic Value of Mangroves: A Manual for Researchers. Economy and Environment Program for Southeast Asia, Singapore

Bann C (1999) A contingent valuation of the mangroves of Benut, Johor State. Johor State Forestry Department. http://www.soc.uoc.gr/calendar/2000EAERE/papers/PDF/A6-Bann.pdf. Accessed 4 Apr 2016

Barbier EB (2006) Natural barriers to natural disasters: replanting mangroves after the tsunami. Front Ecol Environ 4:124–131
Barbier EB (2007a) Natural capital and labor allocation mangrove-dependent households in Thailand. The Journal of Environment & Development 16:398–431. https://doi.org/10.1177/1070496507307458
Barbier EB (2007b) Valuing ecosystem services as productive inputs. Econ Policy 22:177–229
Barbier EB (2010) Poverty, development, and environment. Environ Dev Econ 15(Special Issue 06):635–660. https://doi.org/10.1017/S1355770X1000032X
Barrett CB, Travis AJ, Dasgupta P (2011) On biodiversity conservation and poverty traps. Proc Natl Acad Sci 108(34):13907–13912
Binh TNKD, Vromant N, Nguyen TH, Hens L, Boon EK (2005) Land cover changes between 1968 and 2003 in Cai Nuoc, Ca Mau Peninsula, Vietnam. Environ Dev Sustain 7:519–536
Boyd CE, Clay JW (1998) Shrimp aquaculture and the environment. Sci Am 278(6):58–65
Bravo M (2007) Análisis de los *acuerdos de uso sustentable* otorgados a usuarios tradicionales para la protección y manejo del ecosistema manglar: Estudios de caso de las Concheras 18 de Octubre (Provincia Esmeraldas), Cangrejeros Seis de Julio (Provincia Guayas) y Mariscadores Costa Rica (Provincia el Oro). In: Programa de manejo de recursos costeros, Procedimiento CCI No. PMRC-035-2005, Actualización de los planes de manejo de las concesiones del manglar. Guayaquil, pp 1–50
Bravo M (2010) Interpretación del estudio multitemporal (clirsen 1969–2006) de las coberturas de manglar, camaroneras y áreas salinas en la franja costera del Ecuador Continental. Master's Thesis, University of Guayaquil
Briggs M (2006a) Cultured aquatic species information programme, *Penaeus vannamei*. Food and Agriculture Organization, Rome. www.fao.org/fishery/culturedspecies/Litopenaeus_vannamei/en. Accessed 29 Sept 2013
Briggs M (2006b) National aquaculture sector overview, Ecuador. Food and Agriculture Organization, Rome. www.fao.org/fishery/countrysector/naso_ecuador/en. Accessed 29 Sept 2013
Chen L, Wang W, Zhang Y, Lin G (2009) Recent progresses in mangrove conservation, restoration and research in China. J Plant Ecol 2:45–54
Chen B, Song Y, Zan Q, Tam NFY, Li Y, Yue Y, Tian L, Yu S (2012) Dynamics of the main landscape types at Shenzhen Bay during past three decades. Acta Scientiarum Naturalium Universitatis Sunyatseni 51:86–92
Dahdouh-Guebas F, Mathenge C, Kairo JG, Koedam N (2000) Utilization of mangrove wood products around Mida Creek (Kenya) amongst subsistence and commercial users. Econ Bot 54:513–527. https://doi.org/10.1007/BF02866549
Dahdouh-Guebas F, Zetterström T, Rönnbäck P, Troell M, Wickramasinghe A, Koedam N (2002) Recent changes in land-use in the Pambala-Chilaw Lagoon Complex (Sri Lanka) investigated using remote sensing and GIS: Conservation of mangroves versus development of shrimp farming. Environ Dev Sustain 4:185–200
Dahdouh-Guebas F, Jayatissa LP, Di Nitto D, Bosire JO, Lo Seen D, Koedam N (2005) How effective were mangroves as a defence against the recent tsunami? Curr Biol 15:R443–R447
Danielsen F, Sørensen MK, Olwig MF, Selvam V, Parish F, Burgess ND, Hiraishi T, Karunagaran VM, Rasmussen MS, Hansen LB (2005) The Asian tsunami: A protective role for coastal vegetation. Science 310:643
Das S, Crépin AS (2013) Mangroves can provide protection against wind damage during storms. Estuar Coast Shelf Sci 134(December):98–107. https://doi.org/10.1016/j.ecss.2013.09.021
Dewalt BR, Vergne P, Hardin M (1996) Shrimp aquaculture development and the environment: People, mangroves and fisheries on the Gulf of Fonseca, Honduras. World Dev 26:1193–1208
Ecuadorian National Aquaculture Chamber (2016) Shrimp production report 2016 (online). Consulted on March 13, 2017. Available at http://www.cnaecuador.com/images/comercio_exterior/estadisticas_camaron/2016
Ewel K, Twilley R, Ong J (1998) Different kinds of mangrove forests provide different goods and services. Glob Ecol Biogeogr Lett 7:83–94

Feka NZ, Chuyong GB, Ajonina GN (2009) Sustainable utilization of mangroves using improved fish-smoking systems: A management perspective from the Douala-Edea wildlife reserve, Cameroon. Tropical Conservation Science 2:450–468

FAO (2016) The State of World Fisheries and Aquaculture 2016. Contributing to food security and nutrition for all. Rome, pp 200

Food and Drug Administration (2013) 2013 FDA Seafood List. www.accessdata.fda.gov/scripts/SEARCH_SEAFOOD/index.cfm?other=complete. Accessed 29 Sept 2013

France-Lanord C, Derry LA (1997) Organic carbon burial forcing of the carbon cycle from Himalyan erosion. Nature 390:65–67

Geist HJ, Lambin EF (2002) Proximate causes and underlying driving forces of tropical deforestation. Bioscience 552:143–150

Giri C, Ochieng E, Tieszen LL, Zhu Z, Singh A, Loveland T, Masek J, Duke N (2011) Status and distribution of mangrove forests of the world using earth observation satellite data. Glob Ecol Biogeogr 20:154–159

Gordon C, Tweneboah E, Mensah AM, Ayivor JS (2009) The application of the ecosystem approach to mangrove management: lessons for Ghana. Nature & Faune 24:30–41

Hamilton SE (2011) The impact of shrimp farming on mangrove ecosystems and local livelihoods along the Pacific Coast of Ecuador. Dissertation, University of Southern Mississippi

Hamilton SE (2012) The Commodification of Ecuador's Estuary Environments from 1970–2006. Congress of the Latin American Studies Association, San Francisco, California

Hamilton SE, Collins S (2013) Livelihood responses to mangrove deforestation in the Northern provinces of Ecuador. Bosque 34(2):143–153

He Q, Bertness MD, Bruno JF, Li B, Chen G, Coverdale TC, Altieri AH, Bai J, Sun T, Pennings SC, Liu J, Ehrlich PR, Cui B (2014) Economic development and coastal ecosystem change in China. Sci Rep 4:5995

Hussain SA, Badola R (2008) Valuing mangrove ecosystem services: Linking nutrient retention function of mangrove forests to enhanced agroecosystem production. Wetl Ecol Manag 16:441–450. https://doi.org/10.1007/s11273-008-9080-z

Hussain SA, Badola R (2010) Valuing mangrove benefits: Contribution of mangrove forests to local livelihoods in Bhitarkanika Conservation Area, East Coast of India. Wetl Ecol Manag 18:321–331. https://doi.org/10.1007/s11273-009-9173-3

Islam MS, Ikejima K (2009) Gear type, species composition and economic value of fisheries in the mangroves of Pak Phanang, Thailand. Wetl Ecol Manag 18:27–36. https://doi.org/10.1007/s11273-009-9145-7

Kathiresan K (2012) Importance of mangrove ecosystem. International Journal of Marine Science 2:70–89. https://doi.org/10.5376/ijms.2012.02.0010

Kathiresan K, Rajendran N (2005) Coastal mangrove forests mitigated tsunami. Estuar Coast Shelf Sci 65:601–606

Kerr AM, Baird AH (2007) Natural barriers to natural disasters. Bioscience 57:102–103

Kling CL, Phaneuf DJ, Zhao J (2012) From Exxon to BP: Has some number become better than no number? J Econ Perspect 26:3–26

Kuenzer C, Tuan VQ (2013) Assessing the ecosystem services value of Can Gio Mangrove biosphere reserve: combining earth-observation- and household-survey-based analyses. Appl Geogr 45:167–184

Lai S, Loke LHL, Hilton MJ, Bouma TJ, Todd PA (2015) The effects of urbanisation on coastal habitats and the potential for ecological engineering: A Singapore case study. Ocean Coast Manag 103:78–85

Liao B-W, Zhang Q-M (2014) Area, distribution and species composition of mangroves in China. Wetland Science 12:435–440. (in Chinese with English Abstract)

Lightner DV (2011) Virus diseases of farmed shrimp in the Western Hemisphere (the Americas): A review. J Invertebr Pathol 106:110–130

McNally CG, Uchida E, Gold AJ (2011) The effect of a protected area on the tradeoffs between short-run and long-run benefits from mangrove ecosystems. Proc Natl Acad Sci 108(34):13945–13950

Meyfroidt P (2015) Approaches and terminology for causal analysis in land systems science. J Land Use Sci. https://doi.org/10.1080/1747423X.2015.1117530
Mialhe F, Gunnell Y, Mering C (2013) The impacts of shrimp farming on land use, employment and migration in Tumbes, northern Peru. Ocean Coast Manag 73:1–12
Peng Y, Zheng M, Zheng Z, Wu G, Chen Y, Xu H, Tian G, Peng S, Chen G, Lee SY (2016) Virtual increase or latent loss? A reassessment of mangrove populations and their conservation in Guangdong, southern China. Mar Pollut Bull 109(2):691–699
Primavera J (2006) Overcoming the impacts of aquaculture on the coastal zone. Ocean Coast Manag 49(9):531–545
Richards DR, Friess DA (2016) Rates and drivers of mangrove deforestation in Southeast Asia, 2000–2012. Proc Natl Acad Sci U S A 113:344–349
Rönnbäck P (1999) The ecological basis for economic value of seafood production supported by mangrove ecosystems. Ecol Econ 29:235–252
Rosero Moya K, Santillan Salas C (2011) Ministerio del ambiente entrega acuerdo de uso sustentable y custodia de manglar. Boletin No. 498 09-06-2011. http://www.ambiente.gob.ec/?q=node/1952. Accessed: 8/14/2011
Thompson M (2014) White shrimp (*Litopenaeus vannamei)* on coastal ponds of Ecuador. Monterey Bay Aquarium Seafood Watch, Monterey
Thu PM, Populus J (2007) Status and changes of mangrove forest in Mekong Delta: Case study in Tra Vinh, Vietnam. Estuar Coast Shelf Sci 71:98–109
UNEP (United Nations Environment Programme) (2005) After the tsunami: rapid environmental assessment report, February 22, 2005. United Nations Environmental Programme, Nairobi
UNEP (United Nations Environment Programme) (2011) Economic analysis of mangrove forests: a case study in Gazi Bay, Kenya. UNEP, New York
Valiela I, Bowen JL, York JK (2001) Mangrove forests: One of the world's threatened major tropical environments. Bioscience 51:807–815
Vermaat JE, Thampanya U (2006) Mangroves mitigate tsunami damage: A further response. Estuar Coast Shelf Sci 69:1–3
Vosti S, Reardon T (1997) Sustainability, growth and poverty alleviation: A policy and agroecological perspective. Johns Hopkins University Press, Baltimore
Walters BB (2003) People and mangroves in the Philippines: Fifty years of coastal environmental change. Environ Conserv 30:293–303
Wang W, Wang M (2007) The mangroves of China. Science Press, Beijing
World Bank (2008) Poverty and the environment: understanding linkages at the household level. World Bank, Washington DC
World Commission on Environment and Development (WCED) (1987) Our common future. Oxford University Press, New York

Chapter 10
Mangrove Forest Restoration and Rehabilitation

Jorge López-Portillo, Roy R. Lewis III, Peter Saenger, André Rovai, Nico Koedam, Farid Dahdouh-Guebas, Claudia Agraz-Hernández, and Victor H. Rivera-Monroy

10.1 Introduction

The historical loss of mangrove wetland distribution is on a worldwide scale approximately 35–50% of the current area with a variable loss rate of 1–3% per year (i.e., ~150,000 ha/y) (Valiela et al. 2001; Wilkie and Fortuna 2003; Giri et al. 2011). The most recent global coverage estimate for 2014 is 163,925 km^2 down from 173,067 km^2 in 2000, providing an annual loss during that period of 0.4% (Hamilton and Casey 2016). The ongoing wetland loss has triggered an increasing interest in implementing a better management of existing healthy mangrove areas (Ong and Gong 2013). Such management includes the return of key ecological functions in

J. López-Portillo (✉)
Instituto de Ecología, A.C. (INECOL), Red de Ecología Funcional,
Carretera antigua a Coatepec 351, Xalapa, Veracruz 91070, Mexico
e-mail: jorge.lopez.portillo@inecol.mx

R.R. LewisIII
Lewis Environmental Services, Inc., P.O. Box 5430, Salt Springs, FL 32134-5430, USA

P. Saenger
Centre for Coastal Management, Southern Cross University, Lismore, NSW 2480, Australia

A. Rovai
Departamento de Ecologia e Zoologia, Universidade Federal de Santa Catarina,
Florianópolis, SC 88040-900, Brazil

Department of Oceanography and Coastal Sciences, College of the Coast and Environment,
Louisiana State University, Baton Rouge, LA 70803, USA

N. Koedam
Plant Biology and Nature Management (APNA), Vrije Universiteit Brussel,
VUB 1, B-1050 Brussels, Belgium

F. Dahdouh-Guebas
Laboratoire d'Écologie des Systèmes et Gestion des Ressources, Département de Biologie
des Organismes, Faculté des Sciences, Université Libre de Bruxelles – ULB,
Campus de la Plaine, B-1050 Bruxelles, Belgium

V.H. Rivera-Monroy et al. (eds.), *Mangrove Ecosystems: A Global Biogeographic Perspective*, https://doi.org/10.1007/978-3-319-62206-4_10

coastal areas where wetland mortality is widespread and where these valuable ecosystems and their goods and services are beginning to show deterioration because of increasing human activities (Field 1999a, b; Ellison 2000; Lewis et al. 2005, 2009).

Ecosystem restoration is defined as the return from a deteriorated condition to a state similar to a preserved reference site that represents the structural and functional variability within habitats before a devastating natural or human-induced disturbance (Kaly and Jones 1998). For mangrove wetlands, Lewis (1990) defined restoration as "return from a disturbed or totally altered condition by some action of man" underscoring the more active alternative, as opposed to passive restoration through natural secondary succession; the speed of which depends on the ecosystem resilience capacity, past land-use history, and health of the surrounding landscape matrix (Holl and Aide 2011). In contrast, rehabilitation is not defined as a return to previously existing conditions, a view characterized as "the myth of carbon copy" (Hilderbrand et al. 2005), but to a defined "better" or improved state (Lewis 1990). It has been proposed that rehabilitation is aligned with restoration as both management strategies generally take a culturally acceptable original (preanthropogenic era, sensu Crutzen and Stoermer 2000) or historic ecosystem/landscape as a reference for planned initiatives to halt degradation and initiate more sustainable ecosystem trajectories (Aronson et al. 2007). Indeed, there is a recent consensus based on the historical usage of the terms "restoration" and "rehabilitation" in mangrove wetland management programs, where "the use of the term 'rehabilitation' would reduce confusion as it encompasses the widest range of remedial actions for mangrove degradation" (Dale et al. 2014). However, it is also acknowledged that the term "restoration" has a strong ascendancy in the published literature and therefore we maintain this term in our discussion of the state of mangrove restoration/rehabilitation (R/R) approaches (Primavera et al. 2012; Lewis and Brown 2014).

Similarly to the usage and definitions of "restoration" and "rehabilitation", there is also some confusion regarding the meaning of other related terms such as "forestation", "reafforestation", "replanting", and "plantation". For example, the initial planting of mangrove propagules or seedlings is often called "replanting" where it implies that a first planting may have failed and a second one is taking place. Although this might be a minor detail in describing the type of action and timing to initiate a restoration program, such critical steps must be clearly documented when assessing the success or failure of either a mangrove initial planting effort or repeated plantings in a location or set of locations. Thus, clarity on the type of action can help identify problems with site selection that could, as a consequence,

C. Agraz-Hernández
Instituto de Ecología, Pesquerías y Oceanografía del Golfo de México (EPOMEX).
Universidad Autónoma de Campeche – UAC,
Av. Héroe de Nacozari #480. Campus 6 de Investigaciones, 24029 San Francisco de Campeche, Campeche, Mexico

V.H. Rivera-Monroy
Department of Oceanography and Coastal Sciences, College of the Coast and Environment, Louisiana State University, Baton Rouge, LA 70803, USA

potentially increase the costs of restoration programs. Well-defined actions become critical indicators of the applicability of any method of restoration, particularly when planting has been proposed as an alternative after natural seedling recruitment during secondary succession is insufficient to promote mangrove regeneration (Lewis et al. 2005, 2009; Lewis and Brown 2014). Therefore, we encourage the provision of detailed descriptions and implementation of management strategies to be as specific as possible within the context of the definition of both restoration and rehabilitation, especially the description of the actions selected to remedy or improve a specific environmental condition (e.g., geomorphic setting, such as deltaic vs. karstic) in a mangrove wetland.

In this chapter, we explore the main motivations to implement mangrove restoration projects and evaluate R/R projects across latitudinal gradients in the AEP (West Africa and America; Fig. 10.1a–c) and the Indo-West Pacific (IWP: East Africa, Asia, and Australasia; Figs. 10.1d and 10.2a, b) regions. We also identify research gaps and delineate a strategy to improve the implementation of R/R projects using lessons learned in different environmental and social contexts through case studies. Our synthesis contributes to recent analyses aimed at developing best practices when implementing urgently needed science-based mangrove restoration projects.

10.2 Original Motivations and Plans for Implementation

Mangrove resource management should rely on R/R approaches to enhance the full potential of sites, either with complete or cryptic impairment (sensu Dahdouh-Guebas et al. 2005a, 2005b), for the conservation and community-based participation in projects. One of the main attributes of these projects is relying on the knowledge of key ecosystem properties and on documented successes or failures from other R/R endeavors (Primavera and Esteban 2008; Zaldívar-Jiménez et al. 2010). Following on the wealth of data and information, several institutions have developed technical reports with guidelines for restoration programs in mangrove wetlands, which have improved the communication of technical details to evaluate, at least in the short term, project success and/or failures (e.g., Pulver 1976; Field 1995; Saenger 2002; Agraz Hernández et al. 2007; Primavera et al. 2012, 2014; Lewis and Brown 2014).

As a result of the increasing recognition of valuable direct (e.g., wood, carbon, shoreline protection) and indirect (e.g., fisheries maintenance, water quality, carbon storage/sequestration) ecosystem services provided by mangroves (see Chaps. 5, 8, and 9), we identified several R/R projects throughout tropical and subtropical regions. A web search using the ISI Web of Knowledge platform for publications from 1995 through 2015 with the keywords “mangrove”, “restoration”, “rehabilitation”, “reforestation”, “forestation”, and “recovery” in the title produced 136 references with 2273 citations. From this search, supplemented with results from the Google search engine, we selected references that included specific project location data. This combined publication search produced 65 references that provided infor-

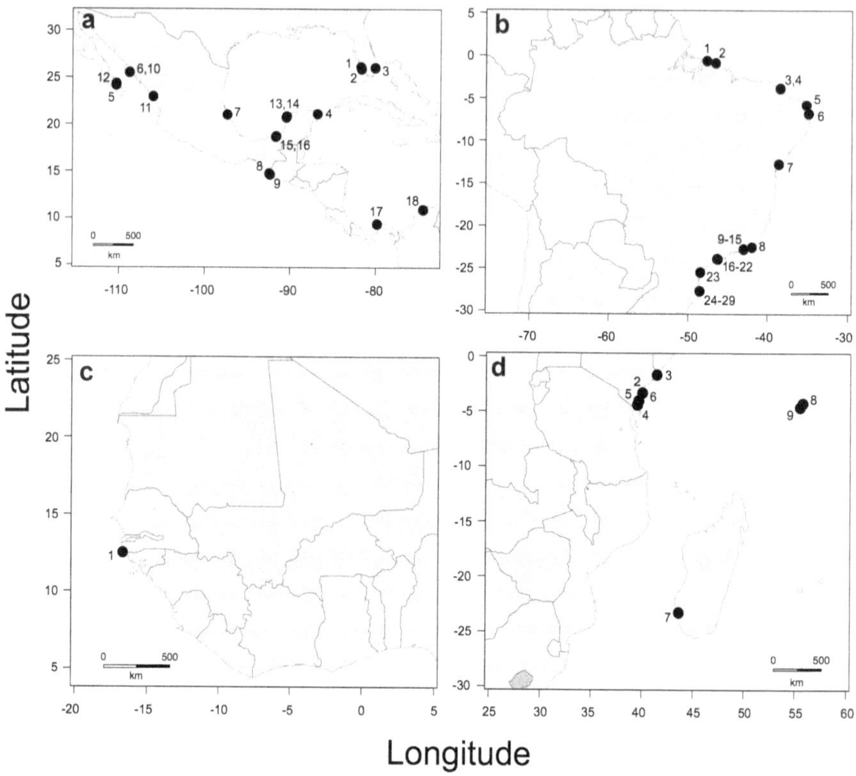

Fig. 10.1 Mangrove R/R projects implemented in the AEP Region (**a–c**) and the Africa sector of the IWP (**d**). Numbers indicating location in each panel are included Tables 10.2 and 10.3. See text for explanation on site identification and selection

mation for our analysis (Table 10.1) and included 90 sites around the world where R/R actions have been implemented (Figs. 10.1 and 10.2). We included each site in a Google Earth KMZ file (available upon request). Given the volume of information in the "gray" literature and other publications not included in the search engines, we acknowledge that this search might not be exhaustive and encourage readers to consult published reports in other coastal regions around the world.

10.2.1 Sources of Mangrove Wetland Damage

The source of damage to mangrove wetlands might be of natural origin (e.g., siltation, erosion, the direct and indirect effect of tropical storms or tsunamis) or induced by anthropogenic activities (e.g., pollution, land use policies, overharvesting, aquaculture, or altered hydrology and hydroperiod; see also Chap. 9). Thus, to

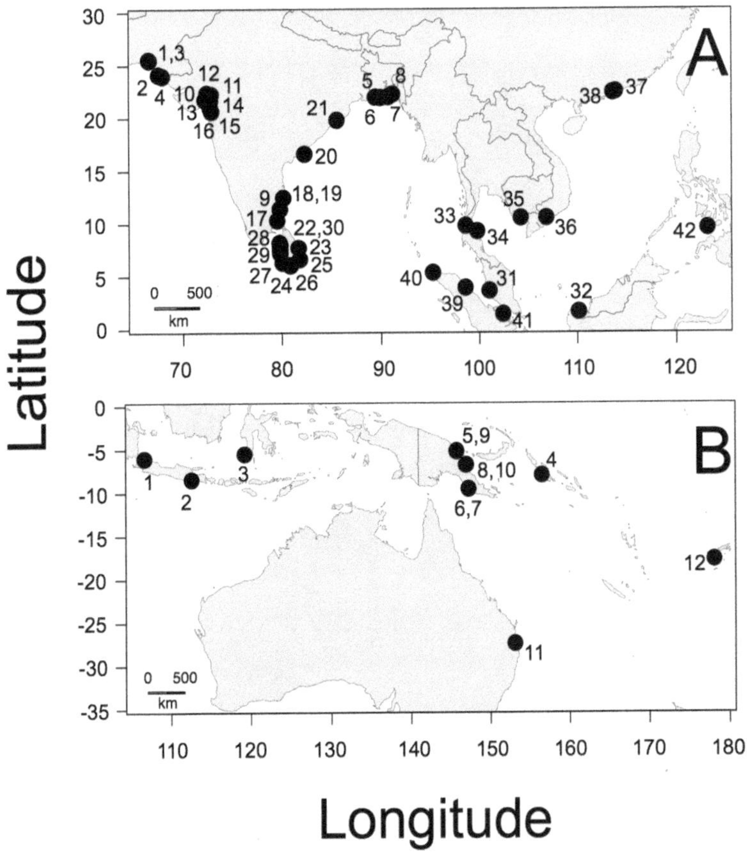

Fig. 10.2 Mangrove R/R projects implemented in the Asia and Australasia sectors of the IWP (**a**, **b**). Numbers indicating location in each panel are listed Tables 10.2 and 10.3 for further information about the sites. See text for explanation on site identification and selection

be effective and efficient, each mangrove wetland project requires a specific R/R approach (i.e., restoration, rehabilitation, or afforestation). There are many causes for mangrove impairment, and because they are frequently mixed and complex, we only assess them according to their frequency in 14 general categories (Table 10.1; percentage [%] of site reports): exposed shores [25%]; impaired hydrological regime [19%]; deforestation [19%]; siltation [11%]; shrimp or fish aquaculture [11%]; conversion to other soil uses, such as palm oil [8%]; blocking of inlets after strong storms such a cyclones/typhoons/hurricanes and tsunamis [7%]; exposure to dredge spoils [5%]; mosquito-preventing dikes [2%]; pollution [2%]; water logging [1%]; soil collapse [1%]; drought [1%]). The quantitative evaluation of the impact by each cause in impairing mangrove wetlands and associated variability in structural and functional properties requires further work at a global scale.

Table 10.1 Mangrove restoration or rehabilitation projects and associated amelioration procedure across Biogeographic regions

Biogeographic region	Project site/country	Cause of impairment	Amelioration procedure	References
Atlantic-East-Pacific (AEP)	Windstar, Florida, USA	Dredge spoil blocked normal tidal flushing	Hydrologic restoration by restoring elevation, Forestation	Stephen (1984), McKee and Faulkner (2000), Proffitt and Devlin (2005)
	West Lake, Florida, USA	Filled wetlands	Excavation of historical fill in mangroves, hydrologic restoration, no planting of mangroves	Lewis (2005), Lewis and Gilmore (2007)
	Florida East Coast, USA	Diked wetlands for mosquito control	Dredged deposits removed, diked mosquito control impoundments breached, very little forestation, natural recovery predominantly	Lewis et al. (1985), Brockmeyer et al. (1997), Rey et al. (2012)
	Rookery Bay, Florida, USA	Incomplete tidal flushing, elevated salinity, waterlogging	Restoring original elevation, excavation of water outlets; Forestation	McKee and Faulkner (2000)
	Bahía de Navachiste, Sinaloa, Mexico	Accumulation of dredging spoils	Channel digging on dredge material terraces and afforestation of nursery plants	Benítez-Pardo et al. (2015)
	Laguna Balandra, Baja California, Mexico	Deforestation	Forestation, natural regeneration	Vovides et al. (2011)
	Laguna de Enfermería, Baja California, Mexico	Block of feeder channel by road	Hydrologic restoration, natural regeneration	Vovides et al. (2011)
	El Mogote, Baja California, Mexico	Hurricane-caused blocking of outlet with a sand dune	Hydrologic restoration	Bashan et al. (2013)
	Huizache-Caimanero, Sinaloa, Mexico	Accumulation of dredging spoils	Forestation with nursery plants	Benítez Pardo et al. (2015)
	Laguna Nichupté, Quintana Roo, Mexico	Hurricane damage, probably including blocking water outlets	Afforestation, Hydrologic restoration	Adame et al. (2014)
	Tampamachoco, Veracruz, Mexico	Water flow obstruction by power line embankments	Hydrologic restoration	López-Portillo et al. (2014)

	Yucatán Peninsula, Mexico	Water flow obstruction by closure of inlets after a strong hurricane	Hydraulic restoration and planting from nursery	Zaldívar-Jiménez et al. (2010)
	Celestún, Yucatán, Mexico	Water flow obstruction by closure of inlets and road construction	Hydraulic restoration and planting from nursery	Miyagi (2013)
	Términos Lagoon, Campeche, Mexico	Water flow obstruction by closure of inlets after a strong hurricane	Hydraulic restoration and planting from nursery	Agraz Hernández et al. (2010)
	Jaina, Petenes BR, Campeche, Mexico.	Water flow obstruction by closure of inlets and road construction	Hydraulic restoration and planting with propagules	Agraz Hernández et al. (2015)
	Isla Arena, Campeche, Mexico	Water flow obstruction by closure of inlets after a strong hurricane	Hydraulic restoration and planting from nursery	Tsuruda (2013)
	Laguna de Cabildo, Chiapas, Mexico	Channel excavation and obstruction of water by bunds	Direct seeding of propagules and nursery plants	Reyes and Tovilla (2002)
	Laguna de Pozuelos, Chiapas, Mexico	Channel excavation and obstruction of water by bunds	Direct seeding of *R. mangle* propagules and nursery plants	Reyes and Tovilla (2002)
	Barra del Río Cahoacán, Mexico	Siltation from upland erosion	Direct sowing of collected propagules and nursery plants	Tovilla et al. (2004)
	Punta Galeta, Panama	Deforestation (?), invasion by *Saccharum spontaneous*	Forestation	Outterson (2014)
	Ciénaga Grande de Santa Marta, Colombia	Interruption of major water flows by road construction	Hydraulic restoration, forestation	Rivera-Monroy et al. (2006), Twilley et al. (1998), Ortiz-Ruiz (2004)
	Parque Nacional Corales del Rosario, Colombia	Unspecified	Seeding and forestation with *R. mangle*	Bohórquez-Rueda and Prada-Triana (1988)-, for other experiments,in Colombia see Álvarez León (2003)
	Two sites, Puerto Rico	Hurricane effects	Natural regeneration by recolonization of *L. racemosa*	Wadsworth (1959)

(continued)

Table 10.1 (Continued)

Biogeographic region	Project site/country	Cause of impairment	Amelioration procedure	References
	Martin Peña Channel, San Juan, Puerto Rico	Urban detritus and siltation, deforestation	Urban renewal, removal of debris, no planting, just natural recolonization	Cintrón-Molero (1992)
	Ajuruteua Peninsula, Bragança, Brazil	Disturbance of hydrological regime by road construction	Natural regeneration by recolonization of *A. germinans*	Vogt et al. (2014)
	Río Jaguaribe, Rio Grande do Norte, Brazil	Deforestation	Seeding and forestation with *R. mangle*	Ferreira et al. (2015)
	Río las Ostras, Río de Janeiro, Brazil	Deforestation	Forestation and natural regeneration	Bernini et al. (2014)
	Baixada Santista, Estuário de Santos, Río Cubatão, Brazil	Deforestation, pollution, dredging	Planting seeds and propagules	Menezes et al. (2005)
Indo-West-Pacific (IWP)	Shenzhen Bay, China	Urban encroachment and pollution, increase in siltation rates	Restoration plan including integration of rustic shrimp ponds (*gei wei*) and mangrove species communities	Ren et al. (2011)
	Qi'ao Reserve, China	Deforestation	Seeding	Chen et al. (2013)
	Barisal, Chitta Gong, Patuakhali, Noakhali, Bangladesh	Newly accreting mudflats	Afforestation	Saenger and Siddiqi (1993)
	Pichavaram, Tamil Nadu, India	Extensive deforestation, soil collapse	Hydraulic rehabilitation by excavating main and secondary channels, forestation	Selvam et al. (2003)
	Nellore, Andhra Pradesh, India	Exposed shores after tsunami or cyclones	Forestation with upland dune plants and some mangroves but no distinction is made between the two at all 18 sites	Mukherjee et al. (2015) (Eighteen [18] sites are reported many apparently without mangrove restoration activities)

	Prakasam, Andhra Pradesh, India	Exposed shores after tsunami or cyclones	Forestation	Mukherjee et al. (2015)
	Guntur, Andhra Pradesh, India	Exposed shores after tsunami or cyclones	Forestation	Mukherjee et al. (2015)
	Krishna, Andhra Pradesh, India	Exposed shores after tsunami or cyclones	Forestation	Mukherjee et al. (2015)
	West Godavari, Andhra Pradesh, India	Exposed shores after tsunami or cyclones	Forestation	Mukherjee et al. (2015)
	East Godavari, Andhra Pradesh, India	Exposed shores after tsunami or cyclones	Forestation	Mukherjee et al. (2015)
	Visakhapatnam, Andhra Pradesh, India	Exposed shores after tsunami or cyclones	Forestation	Mukherjee et al. (2015)
	Pulicat Lake, Andhra Pradesh/Tamil Nadu, India		Planting from nursery	Trump and Gattenlöhner (2015)
	Kannur, Kerala, India	Exposed shores after tsunami or cyclones	Forestation	Mukherjee et al. (2015)
	Kasargod, Kerala, India	Exposed shores after tsunami or cyclones	Forestation	Mukherjee et al. (2015)
	Thiruvallur, Tamil Nadu, India	Exposed shores after tsunami or cyclones	Forestation	Mukherjee et al. (2015)
	Kanchipuram, Tamil Nadu, India	Exposed shores after tsunami or cyclones	Forestation	Mukherjee et al. (2015)
	Nagapattinam, Tamil Nadu, India	Exposed shores after tsunami or cyclones	Forestation	Mukherjee et al. (2015)
	Thiruvarur, Tamil Nadu, India	Exposed shores after tsunami or cyclones	Forestation	Mukherjee et al. (2015)
	Thanjavur, Tamil Nadu, India	Exposed shores after tsunami or cyclones	Forestation	Mukherjee et al. (2015)
	Pudukottai, Tamil Nadu, India	Exposed shores after tsunami or cyclones	Forestation	Mukherjee et al. (2015)

(continued)

Table 10.1 (Continued)

Biogeographic region	Project site/country	Cause of impairment	Amelioration procedure	References
	Ramanathapuram, Tamil Nadu, India	Exposed shores after tsunami or cyclones	Forestation	Mukherjee et al. (2015)
	Tutic, Tamil Nadu, India	Exposed shores after tsunami or cyclones	Forestation	Mukherjee et al. (2015)
	Tirunelveli/Kanyakumari, Tamil Nadu, India	Exposed shores after tsunami or cyclones	Forestation	Mukherjee et al. (2015)
	Trapaing Sangke, Cambodia		Planting from nursery	Trump and Gattenlöhner (2015)
	Kien Giang, Cambodia	Exposure to wave action and erosion	Construction of *Melaleuca* fence and mangrove planting	Cuong et al. (2015)
	Andaman Coast, Thailand		Hydraulic restoration and planting from nursery	Trump and Gattenlöhner, (2015)
	Klong Gnao, Thailand	Wood harvesting, tin mining, aquaculture	Planting of propagules and plants	Macintosh et al. (2002)
	Philippines	Fish/shrimp culture ponds	Multispecies planting	Primavera and Esteban (2008), Primavera et al. (2011, 2012, 2014), Salmo III et al. (2013), Samson and Rollon (2008), Stevenson et al. (1999), Walters (1997)
	Thong Nian, Thailand	Shrimp culture ponds	breaching of banks to rehabilitate water flow, planting of propagules and plants	Matsui et al. (2010)
	Bolgoda Lake, Sri Lanka		Planting from nursery	Trump and Gattenlöhner (2015)
	Madampe Lake, Sri Lanka		Planting from nursery	Trump and Gattenlöhner (2015)
	Pambala-Chilaw lagoon, Sri Lanka	Shrimp aquaculture pond	Remote sensing update, forestation	Dahdouh-Guebas et al. (2002)

	North Sumatra, Aceh Besar, Lhok Nga, Indonesia	Deforestation and erosion due to tsunami	Planting?	Alexandris et al. (2013)
	Banda Aceh, North Sumatra, Indonesia	Deforestation and erosion due to tsunami	Planting?	Alexandris et al. (2013)
	Tanakeke Island, Sulawesi, Indonesia	Shrimp aquaculture pond	Hydrologic restoration of an abandoned shrimp aquaculture pond area followed by limited planting	Brown and Massa (2013)
	Jaring Halus, NE Langkat Wildlife Sanctuary, North Sumatera Province, Sumatra	Shrimp aquaculture pond	Hydrologic restoration in 10 ha of an abandoned shrimp aquaculture pond area followed by limited planting	Brown and Massa (2013)
	Tiwoho Village, North Sulawesi	Shrimp aquaculture pond	Hydrologic restoration in 25 ha of an abandoned shrimp aquaculture pond area followed by limited planting	Brown and Massa (2013)
	Sungai Haji Dorani, Malaysia	Exposed shoreline	Breakwater construction to induce natural establishment of *A. marina*	Tamin et al. (2001), Stanley and Lewis (2009)
	Sungai Haji Dorani, Malaysia	Exposed shoreline	Break water, transplant, natural regeneration	Kamali and Hashim (2011), Stanley and Lewis (2009)
	Sabah, northern Borneo, Malaysia	Areas encroached by oil palms (7 sites) or shrimp ponds (2 sites); five sites are deforested	Forestation in 14 project sites located in five mangrove forest reserves. Additional hydrologic restoration in areas encroached by oil palms (7 sites) or shrimp ponds (2 sites)	Tangah et al. (2015)
	Cabrousse, Senegal	Deforestation, blockage of water flows, droughts	Seeding (propagule planting)	Alexandris et al. (2013)
	Kiunga Marine National Reserve, Kenya	Deforestation, silting	Natural regeneration	Kairo et al. (2001)
	Mida Creek, Kenya	Deforestation, silting	Natural regeneration	Kairo (Kairo et al. 2001)

(continued)

Table 10.1 (Continued)

Biogeographic region	Project site/country	Cause of impairment	Amelioration procedure	References
	Tudor Creek, Kenya	Deforestation, silting	Natural regeneration	Bosire et al. (2014)
	Gazi Bay and Mwache Creek, Kenya	Deforestation, silting	Natural regeneration	Bosire et al. (2003, 2014)
	Gazi, Kenya	Deforestation, silting	Forestation	Kairo et al. (2001)
	Mamelo Honko, Madagascar	Deforestation	Seeding (Propagule planting)	Alexandris et al. (2013)
	Ranongga, Salomon Islands, Melanesia	Deforestation	Replanting to replace vegetation lost after an earthquake	Alexandris et al. (2013)
	Brisbane International Airport, Australia	Filling of main creek and excavation of other channels	Hydrologic restoration by channel digging and forestation	Saenger (1996)

10.2.2 Amelioration Procedures

Forestation practices (Table 10.1) using individual plants from nurseries was the main amelioration procedure ($n = 67$) followed by hydrologic rehabilitation ($n = 29$), although both actions were frequently combined ($n = 22$). Direct seeding or mature propagule planting (mainly the genus *Rhizophora*) was also a frequent action ($n = 11$). Natural regeneration was implemented in 10 sites including cases where it was coupled with transplants ($n = 1$) and forestation ($n = 2$) techniques. We assume that there was afforestation in the 17 sites (covering 43,760 ha) exposed to wave energy and described as "bio-shield" plantations in the states of Kerala, Andhra Pradesh, and Tamil Nadu in peninsular India (Mukherjee et al. 2015).

10.2.3 Spatial Scales of the Amelioration Procedures

The mangrove sites undergoing restoration or just afforestation encompassed a range of area extensions from few square meters to several thousand hectares. The most extensive afforestation sites are located in the Sundarbans, in Bangladesh and India (120,000 ha afforested by 1993, Saenger and Siddiqi 1993), United States (12,605 ha restored, Rey et al. 2012; 500 ha restored, Lewis 2005, Lewis and Gilmore 2007), and other coastal regions in Asia (e.g., Pichavaram Province: >300 ha of restored mangroves, Selvam et al. 2003) and Indonesia at Tanakeke Island (400 ha), where hydrologic restoration was also part of the R/R strategy (Brown and Massa 2013; Brown et al. 2014).

The large mangrove extension in the Sundarbans delta region is characterized by both large spatial scale impacts and management strategies, including erosion, aggradation (i.e., natural sediment accumulation), deforestation, and mangrove rehabilitation programs (Giri et al. 2007). For example, 7300 ha of mangrove wetland were lost to erosion from 1977 to 2000, whereas net aggradation was variable with gains ranging from 2900 ha (1970s) to only 590 ha (2000). Recent estimates show a total loss of 26,200 ha and total gain of 24,000 ha from 1989 through 2014 (Ghosh et al. 2015). Due to the significant new land gains as a result of high sediment deposition, natural mangrove establishment in the newly formed land was combined with active and intense community-based seeding and planting of seedlings to compensate for eroded mangroves (Saenger and Siddiqi 1993; Giri et al. 2007). In contrast to the net gain in mangrove area in this region, a large effort with propagule planting (79 million distributed throughout 7920 ha) in Cabrousse, Senegal, West Africa in 2008, showed no evidence of increase in mangrove coverage as evaluated by changes at the landscape level using remote sensing images obtained up to 2010 (Alexandris et al. 2013).

10.2.4 Mangroves and Aquaculture

Over the last three decades of human impact on mangrove wetlands, shrimp aquaculture and their associated culture ponds have probably been responsible for the greatest losses of mangrove wetland area (see Chap. 9). This activity has been actively encouraged by governments in developing countries (e.g., Brazil, Ecuador, Thailand, Indonesia, and Vietnam) interested in the high earning potential of shrimp as an export product, but also often driven by political patronage (Tobey et al. 1998; Foell et al. 1999; Dahdouh-Guebas et al. 2006; Oliveira-Filho et al. 2016, Table 10.2). A comprehensive work on the total area of mangrove loss to commercial aquaculture indicates that in the eight countries that host about 45% of total world mangrove cover, about 52% of their historic mangrove coverage is lost, including 28% to commercial aquaculture (Hamilton 2013; Hamilton and Casey 2016). Given the proliferation of shrimp farms around the world, many R/R projects have been undertaken in countries where shrimp farms were abandoned due to major disease outbreaks that decimated the industry (e.g., viral diseases) (Stevenson et al. 1999; Matsui et al. 2010; Primavera et al. 2011, 2014; Brown et al. 2014). In fact, some studies have used hydrological models to determine which dikes or artificial barriers should be removed to restore the original hydrology and induce natural mangrove reestablishment and growth (Di Nitto et al. 2013). In other locations, particularly in developed countries (e.g., the USA or Australia), R/R projects were initially used as ecological offsets related to land use and mitigation policies (Teas 1977; Snedaker and Biber 1996; Latif 1996). As an example of this strategy, Brockmeyer et al. (1997) and Rey et al. (2012) reported an accumulated 12,000 ha of successful restoration programs mainly due to reconnection and controlled flooding along the east coast of Florida.

A number of R/R projects have been undertaken to address the problem of extensive abandonment of shrimp ponds due to economic failure in several countries (e.g., Primavera and Esteban 2008; Brown et al. 2014), and as a result, there is growing number of peer-reviewed studies that provides useful insights into designing R/R projects with specific management objectives and goals based on the initial nature of the damage (e.g., Latif 1996; Saenger 1996; Das et al. 1997; Walters 1997;

Table 10.2 Aquaculture pond areas constructed in mangroves in major shrimp producing developing countries (From Tobey et al. 1998)

Country	Pond area (ha)	Number of farms
Indonesia	350,000	60,000
India	200,000	10,000
Vietnam	200,000	2000
Bangladesh	140,000	13,000
Ecuador	130,000	1200
China	127,000	6000
Thailand	70,000	16,000
Philippines	60,000	1000
Mexico	14,000	240
Honduras	12,000	55

Table 10.3 Geographic information of sites where rehabiliation of restoration projects discussed in the text. Figure 10.1 shows the location

Biogeographic region	Country/continent	Site name	Site ID[a]	Latitude	Longitude
Atlantic-East-Pacific (AEP)	Brazil	Pará	A1	−0.551398	−47.735251
		Ajuruteua	A2	−0.8056154	−46.625772
		Sapiranga, Fortaleza	A3	−3.774106	−38.448555
		Sapiranga, Fortaleza	A4	−3.774106	−38.448555
		Jaguaribe	A5	−5.753603	−35.218739
		Barra de Mamanguape, Paraíba	A6	−6.780058	−34.936283
		Baía de Todos os Santos, Bahia	A7	−12.717753	−38.612231
		Estuário do Rio das Ostras	A8	−22.506952	−41.94283
		Angra dos Reis, Rio de Janeiro	A9	−22.733875	−43.018167
		Lagoa Rodrigo de Freitas, Rio de Janeiro	A10	−22.733875	−43.018167
		Ilha do Fundão, Rio de Janeiro	A11	−22.733875	−43.018167
		Ilha do Fundão, Rio de Janeiro	A12	−22.733875	−43.018167
		Ilha do Fundão, Rio de Janeiro	A13	−22.733875	−43.018167
		Ilha do Fundão, Rio de Janeiro	A14	−22.733875	−43.018167
		Ilha do Fundão, Rio de Janeiro	A15	−22.733875	−43.018167
		Baixada Santista, São Paulo	A16	−23.880911	−46.364192
		Baixada Santista, São Paulo	A17	−23.880911	−46.364192
		Baixada Santista, São Paulo	A18	−23.880911	−46.364192
		Baixada Santista, São Paulo	A19	−23.880911	−46.364192
		Baixada Santista, São Paulo	A20	−23.880911	−46.364192
		Baixada Santista, São Paulo	A21	−23.880911	−46.364192
		Baixada Santista, Estuário de Santos	A22	−23.91419	−46.265168
		Baía de Paranaguá, Paraná	A23	−25.463458	−48.475563
		Costeira do Pirajubaé, Florianópolis	A24	−27.652969	−48.539156
		Biguaçu, Santa Catarina	A25	−27.652969	−48.539156

(continued)

Table 10.3 (continued)

Biogeographic region	Country/continent	Site name	Site ID[a]	Latitude	Longitude
		Saco Grande, Florianópolis	A26	−27.652969	−48.539156
		Ratones, Florianópolis	A27	−27.652969	−48.539156
		Itacorubi, Florianópolis	A28	−27.652969	−48.539156
		Saco da Fazenda, Itajaí	A29	−27.652969	−48.539156
	USA	Windstar	B1	26.1196972	−81.782469
		Rookery Bay	B2	25.9102556	−81.703361
		West Lake	B3	26.0384972	−80.119464
	Mexico	Laguna Nichupté	B4	21.099975	−86.793617
		Balandra	B5	24.3234868	−110.32286
		Laguna Enfermería	B6	24.2498611	−110.31276
		Tampamachoco	B7	21.0123861	−97.339694
		Laguna de Cabildo	B8	14.742925	−92.433219
		Laguna de Pozuelos	B9	14.6458278	−92.339764
		Navachiste	B10	25.4980185	−108.79743
		Huizache - Caimanero	B11	22.9458639	−106.00631
		El Mogote	B12	24.1636833	−110.3348
		Celestún	B13	20.8580167	−90.390083
		Isla Arena	B14	20.7124584	−90.44895
		Isla Aguada, Campeche	B15	18.6660821	−91.665588
		Isla Aguada	B16	18.7132933	−91.609765
	Panama	Punta Galeta	B17	9.40270219	−79.862062
	Venezuela	Ciénaga Grande de Santa Marta	B18	10.9371278	−74.541131
	Africa	Cabrousse, Senegal	C1	12.4926172	−16.685131

Indo-West-Pacific (IWP)	Africa	Mida Creek, Watamu, Kenya	D2	−3.3333415	40.0000004
		Kiunga Marine National Reserve, Kenya	D3	−1.668958	41.4066794
		Gazi, Kenya	D4	−4.4273752	39.51063
		Tudor Creek, Kenya	D5	−4.0479108	39.6535163
		Mwache Creek, Mombasa, Kenya	D6	−4.0502697	39.633712
		Mamelo Honko, Madagascar	D7	−23.262529	43.6242508
	Seychelles	Curieuse Island	D8	−4.2791955	55.7277429
		Roche Caiman Sanctuary	D9	−4.6396463	55.4689262
	Pakistan	Sonmiani, Balochistan	E1	25.4890771	66.5182225
		Sha Bandar	E2	23.9882232	67.84664
		Miani Hor	E3	25.5282117	66.4561847
		Keti Bandar	E4	24.1301277	67.4445187
	Bangladesh	Sundarban	E5	22.0026661	89.4464738
		Barguna Sadar	E6	21.9660641	89.9607137
		Char Fasson	E7	22.0397962	90.7422427
		Hatiya	E8	22.2806648	91.1926791
	India	Pichavaram	E9	11.4208443	79.796165
		Ahmedabad	E10	22.3748974	72.4439145
		Bhavnagar	E11	21.7631481	72.2441373
		Anand	E12	22.2613255	72.8892584
		Bharuch	E13	21.6475722	72.8008261
		Surat	E14	21.0542686	72.7628816
		Valsad	E15	20.6380196	72.9119655
		Navsari	E16	20.9294833	72.79864
		Muthupet	E17	10.3408316	79.5378549
		Chidabaram	E18	11.390341	79.8137706
		Krishna	E19	12.4698823	80.1501882

(continued)

Table 10.3 (continued)

Biogeographic region	Country/continent	Site name	Site ID[a]	Latitude	Longitude
		Godavari	E20	16.6170396	82.2825575
		Chilika	E21	19.8101156	85.5365084
	Sri Lanka	Pambala-Chilaw	E22	7.50002222	79.8167167
		Batticaloa	E23	7.73376131	81.6668314
		Kumana National Park	E24	6.64617297	81.7750119
		Rekawa	E25	6.05588762	80.852921
		Madu Ganga	E26	6.31203664	80.0667239
		Negombo	E27	7.19265861	79.8300512
		Arachchikattuwa	E28	7.66656578	79.8014598
		Puttalam	E29	8.00588689	79.832815
		Kalpitiya	E30	8.21745625	79.7638786
	Malaysia	Sungai Haji Dorani	E31	3.65576667	101.009853
	Thailand	Matang	E32	1.66743216	110.121648
		Klong Ngao	E33	9.83335833	98.5833611
		Thong Nian	E34	9.30786852	99.7815087
	Vietnam	Kien Giang	E35	10.5688691	104.230806
		Xuan Thuy National Park	E36	10.576852	106.846581
	China	Shenzhen Bay	E37	22.5045	113.898844
		Reserva Qiao	E38	22.4219186	113.622618
	Indonesia	Jaring Halus, NE Langkat Wildlife Sanctuary	E39	3.94296529	98.5650101
		Bengkalis Island, Riau Province	E40	1.4476312	102.392214
		North Sumatra, Aceh Besar, Lhok Nga	E41	5.36595278	95.2519
	Philipines	Filipinas	E42	9.60581111	123.128139
	Indonesia	Tanjung Pasir	F1	−6.0227622	106.667057
		Segara Anakan	F2	−8.4417594	112.669275
		Tanakeke Island, South Sulawesi Province	F3	−5.4936601	119.307603

	Papua New Guinea	Ranongga	F4	−7.9369326	156.541642
		Madang	F5	−5.2000636	145.784309
		Motupore	F6	−9.5244443	147.285483
		Bottless Bay	F7	−9.4998877	147.283054
		Labu	F8	−6.7546354	146.953832
		Riwo	F9	−5.1321679	145.78296
		Wangang	F10	−6.7334149	147.016566
	Australia	Brisbane airport	F11	−27.353787	153.107414
	Fiji	Fiji	F12	−17.713372	178.065031

[a]As depicted in Fig. 10.1

McKee and Faulkner 2000; Macintosh et al. 2002; Lewis et al. 2005; Darkwa and Smardon 2010; Matsui et al. 2010; Lewis and Brown 2014). Indeed, specific outcomes of mangrove R/R implemented on abandoned shrimp farm locations have been critically reviewed with major emphasis on case studies in the Philippines (Primavera and Esteban 2008) and Costa Rica (Stevenson et al. 1999) and have provided essential and useful practical guidelines (e.g., Brown and Lewis 2006; Lewis and Brown 2014).

10.2.5 Monitoring of R/R Projects

Most R/R projects consist of planting propagules, wildings, or saplings reared in nurseries close to or away from the target site. Few of these projects have detailed monitoring plans, and in most instances, there is no documentation of either positive/negative outcomes or recommendations for modifications of the original planting design (Lewis et al. 2005; Kodikara et al. 2017). An exception is the Ciénaga Grande de Santa Marta (CGSM), Colombia monitoring project (1995–2001), which was carried out after the construction of box culverts to reestablish hydraulic flow in a mangrove area representing the largest restoration project in Latin America (~350 km^2, including freshwater and mangrove wetlands and natural water bodies). The hydrological rehabilitation of the area consisted of dredging and reopening previous tributaries to conduct freshwater from the Magdalena River to the eastern region of the CGSM system, where mangrove mortality was extensive due to hypersalinity (>80 ppt) (Botero and Salzwedel 1999). There was a significant reduction in soil and water column salinity (<30 ppt) in all sampling stations following the hydraulic reconnection, which resulted in a major increase in mangrove forest regeneration promoting a net gain of 99 km^2 from 1995 to 1999 (Rivera-Monroy et al. 2006). Unfortunately, the lack of economic investment in the maintenance of the diversion structures from 2001 to the present has reverted the system to pre-project ecological conditions causing an increase in soil salinity, which has negatively affected the already restored vegetation (Elster 2000; Rivera-Monroy et al. 2006; Rivera-Monroy et al. 2011; Vilardy et al. 2011; Roderstein et al. 2014). In addition, areas where *Avicennia germinans* propagules established and developed into saplings were heavily impacted by the butterfly *Junonia evarete*, further increasing plant mortality rates; yet, some survived and increased plant density in areas with previously extensive mangrove mortality (Elster 2000). Overall, herbivory has not been explicitly addressed as a negative factor in mangrove R/R, but it is probably significant based on reports from other mangrove wetlands (Nagelkerken et al. 2008; Fernandes et al. 2009). Although there are fewer mangrove species in the AEP region (West Africa and Americas; see Chap. 2), such R/R failures still provide essential knowledge on biological, ecological, and hydrological variables that should be considered during forestation or afforestation projects, including the direct impact of trampling, barnacle colonization, and flotsam (Kodikara et al. 2017).

10.3 Geographical Distribution of R/R Projects in Mangrove Habitats

Assessing the geographical distribution of R/R projects (Figs. 10.1 and 10.2) contributes to our understanding of the causes triggering mangrove wetland conversion and its relative impact and how current R/R practices are related to economic or social failure. Indeed, there are some geographical differences (and similarities) concerning the causes of mangrove degradation. In the United States, most of the damage in mangroves and other wetlands was caused by dikes and draglines (which include ditching, dredging, filling, and impounding for land development) to control mosquito and biting midge populations in South East and West Florida and the Florida Keys (Fig. 10.1a). These hydrological modifications at the landscape level had negative consequences by reducing wetland productivity and fisheries abundance (McKee and Faulkner 2000; Rey et al. 2012). In mid-latitudes across the AEP region (Fig. 10.1 a–c), mangrove degradation is generally caused by the construction of highways and embankments that interrupt water (fresh and marine) flow; the opening of artificial inlets, dredging of navigation channels, and deposition of this dredged materials over or nearby mangrove forests; conversion to shrimp farms and the pumping of estuarine/coastal water during operations of shrimp aquaculture (Teas 1977; Twilley et al. 1998; Chargoy Reyes and Tovilla Hernández 2002; Menezes et al. 2005; Primavera 2006; Rivera-Monroy et al. 2006; Pagliosa et al. 2012; Hamilton 2013; Miyagi 2013; Benítez-Pardo et al. 2015; Ferreira et al. 2015).

In West Africa (Fig. 10.1c), the causes of mangrove degradation are related to expansion of agriculture and aquaculture, construction of embankments and access roads, unsustainable wood extraction for fuel wood and charcoal, and fishing and hunting, among other causes (Corcoran et al. 2007). Although mangrove extension and causes of mangrove mortality in these coastal regions are yet to be documented, extensive R/R efforts are implemented at different stages in several sites where most of the same causes of degradation are similar to those observed at the global scale (see Chaps. 8 and 9; Table 10.1; Figs. 10.1 and 10.2). For example, in the IWP region (East Africa, Asia, and Australasia), planting efforts in Gazi Bay, Kenya, were implemented in response to a lack of natural regeneration after the synergetic impact of clear-cut felling of trees about 40 years ago and heavy silting due to major upland deforestation in the middle and upper river basins. This synergy of human impacts along river watersheds from upstream to coastal regions seems to be common for other mangrove forests throughout East Africa (Kairo et al. 2001, Bosire et al. 2003; Dahdouh-Guebas et al. 2004; Fig. 10.1d). Considering mangrove reforestation as an R/R approach, the Payment for Ecosystem Services and REDD+ in Gazi Bay through the Mikoko Pamoja project is a prime example of how important the recognition of mangrove ecosystem services is and how essential it is to clearly identify the social need and economic value of mangrove wetlands (http://www.planvivo.org/project-network/mikoko-pamoja-kenya/; Jerath et al. 2016; see Chaps. 8 and 9).

Human impacts on mangrove-dominated ecosystems in India also include clear cutting and deforestation, fresh water diversions and intensive shrimp farming (Table 10.2, Fig. 10.2a; see also Chap. 9). Mangrove forests in the Pichavaram and Muthupet regions of India have been historically affected by major clear-cut logging (Selvam et al. 2003). In contrast, the impacts of land use changes in the Sundarbans National Park, one of the largest mangrove protected areas in the world (10,000 km^2), seem to be relatively minor; yet, turnover rates "due to erosion, aggradation, reforestation, and deforestation" are apparently significantly greater than the net change estimated using remote sensing techniques (Giri et al. 2007). The estimated actual mangrove wetland area in the vast Sundarbans ecosystem in the year 2000 was 5816 km^2 (Giri et al. 2007). This value includes an area of 1200 km^2 that have been afforested from 1973 to 1990 within the park limits, primarily on new accreting mud deposits as a protection against tropical cyclones (Saenger and Siddiqi 1993). Recent estimates report 1852 km^2 of mangrove cover in 2014 in the Indian Sundarbans (Ghosh et al. 2015); adding this area to the area determined for the Bangladesh Sundarbans (3745 km^2), a total of 5327 km^2 is obtained, which is slightly lower than it has been previously reported (i.e., 5816 km^2 for a deficit of 489 km^2; see Giri et al. 2007). Similar patterns in extensive mangrove loss are also observed in the Seychelles, Sri Lanka, Pakistan, Bangladesh, Myanmar, Thailand, Cambodia, Vietnam, Sumatra, and Java (Macintosh et al. 2012; Alexandris et al. 2013).

Specifically, for the Indian Ocean area, the devastating tsunami of 2004 has been an incentive for mangrove restoration programs through international and national funding initiatives. Unfortunately, most of the funding opportunities do not translate into science-based plans and are often ill prepared and unsuccessful (Jayatissa et al. 2016). A colloquium held in the coastal town of Mamallapuram, India, listed 52 sites where restoration efforts have been implemented, especially in the wake of the tsunami (Macintosh et al. 2012). Similarly, guidelines have been prepared for R/R projects after the tsunami damage to mangroves and coastal forests in Southeast Asia (Chan and Ong 2008; Chan and Baba 2009), or following oil pollution reclamation and camel grazing in the Middle East (Protection of the Environment of the Red Sea and the Gulf of Aden; Saenger and Khalil 2011).

10.3.1 Current Motivations for the R/R projects

Among the main motives identified for the implementation of R/R projects include ecological problems caused by the operation or abandonment of shrimp ponds, altered hydroperiod and tidal circulation patterns, water pollution, loss of habitat (particularly for fisheries of local and regional social and economic value), and significant decrease of soil pH (acid sulfate). In the latter case, some mangrove soils contain pyrite (potential acid-sulfate soils), which remain immobile while waterlogged (see Chap. 6). However, when these soils are used to build pond walls, where they partially dry out, sulfuric acid is produced, which lowers pond water pH values and releases Al^{3+} (Saenger 2002; see Chap. 6). As a consequence, shrimp farms often do not function well in the long term, and shrimp/prawn production dramatically declines leading to bankruptcy of aquaculture farms. In the aftermath of such local/regional socioeconomic failure, soil quality problems are left behind. Pond water acidity and toxic concentration of Al^{3+} must be dealt with before effective restoration or rehabilitation can be implemented, increasing overall R/R project costs. More recently, the motives for the implementation of R/R projects have expanded to include shoreline protection, channel stabilization, fisheries and wildlife enhancement, biodiversity conservation, legislative compliance, or socioeconomic improvement of local communities (Stubbs and Saenger 2002; Mukherjee et al. 2015).

10.3.2 *Effective R/R Projects Goal Setting*

Based on the experiences described above, it is essential that R/R project objectives are clearly defined and prioritized as a first step. A coastal afforestation project in Bangladesh, for example, had several objectives that included the production of commercial timber, acceleration of the accretion rate to form new land areas, and protection of nearshore agricultural and residential land from storm damage (Saenger 2011). These objectives were gradually achieved, but in some cases, there were conflicts in achieving success for each specific objective. For instance, in planting sites where very high sedimentation rates occurred, trees were buried and timber production was negligible. Thus, when assessing the significance of high sedimentation rates at specific sites in such cases, consideration must be given for both well-prepared and managed production of timber and coastal protection as those objectives were of highest priority, giving way to best practices for mangrove restoration and management.

Other examples in the complex implementation of R/R projects include sites in the states of Tamil Nadu and Andhra Pradesh, India (Selvam et al. 2005) and in Celestún, Campeche, Mexico (Miyagi 2013). In some locations in India, soil collapse was a consequence of extensive forest clear felling (wood revenue) of vast mangrove wetland extensions from 1935 to 1975 (Selvam et al. 2003; for other location, see Cahoon et al. 2003). As a result of direct cutting, trough-shaped areas resulted from soil exposure after tree felling causing water stagnation and high soil salt concentration. The proposed solution was to excavate artificial channels (1 m deep, 1.5 m wide at the base and 3 m wide at the soil surface) and connect them to natural adjacent channels (Fig. 10.3). Feeder channels (0.75 m deep, 0.6 wide at the base, and 1.5 m wide at the soil surface) were also excavated throughout the die-back mangrove area, following a "fish bone" spatial pattern (Fig. 10.3). The excavated sediments were deposited next to the channels, increasing soil relative elevation. This strategy was designed to reestablish water exchange between the mangrove die-back areas and the natural channels with the goal of increasing the survival rate of planted and naturally established seedlings, The technique (i.e.,

Fig. 10.3 Hydrological restoration implemented in mangrove wetlands in Pichavaram, Tamil Nadu, India, showing original main and feeder channels excavated circa 1996. (**a**): March 3, 2003; (**b**): January 29, 2016 (Source: Google Earth Pro; image area: 55.5 ha; eye altitude 881 m; Latitude: 11°25′59.86″ N, longitude: 79°47′28.89″ E at the center of the images

feeder channels) was first tested around 1996 in a pilot study involving 10 ha of dead mangrove wetland and resulted in the recovery of an extensive mangrove forest area (Fig. 10.3). After it was demonstrated to be successful, it was used in other areas covering at least 1200 ha impacted mangrove sites in the states of Tamil Nadu and Andhra Pradesh, India (Selvam et al. 2005). One of the main attributes of the R/R project described above (Fig. 10.3) involved an initial diagnostic and a pilot study to test the proposed solution. The implementation of this approach involved the acquisition of permits before and after project implementation, as well as securing funding from government agencies. Additional critical steps included (1) plan-

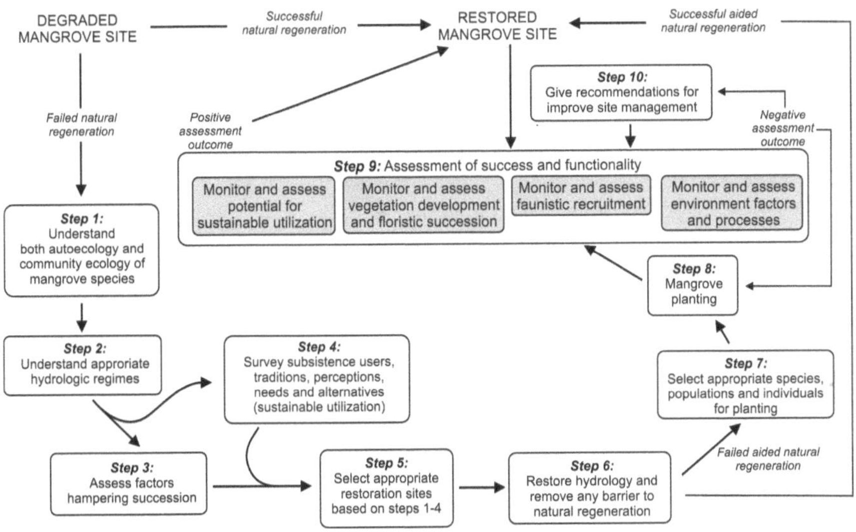

Fig. 10.4 Decision tree showing recommended steps and tasks to restore a mangrove wetland based on original site conditions (From Bosire et al. 2008)

ning to excavate during the period of lowest water level, (2) organizing and working closely in a community-based restoration effort, (3) maintaining nurseries to raise seedlings of several mangrove species for planting in the modified areas, (4) channel maintenance (mainly silt dredging) when required, and (5) monitoring the success or failure of restored areas by means of GIS and ground truthing (Selvam et al. 2003). A similar success history following essentially the same steps was implemented in Celestún, Campeche, and Mexico (Miyagi 2013).

Prioritized objectives underpin the development and implementation of R/R projects as they force the operational identification of the processes that must be included to provide a clear framework that warrant project success. Among other alternatives to ensure a logical selection of steps and clear objectives, we propose the implementation of the Ecological Mangrove Rehabilitation (EMR) protocol as outlined in Lewis and Brown (2014) that includes monitoring and reporting tasks (Fig. 10.4). For example, if the objective is to restore harvestable fish and shellfish habitat, the life history of the target species should be fully understood while monitoring species-specific requirements to document an actual increase in species population density in the restored area (Lewis et al. 1985; Brockmeyer et al. 1997; Lewis and Gilmore 2007). A unique design criterion, such as the restoration of the historical hydrological patterns (e.g., water flow, net volume), and attributes (e.g., cross section area, length) of tidal creeks may also be essential to provide accessibility for migration and reproduction cycles for those targeted species.

An interdisciplinary framework has also been proposed to evaluate coastal "bioshield" plantations (some with mangroves) and involves the consideration of several preplantation, plantation, and postplantation procedures (Mukherjee et al. 2015). In this scheme, one of the major drivers defining the objectives and requirements to ensure success, but usually neglected, is land tenure rights. This consideration is especially critical in plantations established on land under the jurisdiction of the Revenue Department or similar country/regional governance bodies or long-term land grants where projects could become high economic risks if changes in policy occur after project implementation (Primavera 2000; Primavera and Esteban 2008; Mukherjee et al. 2015). In fact, land use change, either in private and public lands, is perhaps the major threat to the implementation of R/R projects given the uncertainty in the change of regional and national policies and economic interests associated to urban and industrial development, particularly in developing countries (see Chap. 9).

10.3.3 Critical Questions: What Were the Ecological Services Sought? What Were the Societal Priorities?

Mangroves have well-defined economic and social values referred to as "instrumental values", "free services", "ecological functions", or "ecological services" (see Chaps. 8 and 9). These values include the provision of habitat and biodiversity conservation, food and wood production, shoreline protection, chemical buffering, water quality maintenance, provision of recreational, aesthetic and education opportunities, and reservoirs of genetic materials. Indeed, coastal protection and socio-economic factors are the main drivers of coastal bio-shield projects in India (Mukherjee et al. 2015). Therefore, in each R/R project it must be decided which of these ecological functions, goods, and services is (are) the most appropriate to be sustainable, including the need to make decisions that are congruent with the priorities of both national governments and local communities.

10.3.4 Implementation Plans

In earlier steps in the implementation of R/R projects, a questionnaire survey is a useful tool for the evaluation of site conditions to compare potential sites. This tool is also necessary in the development of a detailed implementation plan based on the natural conditions of each site (Saenger et al. 1996). Furthermore, this assessment should include a synoptic account of the biotic and abiotic site conditions and, critically important, practical considerations as access, travel time, and land-use status. Since the early 1980s, it has been advocated that the planting of mangroves specifically should occur for the environmental services these wetlands can provide (i.e., Lewis 1982). One of the requirements to implement such an approach is to avoid, as much as possible, the monoculture of mangroves that frequently characterizes restoration projects devoted to timber production. Despite this limitation, few restoration programs have achieved a degree of ecological functioning similar to natural mangrove systems (Latif 1996; McKee and Faulkner 2000; Lewis and Gilmore

2007; Bosire et al. 2008). Based on these experiences, the following conditions should be met to increase the success of a specific mangrove R/R project: (1) it should be viewed by the local people as an economic opportunity and/or offer other tangible benefits; (2) it is compatible with local patterns of resource use and land tenure; (3) local knowledge and skills relevant to restoration are successfully embedded into the project; (4) local groups and organizations are effectively mobilized to support and implement restoration activities; and (5) relevant policies and political factors are supportive of restoration efforts at the local, regional, and national levels (Walters 1997).

10.4 Major Limitations in the Implementation of R/R: Funding Availability and Current Ecological Theory

Funding availability for the implementation of R/R project is generally based on the realization by different countries that a high proportion of mangrove wetlands have been damaged by a complex interaction of human impacts including aquaculture, agriculture, livestock, urban/rural/industrial and touristic development, and misguided practices concerning the construction of roads, extensive dredging and the opening of sand bar inlets along vulnerable coasts. Some of these activities have caused irreversible damage, requiring the implementation of mangrove R/R projects, which may be funded by government agencies/departments and/or Nongovernment Organizations. However, financial support for most of these coastal management projects is limited due, in most instances, to the high cost for implementation. Even when economic resources are available, they are often not appropriately allocated and spent (Kodikara et al. in press). Therefore, current ecological theory and the experience gained through frequent failures, and less frequent successes, must be incorporated in current and future R/R projects to help define the short- and long-term goals and strategies to promote cost-effective small and large-scale mangrove R/R projects (Lewis et al. 2005; Primavera and Esteban 2008; Saenger 2011; Twilley et al. 1998; Twilley and Rivera-Monroy 2005).

10.4.1 Selection of Easily Manageable Species

Among the taxonomic selection of individual for R/R projects, the genus *Rhizophora* has been the preferred taxon used in planting-oriented restoration projects (Ellison 2000). The species within this genus have a worldwide distribution (Tomlinson 1986; Giri et al. 2011; see Chap. 2). Two of the major reasons this genus is used in planting programs are its large hypocotyl nutrient storage that increases survival rates at early developmental stages, even for long-term wood production in natural environments, and its handling versatility (Shamsudin et al. 2008; Goessens et al. 2014).

10.4.2 Planting Seedlings or Saplings from Local or Distal Genetic Sources

Although much is yet to be understood about the effects of planting *Rhizophora* propagules or saplings in a site that is far away from the germplasm source, even when planting the same species, current studies show that genetic diversity decreases toward higher latitudes and under isolation conditions (Sandoval-Castro et al. 2014; De Ryck et al. 2016; Ngeve et al. 2016). This decrease is due to the genetic attenuation (e.g., loss of unique alleles) and an increase in selfing. These findings suggest that genetic recovery of large impacted wetlands areas in tropical latitudes may require more than 30 years (Arnaud-Haond et al. 2009). Similarly, the effect of habitat fragmentation might not influence the genetic makeup of adult populations, although it can occur in cases of higher inbreeding in smaller populations (Hermansen et al. 2015). Perhaps a rule of thumb would be to use, if available, genetic resources from the nearest possible populations, such as transplanting wildings from nearest mangrove wetlands under good or optimal environmental conditions (Ellison and Fiu 2010).

10.4.3 Have Native Species Been Always Used in Restoration Programs?

R/R projects using exotic species in species-rich biogeographic regions have been recently reported in the scientific literature. For instance, the mangrove species *Sonneratia apetala* (originally from India, Sri Lanka, and the Bengal coastal region) has been used in the restoration of physically altered environments lacking natural propagule sources in China (Ren et al. 2008). Over the first decade, the growth performance of the mangrove species *S. apetala* was higher than those of the native species, *Rhizophora stylosa* and *Kandelia candel* (now *K. obovata*); and in some cases, *S. apetala* facilitated the recolonization of native mangrove species (Ren et al. 2008; Peng et al. 2012). However, due to the ecological risk of invasion at broader spatial scales, recent assessments are now recommending that restoration efforts should include competitive control mechanisms and removal of alien plant species once the populations of native species are established (Chen et al. 2013; Ren et al. 2009, 2014). Moreover, the use of exotic species in restoration programs started relatively recently (two decades ago) and was restricted to site-specific experiments. Unfortunately, the lack of adequate monitoring of multilevel performance measures makes it extremely difficult to infer whether these actions will sustain themselves without further human intervention and at higher ecological and economic cost.

The few experiments designed to assess the effects of exotic species on ecosystem functionality include evaluations of macrobenthic faunal communities (Tang et al. 2012; Leung and Tam 2013). These studies revealed that although the exotic mangrove species *S. apetala* seems to be innocuous to the macrobenthic fauna, its presence and dispersion could have negative impacts on other functional groups. For instance, afforestation of mudflats with alien species reduces the feeding ground for water birds (Leung and Tam 2013). Due to the lack of data and information together with an insufficient monitoring timeframe, including the lack of proper spatial and temporal replication, management plans aiming to regulate the use of exotic species and prevent adverse impacts to the estuarine ecosystem are yet to be implemented. Thus, a consensus regarding the use of exotic mangrove species as a good restoration practice remains to be evaluated.

10.5 Implementing R/R Projects in the Context of Climate Change: Carbon Markets and Greenhouse Emissions

R/R projects could be considered a long-term strategy to mitigate carbon emissions given the current estimates of potential carbon storage ("blue carbon") in mangrove wetlands (Donato et al. 2011; Caldeira 2012; Siikamäki et al. 2012). The assessment of carbon stocks in the wide range of mangrove ecotypes (sensu Lugo and Snedaker 1974) throughout tropical and subtropical latitudes confirm that mangrove forests are among the ecosystems with the highest C storage capacity per unit area (e.g., Mcleod et al. 2011; Donato et al. 2011; Alongi 2014; Lovelock et al. 2014; Adame et al. 2015; see Chap. 5). This storage capacity is due to slow decomposition and rapid organic matter accumulation through time in flooded soils. For example, soil carbon sequestration rates in mangroves growing in arid tropical coastal regions (Pacific coast of Mexico) range from 0.1 and 6.9 Mg C ha^{-1} $yr.^{-1}$ in the last 100 years (Ezcurra et al. 2016). However, actual emission rates of previously stored blue carbon into the atmosphere in deforested mangrove areas have not been directly and comprehensively assessed. For example, Kauffman et al. (2015) indirectly estimated a loss of 1464 Mg CO_2 equivalents per ha for the top 1 m soil depth when mangrove forests were converted to pastures in Tabasco, Mexico, representing seven and three times greater emissions than those reported for a tropical dry forest and a tropical forest in the Amazons, respectively. In that study, the carbon stock was lower in older (30-year) than younger (7-year) pasturelands previously occupied by mangroves, suggesting continuous loss to the atmosphere through time (Kauffman et al. 2015), especially when flooded soils are drained and exposed to fast aerobic decomposition (Couwenberg et al. 2010).

It is assumed that some of the carbon emitted could be sequestered again from the atmosphere after these impacted sites are restored; this response has been observed in mangrove forests where superficial soil horizons were similar to preserved forests after 35 years of mangrove tree planting or natural regeneration (Lunstrum and Chen 2014; Nam et al. 2016). Although more information is needed to evaluate the potential sequestration and storage in restored mangrove wetlands, studies suggest that R/R projects could be an efficient strategy to capture carbon from the atmosphere at a relatively low cost (Siikamäki et al. 2013;

Thomas 2014) considering the potentially high estimated economic values of carbon sequestration as an ecosystem service (e.g., Estrada et al. 2015; Jerath et al. 2016). However, adequate species selection and suitable (e.g., middle to upper intertidal) environments must be selected for successful mangrove restoration in contrast to the selection of unsuitable (e.g., lower intertidal) environments, as it has been the case in some coastal regions (Lewis et al. 2005; Primavera and Esteban 2008). Additionally, the economic and social dimension of carbon sequestration valuation and carbon market development require not only community-based mangrove management schemes to achieve restoration goals, but also that local governments are directly aligned to international economic incentives related to carbon markets in the context of climate change (Beymer-Farris and Bassett 2012; Jerath et al. 2016).

10.6 Global, Regional, and Local Perspectives in Mangrove R/R Programs: Beyond Planting Trees

10.6.1 Factors Controlling Long-Term Sustainability of Restored Mangroves

Mangrove R/R strategies have historically been scrutinized to identify both information gaps and operational pitfalls. Despite the broad geographic range of implemented mangrove restoration projects, an analysis of project outcomes from the 1800s until 1999 (Ellison 2000) indicated that the methods used are mainly based on planting of single mangrove species and that the primary focus remained on a silviculture-oriented approach (e.g., fuelwood, charcoal, Lewis 1982). Recently, a number of assessments of R/R practices and methods indicate a limited advance in improving R/R strategies and confirm that planting, rather than eliminating the stressors and assisting natural regeneration, remains the main strategy used worldwide (Bosire et al. 2008; Dale et al. 2014).

Effective mangrove restoration can only be achieved by eliminating environmental stressors, a strategy proposed more than 30 years ago (e.g., Cintrón and Schaeffer-Novelli 1983; Cintrón-Molero 1992). A stressor is any factor or situation that diverts potential energy flows that could be used for the system's own maintenance, stability, and resilience (Odum 1967; Lugo and Snedaker 1974; Twilley and Rivera-Monroy 2005). The ecosystem response to a stressor depends on its effect/impact on the system (e.g., physiological mechanisms, structure, and composition) that influence the recovery rates depending on the type, persistence, and synergy among natural and human-induced stressors (Lugo 1978; Lugo et al. 1981). If we consider that environmental stressors can impair the system's recovery capacity, it is important to prioritize ecological-based restoration strategies over single species planting (Lewis 2000).

Mangroves, as is the case for other wetlands, are flow-through ecosystems. Thus, an understanding of their ecology and hydrology is a critical step in designing successful mangrove restoration plans (Lewis et al. 2005). There are successful wetland restoration projects based on hydrologic restoration (Turner and Lewis 1997; Selvam et al. 2003; Miyagi 2013). In mangrove forests, the hydroperiod (flooding frequency, duration, and depth) regulates biogeochemical processes such as gas exchange (O_2 and CO_2) between plants and the environment, metabolic turnover rates, and the accumulation of sulfide in soil (Twilley and Rivera-Monroy 2005; Lugo and Medina 2014; see Chaps. 5 and 6). Mangrove forests are very sensitive to edaphic modifications, mainly due to shifts in substrate elevation relative to water level; and their ability to return to a more complex level of organization is strongly affected by the intensity and frequency of the stressor (Cintrón and Schaeffer-Novelli 1983). In fact, regrading sites to previous relative elevation is recommended for restoration projects and ignoring this step has led to numerous failures (Lewis et al. 2005 and references therein).

On a mangrove forest scale, the environmental gradient created by the microtopography sets ecological patterns relevant to restoration strategies such as species distribution in response to hydroperiod (Lugo and Snedaker 1974; Twilley et al. 1998; Twilley and Rivera-Monroy 2005; Flores Verdugo et al. 2007; Flores-de-Santiago 2017; see Chaps. 6 and 9), as well as to other regulators (salinity, sulfide, pH, redox potential) and resources (nutrients, light, space) (Twilley and Rivera-Monroy 2005). Moving up one level to the landscape scale, mangrove stands are nested within environmental settings (e.g., deltas, coastal lagoons, oceanic islands) and are necessarily subjected to environmental variability as a result of major changes in hydrology or sediment input and deposition rates (Twilley et al. 1998; Schaeffer-Novelli et al. 2005). Therefore, restoration strategies should not be limited to the local site, but also consider the interconnectedness with regional and global process (Twilley et al. 1998; Twilley and Rivera-Monroy 2005). This is particularly important when considering recurrent large-scale climate phenomena (e.g., El Niño Southern Oscillation) and changes triggered by events that can affect site-level management strategies as shown in large mangrove restoration projects in the Americas (Blanco et al. 2006; Rivera-Monroy et al. 2006; Rivera-Monroy et al. 2011). These hierarchical levels should be considered in mangrove R/R projects to capture the combined effects of geophysical, geomorphic, and ecological processes that control the mosaic and development of mangrove wetlands (Twilley et al. 1998).

In the context of adaptive management of natural resources, there is no "one-size-fits-all solution". Thus, the studies discussed here underscore the constraints and opportunities for successful mangrove restoration. A large body of evidence shows that neglecting ecological baselines is the main factor hindering effective restoration initiatives worldwide, and when appropriate hydrological conditions are restored, mangroves can fully develop and function as natural stands with no further human intervention required (Twilley et al. 1998; Ellison 2000; Lewis et al. 2005; Rivera-Monroy et al. 2006; Lewis and Gilmore 2007; Bosire et al. 2008; Rovai et al. 2012; Rovai et al. 2013; Dale et al. 2014).

10.6.2 Monitoring the Functionality of Restored Mangroves

A number of variables have been proposed to assess mangrove restoration outcomes (Twilley and Rivera-Monroy 2005; Bosire et al. 2008; Dale et al. 2014). Issues related to monitoring of restoration projects are coupled to the economic priorities, timeframe, and diversity of methods. In addition to the lack of standardized methods to monitor mangrove restoration outcomes, assessments often limit their analyses to one specific indicator species or group. This approach does not provide an overview of the functionality, which should reflect the system's capacity to maintain an effective energy flow as well as structural and functional properties considering the multiple pathways and mechanisms by which ecological services are delivered (see Chaps. 8 and 9). Again, because environmental stressors can affect the target ecosystem at different levels of organization, it is important to define and consider multiple functional indicators as performance measures in mangrove restoration strategies (Twilley and Rivera-Monroy 2005).

Most projects are short in duration (<3 years) and do not devote funding for adequate maintenance and monitoring periods (Rivera-Monroy et al. 2006; Lewis et al. 2005; Roderstein et al. 2014). Periods ranging from 2 to 16 years (Bosire et al. 2008 and references therein) and 10 to 50 years (Crewz and Lewis 1991; Lugo 1992; Shafer and Roberts 2008; Luo et al. 2010; Rovai et al. 2012, 2013) may be required to fully ascertain mangrove restoration success based on faunal diversity and vegetation structural (e.g., basal area, species diversity) as well as functional (e.g., net primary productivity, carbon storage, resilience) properties. Based on these studies, we recommend that the monitoring and maintenance of R/R projects cover at least 5 years after project implementation. For example, one functional ecosystem property might be an assessment of the abundance and diversity of fish populations to ensure that both keystone and of economic important species to return to reference condition within 5 years (Lewis and Gilmore 2007). However, depending on the intensity of the damage, ecosystem functionality in wetlands can take over a century to be restored. Moreno-Mateos et al. (2012) found that only 7 out of the 124 references used in their analysis corresponded to mangrove ecosystems with restoration ages ranging from 22 months to 14 years. Appropriate spatial and temporal replication incorporating key and multilevel functional indicators is needed to draw conclusions at a range of population, community, or ecosystem dynamics.

The key set of functional indicators used as performance measures to evaluate the success of a mangrove R/R projects should include physiological and structural attributes as response variables to gradients of environmental factors. These include resources (light and nutrients), regulators (salinity, pH, soil sulfide, redox potential), and hydroperiod (water depth, frequency and duration of flooding; Twilley and Rivera-Monroy 2005; Rivera-Monroy et al. 2011) that account for the main stressors to mangrove development and long-term sustainability. The performance measures should provide information about the restoration trajectory of the ecosystem at specific sites, thus describing the degree and timing of changes anticipated in both structural and functional characteristics and enabling adaptive management actions. The integration of multilevel performance measures, including abiotic and biotic compartments, allows for the identification of cause and effect relationships, documenting the effectiveness of restoration strategies and testing assumptions concerning the stressors that are associated with the system's degradation (Twilley and Rivera-Monroy 2005).

The difficulty and utility of monitoring performance measures in R/R mangrove projects can be illustrated by some examples. The trajectories of vegetation and soil properties of a mangrove rehabilitation project by reconnecting water bodies in the Ciénaga Grande de Santa Marta lagoon complex (Colombia), one of the largest restoration efforts ever implemented (mangrove area: 99 km^2) in the AEP region, indicated a reversal of the initial success (Rivera-Monroy et al. 2006). After a successful response to the large spatial scale hydrological modifications by widespread natural regeneration in 1996 and 1999, the mangrove forest in the region began to show potentially irreversible deterioration due to a lack of a long-term economic strategy that included maintenance of the originally dredged channel to maintain freshwater exchange between the mangrove die-back areas and the natural creeks and estuary (Roderstein et al. 2014). Similarly, extensive canal digging toward river and tidal water sources was carried out in the Pichavaram mangrove area in South India (Selvam et al. 2003) that resulted in the recovery of an extensive area (~300 ha), visible form space (Fig. 10.3) and originally lost due to clear-cutting and soil subsidence. In contrast to the case in Colombia, canal maintenance to avoid siltation is currently performed in this location with the participation of local communities and adequate technical and economic support. Another successful hydrological rehabilitation implemented at both Términos Lagoon and Jaina Island in Campeche, Mexico, has promoted a maintenance-free mangrove restoration areas, enhancing further recovery of vegetation cover and ecosystem services at low investment cost (Agraz-Hernández and Arriaga 2010; Agraz-Hernández et al. 2015).

Another R/R project in the AEP region (Brazil) coupled structural and physiological properties of mangrove vegetation with edaphic conditions to assess the success of different mangrove restoration projects (Rovai et al. 2012, 2013). Those studies demonstrated that although restoration sites did not differ from reference stands in terms of forest structural characteristics, there was impaired photosynthetic performance due to stress caused by soil elevation changes and heavy metal inputs, thus making it difficult to infer possible restoration trajectories. This study shows the advantage of using hierarchical performance measures in restoration strategies, since ecological responses at lower levels of organization may anticipate threats to the system's structure, and reveal critical trends in ecosystem development (Twilley et al. 1998). For example, nitrogen fixation, a functional ecosystem service, has been used successfully as an indicator of success in reforested and naturally regenerated mangroves in Mexico (Vovides et al. 2011)

The mangrove fauna plays indeed a significant role in the functioning of mangrove ecosystems and can thus be a useful indicator of integrity of managed mangroves (Lewis 1982; Lewis and Gilmore 2007; Bosire et al. 2008; Cannicci et al. 2008; Ellison 2008; see Chaps. 3 and 6). The assessment of trends in recolonization of epibiotic, macrobenthic, and sediment-infauna communities and the distribution patterns of benthic macrofauna, fish, and shrimp in R/R stands across the world show significant and short-term response (Bosire et al. 2008). Although selected biota groups seem to be more responsive to mangrove restoration, there are still only few studies on the spatial and temporal changes in biodiversity in restored mangroves (see Chap. 3); the scant information on age range, species composition, and hydroperiod in restored sites make generalizations highly uncertain.

We underscore the premise that there is no "*one-size-fits-all*" solution in restoration ecology. Mangrove restoration monitoring programs should include as many indicators as the budget and timeframe allow and may be amended as required by the specific goals of the initial restoration plan (i.e., adaptive management). An empirical framework that models mangrove restoration trajectories by integrating indicators that reflect ecological processes at different time and spatial scales is strongly recommended (Twilley and Rivera-Monroy 2005). This framework should highlight the opportunities and constraints of monitoring programs and operationally define the basic performance measures that should assist in the advancement of mangrove restoration in all biogeographic regions.

10.7 Future Directions: Lessons Learned and Research Agenda

To advance mangrove R/R efforts worldwide, data sharing and exchange of experiences should be promoted and orchestrated at a comparative level in different geomorphological settings and latitudes within and across the IWP and AEP regions. Below we discuss four proposed R/R protocols that could be considered as a general research agenda to be implemented given the inclusion of critical ecological processes and operational tasks to improve the success of mangrove R/R projects. A critical step is to develop a decision tree that could serve as a guide to optimize the use of available funding in the development, implementation, and monitoring of R/R projects (Fig. 10.4). Future protocols should list clear objectives, goals and deadlines, a robust research agenda that include specific questions (and hypotheses) based on sound ecological theory, and reliable monitoring practices that maximize the usefulness of current and past R/R project experiences (Ellison 2000; Bosire et al. 2008). We propose that these initial steps could be based on the current available protocols for mangrove R/R projects that could be further developed under the specific conditions at each individual location.

The first, and most commonly used protocol, emphasizes that if natural recolonization after site selection or improvement (secondary succession) does not occur or is too slow (Field 1996b; Primavera et al. 2012) a mangrove nursery should be set up as sites for possible planting or out-planting (sensu Primavera et al. 2012) are identified primarily based on the current lack of mangrove cover or on evidence of their historical cover loss. A very large part of this protocol is devoted to successful nursery practices including seed or seedling collection and planting, and the use of some natural seedlings transplants (i.e., wildlings) from healthy forests (Field 1996a, b; Primavera et al. 2012). However, this approach does not emphasize steps to clearly identify the drivers causing mangrove mortality in the first place or factors hindering the lack of natural mangrove regeneration and growth in the proposed planting site. Indeed, Samson and Rollon (2008) documented the failure of a similar mangrove restoration protocol implemented over 40,000 ha during a 20-year period in the Philippines.

The second protocol, called Ecological Mangrove Rehabilitation (or Restoration) (EMR, Lewis and Marshall 1998; Stevenson et al. 1999), was initially described as a five-step process (Brown and Lewis 2006), and later expanded to six steps (Lewis 2009, which have been implemented at a number of sites around the world (Lewis and Brown 2014). For example, Rey et al. (2012) described the success of this "hydrologic restoration" approach (Lewis et al. 1985; Brockmeyer et al. 1997; Turner and Lewis 1997) when implemented in 12,605 ha out of the original 16,185 ha mangrove area that was diked and filled in the East Coast of Florida, USA. The localities were hydrologically reconnected, breached, or restored for the rehabilitation of formerly diked mosquito control impoundments. Nursery establishment and planting of mangroves is only used under this protocol if natural propagule recruitment does not occur after site preparation and monitoring (i.e., "propagule limitation"; Lewis et al. 2005). Thus, planting of mangroves is not precluded under EMR, but is based upon a documented lack of natural establishment of propagules (i.e., secondary succession).

The six steps of EMR (sensu Lewis and Brown 2014) are as follows.

1. Understand the autecology (individual species ecology) of the mangrove species at the site, the patterns of reproduction, propagule distribution, and successful seedling establishment.
2. Understand the normal hydrologic patterns that control the distribution and successful establishment and growth of targeted mangrove species.
3. Assess the modifications of the previous mangrove environment that currently prevent natural secondary succession.
4. Select appropriate mangrove restoration sites through application of Steps 1–3. These steps increase the likelihood of success in restoring a sustainable mangrove forest ecosystem, and are cost-effective given the available funds and manpower to implement projects, including adequate monitoring to assess quantitative goals established prior to restoration. This step includes resolving land ownership/use issues necessary for ensuring long-term access to and conservation of the site.
5. Design the restoration program at appropriate sites selected in Step 4 to initially restore the appropriate hydrology and utilize natural mangrove propagule recruitment for plant establishment.
6. Only utilize actual planting of propagules, collected seedlings, or cultivated seedlings after determining through steps 1–5 that natural recruitment will not provide the quantity of successfully established seedlings, rate of stabilization,

or rate of growth of saplings established as quantitative goals for the restoration project.

In a third protocol proposed for mangrove restoration, Bosire et al. (2008) present a ten-step flow diagram that expands even further on the six steps from EMR and that can be used as a decision tree for restoration programs (Fig. 10.4). These steps integrate the essential procedure of consulting with the local communities (Step 4) and post-plantation phases, similar to those discussed by Mukherjee et al. (2015). The step 9 in this approach underscores the need to monitor ecological succession in all main biological groups as well as resource use by local people, which is a much-desired step toward functional integrity when the goods and services mangrove forest provide directly benefit local communities (see Chap. 8).

The fourth protocol explicitly adds economic and social issues and emphasizes the use of local ecological knowledge to substitute for baseline information gaps (e.g., detailed reference site topography and hydrology) (Biswas et al. 2009). This approach is akin to "community based rehabilitation" (Primavera et al. 2012) or "community based ecological mangrove rehabilitation" (CBMER) (Brown and Lewis 2006; Lewis and Brown 2014) and was tested in four R/R projects (Biswas et al. 2009) with "minimum" success for two projects and "uncertain" success for the other two. A major problem when relying on community support to implement R/R project is that funding for the participation of volunteer planting and monitoring is limited, thus "[…] it is not uncommon that the whole effort collapses as soon as the external support is withdrawn" (Biswas et al. 2009; p. 379). This limitation does not invalidate the general approach, but introduces a potential problem by not emphasizing enough ecological engineering considerations such as the assessment of hydrology and topography as important initial step in data gathering efforts before project implementation. An integrated approach similar to that of CBEMR have been implemented in Indonesia relying on community-based data gathering on hydrology and topography, underlining adequate funding and training as key to the overall success of that rehabilitation project (Brown et al. 2014).

Finally, it is paramount to include in any monitoring and reporting program both spatial and temporal replication (Underwood 1997), including reference sites within the restoration site or nearby (see Rovai et al. 2012, 2013 for a detailed spatial and built-in time sampling strategy). In addition, the program should consider establishment of long-term research plots and multiple sequential research programs when and where possible. The results, whether successful or not, should be published, as it is the only sound alternative to learn from past experiences, and further advance mangrove restoration ecological science based on the actual successes and failures of the four protocols previously described. We urge the continental level implementation of these guidelines to advance international initiatives aimed to protect and conserve one of the most productive and threaten coastal ecosystems in the world.

Acknowledgments JLP was funded by the Comisión Nacional para el Conocimiento y Uso de la Biodiversidad-CONABIO (project nos. HH05 and MN001). The Louisiana Sea Grant College Program (NOAA) and the CAPES/CNPq Science without Borders Program (grant no. BEX1930/13-3) provided funding for ASR. VHRM was partially funded by The Florida Coastal

Everglades Long-Term Ecological Research program (grant nos. DBI-0620409 and DEB-1237517), NASA-JPL project "Vulnerability Assessment of Mangrove Forest Regions of the Americas" (LSU Subcontract no. 1452878), and the Department of the Interior South Central Climate Science Center through Cooperative Agreement # G12 AC00002. We thank A. F. Zaragoza-Méndez for help in locating the R/R sites in Google Earth.

References

Adame MF, Hermoso V, Perhans K, Lovelock E, Herrera Silveira JA (2014) Selecting cost effective areas for restoration of ecosystem services. Conserv Biol 29(2):493–502

Adame MF, Santini NS, Tovilla C, Vázquez-Lule A, Castro L, Guevara M (2015) Carbon stocks and soil sequestration rates of tropical riverine wetlands. Biogeosciences 12(12):3805–3818

Agraz Hernández CM, Arriaga V (2010) Restauración del manglar en la laguna de Términos. In: Carabias J et al (coords). Patrimonio natural de México. Cien casos de éxito. México, CONABIO, p. 151

Agraz Hernández CM, Osti Sáenz J, Reyes Castellano J, Chan Keb C, Expósito Díaz, G, Conde Medina KP (2015) Restauración ecológica en el ecosistema de manglar en Isla Jaina, mediante el empleo de modelos de circulación hidrodinámica en el Municipio de Hecelchakán, Campeche en 600 ha. First Report, CONAFOR-UAC. 120 p

Agraz Hernández CM; Osti-Sáenz J, Jiménez-Zacarías C, García Zaragoza, Chan-Canul E, González Durán L, Palomo Rodríguez A (2007) Restauración con manglar: criterios y técnicas hidrológicas de reforestación y forestación. Universidad Autónoma de Campeche, Comisión Federal de Electricidad, Comisión Nacional Forestal. 132 p

Alexandris N, Chatenoux B, Lopez Torres L, Peduzzi P (2013) Monitoring mangrove restoration from space. UNEP/GRID-Geneva. 48 p

Álvarez-León R (2003) Los manglares de Colombia y la recuperación de sus áreas degradadas: revisión bibliográfica y nuevas experiencias. Madera y Bosques 9:3–25

Alongi DM (2014) Carbon cycling and storage in mangrove forests. Annu Rev Mar Sci 6:195–219

Arnaud-Haond S, Duarte CM, Teixeira S, Massa SI, Terrados J, Tri NH, Hong PN, Serrão E (2009) Genetic recolonization of mangrove: genetic diversity still increasing in the Mekong Delta 30 years after Agent Orange. Mar Ecol Prog Ser 390:129–135

Aronson J, Milton SJ, Blignaut JN (2007) Restoring natural capital. Science, business, and practice. Society for Ecological Restoration International-Island Press. 384 p

Bashan Y, Moreno M, Salazar G, Álvarez L (2013) Restoration and recovery of hurricane-damaged mangroves using the knickpoint retreat effect and tides as dredging tools. J Environ Manag 116:196–203

Benítez-Pardo D, Flores-Verdugo FJ, Casas-Valdéz M, Hernández-Carmona G, Valdez-Hernández JI, Gómez-Muñoz V (2015) Forestación de isletas de dragado utilizando dos especies de mangles, en una laguna costera del Golfo de California, México. Bot Sci 93:165–174

Bernini E, Cristina de Fátima NRS, Lage-Pinto F, Paiva Chagas G, Rezende CE (2014) Fitossociologia de florestas de mangue plantadas e naturais no estuário do Rio das Ostras, Rio de Janeiro, Brasil. Biotemas 27:37–48

Beymer-Farris BA, Bassett TJ (2012) The REDD menace: resurgent protectionism in Tanzania's mangrove forests. Glob Environ Chang 22:332–341

Biswas SR, Malik AU, Choudhury JK, Nishat A (2009) A united framework for the restoration of Southeast Asian mangroves – bridging ecology, society and economics. Wetl Ecol Manag 17:365–383

Blanco JA, Viloria EA, Narvaez JC (2006) ENSO and salinity changes in the Ciénaga Grande de Santa Marta coastal lagoon system, Colombian Caribbean. Estuar Coast Shelf Sci 66:157–167

Bohórquez-Rueda C, Prada-Triana MC (1988) Transplante de plántulas de Rhizophora mangle (Rhizophoraceae) en el Parque Nacional Corales del Rosario, Colombia. Rev Biol Trop 36 (2B): 555-557

Bosire JO, Dahdouh-Guebas F, Kairo JG, Koedam N (2003) Colonization of non-planted mangrove species into restored mangrove stands in Gazi Bay, Kenya. Aquat Bot 76:267–279
Bosire JO, Dahdouh-Guebas F, Walton M, Crona BI, Lewis RR, Field C, Kairo JG, Koedam N (2008) Functionality of restored mangroves: a review. Aquat Bot 89:251–259
Bosire JO, Kaino JJ, Olagoke AO, Mwihaki LM, Ogendi GM, Kairo JG, Berger U, Macharia D (2014) Mangroves in peril: unprecedented degradation rates of peri-urban mangroves in Kenya. Biogeosciences 11:2623–2634
Botero L, Salzwedel H (1999) Rehabilitation of the Ciénaga Grande de Santa Marta, a mangrove-estuarine system in the Caribbean coast of Colombia. Ocean Coast Manag 42:243–256
Brockmeyer RE Jr, Rey JR, Virnstein RW, Gilmore RG, Ernest L (1997) Rehabilitation of impounded estuarine wetlands by hydrologic reconnection to the Indian River Lagoon, Florida (USA). Wetl Ecol Manag 4:93–109
Brown B, Lewis RR (2006) Five steps to successful ecological restoration of mangroves. In: Quarto A, Enright J, Corets E, Primavera J, Ravishankar T, Stanley OD, Djamaluddin R (eds) Mangrove action project – Yayasan Akar Rumput Laut (YARL), Yogyakarta, Indonesia, p 64. www.mangroverestoration.com
Brown B, Fadilla R, Nurdin Y, Soulsby I, Ahmad R (2014) Case study: Community based ecological mangrove rehabilitation (CBEMR) in Indonesia. S.A.P.I.E.N.S. 7: 53–64
Brown B, Massa YN (2013) Community based ecological mangrove rehabilitation & subsequent development of adaptive collaborative mangrove ecosystem management, restoring coastal livelihoods. Mainstreaming Mangrove Indonesian Ministry of Forestry, Jakarta, p 35
Cahoon DR, Hensel P, Rybczyk J, McKee KL, Proffitt CE, Perez BC (2003) Mass tree mortality leads to mangrove peat collapse at Bay Islands, Honduras after Hurricane Mitch. J Ecol 91:1093–1105
Caldeira K (2012) Avoiding mangrove destruction by avoiding carbon dioxide emissions. Proc Natl Acad Sci USA 109:14287–14288
Cannicci S, Burrows D, Fratini S, Lee SY, Smith TJ III, Offenberg J, Dahdouh-Guebas F (2008) Faunal impact on vegetation structure and ecosystem function in mangrove forests: a review. Aquat Bot 89(2):186–200
Chan HT, Baba S 2009 Manual on guidelines for rehabilitation of coastal forests damaged by natural hazards in the Asia-Pacific Region. International Society for Mangrove Ecosystems and International Tropical Timber Organization, Okinawa http://www.mangrove.or.jp
Chan HT, Ong JE (eds) (2008) Guidelines for the rehabilitation of mangroves and other coastal forests damaged by tsunamis and other natural hazards in the Asia-Pacific region. ISME Mangrove Ecosystem Proceedings Number 5. International Society for Mangrove Ecosystems and International Tropical Timber Organization, Okinawa http://www.mangrove.or.jp
Chargoy Reyes MA, Tovilla Hernández C (2002) Restauración de áreas alteradas de manglar con *Rhizophora mangle* en la Costa de Chiapas. Madera y Bosques, Número especial:103–114
Chen L, Peng S, Li J, Lin Z, Zeng Y (2013) Competitive control of an exotic mangrove species: restoration of native mangrove forests by altering light availability. Restor Ecol 21:215–223
Cintrón G, Schaeffer-Novelli Y (1983) Introducción a la ecología del manglar. Oficina regional de ciencia y tecnología de la UNESCO para América Latina y el Caribe-ROSTLAC, Montevideo, p 109
Cintrón-Molero G (1992) Restoring mangrove systems. In: Thayer GW (ed) Restoring the nation's marine environment. Maryland Sea Grant, College Park, Chapter 6, pp 223–277
Corcoran E, Ravilious C, Skuja M (2007) Mangroves of Western and Central Africa (No. 26). UNEP/Earthprint. 88 p
Couwenberg J, Dommain R, Joosten H (2010) Greenhouse gas fluxes from tropical peatlands in south-east Asia. Glob Chang Biol 16:1715–1732
Crewz DW, Lewis RR 1991 An evaluation of historical attempts to establish emergent vegetation in marine wetlands in Florida. Technical paper no. 60. Florida Sea Grant College, Gainesville
Crutzen PJ, Stoermer EF (2000) The "Antropocene". Global Change Newsletter 41:12–13
Cuong CV, Brown S, Huu To H, Hockings M (2015) Using Melaleuca fences as soft coastal engineering for mangrove restoration in Kien Giang, Vietnam. Ecol Eng 81:256–265

Dahdouh-Guebas F, Collin S, Lo Seen D, Rönnbäck P, Depommier D, Ravishankar T, Koedam N (2006) Analysing ethnobotanical and fishery-related importance of mangroves of the East-Godavari Delta (Andhra Pradesh, India) for conservation and management purposes. J Ethnobiol Ethnomed 2:24

Dahdouh-Guebas F, Hettiarachchi S, Seen DL, Batelaan O, Sooriyarachchi S, Jayatissa LP, Koedam N (2005a) Transitions in ancient inland freshwater resource management in Sri Lanka affect biota and human populations in and around coastal lagoons. Curr Biol 15:579–586

Dahdouh-Guebas F, Jayatissa LP, Di Nitto D, Bosire JO, Seen DL, Koedam N (2005b) How effective were mangroves as a defence against the recent tsunami? Curr Biol 15:R443–R447

Dahdouh-Guebas F, Van Pottelbergh I, Kairo JG, Cannicci S, Koedam N (2004) Human-impacted mangroves in Gazi (Kenya): predicting future vegetation based on retrospective remote sensing, social surveys, and distribution of trees. Mar Ecol Prog Ser 272:77–92

Dahdouh-Guebas F, Zetterström T, Rönnbäck P, Troell M, Wickramasinghe A, Koedam N (2002) Recent changes in land-use in the Pambala-Chilaw Lagoon complex (Sri Lanka) investigated using remote sensing and GIS: conservation of mangroves vs. development of shrimp farming. Environ Dev Sustain 4:185–200

Dale PER, Knight JM, Dwyer PG (2014) Mangrove rehabilitation: a review focusing on ecological and institutional issues. Wetl Ecol Manag 22:587–604

Darkwa S, Smardon R (2010) Ecosystem restoration: evaluating local knowledge and management systems of fishermen in Fosu Lagoon, Ghana. Environ Pract 12:202–213

Das P, Basak UC, Das AB (1997) Restoration of the mangrove vegetation in the Mahanadi delta, Orissa, India. Mangrove Salt Marshes 1:155–161

De Ryck DJ, Koedam N, Van der Stocken T, van der Ven RM, Adams J, Triest L (2016) Dispersal limitation of the mangrove *Avicennia marina* at its South African range limit in strong contrast to connectivity in its core East African region. Mar Ecol Prog Ser 545:123–134

Di Nitto D, Erftemeijer PLA, van Beek JKL, Dahdouh-Guebas F, Higazi L, Quisthoudt K, Jayatissa LP, Koedam N (2013) Modelling drivers of mangrove propagule dispersal and restoration of abandoned shrimp farms. Biogeosciences 10:5095–5113

Donato DC, Kauffman JB, Murdiyarso D, Kurnianto S, Stidham M, Kanninen M (2011) Mangroves among the most carbon-rich forests in the tropics. Nat Geosci 4:293–297

Ellison AM (2000) Mangrove restoration: do we know enough? Restor Ecol 8:219–229

Ellison AM (2008) Managing mangroves with benthic biodiversity in mind: moving beyond roving banditry. J Sea Res 59:2–15

Ellison JC, Fiu M (2010) Vulnerability of Fiji's mangroves and associated coral reefs to climate change. A review. WWF South Pacific Office, Suva, p 50

Elster C (2000) Reasons for reforestation success and failure with three mangrove species in Colombia. For Ecol Manag 131:201–214

Estrada GCD, Soares MLG, Fernadez V, de Almeida PMM (2015) The economic evaluation of carbon storage and sequestration as ecosystem services of mangroves: a case study from southeastern Brazil. Int J Biodiv Sci Ecosyst Serv Manage 11:29–35

Ezcurra P, Ezcurra E, Garcillán PP, Costa M, Aburto-Oropeza M (2016) Coastal landforms and accumulation of mangrove peat increase carbon sequestration and storage. Proc Natl Acad Sci USA 113(16):4404-4409

Fernandes MEB, Nascimento AAM, Carvalho ML (2009) Effects of herbivory by *Hyblaea puera* (Hyblaeidae: Lepidoptera) on litter production in the mangrove on the coast of Brazilian Amazonia. J Trop Ecol 25:337–339

Ferreira AC, Ganade G, de Attayde JL (2015) Restoration versus natural regeneration in a neotropical mangrove: effects on plant biomass and crab communities. Ocean Coast Manag 110:38–45

Field CD (1995) Impact of expected climate change on mangroves. Hydrobiologia 295:75–81

Field CD (ed) (1996a) Restoration of mangrove ecosystems. International Society for Mangrove Ecosystems, Okinawa, p 250

Field CD (1996b) General guidelines for the restoration of mangrove ecosystems. In: Field CD (ed) Restoration of mangrove ecosystems. International Society for Mangrove Ecosystems, Okinawa, pp 233–250

Field CD (1999a) Rehabilitation of mangrove ecosystems: an overview. Mar Pollut Bull 37:383–392
Field CD (1999b) Mangrove rehabilitation: choice and necessity. Hydrobiologia 413:47–52
Flores Verdugo FJ, Moreno Casasola P, Agraz Hernández CM, López Rosas H, Benítez Pardo D, Travieso Bello AC (2007) La topografía y el hidroperiodo: dos factores que condicionan la restauración de los humedales costeros. Bol Soc Bot Mex suppl 80:33–47
Flores-de-Santiago F, Serrano D, Flores-Verdugo F, Monroy-Torres M (2017) Application of a simple and effective method for mangrove afforestation in semiarid regions combining nonlinear models and constructed platforms. Ecol Eng 103:244–255
Foell J, Harrison E, Stirrat RL (1999) Participatory approaches to natural resource management – the case of coastal zone management in the Puttalam District. Summary findings of DFID-funded research 'participatory mechanisms for sustainable development of coastal ecosystems' (Project R6977), School of African and Asian studies, University of Sussex, Falmer, 48 p
Giri C, Ochieng E, Tieszen LL, Zhu Z, Singh A, Loveland T, Masek J, Duke N (2011) Status and distribution of mangrove forests of the world using earth observation satellite data. Glob Ecol Biogeogr 20:154–159
Ghosh A, Schmidt S, Fickert T, Nüsser M (2015) The Indian Sundarban mangrove forests: history, utilization, conservation strategies and local perception. Diversity 7:149–169
Giri C, Pengra B, Zhu Z, Singh A, Tieszen LL (2007) Monitoring mangrove forest dynamics of the Sundarbans in Bangladesh and India using multi-temporal satellite data from 1973 to 2000. Estuar Coast Shelf Sci 73:91–100
Goessens A, Satyanarayana B, Van der Stocken T, Quispe Zuniga M, Mohd-Lokman H, Sulong I, Dahdouh-Guebas F (2014) Is Matang mangrove forest in Malaysia sustainably rejuvenating after more than a century of conservation and harvesting management? PLoS One 9(8):e105069
Hamilton S (2013) Assessing the role of commercial aquaculture in displacing mangrove forest. Bull Mar Sci 89(2):585–601
Hamilton S, Casey D (2016) Creation of a high spatio-temporal resolution global database of continuous mangrove forest cover for the 21st century. Glob Ecol Biogeogr 25:729–738
Hermansen TD, Roberts DG, Toben M, Minchinton TE, Ayre DJ (2015) Small urban stands of the mangrove *Avicennia marina* are genetically diverse but experience elevated inbreeding. Estuar Coast 1–10. doi:https://doi.org/10.1007/s12237-015-9955-1
Hilderbrand RH, Watts AC, Randle AM (2005) The myths of restoration ecology. Ecol Soc 10:19
Holl KD, Aide TM (2011) When and where to actively restore ecosystems? Forest Ecol Manag 261:1558–1563
Jayatissa LP, Kodikara KAS, Dissanayaka NP, Satyanarayana B (2016) Post-Tsunami assessment of coastal vegetation, with the view to protect coastal areas from ocean surges in Sri Lanka. In: Santiago-Fandiño V, Tanaka H, Spiske M (eds) Tsunamis and earthquakes in coastal environments, Coastal Research Library. Springer International Publishing, pp 47–64
Jerath M, Bhat M, Rivera-Monroy VH, Castaneda-Moya E, Simard M, Twilley RR (2016) The role of economic, policy, and ecological factors in estimating the value of carbon stocks in Everglades mangrove forests, South Florida, USA. Environ Sci Pol 66:160–169
Kairo JG, Dahdouh-Guebas F, Bosire J, Koedam N (2001) Restoration and management of mangrove systems – a lesson for and from the East African region. S Afr J Bot 67:383–389
Kaly UL, Jones GP (1998) Mangrove restoration: a potential tool for coastal management in tropical developing countries. Ambio 27:656–661
Kamali B, Hashim R (2011) Mangrove restoration without planting. Ecol Eng 37:387–391
Kauffman JB, Trejo HH, García MDCJ, Heider C, Contreras WM (2015) Carbon stocks of mangroves and losses arising from their conversion to cattle pastures in the Pantanos de Centla, Mexico. Wetl Ecol Manag:1–14
Kodikara KAS, Mukherjee N, Jayatissa LP, Dahdouh-Guebas F, Koedam N (2017) Have mangrove restoration projects worked? An in-depth study in Sri Lanka. Restor Ecol. https://doi.org/10.1111/rec.12492
Latif MA, 1996. An evaluation of mangrove restoration projects in eastern Australia. Unpublished thesis, Southern Cross University, Lismore

Leung JYS, Tam NFY (2013) Influence of plantation of an exotic mangrove species, *Sonneratia caseolaris* (L.) Engl. on macrobenthic infaunal community in Futian Mangrove National Nature Reserve, China. J Exp Mar Biol Ecol 448:1–9
Lewis RR 1982 Mangrove forests. pp 153–172 in RR Lewis (ed), Creation and restoration of coastal plant communities. CRC Press, Boca Raton. 219 p
Lewis RR, Gilmore RG Jr, Crewz DW, Odum WE (1985) Mangrove habitat and fishery resources of Florida. In: Seaman W Jr (ed) Florida aquatic habitat and fishery resources. Florida chapter. American Fisheries Society, Kissimmee, pp 281–233
Lewis RR (1990) Wetlands restoration/creation/enhancement terminology: suggestions for standardization. In: Kusler JA, Kentula ME (eds) Wetland creation and restoration: The status of the science. Island Press, Washington, pp 417–419
Lewis RR, Marshall MJ (1998) Principles of successful restoration of shrimp aquaculture ponds back to mangrove forests (abstract). Book of Abstracts of the Aquaculture '98 Meeting, 15–19 Feb 1998, Las Vegas, p 327. World Aquaculture Society
Lewis RR (2000) Ecologically based goal setting in mangrove forest and tidal marsh restoration. Ecol Eng 15:191–198
Lewis RR (2005) Ecological engineering for successful management and restoration of mangrove forests. Ecol Eng 24:403–418
Lewis RR, Hodgson AB, Mauseth GS (2005) Project facilitates the natural reseeding of mangrove forests (Florida). Ecol Restor 23:276–277
Lewis RR (2009) Methods and criteria for successful mangrove forest restoration. In: GME P, Wolanski E, Cahoon DR, Brinson MM (eds) Coastal Wetlands: An Integrated Ecosystem Approach. Elsevier Press, Amsterdam, pp 787–800
Lewis RR, Brown B (2014) Ecological mangrove rehabilitation – a field manual for practitioners. Version 3. Mangrove Action Project Indonesia, Blue Forests, Canadian International Development Agency, and OXFAM. p 275. www.mangroverestoration.com
Lewis RR, Gilmore RG (2007) Important considerations to achieve successful mangrove forest restoration with optimum fish habitat. Bull Mar Sci 80(3):823–837
López-Portillo J, Lara-Domínguez AL, Sáinz-Hernández E, Vásquez VM, Rodríguez-Rivera M, Martínez-García MC, Bartolo-Mateos O, Ortiz-Vela II, Alvarado G (2014) Restauración hidráulica en la laguna de Tampamachoco en el estado de Veracruz para la rehabilitación del manglar y de sus servicios ambientales. Instituto de Ecología A.C. Final Report SNIB-CONABIO, Proyect HH025, Mexico
Lovelock CE, Adame MF, Bennion V, Hayes M, O'Mara J, Reef R, Santini NS (2014) Contemporary rates of carbon sequestration through vertical accretion of sediments in mangrove forests and saltmarshes of South East Queensland, Australia. Estuar Coasts 37:763–771
Lugo AE (1978) Stress and ecosystems. In: Thorp JH, Gibbons JW (eds) Energy and environmental stress in aquatic systems. National Technical Information Service, Springfield, pp 62–98
Lugo AE (1992) Comparison of tropical tree plantations with secondary forests of similar age. Ecol Monogr 62:1–41
Lugo AE, Cintrón G, Goenaga C (1981) Mangrove ecosystems under stress. In: Barret GW, Rosenberg R (eds) Stress and natural ecosystems. Wiley, Chichester, pp 129–153
Lugo AE, Medina E (2014) Mangrove forests. In: Encyclopedia of natural resources: land. Taylor and Francis, New York, pp 343–352
Lugo AE, Snedaker SC (1974) The ecology of mangroves. Annu Rev Ecol Syst 5:39–64
Lunstrum A, Chen L (2014) Soil carbon stocks and accumulation in young mangrove forests. Soil Biol Biochem 75:223–232
Luo Z, Sun OJ, Xu H (2010) A comparison of species composition and stand structure between planted and natural mangrove forests in Shenzhen Bay, South China. J Plant Ecol 3:165–174
Macintosh DJ, Ashton EC, Havanon S (2002) Mangrove rehabilitation and intertidal biodiversity: a study in the Ranong mangrove ecosystem, Thailand. Estuar Coast Shelf Sci 55:331–345
Macintosh DJ, Mahindapala R, Markopoulos M (eds) (2012) Sharing lessons on mangrove restoration. Bangkok, Thailand: mangroves for the future and gland. IUCN, Gland, p 300

Matsui N, Suekuni J, Nogami M, Havanond S, Salikul P (2010) Mangrove rehabilitation dynamics and soil organic carbon changes as a result of full hydraulic restoration and re-grading of a previously intensively managed shrimp pond. Wetl Ecol Manag 18:233–242

McKee K, Faulkner P (2000) Restoration of biogeochemical function in mangrove forests. Restor Ecol 8:247–259

Mcleod E, Chmura GL, Bouillon S, Salm R, Björk M, Duarte CM, Lovelock CE, Schlesinger WH, Silliman BR (2011) A blueprint for blue carbon: toward an improved understanding of the role of vegetated coastal habitats in sequestering CO_2. Front Ecol Environ 9:552–560

Menezes GV, Schaeffer-Novelli Y, Poffo IRF, Eysink GGJ (2005) Recuperação de manguezais: um estudo de caso na Baixada Santista de São Paulo, Brasil. Braz J Aquat Sci Technol 9(1):67–74

Miyagi T (2013) Environmental characteristics of mangroves for restoration in the Yucatan Peninsula, Mexico. ISME Mangrove Ecosystems Occasional Papers No. 4. Tohoku Gakuin University, Action for Mangrove Reforestation (ACTMANG), Japan International Cooperation Agency (JICA), United Mexican States Ministry of the Environment and International Society for Mangrove Ecosystems (ISME)

Moreno-Mateos D, Power ME, Comín FA, Yockteng R (2012) Structural and functional loss in restored wetland ecosystems. PLoS Biol 10:e1001247

Mukherjee N, Dahdouh-Guebas F, Koedam N, Shanker K (2015) An interdisciplinary framework to evaluate bioshield plantations: Insights from Peninsular India. Acta Oecol 63:91–100

Nagelkerken I, Blaber SJM, Bouillon S, Green P, Haywood M, Kirton LG, Meynecke J-O, Pawlik J, Penrose HM, Sasekumar A, Somerfield PJ (2008) The habitat function of mangroves for terrestrial and marine fauna: a review. Aquat Bot 89:155–185

Nam VN, Sasmito SD, Murdiyarso D, Purbopuspito J, MacKenzie RA (2016) Carbon stocks in artificially and naturally regenerated mangrove ecosystems in the Mekong Delta. Wetl Ecol Manag 24:231–244

Ngeve MN, Van der Stocken T, Menemenlis D, Koedam N, Triest L (2016) Contrasting effects of historical sea level rise and contemporary ocean currents on regional gene flow of *Rhizophora racemosa* in Eastern Atlantic Mangroves. PLoS One 11(3):e0150950

Odum HT (1967) Work circuits and system stress. In: Young HE (ed) Symposium on primary productivity and mineral cycling in natural ecosystems. University of Maine Press, Orono, pp 81–138

Oliveira-Filho RR, Rovai AS, Menghini RP, Coelho-Jr C, Schaeffer-Novelli Y, Cintron-Molero G (2016) On the impact of the Brazilian Forest Code on mangroves: a comment to Ferreira and Lacerda (2016). Ocean Coast Manag 132:36–37

Ong JE, Gong WK (2013) Structure, function and management of mangrove ecosystems. ISME Mangrove Educational Book Series No. 2. International Society for Mangrove Ecosystems (ISME), Okinawa, Japan, and International Tropical Timber Organization (ITTO), Yokohama, Japan

Ortiz-Ruiz JJ (2004) Evaluación de la regeneración artificial de manglar en cinco zonas de la Ciénaga Grande de Santa Marta. Thesis, Jorge Tadeo Lozano Univ., Bogotá, Colombia 97 p

Outterson AH (2014) Evaluating the Progress of a Mangrove Reforestation Project on Isla Galeta, Colon. Independent Study Project (ISP) Collection. Paper 1997. http://digitalcollections.sit.edu/isp_collection/1997

Pagliosa PR, Rovai AS, Fonseca AL (2012) Carbon mismanagement in Brazil. Nat Clim Chang 2:764

Peng YG, Xu ZC, Liu MC (2012) Introduction and ecological effects of an exotic mangrove species *Sonneratia apetala*. Acta Ecol Sin 32:2259–2270

Primavera JH (2000) Development and conservation of Philippine mangroves: institutional issues. Ecol Econ 35:91–106

Primavera J (2006) Overcoming the impacts of aquaculture on the coastal zone. Ocean Coast Manag 49:531–545

Primavera JH, Esteban JMA (2008) A review of mangrove rehabilitation in the Philippines: successes, failures and future prospects. Wetl Ecol Manag 16:345–358

Primavera JH, Rollon RN, Samson MS (2011) The pressing challenges of mangrove rehabilitation: pond reversion and coastal protection. In: Chicharo L, Zalewski M (eds) Ecohydrology and restoration. Elsevier, Amsterdam, pp 217–244

Primavera JH, Savaris JD, Bajoyo B, Coching JD, Curnick DJ, Golbeque R, Guzman AT, Henderin JQ, Joven RV, Loma RA, Koldewey HJ (2012) Manual on community-based mangrove rehabilitation –Mangrove Manual Series 1. Zoological Society of London, London. viii + 240 pp.

Primavera JH, Yap WG, Savaris JP, Loma RA, Moscoso ADE, Coching JD, Montilijao CL, Poignan RP, Tayo ID (2014) Manual on mangrove reversion of abandoned and illegal brackishwater fishponds – Mangrove Manual Series 2. Zoological Society of London, London. ZSL. xii + 108 p

Proffitt CE, Devlin DJ (2005) Long-term growth and succession in restored and natural mangrove forests in southwestern Florida. Wet Ecol Manage 13:531–551

Pulver TR 1976 Transplant techniques for sapling mangrove trees, *Rhizophora mangle*, *Laguncularia racemosa*, *Avicennia germinans* in Florida. St. Petersburg, FL, Florida Department of Natural Resources, Marine Research Laboratory (Florida Marine Research Publications, 22)

Ren HR, Wu X, Ning T, Huang G, Wang J, Jian S, Lu H (2011) Wetland changes and mangrove restoration planning in Shenzhen Bay, Southern China. Landsc Ecol Eng 7:241–250

Ren H, Guo Q, Liu H, Li J, Zhang Q, Xu H, Xu F (2014) Patterns of alien plant invasion across coastal bay areas in southern China. J Coast Res 30:448–455

Ren H, Jian S, Lu H, Zhang Q, Shen W, Han W, Yin Z, Guo Q (2008) Restoration of mangrove plantations and colonization by native species in Leizhou Bay, South China. Ecol Res 23:401–407

Ren H, Lu HF, Shen WJ, Huang C, Guo QF, Li ZA, Jian SG (2009) *Sonneratia apetala* Buch. Ham in the mangrove ecosystems of China: an invasive species or restoration species? Ecol Eng 35:1243–1248

Rey JR, Carlson DB, Brockmeyer RE Jr (2012) Coastal wetland management in Florida: environmental concerns and human health. Wetl Ecol Manag 20:197–211

Reyes Cargoy MA and Tovilla Hernández C (2002) Restauración de áreas alteradas de manglar con Rhizophora mangle en la costa de Chiapas. Madera y Bosques. Número especial. 103–114

Rivera-Monroy VH, Twilley RR, Mancera E, Alcántara-Egurren A, Castañeda-Moya E, Casas Monroy O, Reyes P, Restrepo J, Perdomo L, Campos E, Cotes G, Viloria E (2006) Aventuras y desventuras en Macondo: rehabilitación de la Ciénaga Grande de Santa Marta, Colombia. Ecotropicos 19:72–93

Rivera-Monroy VH, Twilley RR, Mancera-Pineda JE, Madden CJ, Alcantara-Eguren A, Moser EB, Jonsson BF, Castañeda-Moya E, Casas-Monroy O, Reyes-Forero P, Restrepo J (2011) Salinity and Chlorophyll a as performance measures to rehabilitate a mangrove-dominated deltaic coastal region: the Ciénaga Grande de Santa Marta-Pajarales Lagoon Complex, Colombia. Estuar Coasts 34:1–19

Roderstein M, Perdomo L, Villamil C, Hauffe T, Schnetter ML (2014) Long-term vegetation changes in a tropical coastal lagoon system after interventions in the hydrological conditions. Aquat Bot 113:19–31

Rovai AS, Barufi JB, Pagliosa PR, Scherner F, Torres MA, Horta PA, Simonassi JC, Quadros DPC, Borges DLG, Soriano-Sierra EJ (2013) Photosynthetic performance of restored and natural mangroves under different environmental constraints. Environ Pollut 181:233–241

Rovai AS, Soriano-Sierra EJ, Pagliosa PR, Cintrón G, Schaeffer-Novelli Y, Menghini RP, Coelho-Jr C, Antunes Horta P, Lewis RR, Simonassi JC, Andrade Alves JA, Boscatto F, Dutra SJ (2012) Secondary succession impairment in restored mangroves. Wetl Ecol Manag 20:447–459

Saenger P (1996) Mangrove restoration in Australia: a case study of Brisbane International Airport. In: Field C (ed) Restoration of mangrove ecosystems. ISME/ITTO, Okinawa, pp 36–51

Saenger P (2002) Mangrove ecology, silviculture and conservation. Kluwer Academic Publishers, Dordrecht, p 360

Saenger P (2011) Mangroves: sustainable management in Bangladesh. In: Günter S, Weber M, Stimm B, Mosandl R (eds) Silviculture in the tropics. Tropical forestry 8. Springer, Berlin/Heidelberg, pp 339–347

Saenger P, Khalil A (2011) Regional guidelines for mangrove restoration in the Red Sea and Gulf of Aden. PERSGA Guidelines No. GD-0010, PERSGA, Jeddah
Saenger P, Sankaré Y, Perry T (1996) Review of selection criteria and ecological guidelines for mangrove restoration studies. Large Marine Ecosystem Project for the Gulf of Guinea. UNIDO, UNDP, NOAA and UNEP, Abidjan
Saenger P, Siddiqi NA (1993) Land from the sea: the mangrove afforestation program of Bangladesh. Ocean Coast Manag 20:23–39
Salmo SG III, Lovelock C, Duke NC (2013) Vegetation and soil characteristics as indicators of restoration trajectories in restored mangroves. Hydrobiologia 720:1–18
Samson MS, Rollon RN (2008) Growth performance of planted red mangroves in the Philippines: revisiting forest management strategies. Ambio 37:234–240
Sandoval-Castro E, Dodd RS, Riosmena-Rodríguez R, Enríquez-Paredes LM, Tovilla-Hernández C, López-Vivas JM, Aguilar-May B, Muñiz-Salazaret R (2014) Post-glacial expansion and population genetic divergence of mangrove species *Avicennia germinans* (L.) Stearn and *Rhizophora mangle* L. along the Mexican Coast. PLoS One 9:e93358
Schaeffer-Novelli Y, Cintrón-Molero G, Cunha-Lignon M, Coelho-Jr C (2005) A conceptual hierarchical framework for marine coastal management and conservation: a "Janus-Like" approach. J Coast Res SI 42:191–197
Selvam V, Ravichandran KK, Gnanappazham L, Navamuniyammal M (2003) Assessment of community-based restoration of Pichavaram mangrove wetland using remote sensing data. Curr Sci India 85:794–798
Selvam V, Ravishankar T, Karunagaran VM, Ramasubramanian R, Eganathan P, Parida AK (2005) Toolkit for establishing coastal bioshield. MS Swaminathan Research Foundation Centre for Research on Sustainable Agriculture and Rural Development, Chennai, p 117
Shafer DJ, Roberts TH (2008) Long-term development of tidal mitigation wetlands in Florida. Wetl Ecol Manag 16:23–31
Shamsudin I, Raja Barizan RS, Azian M, Wan Nurzalia WS (2008) Rehabilitation of mangroves in Peninsular Malaysia after the 2004 Indian Ocean tsunami. In: Chan HT, Ong JE (eds) Guidelines for the rehabilitation of mangroves and other coastal forests damaged by Tsunamis and other natural hazards in the Asia-Pacific Region, ISME-ITTO, pp 9–14
Siikamäki J, Sanchirico JN, Jardine S, McLaughlin D, Morris D (2013) Blue carbon: coastal ecosystems, their carbon storage, and potential for reducing emissions. Environ Sci Policy Sustain Devel 55(6):14–29
Siikamäki J, Sanchirico JN, Jardine SL (2012) Global economic potential for reducing carbon dioxide emissions from mangrove loss. Proc Natl Acad Sci USA 109:14369–14374
Snedaker SC, Biber PD (1996) Restoration of mangroves in the United States of America. In: Field C (ed) Restoration of mangrove ecosystems. ISME/ITTO, Okinawa, pp 170–188
Stanley OD, Lewis RR (2009) Strategies for mangrove rehabilitation in an eroded coastline of Selangor, peninsular Malaysia. J Coastal Devel 12:144–156
Stephen MF 1984 Mangrove restoration in Naples, Florida. In: FJ Webb (ed) Proceedings of the 10th annual conference on wetland restoration and creation. Hillsborough Community College, Tampa, pp 201–216
Stevenson NJ, Lewis RR, Burbridge PR (1999) Disused shrimp ponds and mangrove rehabilitation. In: Streever WJ (ed) An international perspective on wetland rehabilitation. Kluwer Academic Publishers, Dordrecht, pp 277–297
Stubbs BJ, Saenger P (2002) The application of forestry principles to the design, execution and evaluation of mangrove restoration projects. Bois et Forêts des Tropiques 273:5–21
Tamin NM, Zakaria R, Hashim R, Yin Y (2001) Establishment of *Avicennia marina* mangroves on accreting coastline at Sungai Haji Dorani, Selangor, Malaysia. Estuar Coast Shelf Sci 94:334–342
Tang Y, Fang Z, Chen K, Zhang Z, Zhong Y, An D, Yang X, Liao B (2012) Ecological influence of exotic plants of *Sonneratia apetala* on understory macrofauna. Acta Oceanol Sin 31:115–125
Tangah J, Bajau FE, Jilimin W, Baba S, Chan HT, Kezuka M (2015) Rehabilitation of mangroves in Sabah – The SFD-ISME Collaboration (2011–2014). Sabah Forestry Department,

International Society for Mangrove Ecosystems and Tokio Marine & Nichido Fire Insurance Co., Ltd. 56 pp. http://www.mangrove.or.jp/isme/english/books/Rehabilitation-in-Sabah.pdf
Teas HJ (1977) Ecology and restoration of mangrove shorelines in Florida. Environ Conserv 4:51–58
Thomas S (2014) Blue carbon: Knowledge gaps, critical issues, and novel approaches. Ecol Econ 107:22–38
Tobey J, Clay J, Vergne P (1998) The economic, environmental and social impacts of shrimp farming in Latin America. Coastal Resources Center, University of Rhode Island
Tomlinson PB (1986) The Botany of Mangrove. Cambridge University Press, Cambridge, 413 pp
Tovilla C, Román AV, Simuta GM, Linares RM (2004) Recuperación del manglar en la Barra del Río Cahoacán, en la costa de Chiapas. Madera y Bosques, Special no 2:77–91
Trump K, Gattenlöhner U (2015) Mangrove restoration guide: best practices and lessons learned from a community-based conservation project. Global Nature Fund, Radolfzell, 60 pp
Tsuruda K (2013) Silviculture manual for mangrove restoration in the Yucatan Peninsula, Mexico. Mangrove Ecosystems Occasional Papers ISME 4:23–34
Turner R, Lewis RR (1997) Hydrologic restoration of coastal wetlands. Wetl Ecol Manag 4:65–72
Twilley RR, Rivera-Monroy V (2005) Developing performance criteria using simulation models of mangrove ecosystem restoration: a case study of the Florida Coastal Everglades. J Coast Res 40:79–93
Twilley RR, Rivera-Monroy VH, Chen R, Botero L (1998) Adapting an ecological mangrove model to simulate trajectories in restoration ecology. Mar Pollut Bull 37:404–419
Underwood AJ (1997) Experiments in ecology: their logical design and interpretation using analysis of variance. Cambridge University Press, Cambridge
Valiela I, Bowen JL, York JK (2001) Mangrove forests: one of the world's threatened major tropical environments at least 35% of the area of mangrove forests has been lost in the past two decades, losses that exceed those for tropical rain forests and coral reefs, two other well-known threatened environments. Bioscience 51:807–815
Vilardy SP, Gonzalez JA, Martin-Lopez B, Montes C (2011) Relationships between hydrological regime and ecosystem services supply in a Caribbean coastal wetland: a social-ecological approach. Hydrolog Sci J 56:1423–1435
Vogt J, Lin Y, Pranchai A, Frohberg P, Mehlig U, Berger U (2014) The importance of conspecific facilitation during recruitment and regeneration: a case study in degraded mangroves. Basic Appl Ecol 15:651–660
Vovides AG, Bashan Y, López-Portillo J, Guevara R (2011) Nitrogen fixation in preserved, reforested, naturally regenerated and impaired mangroves as an indicator of functional restoration in mangroves in an arid region of Mexico. Restor Ecol 19:236–244
Wadsworth FH (1959) Growth and regeneration on the white mangrove *Laguncularia racemosa* in Puerto Rico, Caribbean. Forester 20:59–71
Walters BB (1997) Human ecological questions for tropical restoration: experiences from planting native upland trees and mangroves in the Philippines. Forest Ecol Manag 99:275–290
Wilkie ML, Fortuna S (2003) Status and trends in mangrove area extent worldwide. Forest Resources Assessment Working Paper 63. Forestry Department, Food and Agriculture Organization of the United Nations, Rome
Zaldívar-Jiménez MA, Herrera Silveira JA, Teutli Hernández C, Comin FA, Andrade JL, Coronado Molina C, Pérez Ceballos R (2010) Conceptual framework for mangrove restoration in the Yucatán Peninsula. Ecol Restor 28:333–342

Chapter 11
Advancing Mangrove Macroecology

Victor H. Rivera-Monroy, Michael J. Osland, John W. Day, Santanu Ray, Andre Rovai, Richard H. Day, and Joyita Mukherjee

11.1 Introduction

Macroecology is broadly defined as a discipline that uses statistical analyses to investigate large-scale, universal patterns in the distribution, abundance, diversity, and organization of species and ecosystems (Brown 1995; Enquist et al. 1995; Smith et al. 2008), including the scaling of ecological processes and structural and functional relationships (e.g., McGill and Collins 2003). Though related to biogeography (covered in Chap. 2), macroecology differs in that it also addresses associations between organisms and their environment (i.e., traditional ecological questions) at large spatial and temporal scales (Keith et al. 2012). Thus, macroecology as a transdiscipline explores the boundaries where ecology, biogeography, paleontology, landscape ecology, and macroevolution come together (Brown 1995).

Macroecology attempts to provide an explicit mechanistic ecological understanding about questions that deal with the distribution, abundance, energetics, and

V.H. Rivera-Monroy (✉) • J.W. Day
Department of Oceanography and Coastal Sciences, College of the Coast and Environment, Louisiana State University, Baton Rouge, LA 70803, USA
e-mail: vhrivera@lsu.edu

M.J. Osland • R.H. Day
U.S. Geological Survey, 700 Cajundome Blvd., Lafayette, LA 70506, USA

S. Ray • J. Mukherjee
Department of Zoology, System Ecology and Ecological Modelling Laboratory, Visva-Bharati University, Santiniketan 731235, India

A. Rovai
Departamento de Ecologia e Zoologia, Universidade Federal de Santa Catarina, Florianópolis, SC 88040-900, Brazil

Department of Oceanography and Coastal Sciences, College of the Coast and Environment, Louisiana State University, Baton Rouge, LA 70803, USA

V.H. Rivera-Monroy et al. (eds.), *Mangrove Ecosystems: A Global Biogeographic Perspective*, https://doi.org/10.1007/978-3-319-62206-4_11

interaction networks of individuals and species across multiple spatial and temporal scales (Keith et al. 2012; Peters et al. 2014). Macroecological approaches have been used in ecological studies of coastal and oceanic environments (Witman and Roy 2009), although not as often as for terrestrial ecosystems (Blackburn and Gaston 2003). The approach explicitly recognizes the importance of using models of the spatial and temporal variation in ecosystem function and structure to better understand complex interactions among entities composing these systems, no matter what the scale of the analysis (Marquet et al. 2004). For example, the urgency to assess the effects of global change upon the interactions between ecological, social, and economic drivers has underscored the potential utility of conceptual frameworks such as macroecology (Fowler 2008; Brown et al. 2011, 2014; Burger et al. 2012; Nekola et al. 2013; McBride et al. 2014; Peters et al. 2014). In coastal regions, macroecological approaches can advance the understanding of how non-linear responses in natural systems can be triggered by human impacts at local, regional, and global scales (Kerr et al. 2007; Day et al. 2008; Barbier et al. 2011).

There are many examples of studies that have applied a macroecological approach without explicitly acknowledging its connection to the realm of macroecology (e.g., Simberloff and Wilson 1969). In the first paper devoted explicitly to the application of macroecology to mangrove ecosystems, Ellison (2002) pointed out that "a focus on the larger-scale contexts that constrain local processes (a macroecology of mangroves) will provide us with new insights into the structure and function of mangrove ecosystems." Using relationships between local species richness and latitude, longitude, regional diversity, and structural properties of leaf traits, he proposed that ecological theory developed in upland forests could be directly applied to mangrove forests. Since that first paper on mangrove macroecology, no further work has been published that explicitly uses this conceptual framework to advance our understanding of mangrove-dominated ecosystems. An ISI Web of Science search identifying published work explicitly listing the words "macroecology" and "mangrove" in the title, keywords, and abstract produced only two papers (i.e., Whittaker et al. 2001; Ellison 2002). In contrast, macroecology is a term and an approach that has been frequently incorporated in analyses of terrestrial and marine ecosystems (Fig. 11.1). We acknowledge that there are many mangrove-focused studies that have used macroecological approaches without ever explicitly mentioning the term macroecology (see examples provided in this chapter); however, in general, there is a need to more explicitly incorporate macroecology into the study of mangroves.

We suggest that macroecology can be used to embrace ecological theory regarding complex structural and functional spatiotemporal patterns. Macroecology can even be used to assess and anticipate human impacts to mangrove ecosystems. Although mangrove forests provide a wide range of ecosystem services to society, they are among the most anthropogenically impacted coastal ecosystems in the world (Ewel et al. 1998; Alongi 2009; Barbier et al. 2011; Twilley and Day 2012; Mukherjee et al. 2012) (see Chaps. 8 and 9). In some cases, there is a lack of data, information, and knowledge (Baker and Bowker 2007) needed to thoroughly

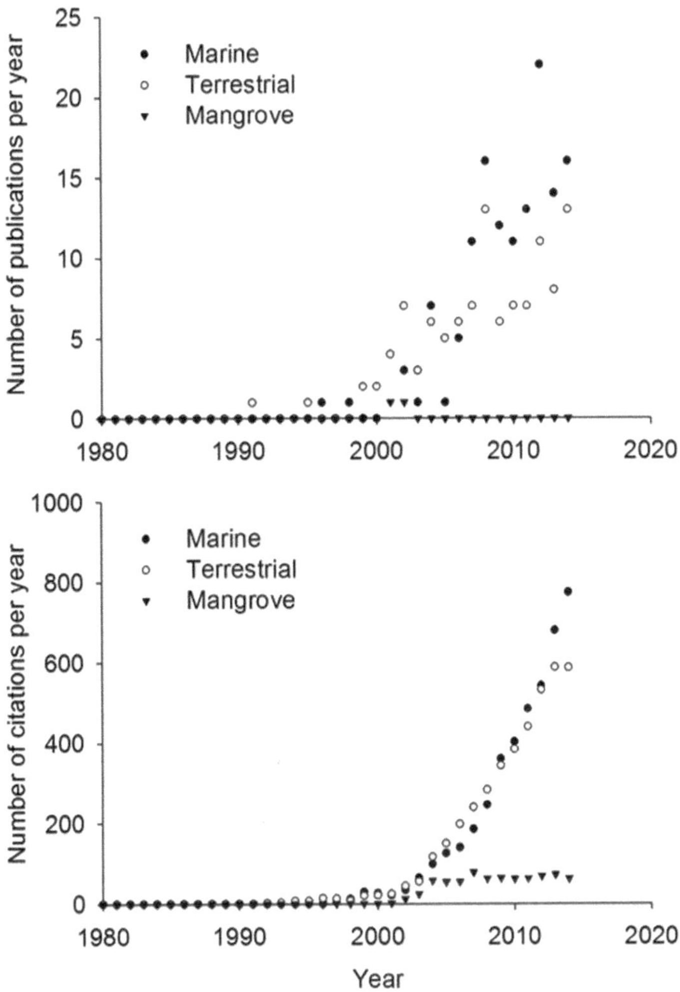

Fig. 11.1 The number of macroecology-related publications (*top*) and citations (*bottom*) for marine, terrestrial, and mangrove ecosystems for the period 1980–2014. These data come from an ISI Web of Science search for publications that used the following search topic terms: (1) macroecology and marine; (2) macroecology and terrestrial; or (3) macroecology and mangrove. The mangrove-focused search revealed only two papers (Ellison 2002; Whittaker et al. 2001)

evaluate how direct and indirect human impacts affect mangrove survival and persistence. Unfortunately, the indirect impacts are typically subtler than the direct impacts (e.g., deforestation) since major environmental changes are not usually immediate, as is the case, for example, with incremental and irreversible hydrological changes in coastal and estuarine environments. Indeed, the mangrove literature abounds with examples of how changes in river discharge or coastal hydrology, such as levee construction for flood protection or road construction, modify critical environmental processes, including sediment delivery, hydroperiod, and salinity, which can cause major mangrove diebacks (Hicks and Burns 1975; Odum and Tohannes 1975; Ellison 1998; Rivera-Monroy et al. 2011). The causes and rates of mangrove mortality due to indirect impacts at the global scale are not well known. However, it is understood that assessing the effects of management decisions requires a comprehensive approach that explicitly identifies the interactions between environmental variables regulating mangrove ecosystems and human impacts at different spatiotemporal scales. Since macroecology is a discipline of synthesis, it can offer both an operational approach and foundation for designing and implementing comparative ecological studies at the global scale to conserve mangrove resources in the long term.

In this chapter, we provide examples of how a macroecological conceptual framework can be used in mangrove ecological studies. Our examples include evaluations of the variation in mangrove forest ecosystem structure and function in relation to macroclimatic drivers (e.g., temperature and rainfall regimes) and climate change. Additional examples include work focused upon the continental distribution of aboveground net primary productivity (see also Chap. 5) and carbon storage, which are research areas that have been rapidly advancing in recent years. These research priorities are good examples of the value of a macroecological perspective for the understanding of global- and regional-scale effects of both changing environmental conditions and management actions on ecosystem structure, function, and the supply of goods and services. Beginning with mechanistic hypotheses and a local-scale understanding of important processes, macroecology has used modeling as a critical component in the synthesis of information. We therefore present current trends in mangrove modeling approaches and their potential utility to test hypotheses about mangrove structural and functional properties. Given the importance of experimental work at intermediate (i.e., regional) scales, we also discuss the potential integration of restoration and rehabilitation projects (see also Chap. 10) in macroecology studies. These projects could help to advance the critical selection and conservation of ecosystem services when managing mangrove resources. We by no means provide an exhaustive review of potential mangrove ecological studies within a macroecological framework. Our goal here is to show the prospective utility of macroecology-based studies that could answer process-based ecological questions and help expand long-term ecological studies at regional and continental scales. We also aim to show how such studies could help improve conservation outcomes and ensure that mangroves will continue to provide ecosystem goods and services for current and future generations.

11.2 Macroecology of Mangrove-Dominated Ecosystems

As mentioned above, macroecology addresses associations between organisms and their environment based on statistical patterns of abundance, distribution, and diversity (Brown 1995). To discern the relative importance of these interactions, it is necessary to conceptually separate the entities and mechanisms involved to elucidate these patterns at different spatiotemporal scales. Studies on the biogeography of mangroves have provided valuable information on the global distribution and taxonomic variability of mangrove species in relation to both sea and atmospheric temperatures (see Chap. 2). However, with the advent of improved and more relevant climate data, further work is needed to better understand the relative importance of different climatic drivers in controlling mangrove ecosystem structure and function at the global scale, especially tropical-temperate and/or humid-arid climatic transitions zones (e.g., Mandal et al. 2009; Osland et al. 2013, 2014b; Cavanaugh et al. 2015). Also, we still lack general models, for example, ranking the statistical and mechanistic significance of local environmental variables (e.g., salinity, tidal regimes, phosphorus and nitrogen limitation, hydrogen sulfide) in determining mangrove zonation, net primary productivity, or interspecific competition at local and regional scales within and across biographical regions (see modeling Sect. 11.3 of this chapter). Comparative studies assessing how species diversity and associated species-specific adaptations respond to changing coastal environmental conditions in the species-rich Indo-west-Pacific and the depauperate Atlantic-east-Pacific regions are needed given the large differences in phylogeny, taxa hybridization, and endemic origin (see Chap. 2).

Understanding the processes that define how patterns of biological diversity vary in space and time is one of the key questions in the field of ecology. Although major advances have been achieved in a number of terrestrial ecosystems during the last 200 years (Whittaker et al. 2001; Pfeifer et al. 2012), mangrove ecological studies at the macro-scale have not kept pace. This knowledge gap is understandable because of the historical trend in the number of mangrove studies around the world, where most of the focus has centered on assessing mangrove species diversity and spatial distribution while the focus on ecosystem functional studies (e.g., nutrient and carbon cycling, species physiological thresholds) has begun relatively recently (Rivera-Monroy et al. 2004; Feller et al. 2007, 2010; McKee 2011; Quisthoudt et al. 2012; Record et al. 2013) in comparison to tropical and temperate terrestrial forests (Schaeffer-Novelli and Cintron 1990). Among the largest challenges to understanding the variation in functional properties of mangrove ecosystems (e.g., biodiversity, primary and secondary productivity) is our ability to scale up from local-scale process-based knowledge to global- or region-scale ecological patterns (Kennedy et al. 2014). The solution to this problem is not easy and remains one of the most active topics of research in ecology overall. Because ecological patterns depend on the selection of the spatial scale of local studies, mangrove ecology advancement is still limited to extrapolation from location-specific small-scale studies to larger landscape scales; a challenge already addressed in other type of ecosystems (e.g., arid ecosystems, tropical and temperate forests) and in species-specific studies at the global scale (Pfeifer et al. 2012; Peters et al. 2014).

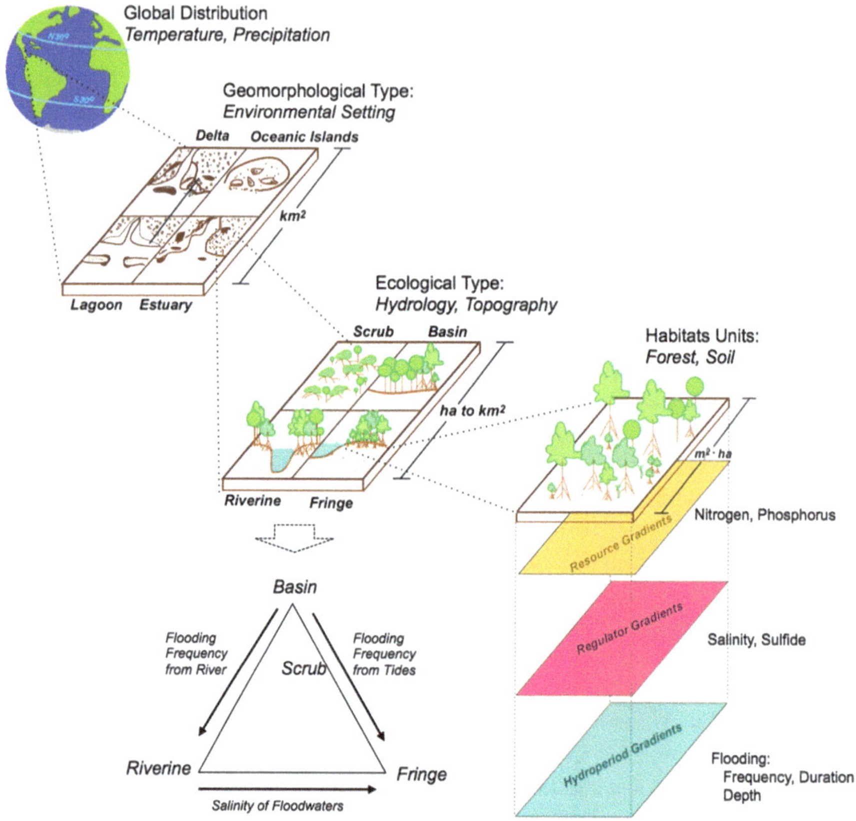

Fig. 11.2 Hierarchical classification for mangrove-dominated ecosystems describing abiotic controls on mangrove structural and functional properties at the global, regional and local scale. Factors in each spatial level control processes defining soil properties, fertility gradients, and mangrove productivity (Modified from Twilley and Rivera-Monroy 2005; Ewel et al. 1998)

11.2.1 Linking Local and Regional Scales to the Global Dimension

The hydrogeomorphic classification of mangrove wetlands proposed by Lugo and Snedaker (1974) in the mid-1970s brought the local and regional scales together to improve our understanding of the relative importance of some environmental factors in determining functional attributes (Fig. 11.2), especially for the Atlantic-east-Pacific region. Lugo and Snedaker's classification represented an operational link between local and regional spatial scales allowing an explicit linkage between forest structure and function and regional environmental settings as represented by four distinct ecotypes (riverine, basin, fringe, scrub). Further, because mangrove forests occupy the intertidal and supratidal zones of diverse coastal settings (river deltas, muddy coasts, coastal lagoons, rias, continental/oceanic island carbonate settings), these four basic ecotypes are useful in defining a macroecological approach that can be used to identify mangrove structural and functional patterns (Thom 1982; Twilley 1997; Woodroffe 2002). Thus, recent studies on the interactions between coastal settings and mangrove ecological function (e.g., net primary productivity) show a correlation that reflects the importance of site-specific spatial patterns of nutrient fertility, environmental stressors (Twilley and Rivera-Monroy 2005), and hydroperiod (Twilley 1997; Twilley and Rivera-Monroy 2009; Castañeda-Moya et al. 2011). Figure 11.2 shows a hierarchical classification of spatial scales and processes controlling an array of factors, from the global distribution of mangroves (i.e., temperature and precipitation) and the development of different geomorphic settings (Woodroffe 2002), to colonization by different mangrove ecotypes as they respond to diverse gradients in nutrients (resource), salinity (regulators/stressors), and hydroperiod (frequency, duration, and depth of inundation). We use this hierarchical model as a tool to further propose the development of a macroecological approach and to show examples of knowledge gaps to be addressed among hierarchical scales (Rivera-Monroy et al. 2004; Twilley and Rivera-Monroy 2005; Reef et al. 2010).

In terrestrial and marine systems, macroecology is relatively advanced in its use of statistical methods (Nee 2002; Urban 2005). However, the theoretical basis of the identified statistical relationships is sometimes poorly developed (Blackburn and Gaston 2003; Smith et al. 2014). The lack of macroecological studies in mangrove systems is partially due to the spatial scale employed by researchers and the quality and quantity of the information used to develop and validate statistical models. For example, in predictions of aboveground biomass, carbon sequestration, and species dominance at the continental scale (Peters et al. 2014), explicit formulations of theoretical models and the robust derivation of statistical expectations from those models is one of macroecology's major challenges (e.g., Gaston and Blackburn 1999). In the case of the few published macro-level mangrove studies, the challenge is underscored by the lack of global hypotheses, especially when related to the degree of interaction among specific biotic and abiotic factors in the context, for example, of niche theory and assembly rules development (i.e., local scale) (Clarke 2014). Macroecological studies in other ecosystems already show how the incorporation of local processes can influence large-scale patterns of species distribution and productivity; hence, in the case of mangroves, it is paramount to assess the relative importance of biotic versus abiotic responses in experimental work at different scales (Feller et al. 2010; Reef et al. 2010). This comparison is not an easy task given, for example, the confounding effects of the large interactions among salinity gradients, species dominance, and in situ topography, to name just a few critical factors. Testing hypotheses at the local and regional scale is necessary if robust predictions at the global scale are to be expected. In fact, implementing hypothesis-driven studies in the context of mangrove rehabilitation and restoration efforts presents a promising option for scaling up results from the local and regional to the global scale (see chapter Sect 11.4 below).

11.2.2 Two Examples: Carbon Storage and Response to Climate Change

As illustrations of the issues discussed above, we describe here the advances and limitations in developing a macroecological approach for mangrove ecosystems using two current issues of economic and ecological relevance in the context of global climate change. These examples help underscore the importance of linking potential large-scale predictions with small-scale processes to generate a consistent signal in macroecological patterns and test hypotheses at different spatial scales as performed in other type of ecosystems (Smith et al. 2014). Whereas the first example in this section focuses on soil carbon (Jardine and Siikamäki 2014), the second example focuses on mangrove forest responses to climate change (Osland et al. 2013, 2016).

11.2.2.1 Global Controls of Carbon Storage in Mangroves

The discrepancy in operationally linking small to large spatial scales in mangrove ecosystems is currently more evident as mangrove wetlands are recognized as large reservoirs of organic carbon in subtropical and tropical coastal latitudes (e.g., Donato et al. 2011; Suárez-Abelenda et al. 2014). This ecosystem attribute, widely recognized as a desired ecosystem service, can play a potential role not only in regulating climate change, but also in halting potential mangrove deforestation trends given its potential economic importance in carbon markets (Alongi 2011). As a result, recent studies now extrapolate site-specific soil carbon and aboveground biomass values to the continental scale using statistical modeling, thus offering a "snap-shot," along with associated uncertainties, of global carbon storage in mangrove ecosystems (see also Chaps. 5 and 6). Paradoxically, comprehensive carbon cycling studies assessing the seasonal and inter-annual variation of the role of mangrove forests as actual sinks or sources of carbon at different latitudes and within climatic zones are still lacking (Alongi 2009; Rivera-Monroy et al. 2013). The need to evaluate net fluxes of organic and inorganic carbon across mangrove forest boundaries at large spatial scales is critical for determining the impact of increasing ambient temperatures and atmospheric CO_2 concentrations on the complex interactions among net ecosystem production, heterotrophic respiration, and net primary production (NPP) (Alongi 2014). Discerning the relative importance of these net fluxes is necessary since key ecological mechanisms, such as microbial decomposition, could limit soil carbon storage in the long term (e.g., van Groenigen et al. 2014). Currently there are no studies in a single location that could help "construct" whole mangrove ecosystem carbon budgets in tropical and subtropical latitudes, although research is ongoing in some regions (Bouillon et al. 2008; Rivera-Monroy et al. 2013; Troxler et al. 2013) (see also Chaps. 5 and 6).

Carbon sequestration and carbon storage values are constrained by the strong interaction between NPP and environmental factors in coastal ecosystems, including

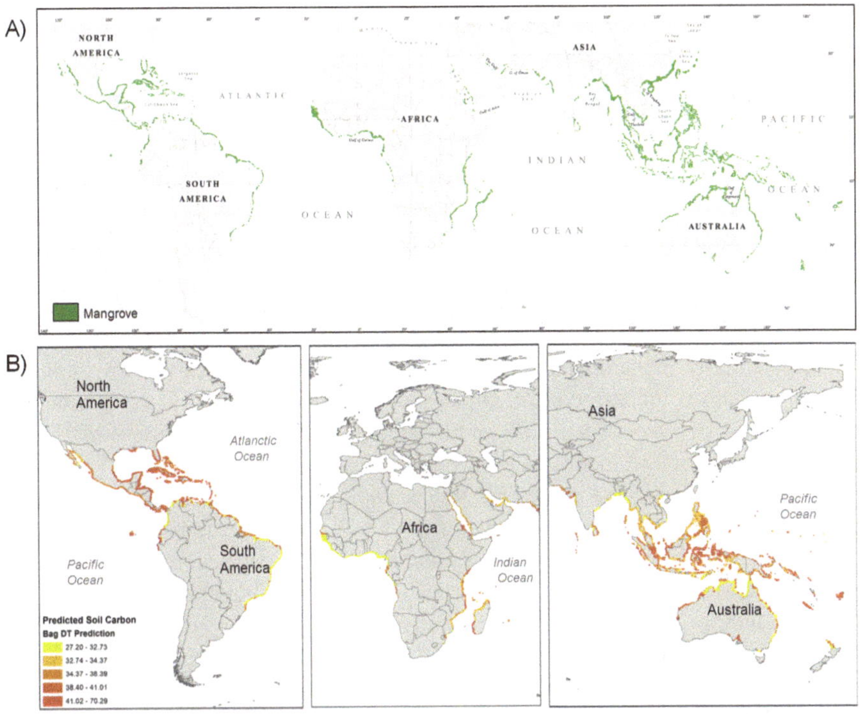

Fig. 11.3 The global distribution of: (**a**) mangrove wetlands (modified from Giri et al. 2011) and (**b**) mangrove soil carbon density (mg cm^{-3}) extrapolated at the global scale using statistical models (modified from Jardine and Siikamäki 2014)

mangrove wetlands. Differences in NPP can be readily observed among riverine, fringe, and scrub forests due to dissimilarities in freshwater sources, nutrient input from both upland and estuarine sources and salinity controlled by tidal flushing, precipitation, and evaporation (Fig. 11.2). Differences in NPP can also be readily observed across climatic gradients (Lot-Helgueras et al. 1975). Some of the research issues related to carbon cycling that can be better addressed via a macroecological perspective include: (1) the partitioning of organic carbon aboveground versus belowground; (2) the differences in carbon storage among forests with different structural and functional properties; and (3) biotic and abiotic factors (e.g., climatic, hydrologic, geomorphic, species composition) controlling the observed patterns.

Current soil carbon storage estimates in mangrove forests illustrate the limitations in trying to address those issues at local, regional, and global scales. Recently, Jardine and Siikamäki (2014) produced a global map of mangrove soil carbon (Fig. 11.3). In that study, carbon concentration values previously compiled by other literature reviews (Chmura et al. 2003; Kristensen et al. 2008; Donato et al. 2011) were used to fit statistical models to extrapolate values at a scale of 5 arc minute (i.e., a cell size of approximately 10 km by 10 km). The statistical models (parametric predictive and machine learning algorithms) fitted 932 data points obtained in field studies in 28 countries. The independent variables included in the regression models were latitude of each location, an array of variables describing climate conditions at the sampling locations, and other regional indicators. The models showed an uneven regional and latitudinal soil carbon spatial distribution in four carbon storage categories (~27–32, 32–34, 34–38, 38–41, 41–70 mg cm^{-3}) (Fig. 11.3). In addition to differences in local and regional climate factors, some of the variability was explained as the uncertainty of oversampling "pristine mangrove forests" in the original studies, although this was not quantified.

The information provided by Jardine and Siikamäki (2014) is useful when attempting to obtain a global estimate of the potential range of mangrove carbon storage values and when aiming to spatially identify associated environmental variables potentially explaining those patterns at the selected scale (5 arc minute). However, there is much room for improvement in these models. Macroecological analyses are often most successful and useful when they test hypotheses that stem from a combination of local-scale mechanistic ecological knowledge and rigorous data synthesis to evaluate ecological processes at different spatial scales. Indeed, macroecology aims to achieve synthesis to uncover broad-scale patterns and processes that often span local to global spatial scales. For instance, metabolic theory proposes specific predictions for the non-linear relationships between biological rates, including productivity, and body size and temperature (e.g., Brown et al. 2004). These relationships are considered "scale free" since observed constants (exponents) encompass all system sizes (e.g., Anderson-Teixeira and Vitousek 2012; Schramski et al. 2015).

In the case of the Jardine and Siikamäki (2014) analysis, the ecological processes responsible for the global-scale patterns observed were not examined in depth and there was very little discussion of the processes underlying the variables included in the models. Some of the variables are correlated (e.g., latitude and temperature), and some of the results could even be interpreted as contradictory when they are viewed in the context of temperature- and rainfall-dependent global-scale patterns for other ecosystem attributes and processes (e.g., aboveground biomass and carbon storage, litterfall, NPP, and tree height) (Twilley et al. 1992; Simard et al. 2006). The carbon storage pattern reported for cells at resolution of ~10 km would then suggest that, at latitudes close to the equator, there is an inverse relationship between carbon storage aboveground (e.g., trunks and canopy) and belowground (e.g., soil and roots). However, this relationship is counterintuitive when considering that (1) previous estimates of litterfall production show higher rates close to the equator in the Neotropics and (2) tree height, aboveground biomass, and litterfall are correlated (Twilley et al. 1992; Saenger and Snedaker 1993).

Just as there are limitations for approaches that use results from global-level data to explain local ecological patterns, there are also constraints to extrapolating findings from local studies to the global scale. Certainly, since the seminal work by Thom (1967) linking geomorphology to mangrove forest zonation in the 1960s and further advanced by Woodroffe (1992), there has been a lack of studies explicitly linking the local to the regional scale in mangrove function and structure (e.g., Lovelock et al. 2007). The inclusion of this hierarchical level is necessary to correctly interpreting the wide spatiotemporal variability due to the effects of macroclimatic drivers (e.g., temperature, precipitation), resources (e.g., nitrogen, phosphorus), regulators (e.g., salinity, sulfide), and hydroperiod (e.g., flooding frequency and duration) (Fig. 11.2) upon ecosystem structure and function. This variability at local scales is well represented by species zonation along fertility gradients, productivity levels, and carbon sequestration rates in response to pulsing flooding regimes. We believe that the lack of inclusion of this geomorphological scale has greatly limited the interpretation of mangrove structural and functional patterns when upscaling (e.g., to the region and/or global scale) and downscaling (e.g., to the local-scale) currently available data sets. Further, studies attempting to extrapolate results to the global scale should recognize the presence of distinct geomorphological settings within similar climatic zones (Castañeda-Moya et al. 2013). This perspective is necessary for correctly characterizing regional properties due to the potential presence of an array of mangrove ecotypes that reflect complex mosaics of forest ecological properties emerging from geomorphological development as discussed above (Fig. 11.2).

One example illustrating the relative importance of intermediate scales (geomorphic) to interpret local processes and explain global patterns is the recent publication by Crase et al. (2013). They evaluated the relative role of hydroperiod and salinity for explaining the dominance and spatial distribution of three mangrove species (*Sonneratia alba, Rhizophora stylosa, Ceriops tagal*) in Northern Australia and concluded that hydroperiod was the major variable explaining species dominance. Given the level of generality and extrapolation reflected in the conclusions, Clarke (2014) wrote a response critiquing the bold generalization and pointed out the partial selection of species along hydrological gradients and the lack of other potential explanations related to interactions of other variables such as competition, propagule dispersal, soil texture, or nutrient availability. Although their arguments differ, both perspectives offer very valid points once the targeted spatial and temporal scales are explicitly considered. For example, another argument against the broad generalization regarding the role of hydroperiod as a single cause determining mangrove species zonation in Australian mangroves is contained in the work of Castañeda-Moya et al. (2013) in neotropical carbonate settings. *Rhizophora mangle* is the dominant species in locations experiencing very different hydroperiods due to this species' ecophysiological adaptations to withstand a long duration of inundation. Since phosphorus is a limiting nutrient in this coastal carbonate-rich setting, very distinct ecotypes (e.g., scrub vs. fringe) of *R. mangle* develop. However, in the case of the fringe mangrove forest, even under an optimal interannual and seasonal soil pore water salinity (i.e., <35), *R. mangle* is replaced by the mangrove species *Laguncularia racemosa* as a result of large-scale disturbances (hurricanes) that trigger forest mortality, cause defoliation, and provide ecologically relevant phosphorus inputs (Castañeda-Moya et al. 2010). These studies, from a particular setting within the neotropical coastal region, emphasize the dynamic role of different processes (e.g., phosphorus availability, hydroperiod, tropical cyclone disturbance) interacting at different spatial and temporal scales; similar relationships have been discerned by several studies over several decades (e.g., Duke 2001; Fromard et al. 2004; López-Hoffman et al. 2007; Smith et al. 2009).

Regardless of the relative importance of other environmental variables that influence mangrove zonation (i.c., tropical storms, geomorphology), hydroperiod is still a critical variable needed to understand and explain not only mangrove zonation and productivity patterns, but also biogeochemical cycling, including carbon cycling (see Chap. 6). In contrast to hydrological studies in other type of wetlands (e.g., marshes, freshwater forested wetlands), published hydroperiod measurements in mangrove forest are limited (Twilley and Chen 1998; Krauss et al. 2006). Although there are hydrodynamic studies assessing diverse hydrological processes (e.g., seasonal riverine inputs, tidal prism, water circulation, water residence) in coastal lagoons and estuaries adjacent to mangrove wetlands, there is comparatively little information on the variation in frequency and duration of inundation and depth inside mangrove forests. These variables define the hydroperiod and greatly influence soil properties (e.g., salinity, redox, sulfide, oxygen, organic matter content). These hydrologic variables also affect a wide range of ecological processes, including propagule dispersal, plant community development, CO_2 fluxes, and organic matter and nutrient export (Lovelock et al. 2004; Feller et al. 2007; Lovelock 2008; Twilley and Rivera-Monroy 2009). However, most studies do not directly measure hydroperiod, which limits the explanation and scope of observed patterns at regional and global scales.

The limited number of direct measurements of hydroperiod inside mangrove forests is the result of a combination of financial, logistical, and technical constraints. If the research goal is to correctly characterize functional and structural processes in mangrove wetlands at the scale demanded by societal problems, a strategic effort has to be launched to expand current hydroperiod and hydrological measurements across climatic gradients. This could be associated with large-scale comparative efforts underway to understand, for example, the impact of sea-level rise in different tropical ecogeomorphic settings (Webb et al. 2013). Clarke's (2014) criticism arguing for the need for a macroecology perspective when assessing mangrove zonation is a good reminder to avoid excessive generalizations when data are lacking. Macroecological research in mangroves can be challenging due to the lack of regional data (ecogeomorphic scale), but also due the lack of long-term studies. The work by Crase et al. (2013), underscoring the role of hydroperiod in species zonation, is a good example of the type of local information needed to establish robust patterns at higher spatial scales and offer mechanistic hypotheses to explain emerging patterns. Care should be exercised to correctly define the relative role of each variable when partitioning their relative contribution in a multivariable world represented by mangrove-dominated coastlines (Adame et al. 2010).

11.2.2.2 Mangrove Forest Responses to Climate Change: The Contributions of Macroecology

In terrestrial and marine ecosystems, macroecological studies have provided an important foundation for better understanding the ecological importance of macroclimatic drivers (i.e., temperature and rainfall regimes) (Holdridge 1967; Whittaker 1970; Davis and Shaw 2001; Staver et al. 2011). This is an important first step for predicting and preparing for the ecological effects of climate change (Shafer et al. 2001; Lawler et al. 2009; Araújo et al. 2011). Macroecological analyses, especially those conducted across large climatic gradients, are valuable tools for placing local field observations and experimental studies within regional and global contexts. Our aim in this subsection is to illustrate, via several examples, how climate-focused macroecological studies can improve our understanding of mangrove forest responses to climate change.

As in all wetland and terrestrial ecosystems, the structure and functioning of mangrove forests is greatly influenced by temperature and rainfall regimes (Alongi 2009; Twilley and Day 2012; Alongi 2015; Lovelock et al. 2016). Field observations and experimental studies conducted in diverse climatic settings have helped to characterize the important role of temperature and rainfall, and the mangrove literature is replete with information on the effects of temperature and rainfall upon the distribution and performance of mangrove forests. However, much of this information is qualitative because, until recently, climate data has not been readily available at the relevant scales. In some cases, climate proxies have been employed. For example, latitude has frequently been used as a proxy in analyses of the importance of temperature regimes, and correlative relationships have been developed between latitude and mangrove forest aboveground biomass (Twilley et al. 1992; Saenger and Snedaker 1993; Alongi 2009; Twilley and Day 2012), litterfall (Saenger and Snedaker 1993), net primary productivity (Alongi 2009), and species richness (Ellison et al. 1999; Ellison 2002). With the increasing availability and accessibility of climate data combined with field observations, experimental studies and/or remotely sensed data, large advances will be made in the next decade to improve our understanding and ability to model the effects of climate change on mangrove ecosystem structure and function. For example, temperature and precipitation data were recently used to develop models predicting aboveground biomass (Hutchison et al. 2014; Rovai et al. 2016) and soil carbon stocks (Jardine and Siikamäki 2014; Ouyang et al. 2017). Though simplistic, these models provide a foundation for better quantifying the role of macroclimatic drivers in determining changes in mangrove forest ecosystem structure and function due to climate change.

At the poleward mangrove-marsh ecotone, there are several recent examples of how a macroecological perspective can advance understanding of climate change effects on mangrove forests. On multiple continents, climate change will probably cause poleward migration of mangrove forests, often at the expense of salt marshes (Saintilan et al. 2014; Alongi 2015; Lovelock et al. 2016). As a result, there is a pressing need to better understand the effects of changing winter climate extremes (i.e., freeze events) on mangrove-salt marsh interactions, as well as the implications for range expansion and salt marsh displacement. In subtropical and warm-temperate climate zones, low air temperatures can damage and/or kill mangrove trees (Lugo and Patterson-Zucca 1977; West 1977; Stuart et al. 2007; Osland et al. 2014a; Lovelock et al. 2016). However, a reduction in the intensity, duration, and frequency of extreme winter air temperature events may lead to mangrove expansion. Until recently, the poleward distribution of mangrove forests was often attributed in the literature to mean monthly air, mean annual air, or sea surface temperatures

(Chapman 1976; Tomlinson 1986; Duke et al. 1998; Quisthoudt et al. 2012; Record et al. 2013). However, in some locations (e.g., the southeastern United States and China), minimum air temperature extremes are much more important than mean winter air temperatures (Osland et al. 2013; Cavanaugh et al. 2014).

Recent advances and improved access to historical daily air temperature data and models have enabled researchers to better quantify minimum temperature-based thresholds that control the northern range limit of mangrove forests in North America (Osland et al. 2013; Cavanaugh et al. 2014, 2015). Gabler et al. (2017) recently used field data collected across the northern Gulf of Mexico to: (1) quantify the relationship between climatic drivers and plant functional groups (i.e., mangroves, graminoids, succulents, and algal mats) and (2) investigate the effects of climate change on coastal wetland foundation species. These models can be incorporated into climate change vulnerability assessments (sensu Glick et al. 2011) and be used to better predict future expansion of mangrove forests at the expense of salt marshes. In an evaluation of vulnerability in the southeastern United States, Osland et al. (2013) used historical climate data, mangrove abundance data, and future climate projections to develop models and show that salt marshes in Texas, Louisiana, and parts of Florida are particularly vulnerable to mangrove forest range expansion under climate change. Osland et al. modeled future mangrove forest expansion throughout the region under alternative future climate change scenarios. Along the Atlantic coast of Florida, Cavanaugh et al. (2015) developed species-specific models that were used to predict poleward migration of mangrove forests in response to future climate change. These efforts have helped identify sensitive coastal reaches (e.g., specific areas in Texas, Louisiana, and Florida) where small changes in winter air temperature extremes are expected to lead to landscape-scale changes in coastal wetland ecosystem structure and function (Osland et al. 2013). These approaches could be refined, expanded, and applied to identify other climate-sensitive zones across the globe. For example, Osland et al. (2017) recently quantified temperature and rainfall controls on the global distribution, abundance, and species richness of mangrove forests. Those analyses show that temperature and rainfall thresholds for mangrove distribution, abundance, and richness are range limit specific.

In addition to improving our understanding of the role of temperature regimes, we expect that, in the coming decade, macroecological analyses will also greatly improve our understanding of the effects of precipitation, aridity, and freshwater availability upon mangrove forests. The ecological influence of freshwater availability on coastal wetlands is particularly large in arid and semiarid climates (Cintrón et al. 1978; Zedler 1982; Semeniuk 2013; Osland et al. 2014b, Lovelock et al. 2016). In Australia, region-scale analyses have shown that precipitation greatly influences the abundance and diversity of mangrove forest and salt marsh plants (Smith and Duke 1987; Bucher and Saenger 1994; Castañeda-Moya et al. 2006; Saintilan 2009; Semeniuk 2013). Across a freshwater availability gradient in the northwestern Gulf of Mexico (range of mean annual precipitation: ~700–1800 mm), the coverage of wetland plants (including mangroves) is positively correlated with precipitation and inversely correlated with salinity (Longley 1994; Montagna et al. 2011; Osland et al. 2014b, Gabler et al. 2017). Interactions between precipitation and winter temperatures within this region greatly determine the structural attributes and physiology of mangrove forests (Lot-Helgueras et al. 1975; Lugo and Patterson-Zucca 1977; Méndez-Alonzo et al. 2008; Madrid et al. 2014). These regional and global climate-coastal wetland linkages can only be revealed via large-scale analyses that incorporate diverse data sources and employ a macroecological approach. Though simple, such analyses have helped quantify important climate-coastal wetland linkages and provide an important foundation for improving our ability to predict and prepare for the effects of climate change upon mangrove forests and other coastal wetland ecosystems.

11.3 Mangrove Modeling of Ecological Processes and Function Within a Macroecological Approach

Clarke (2014) provides a powerful argument for the role and contribution of ecological modeling in mangrove ecological research. One of Clarke's main points is that critical care is needed when developing statistical models assessing the inter-correlated and multi-factorial nature of abiotic and biotic variables that limit mangrove spatial distribution. Care is especially important when models are developed from a small amount of observational data and extrapolated to the global scale. This observation is important since one of the major steps when modeling ecosystems is to focus special attention upon model calibration that closely considers the original model's objectives and architecture, as well as the availability of enough data for both model validation and uncertainty analyses (Jørgensen and Nielsen 2013; Ulanowicz et al. 2014). Failure to strictly follow these steps seriously limits the reliability of the model to simulate site-specific process, as has been the case for defining global and latitudinal patterns of mangrove ecosystem processes such as aboveground biomass, carbon storage, or species extinction (Record et al. 2013; Hutchison et al. 2014; Jardine and Siikamäki 2014).

We consider the utility of modeling (either statistical or dynamic/process-based) as a paramount step in advancing the macroecology of mangroves. There is no question that a model's predictive capacity based solely on correlative or regressive statistical approaches is greatly limited when seeking an explanatory interpretation, especially when it goes beyond the spatial or temporal scale used to build the model (see Sect. 11.2.2.2) (Berger et al. 2008). In some cases, this limitation is ignored or overlooked, paradoxically, when management decisions are urgently needed or not enough data is available to evaluate model uncertainty. This is usually the case in ecosystem restoration and rehabilitation programs (see Chap. 10 and section below), environmental pollution impact assessments, and even climate change impact predictions (Kerr et al. 2007; Peyronnin et al. 2013). Despite the limitation of models for explaining a wide range of emerging ecosystem properties regulated by non-linear complex and multifactorial interactions, they are powerful tools for synthesizing knowledge and advancing hypotheses about the functional and structural attributes of a wide range of ecosystems (Yue et al. 2011), which is precisely one of the main goals of macroecology (Brown 1995).

The number of available static and dynamic mangrove models, including spatially explicit models, is relatively small when compared to existing models developed for other ecosystems (e.g., tropical and temperate forests) (Yue et al. 2011). Yet, these modeling efforts have been very useful for identifying processes and data gaps where more experimental and observational information is needed (Chen and Twilley 1998a, b; Twilley and Chen 1998; Twilley et al. 1998; Berger et al. 2008; Piou et al. 2008; Fontalvo-Herazo et al. 2011; Grueters et al. 2014). Since the first conceptual model proposed by Lugo (1980) (Fig. 11.4), depicting critical mangrove ecosystem components, successional trajectories, interactions and connections to adjacent ecosystems, a number of mangrove models have been developed as indicated by the increasing number of publications on the subject. The models have taken different statistical and computational approaches to simulate biotic and abiotic interactions. These range from trophodynamic and biogeochemical models showing energy and material flows among different ecosystem levels (e.g., Ray et al. 2000; Ray and Straškraba 2001; Vega-Cendejas and Arreguín-Sánchez 2001; Ray 2008; Roy et al. 2008; Mandal et al. 2009; Das and Ray 2010; Mukherjee et al. 2014) to individual-based models depicting the effect of regulators (e.g., salinity) and resources (e.g., phosphorus, light) on tree growth (Chen and Twilley 1998a, b). Some of the models include the spatially explicit degree of competition for space among trees and the influence of tree architecture in biomass allocation (Grueters et al. 2014; Vovides et al. 2014). In the latter examples, new and improved computational methods (e.g., Kautz et al. 2011) and computer architecture (e.g., parallel processing) have opened a wide range of possibilities, for example, in tracking individual trees and species-specific canopy architecture (Fontalvo-Herazo et al. 2011; Vogt et al. 2013; Vovides et al. 2014) to capturing emerging forest and potentially whole ecosystem properties (e.g., Clarke 1995; Berger et al. 2008; Mukherjee et al. 2013). In particular, mangrove individual-based models hold promise for exploring non-linear interactions between environmental factors and species-specific physiological adaptations that could help define complex niche partitions along environmental gradients (e.g., fertility, elevation, hydrological) (e.g., Chen et al. 2013). Similarly, individual-based models can contribute to the identification and refinement of assembly rules (Keddy 1992; Belyea and Lancaster 1999; Keogh et al. 1999; Hammond and Niklas 2009, 2011) that explain potential trajectories in primary and secondary succession. This is a key ecological mechanism that can inform the definition and assessment of performance measurements in mangrove rehabilitation and restoration projects (see Chap. 10; Twilley et al. 1998). Another approach with promising utility for advancing mangrove macroecology is ecological network analysis, particularly given its solid foundation in input-output analysis, information theory, and thermodynamics (Ray et al. 2000; Ray 2008; Mukherjee et al. 2015) as demonstrated by its historical applications in the analysis of food chains/webs and species richness and diversity (Xiao et al. 2015). In contrast to some of the mechanistic models listed above, ecological network analysis includes more than

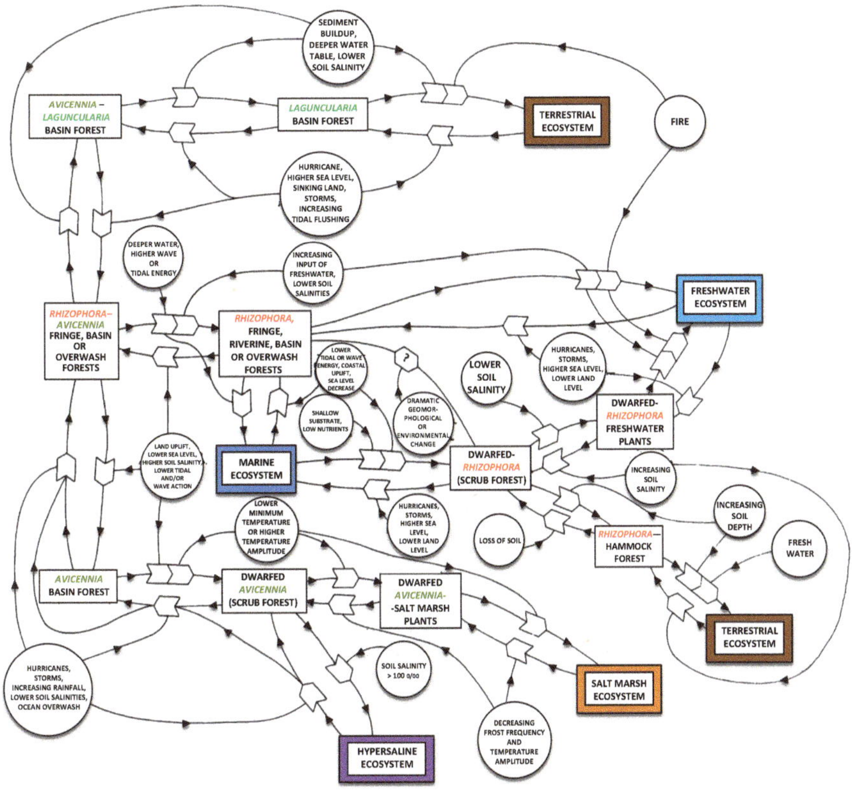

Fig. 11.4 Early conceptual model showing mangrove successional pathways and ecosystem and climatic drivers (*circles*) (modified from Lugo 1980). The actual species composition and forest ecotype in these systems depend on the interaction among climate, substrate, and hydroperiod/hydrology

one biological process into subsets of categories with the aim of expanding explanatory power and characterizing the influences of individual species upon one another (Ulanowicz et al. 2014). In this system network approach, trophic groupings and clusters of cycling can potentially be identified and quantified (Ulanowicz et al. 2014). Unfortunately, data limitations have constrained the development and application of ecological network analyses and individual-based models in mangrove ecosystems (Berger et al. 2008; Brolly et al. 2012; Niklas and Hammond 2013).

Our general description here of some of the most relevant modeling approaches underscores the need for further work to identify major gaps and limitations in the development, validation, and application of mangrove ecosystem models. As mentioned earlier, ecosystem modeling is a data-intensive endeavor that requires clear definition of spatiotemporal scales. Thus, mangrove modeling efforts in the context of macroecological research will require, as a first step, identification of suitable data sets and metadata from diverse geographic regions and geomorphic settings to advance the development and validation of existing models at the global scale (Yue et al. 2011). Keeping in mind that there is not a single best modeling approach for all situations, a strategic use of diverse modeling tools based on well-defined objectives and hypotheses should guide future research in mangrove macroecology.

11.4 Using Mangrove Restoration Projects to Advance a Macroecological Approach

In previous sections, we have emphasized how, despite the lack of information at the local and regional scale (e.g., geomorphic, Fig. 11.2), there have been limited and uncertain extrapolations of local data to biogeographical and latitudinal patterns of mangrove functional and structural attributes. Although some of these studies consider a ~10-km cell (i.e., 5-arc minute) as a fine spatial resolution, the patchy spatial distribution of mangrove ecotypes, partially determined by the presence or absence of watershed and groundwater sources, underlines the need for a better characterization (Fig. 11.2, see Figure 4.9 in Chap. 4 Remote Sensing). We propose that the spatial scale of interest could be selected based upon the spatial extent generally used in mangrove restoration and rehabilitation projects. These are implemented with the purpose of recuperating extensive mangrove dieback areas ranging from few hectares to one or a few km^2 (see Chap. 10). Since the causes of mangrove mortality justifying the implementation of these projects are numerous (e.g., hypersalinity, alterations in hydroperiod) and generally overlap with both local and regional scales, ecological information about both project success and failure could contribute to fill data gaps.

Mangrove zonation and succession represent key ecological processes that occur at a local scale and have a major impact (bottom-up) on shaping ecosystem functional and structural properties at regional and global scales (Fig. 11.2). The dynamic role of these processes is one of the reasons why secondary succession in mangrove ecosystems is one of the major ecological processes targeted for manipulation in restoration ecology. Historically, understanding and forecasting mangrove zonation has been one of the primary research priorities in mangrove ecology (Lugo 1980) (Fig. 11.4), yet there are major challenges in defining successional trajectories to advance a macroecology-based conceptual framework. One of the challenges is the lack of regional long-term data to help characterize these trajectories using a latitudinal comparative approach (Twilley et al. 1999). For example, although neotropical mangroves encompass ~10 species in contrast to >30 species in the Old World tropics (see Chap. 2), this low diversity has not been translated into mechanisms to explain species dominance or replacement patterns along fertility and stressor gradients at a global scale.

One way to use data obtained from restoration projects is to define species-specific environmental constraints that are indirectly manifested in plants' competitive ability to reproduce, grow, and expand when environmental conditions are modified as result of restoration measures (e.g., see Chap. 10). Defining optimal growth and reproductive rates under field conditions is generally difficult, not only in mangrove forests, but also wetlands in general (Mitsch and Gosselink 2015), due to a complex mosaic of interactions among abiotic and biotic factors. For example, our current understanding about the ecophysiological tolerance of mangrove plants to salinity regimes is based primarily on greenhouse studies where seedlings are raised under controlled experimental conditions over short periods of time (days to weeks). However, extrapolating findings from the responses of propagules and seedlings in experimental "pots" to responses by adult trees in field conditions present several limitations. Observational studies are often needed to complement smaller scale experimental findings (Ball 1988). Defining abiotic limits is paramount for evaluating if plants are under quasi-optimal or stressful conditions (Wang et al. 2011; Krauss and Ball 2013). Further, soil salinity stress in mangrove forests is not only confounded by hydroperiod (frequency and duration of inundation) in the long term, but also by nutrient availability (particularly phosphorus and nitrogen) (Twilley and Rivera-Monroy 2009; Feller et al. 2010).

Therefore, information such as mortality, biomass, net primary productivity, growth, or plant height distribution acquired in restoration projects could contribute to the definition of complex niche partition by explaining species net primary productivity or maximal basal area and tree height values. Gathering these types of data sets is often possible via long-term post-restoration monitoring projects, especially where plantings are implemented across a wide range of ecogeomorphic settings within biogeographical regions (see Chap. 10). Although mangrove plantations are not a common management strategy in all continents due to a diverse legal framework for mangrove conservation and protection, these locations offer valuable data to learn, for example, about plant population dynamics (e.g., reproduction rates, demography patterns) or obtain data on biogeochemical transformations (e.g., nutrient demands, carbon storage) (Fontalvo-Herazo et al. 2011; Vogt et al. 2013; Lunstrum and Chen 2014; Manna et al. 2014). Studies that take advantage of restoration-based and natural mangrove forest chronosequences (i.e., space-for-time substitutions) in multiple settings have the potential to improve our understanding of mangrove forest ecosystem development in novel settings (Lovelock et al. 2010; Osland et al. 2012; Salmo et al. 2013; Lunstrum and Chen 2014). Given the logistical and economic constraints to launch large-scale ecological studies at the regional scale, both restoration sites and mangrove plantations represent an opportunity to acquire valuable information to advance our understanding of mangrove ecological processes at the local and regional scales to explain observed complex global patterns.

11.5 Macroecology and the Complexity of Mangrove Ecosystem Services at the Global Scale

Defining, classifying, and valuing ecosystem services is currently a highly dynamic research field due to the urgent need to identify tradeoffs in natural resource management and policy decisions (Fisher et al. 2009). Although ecosystem services are typically not accounted for in economic and land use-relevant decision-making processes, ecologists and environmental economists are increasingly challenged to quantify, compare, and communicate their value (Costanza et al. 1997; Millenium Ecosystem Assessment 2005; Daily et al. 2009; Polasky and Segerson 2009; Jenkins et al. 2010). In fact, the economic value of ecosystems, including mangroves, has been emphasized in several contexts since the early 1990s (Costanza et al. 1997), but there are still discussions about the best approach for quantifying ecosystem services due to the complex suite of benefits that are closely linked to human welfare (see Chap. 8; Barbier et al. 2011). However, there is a general agreement that any attempt at classifying ecosystem services should be based on both the characteristics of the ecosystems of interest and a decision context for which the concept of ecosystem services is being implemented (Fisher et al. 2008). This criterion is particularly important in the case of mangrove wetlands, given the high spatial variability in ecosystem services and the role of subtle hydrological gradients within multiple arrays of geomorphological settings (see Fig. 11.2).

Macroecological approaches can be used to link spatial and temporal scales of ecosystem function and structure in order to explain complex interactions among entities within the systems (Peters et al. 2014). Mangrove forests support many ecological functions that are valuable to society; in addition to supporting habitats for fish and wildlife species, mangrove forests improve water quality, protect coastlines from erosion and storms, maintain coastal food webs, sequester carbon, provide food and timber, and support recreation opportunities (Ewel et al. 1998; Alongi 2009; Barbier et al. 2011; Twilley and Day 2012). Because ecosystem services are dependent upon ecosystem functions, understanding the provision of ecosystem services is closely linked to one of macroecology's major goals. Overall, ecosystem services are defined as the benefits of nature to households, communities, and economies (Boyd and Banzhaf 2007). Since different mangrove ecotypes provide different goods and services (Ewel et al. 1998), mangrove macroecology can help operationally elucidate the drivers of spatial variation in ecosystems services currently observed between continents, regions, and individual mangrove stands, including ecotypes (see Sect. 11.2).

Ewel et al. (1998) proposed a functional classification of mangrove forests that evaluates the ecosystem goods and services provided by distinct types of mangrove forest. The approach of Ewel et al. (1998) largely draws from the conceptual framework described in Sect. 11.2.1 by blending the ecotype categories readily recognized in the low diversity Atlantic-east-Pacific region (overwash, fringe, riverine, basin, and scrub; Lugo et al. 1976) (Fig. 11.2) with a system developed to explain mangrove development for the highly diverse Indo-west-Pacific region (Woodroffe

1992). The scheme highlights three of the hydrogeomorphic categories (sensu Brinson 1993): fringe, riverine, and basin mangroves (Fig. 11.2) (for further description of these categories, see Ewel et al. 1998). In general, fringe forests are strongly influenced by marine tidal waters that are attenuated by modified tree structures (pneumatophores, buttresses, and prop roots). Riverine forests receive mixed flooding from tides and river discharge that control and moderate porewater salinities. In contrast, salinity in basin forests is higher, caused by low frequency of inundation and higher elevations, occasionally developing hypersaline conditions due to high evapotranspiration rates and seasonally low rainfall. Basin forests are generally located behind fringe and riverine forest and can potentially have moderate salinities if flooding frequency or precipitation is high and/or groundwater inputs are large (Fig. 11.2) (Ewel et al. 1998). One of the main functional ecosystem attributes that defines these categories is high NPP that diminishes with decreasing water turnover in the sequence riverine>fringe>basin (Twilley 1995). High water turnover in wetland ecosystems is directly correlated with higher sediment (e.g., silts and clays) and nutrient (e.g., nitrogen, phosphorus) inputs and higher oxygen availability to anoxic sediments that restrict the accumulation of toxic compounds in pore waters (e.g., sulfide) (Alongi 2010; Mitsch and Gosselink 2015). The distinction among hydrogeomorphic types, although not readily apparent in some cases depending on the aerial extension, could be recognized and enhanced by combining extensive and rapidly acquired field data (e.g., water and soil pore water salinity; duration, frequency, and depth of inundation) and remotely sensed data (see Chap. 4).

Ecogeomorphic-based classifications provide a direct link between the spatially explicit ecological attributes of mangrove hydrogeomorphic types and ecosystem service availability and quantity. Such linkage may potentially allow the mapping of ecosystem services (e.g., carbon storage, fish habitat, shoreline protection extension) at different spatial scales. This connection between types of ecosystem services and location along hydrological gradients starts with the identification of the relative importance of each forest ecotype and the provision of goods and services as shown in Table 11.1 (Ewel et al. 1998). For example, mangroves in muddy coasts influenced by major rivers can, given the water discharge, trap more sediment (rank =1) than fringe forests (rank = 3), where tides and marine sources prevail (Ray 2008; Rivera-Monroy et al. 2011; Mandal et al. 2012, 2013). Similarly, the role of forests as carbon and nitrogen sinks is relatively more important in basin forests (rank=1) than in riverine and fringe forests, due to the differences in water residence time, organic matter accumulation, and aerial extension (Rivera-Monroy et al. 1999) (Table 11.1). Arguably the relative importance of each category should be based on field-based quantitative information to develop robust integrative indices with the aim of reducing uncertainty in the assignment of scales.

The deliverable of this approach resides in converting spatially explicit ecological attributes to a map tool, not only ecosystem services quantity and quality, but also their value. Mapping ecosystem services and associated economic value is a major research priority for ecosystems conservation and management, including mangrove wetlands (See Chaps. 5, 8, 9; Ruggiero and Hawkins 2006). Although

Table 11.1 A comparison of mangrove ecosystem services (rows) based on the functional role of three major mangrove ecotypes (columns). Value are shown in decreasing rank (i.e., 1 is most important) (modified from Ewel et al. 1998) (see Fig. 11.2)

Role	Riverine	Basin	Fringe
Trap sediments	1	2	3
Process nutrients and organic matter			
Provide a source of detritus to nearshore waters	1	3	2
Serve as sink for nutrients and carbon:			
Carbon, Nitrogen	2	1	3
Phosphorus	1	3	2
Improve water quality	2	1	3
Provide food and habitat for animals	1	3	2
Provide aesthetically pleasing environments	1	3	2
Protect shorelines	2	3	1
Provide plant products	2	1	3

there are currently no global maps of the economic value of mangrove ecosystem services, there are maps showing the quantity of selected goods and services (e.g., carbon storage, aboveground biomass, species diversity-extinction, litterfall, productivity) (Hutchison et al. 2014; Record et al. 2013; Jardine and Siikamäki 2014). However, as mentioned before, the resolution of current mangrove resource maps is often too coarse for dynamic evaluation of local and regional vulnerabilities to anthropogenic pressures that may lead to selective replacement and/or deforestation of some mangrove ecotypes over others (Record et al. 2013).

A critical priority in ecosystem service research is to advance understanding and communication between scientists and stakeholders at the boundary between what represents final services and benefits (Fisher et al. 2009; Häyhä and Franzese 2014). Accordingly, it is important to manage, monitor, and make policy to protect services that help maintain specific benefits and interlink ecosystem and socioeconomic variables (Nielsen and Müller 2009). Although historically acknowledged, this explicit linkage between the "socioeconomic" and "ecosystem" realms is just recently being operationally defined (Collins et al. 2011; Häyhä and Franzese 2014) (Fig. 11.5). In the case of mangrove wetlands, there are major knowledge gaps in our understanding of the social, economic, and cultural causes for habitat loss and fragmentation. Although mangrove vulnerability assessments are currently available for some areas, they are often very limited in scope and resolution. For instance, poverty is a major factor contributing to mangrove loss in developing countries, precisely where most of the mangrove area is found (Giri et al. 2011). Thus, a better understanding of poverty-inducing mechanisms (Barrett et al. 2011; McNally et al. 2011) and poverty alleviation in connection to functional mangrove attributes should be a research and conservation priority (see also Chaps. 8 and 9). Such knowledge is especially important in areas where ecosystem service preferences by local communities are determined by single socioeconomic and cultural priorities, leading to the direct and indirect overexploitation of mangrove resources (see

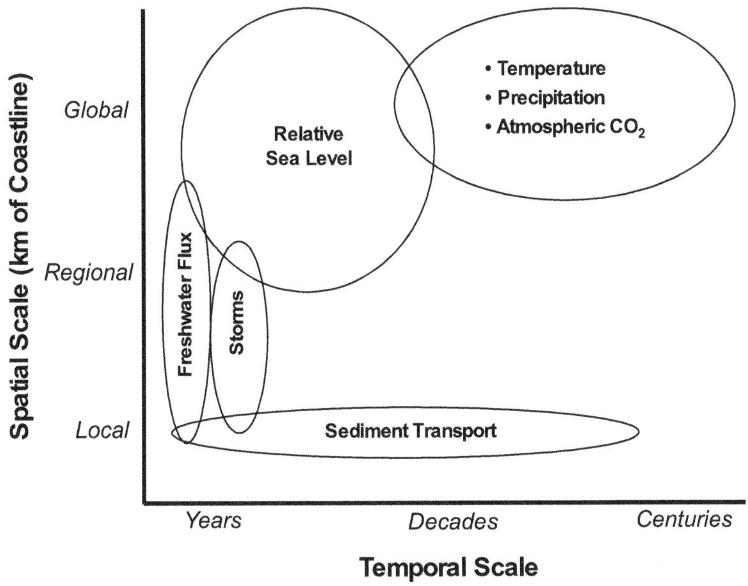

Fig. 11.5 Relative temporal and spatial scales of processes directly and indirectly impacting estuarine geomorphology (modified from Day et al. 2008)

Chaps. 8 and 9). There is much work to be done to improve our understanding of these complex issues. We propose that macroecology can help advance our understanding of the spatial extent of these multifactorial social and environmental interactions. This may contribute to the development of an integrative "socio-ecology" of mangroves that is badly needed to globally conserve and manage mangrove ecosystems (Adger et al. 2005).

11.6 Conclusions

Macroecology is broadly defined as a transdiscipline of ecology that investigates large-scale, universal patterns in the distribution, abundance, diversity, and organization of species and ecosystems (Brown 1995; Enquist et al. 1995; Smith et al. 2008). This includes the scaling of ecological processes and structural and functional relationships (e.g., McGill and Collins 2003; Niklas and Hammond 2013). Macroecology offers a useful conceptual framework to advance mangrove research and to operationally assess processes occurring at different spatiotemporal scales. However, the explicit application of macroecology to mangroves has been limited, which is surprising given the global distribution of mangroves. The major advantage of a macroecology perspective is the explicit integration of spatiotemporal scales to evaluate ecological and socioeconomic processes (e.g., Brown et al. 2014) that drive the quality and quantity of human impacts on mangrove-dominated ecosystems.

Recent mangrove research has focused on providing estimates of contribution to global carbon storage (i.e., "blue carbon") and potential impacts of mangrove poleward range expansion. These findings help to highlight the ecological importance of mangroves at the global scale, but also expand our limited knowledge about processes occurring at local and regional scales. These include the role of nitrogen and phosphorus availability on net carbon fluxes (Rivera-Monroy et al. 2004; Alongi 2010; see Chap. 5), the relative contribution of biotic process versus ambient temperature in regulating mangrove establishment and zonation (e.g., Guo et al. 2013; Clarke 2014), or the common socioeconomic mechanisms driving the degree of vulnerability and overexploitation of mangrove resources.

Researchers must consider and address the increasing trend in mangrove area loss during last 20 years due to land use change (Valiela et al. 2001; Giri et al. 2011). At the global scale, the area loss across diverse coastal regions is tremendously important since the temporal scale at which mangrove deforestation or destruction operates is much smaller (i.e., months, years) than some other aspects of global change (e.g., climate change and sea-level rise) (Figs. 11.2 and 11.5). Thus, realistic global projections of mangrove response to climate change and sea-level rise should include local and regional land-use change. This scenario is illustrated by the concept of the "coastal squeeze" where coastal wetlands are "trapped" between increasing sea-level rise (years/decades; global) and the lack of space to migrate inland, which is occupied by expanding human infrastructure (months/year; local) (Torio and Chmura 2013; Enwright et al. 2016). Of the total global mangrove area registered in 2000 (137,760 km^2), only 6.9% is within the boundary of protected areas (national parks, biosphere reserves). The remainder area is exposed to an exponential increase in local anthropogenic impacts and emerging climate change forcing within the next two decades (Duke et al. 2007; Polidoro et al. 2010; Giri et al. 2011). From a macroecological perspective, the potential area loss has to be included when selecting research priorities and questions at different temporal and spatial scales. This criterion is particularly important when selecting processes that impact coastal geomorphological processes such as sediment transport, freshwater flux, sea-level rise, or changes in atmospheric CO_2 concentrations (Fig. 11.5; Day et al. 2008).

In the Caribbean region (Rivera-Monroy et al. 2004) and West Africa (Tanzania) (Uchida et al. 2014; Gaiser et al. 2015), relevant research questions to compare mangrove functional and structural attributes at the regional scale represent an example to advance the development of a research agenda for mangrove macroecology (Fig. 11.6). These questions emerge from specific links connecting a socioeconomic realm and the environmental signature of a coastal regional setting that results from a combination of geomorphological type, geophysical energies, and levels of disturbance (Figs. 11.2 and 11.6). The interaction of regional climate and topographic features influences the amount of freshwater input into coastal regions that, in the case of mangrove-dominated coastlines, regulate hydrological, sediment transport, and mangrove ecotype development (see Sect. 11.2, Fig. 11.5).

The frequency and magnitude of disturbance, both natural and human, are also important considerations in the environmental signature of coastal settings. Indeed, there is extensive research on the relative impact of tropical cyclones or hurricanes

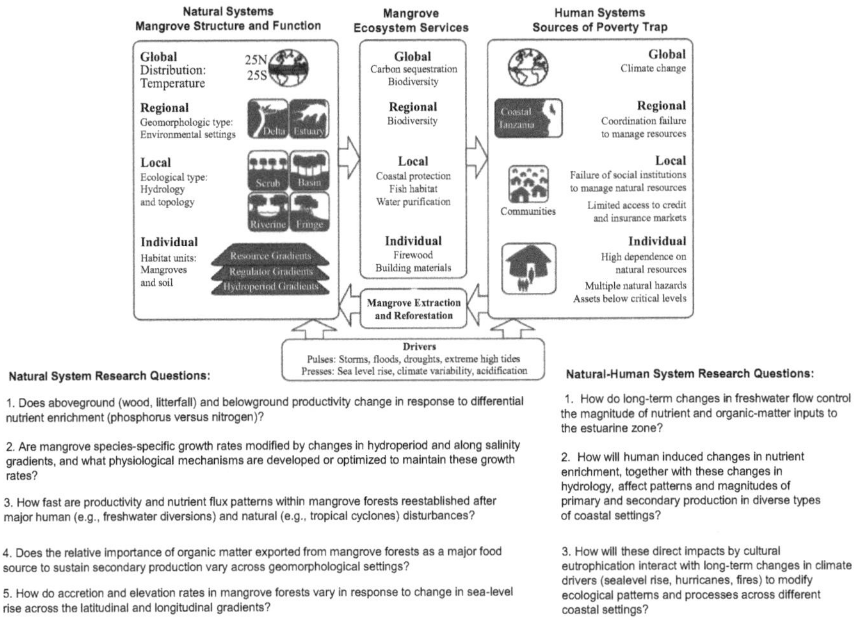

Fig. 11.6 Example of research questions to advance a mangrove socio-ecological research agenda using a macroecological approach with an emphasis on East Africa (modified from Rivera-Monroy et al. 2004; Gaiser et al. 2015)

on the structural and functional ecological conditions of mangrove regions where this climatic event is recurrent (Smith et al. 2009; Farfán et al. 2014; Vogt et al. 2014). Yet, due to the absence of long-term studies in sites where most mangrove area is present, it is not clear how the interactions between human and natural impacts affect mangrove wetland resilience and vulnerability (Badola and Hussain 2005; Vogt et al. 2014; Gaiser et al. 2015). Specific themes that can be advanced within this perspective include (1) linkages between mangrove area and shrimp or fish abundance (reviewed by: Alongi 2009); (2) linkages between climatic drivers and above- and belowground carbon pools (Hutchison et al. 2014; Jardine and Siikamäki 2014; Rovai et al. 2016); and (3) analyses of the effects of mangroves on wave attenuation (reviewed by: Marois and Mitsch 2015) (Fig. 11.6). Future work to advance a mangrove macroecological perspective would benefit from a concerted effort by research groups and institutions to launch research initiatives across biogeographic regions. In an era of unprecedented mangrove degradation and loss (i.e., the Anthropocene; Lugo et al. 2014), macroecology can advance our understanding of mangrove ecosystems and provide information that can be used to maintain goods and services from these unique wetlands for future generations.

Acknowledgments VHRM participation in the preparation of this chapter was partially supported by the Florida Coastal Everglades Long-Term Ecological Research program through US National Science Foundation (NSF) grants (DEB-9910514, DBI-0620409, DEB-1237517),

NASA-JPL (LSU Subcontract# 1452878) project "Vulnerability Assessment of Mangrove Forest Regions of the Americas," the Department of the Interior South Central Climate Science Center Cooperative Agreement #G12AC00002, and the NSF Dynamics of Coupled Natural and Human Systems Program CNH-1518471.

References

Adame MF, Virdis B, Lovelock CE (2010) Effect of geomorphological setting and rainfall on nutrient exchange in mangroves during tidal inundation. Mar Freshw Res 61:1197–1206

Adger WN, Hughes TP, Folke C, Carpenter SR, Rockström J (2005) Social-ecological resilience to coastal disasters. Science 309:1036–1039

Alongi DM (2009) The energetics of mangrove forests. Springer, Dordrecht

Alongi DM (2010) Dissolved iron supply limits early growth of estuarine mangroves. Ecology 91:3229–3241. https://doi.org/10.1890/09-2142.1

Alongi DM (2011) Carbon payments for mangrove conservation: ecosystem constraints and uncertainties of sequestration potential. Environ Sci Pol 14:462–470

Alongi DM (2014) Carbon cycling and storage in mangrove forests. Annu Rev Mar Sci 6:195–219

Alongi DM (2015) The impact of climate change on mangrove forests. Curr Climate Change Rep 1:30–39

Anderson-Teixeira KJ, Vitousek PM (2012) Ecosystems. In: Sibly RM, Brown JH, Kodric-Brown A (eds) Metabolic ecology: a scaling approach. Wiley-Blackwell, Oxford, pp 99–111

Araújo MB, Alagador D, Cabeza M, Nogués-Bravo D, Thuiller W (2011) Climate change threatens European conservation areas. Ecol Lett 14:484–492

Badola R, Hussain SA (2005) Valuing ecosystem functions: an empirical study on the storm protection function of Bhitarkanika mangrove ecosystem, India. Environ Conserv 32:85–92

Baker KS, Bowker GC (2007) Information ecology: open system environment for data, memories, and knowing. J Intell Inf Syst 29:127–144

Ball MC (1988) Ecophysiology of mangroves. Trees 2:129–142

Barbier EB, Hacker SD, Kennedy C, Koch EW, Stier AC, Silliman BR (2011) The value of estuarine and coastal ecosystem services. Ecol Monogr 81:169–193

Barrett CB, Travis AJ, Dasgupta P (2011) On biodiversity conservation and poverty traps. Proc Natl Acad Sci U S A 108:13907–13912

Belyea LR, Lancaster J (1999) Assembly rules within a contingent ecology. Oikos 86:402–416

Berger U, Rivera-Monroy VH, Doyle TW, Dahdouh-Guebas F, Duke NC, Fontalvo-Herazo ML, Hildenbrandt H, Koedam N, Mehlig U, Piou C, Twilley RR (2008) Advances and limitations of individual-based models to analyze and predict dynamics of mangrove forests: a review. Aquat Bot 89:260–274

Blackburn TM, Gaston KJ (2003) Macroecology: concepts and consequences: The 43rd Annual Symposium of the British Ecological Society. Cambridge University Press, Cambridge

Bouillon S, Borges AV, Castañeda-Moya E, Diele K, Dittmar T, Duke NC, Kristensen E, Lee SY, Marchand C, Middelburg JJ, Rivera-Monroy VH, Smith TJ III, Twilley RR (2008) Mangrove production and carbon sinks: a revision of global budget estimates. Glob Biogeochem Cycles 22:GB2013. https://doi.org/10.1029/2007GB003052

Boyd J, Banzhaf S (2007) What are ecosystem services? The need for standardized environmental accounting units. Ecol Econ 63:616–626

Brinson MM (1993) A hydrogeomorphic classification for wetlands. Technical Report WRP-DE-4, August. U.S. Army Corp of Enginners, Washington, DC, p 79. http://www.dtic.mil/docs/citations/ADA270053

Brolly M, Woodhouse IH, Niklas KJ, Hammond ST (2012) A macroecological analysis of SERA derived forest heights and implications for forest volume remote sensing. PLoS ONE 7:e33927. https://doi.org/10.1371/journal.pone.0033927

Brown JH (1995) Macroecology. The University of Chicago Press, Chicago
Brown JH, Gillooly JF, Allen AP, Savage VM, West GB (2004) Toward a metabolic theory of ecology. Ecology 85:1771–1789. https://doi.org/10.1890/03-9000
Brown JH, Burnside WR, Davidson AD, DeLong JP, Dunn WC, Hamilton MJ, Mercado-Silva N, Nekola JC, Okie JG, Woodruff WH, Zuo W (2011) Energetic limits to economic growth. Bioscience 61:19–26
Brown JH, Burger JR, Burnside WR, Chang M, Davidson AD, Fristoe TS, Hamilton MJ, Hammond ST, Kodric-Brown A, Mercado-Silva N, Nekola JC, Okie JG (2014) Macroecology meets macroeconomics: resource scarcity and global sustainability. Ecol Eng 65:24–32
Bucher D, Saenger P (1994) A classification of tropical and subtropical Australian estuaries. Aquat Conserv Mar Freshwat Ecosyst 4:1–19
Burger JR, Allen CD, Brown JH, Burnside WR, Davidson AD, Fristoe TS, Hamilton MJ, Mercado-Silva N, Nekola JC, Okie JG, Zuo W (2012) The macroecology of sustainability. PLoS Biol 10:e1001345
Castañeda-Moya E, Rivera-Monroy VH, Twilley RR (2006) Mangrove zonation in the dry life zone of the Gulf of Fonseca, Honduras. Estuar Coasts 29:751–764
Castañeda-Moya E, Twilley RR, Rivera-Monroy VH, Zhang K, Davis SE, Ross M (2010) Sediment and nutrient deposition associated with Hurricane Wilma in mangroves of the Florida Coastal Everglades. Estuar Coasts 33:45–58
Castañeda-Moya E, Twilley RR, Rivera-Monroy VH, Marx BD, Coronado-Molina C, Ewe SML (2011) Patterns of root dynamics in mangrove forests along environmental gradients in the Florida Coastal Everglades, USA. Ecosystems 14:1178–1195
Castañeda-Moya E, Twilley RR, Rivera-Monroy VH (2013) Allocation of biomass and net primary productivity of mangrove forests along environmental gradients in the Florida Coastal Everglades, USA. For Ecol Manag 307:226–241
Cavanaugh KC, Kellner JR, Forde AJ, Gruner DS, Parker JD, Rodriguez W, Feller IC (2014) Poleward expansion of mangroves is a threshold response to decreased frequency of extreme cold events. Proc Natl Acad Sci 111:723–727
Cavanaugh KC, Parker JD, Cook-Patton SC, Feller IC, Williams AP, Kellner JR (2015) Integrating physiological threshold experiments with climate modeling to project mangrove species' range expansion. Glob Chang Biol 21:1928–1938
Chapman VJ (1976) Mangrove vegetation. J. Cramer, Vaduz
Chen R, Twilley RR (1998a) A gap dynamic model of mangrove forest development along gradients of soil salinity and nutrient resources. J Ecol 86:37–51
Chen R, Twilley RR (1998b) A simulation model of organic matter and nutrient accumulation in mangrove wetland soils. Biogeochemistry 44:93–118
Chen LZ, Tam NFY, Wang WQ, Zhang YH, Lin GH (2013) Significant niche overlap between native and exotic *Sonneratia* mangrove species along a continuum of varying inundation periods. Estuar Coastal Shelf Sci 117:22–28
Chmura GL, Anisfeld SC, Cahoon DR, Lynch JC (2003) Global carbon sequestration in tidal, saline wetland soils. Glob Biogeochem Cycles 17
Cintrón G, Lugo AE, Pool DJ, Morris G (1978) Mangroves of arid environments in Puerto Rico and adjacent islands. Biotropica 10:110–121
Clarke PJ (1995) The population-dynamics of the mangrove *Avicennia marina* demographic synthesis an predictive modeling. Hydrobiologia 295:83–88
Clarke PJ (2014) Seeking global generality: a critique for mangrove modellers. Mar Freshw Res 65:930–933
Collins SL, Carpenter SR, Swinton SM, Orenstein DE, Childers DL, Gragson TL, Grimm NB, Grove JM, Harlan SL, Kaye JP, Knapp AK, Kofinas GP, Magnuson JJ, McDowell WH, Melack JM, Ogden LA, Robertson GP, Smith MD, Whitmer AC (2011) An integrated conceptual framework for long-term social-ecological research. Front Ecol Environ 9:351–357
Costanza R, d'Arge R, De Groot R, Farber S, Grasso M, Hannon B, Limburg K, Naeem S, O'Neill RV, Paruelo J (1997) The value of the world's ecosystem services and natural capital. Nature 387:253–260

Crase B, Liedloff A, Vesk PA, Burgman MA, Wintle BA (2013) Hydroperiod is the main driver of the spatial pattern of dominance in mangrove communities. Glob Ecol Biogeogr 22:806–817
Daily GC, Polasky S, Goldstein J, Kareiva PM, Mooney HA, Pejchar L, Ricketts TH, Salzman J, Shallenberger R (2009) Ecosystem services in decision making: time to deliver. Front Ecol Environ 7:21–28
Das K, Ray S (2010) Delay in the decomposition of dead algae and its impact on benthic food chain of reclaimed mangrove ecosystem: a theoretical approach. Nonlinear Stud 17:387–398
Davis MB, Shaw RG (2001) Range shifts and adaptive responses to Quaternary climate change. Science 292:673–679
Day JW, Christian RR, Boesch DM, Yáñez-Arancibia A, Morris J, Twilley RR, Naylor L, Schaffner L, Stevenson C (2008) Consequences of climate change on the ecogeomorphology of coastal wetlands. Estuar Coasts 31:477–491
Donato DC, Kauffman JB, Murdiyarso D, Kurnianto S, Stidham M, Kanninen M (2011) Mangroves among the most carbon-rich forests in the tropics. Nat Geosci 4:293–297
Duke NC (2001) Gap creation and regenerative processes driving diversity and structure of mangrove ecosystems. Wetl Ecol Manag 9:267–279
Duke NC, Ball MC, Ellison JC (1998) Factors influencing biodiversity and distributional gradients in mangroves. Glob Ecol Biogeogr Lett 7:27–47
Duke NC, Meynecke JO, Dittmann S, Ellison AM, Anger K, Berger U, Cannicci S, Diele K, Ewel KC, Field CD, Koedam N, Lee SY, Marchand C, Nordhaus I, Dahdouh-Guebas F (2007) A world without mangroves? Science 317:41–42
Ellison JC (1998) Impacts of sediment burial on mangroves. Mar Pollut Bull 37:420–426
Ellison AM (2002) Macroecology of mangroves: large-scale patterns and processes in tropical coastal forest. Trees-Struct Func 16:181–194
Ellison AM, Farnsworth EJ, Merkt RE (1999) Origins of mangrove ecosystems and the mangrove biodiversity anomaly. Glob Ecol Biogeogr 8:95–115
Enquist BJ, Jordan MA, Brown JH (1995) Connections between ecology, biogeography, and paleobiology: relationship between local abundance and geographic distribution in fossil and recent molluscs. Evol Ecol 9:586–604
Enwright NM, Griffith KT, Osland MJ (2016) Barriers to and opportunities for landward migration of coastal wetlands with sea-level rise. Front Ecol Environ 14:307–316
Ewel KC, Twilley RR, Ong JE (1998) Different kinds of mangrove forests provide different goods and services. Glob Ecol Biogeogr Lett 7:83–94
Farfán LM, D'Sa E, Liu K, Rivera-Monroy VH (2014) Tropical cyclone impacts on coastal regions: the case of the Yucatán and the Baja California Peninsulas, Mexico. Estuaries and Coasts 37:1388–1402
Feller IC, Lovelock CE, McKee KL (2007) Nutrient addition differentially affects ecological processes of *Avicennia germinans* in nitrogen versus phosphorus limited mangrove ecosystems. Ecosystems 10:347–359
Feller IC, Lovelock CE, Berger U, McKee KL, Joye SB, Ball MC (2010) Biocomplexity in mangrove ecosystems. Annu Rev Mar Sci 2:395–417
Fisher B, Turner K, Zylstra M, Brouwer R, De Groot R, Farber S, Ferraro P, Green R, Hadley D, Harlow J, Jefferiss P, Kirkby C, Morling P, Mowatt S, Naidoo R, Paavola J, Strassburg B, Yu D, Balmford A (2008) Ecosystem services and economic theory: Integration for policy-relevant research. Ecol Appl 18:2050–2067. https://doi.org/10.1890/07-1537.1
Fisher B, Turner RK, Morling P (2009) Defining and classifying ecosystem services for decision making. Ecol Econ 68:643–653
Fontalvo-Herazo ML, Piou C, Vogt J, Saint-Paul U, Berger U (2011) Simulating harvesting scenarios towards the sustainable use of mangrove forest plantations. Wetl Ecol Manag 19:397–407
Fowler CW (2008) Maximizing biodiversity, information and sustainability. Biodivers Conserv 17:841–855
Fromard F, Vega C, Proisy C (2004) Half a century of dynamic coastal change affecting mangrove shorelines of French Guiana. A case study based on remote sensing data analyses and field surveys. Mar Geol 208:265–280

Gabler CA, Osland MJ, Grace JB, Stagg CL, Day RH, Hartley SB, Enwright NM, From AS, McCoy ML, McLeod JL (2017) Macroclimatic change expected to transform coastal wetland ecosystems this century. Nat Clim Chang 7:142–147

Gaiser EE, Anderson EP, Castañeda-Moya E, Collado-Vides L, Fourqurean JW, Heithaus MR, Jaffé R, Lagomasino D, Oehm NJ, Price RM, Rivera-Monroy VH, Chowdhury RR, Troxler TG (2015) New perspectives on an iconic landscape from comparative international long-term ecological research. Ecosphere 6:1–18

Gaston KJ, Blackburn TM (1999) A critique for macroecology. Oikos 84:353–368

Giri C, Ochieng E, Tieszen LL, Zhu Z, Singh A, Loveland T, Masek J, Duke N (2011) Status and distribution of mangrove forests of the world using earth observation satellite data. Glob Ecol Biogeogr 20:154–159

Glick P, Stein BA, Edelson NA (eds) (2011) Scanning the conservation horizon: a guide to climate change vulnerability assessment. National Wildlife Federation, Washington, DC

Grueters U, Seltmann T, Schmidt H, Horn H, Pranchai A, Vovides AG, Peters R, Vogt J, Dandouh-Guebas F, Berger U (2014) The mangrove forest dynamics model mesoFON. Ecol Model 291:28–41

Guo H, Zhang Y, Lan Z, Pennings SC (2013) Biotic interactions mediate the expansion of black mangrove (*Avicennia germinans*) into salt marshes under climate change. Glob Chang Biol 19:2765–2774

Hammond ST, Niklas KJ (2009) Emergent properties of plants competing in silico for space and light: seeing the tree from the forest. Am J Bot 96:1430–1444

Hammond ST, Niklas KJ (2011) Computer simulations of plant biodiversity in stable and unstable environments: A test of the neutral biodiversity theory. J Biol Syst 19:1–17

Häyhä T, Franzese PP (2014) Ecosystem services assessment: a review under an ecological-economic and systems perspective. Ecol Model 289:124–132

Hicks DB, Burns LA (1975) Mangrove metabolic response to alterations of natural freshwater drainage to southwestern Florida estuaries. In: Walsh GE, Snedaker SC, Teas HI (eds) Proceedings of the International Symposium on Biology and Management of Mangroves. East-West Center, Honolulu, pp 238–255

Holdridge LR (1967) Life zone ecology. Tropical Science Center, San Jose

Hutchison J, Manica A, Swetnam R, Balmford A, Spalding M (2014) Predicting global patterns in mangrove forest biomass. Conserv Lett 7:233–240

Jardine SL, Siikamäki JV (2014) A global predictive model of carbon in mangrove soils. Environ Res Lett 9:104013

Jenkins WA, Murray BC, Kramer RA, Faulkner SP (2010) Valuing ecosystem services from wetlands restoration in the Mississippi Alluvial Valley. Ecol Econ 69:1051–1061

Jørgensen SE, Nielsen SN (2013) The properties of the ecological hierarchy and their application as ecological indicators. Ecol Indic 28:48–53

Kautz M, Berger U, Stoyan D, Vogt J, Khan NI, Diele K, Saint-Paul U, Tran T, Vien Ngoc N (2011) Desynchronizing effects of lightning strike disturbances on cyclic forest dynamics in mangrove plantations. Aquat Bot 95:173–181

Keddy PA (1992) Assembly and response rules: two goals for predictive community ecology. J Veg Sci 3:157–164

Keith SA, Webb TJ, Boehning-Gaese K, Connolly SR, Dulvy NK, Eigenbrod F, Jones KE, Price T, Redding DW, Owens IPF, Isaac NJB (2012) What is macroecology? Biol Lett 8:904–906

Kennedy RE, Andréfouët S, Cohen WB, Gómez C, Griffiths P, Hais M, Healey SP, Helmer EH, Hostert P, Lyons MB, Meigs GW, Pflugmacher D, Phinn SR, Powell SL, Scarth P, Sen S, Schroeder TA, Schneider A, Sonnenschein R, Vogelmann JE, Wulder MA, Zhu Z (2014) Bringing an ecological view of change to Landsat-based remote sensing. Front Ecol Environ 12:339–346

Keogh TM, Keddy PA, Fraser LH (1999) Patterns of tree species richness in forested wetlands. Wetlands 19:639–647

Kerr JT, Kharouba HM, Currie DJ (2007) The macroecological contribution to global change solutions. Science 316:1581–1584

Krauss KW, Ball MC (2013) On the halophytic nature of mangroves. Trees-Struct Func 27:7–11
Krauss KW, Doyle TW, Twilley RR, Rivera-Monroy VH, Sullivan JK (2006) Evaluating the relative contributions of hydroperiod and soil fertility on growth of south Florida mangroves. Hydrobiologia 569:311–324
Kristensen E, Bouillon S, Dittmar T, Marchand C (2008) Organic carbon dynamics in mangrove ecosystems: a review. Aquat Bot 89:201–219
Lawler JJ, Shafer SL, White D, Kareiva P, Maurer EP, Blaustein AR, Bartlein PJ (2009) Projected climate-induced faunal change in the Western Hemisphere. Ecology 90:588–597
Longley WL (1994) Freshwater inflows to Texas bays and estuaries: ecological relationships and methods for determination of needs. Texas Water Development Board and Texas Parks and Wildlife Department, Austin
Lot-Helgueras A, Vázquez-Yanes C, Menendez F (1975) Physiognomic and floristic changes near the northern limit of mangroves in the Gulf Coast of Mexico. In: Walsh GE, Snedaker SC, Teas HJ (eds) Proceedings of the International Symposium on Biology and Management of Mangroves. Institute of Food and Agricultural Sciences, University of Florida, Gainesville, pp 52–64
Lovelock CE (2008) Soil respiration and belowground carbon allocation in mangrove forests. Ecosystems 11:342–354
Lovelock CE, Feller IC, Mckee KL, Engelbrecht BMJ, Ball MC (2004) The effect of nutrient enrichment on growth, photosynthesis and hydraulic conductance of dwarf mangroves in Panama. Funct Ecol 18:25–33
Lovelock CE, Feller IC, Ball MC, Ellis J, Sorrell B (2007) Testing the growth rate vs. geochemical hypothesis for latitudinal variation in plant nutrients. Ecol Lett 10:1154–1163
Lovelock CE, Sorrell BK, Hancock N, Hua Q, Swales A (2010) Mangrove forest and soil development on a rapidly accreting shore in New Zealand. Ecosystems 13:437–451
Lovelock CE, Krauss KW, Osland MJ, Reef R, Ball MC (2016) The physiology of mangrove trees with changing climate. In: Goldstein G, Santiago LS (eds) Tropical tree physiology: adaptations and responses in a changing environment. Springer, New York, pp 149–179
Lugo AE (1980) Mangrove ecosystems: successional or steady state? Biotropica 12:65–72
Lugo AE, Sell M, Snedaker SC (1976) Mangrove ecosystem analysis. In: Systems analysis and simulation in ecology. Academic Press, Inc, New York, pp 113–145
Lugo AE, Patterson-Zucca C (1977) The impact of low temperature stress on mangrove structure and growth. Trop Ecol 18:149–161
Lugo AE, Snedaker SC (1974) The ecology of mangroves. Annu Rev Ecol Syst 5:39–64
Lugo AE, Medina E, McGinley K (2014) Issues and challenges of mangrove conservation in the Anthropocene. Madera y Bosques 20:11–38
Lunstrum A, Chen L (2014) Soil carbon stocks and accumulation in young mangrove forests. Soil Biol Biochem 75:223–232
López-Hoffman L, Ackerly DD, Anten NPR, Denoyer JL, Martinez-Ramos M (2007) Gap-dependence in mangrove life-history strategies: a consideration of the entire life cycle and patch dynamics. J Ecol 95:1222–1233. https://doi.org/10.1111/j.1365-2745.2007.01298.x
Madrid EN, Armitage AR, López-Portillo J (2014) *Avicennia germinans* (black mangrove) vessel architecture is linked to chilling and salinity tolerance in the Gulf of Mexico. Front Plant Sci 5:503
Mandal S, Ray S, Ghosh PB (2009) Modelling of the contribution of dissolved inorganic nitrogen (DIN) from litterfall of adjacent mangrove forest to Hooghly-Matla estuary, India. Ecol Model 220:2988–3000
Mandal S, Ray S, Ghosh PB (2012) Modeling nutrient (dissolved inorganic nitrogen) and plankton dynamcis at Sagar Islands of Hooghly-Matla Estuarine system, West Bengal, India. Nat Resour Model 25:629–652
Mandal S, Ray S, Ghosh PB (2013) Impact of mangrove litterfall on nitrogen dynamics of virgin and reclaimed islands of Sundarban mangrove ecosystem, India. Ecol Model 252:153–166
Manna S, Nandy S, Chanda A, Akhand A, Hazra S, Dadhwal VK (2014) Estimating aboveground biomass in *Avicennia marina* plantation in Indian Sundarbans using high-resolution satellite data. J Appl Remote Sens 8:083638

Marois DE, Mitsch WJ (2015) Coastal protection from tsunamis and cyclones provided by mangrove wetlands–a review. Int J Biodivers Sci Ecosyst Serv Manage 11:71–83
Marquet PA, Labra FA, Maurer BA (2004) Metabolic ecology: linking individuals to ecosystems. Ecology 85:1794–1796
McBride PD, Cusens J, Gillman LN (2014) Revisiting spatial scale in the productivity-species richness relationship: fundamental issues and global change implications. AoB Plants 6
McGill B, Collins C (2003) A unified theory for macroecology based on spatial patterns of abundance. Evol Ecol Res 5:469–492
McKee KL (2011) Biophysical controls on accretion and elevation change in Caribbean mangrove ecosystems. Estuar Coastal Shelf Sci 91:475–483
McNally CG, Uchida E, Gold AJ (2011) The effect of a protected area on the tradeoffs between short-run and long-run benefits from mangrove ecosystems. Proc Natl Acad Sci 108:13945–13950
Méndez-Alonzo R, López-Portillo J, Rivera-Monroy VH (2008) Latitudinal variation in leaf and tree traits of the mangrove *Avicennia germinans* (Avicenniaceae) in the central region of the Gulf of Mexico. Biotropica 40:449–456
Millenium Ecosystem Assessment (2005) Ecosystems and human well-being: Synthesis. Island Press, Washington, DC
Mitsch WJ, Gosselink JG (2015) Wetlands, 5th edn. John Wiley & Sons, Inc., Hoboken, p 736
Montagna PA, Brenner J, Gibeaut J, Morehead S (2011) Coastal impacts. In: Schmandt J, North GR, Clarkson J (eds) The impact of global warming on Texas, 2nd edn. University of Texas Press, Austin, pp 96–123
Mukherjee J, Roy M, Ray S, Ghosh PB, Sarkar A (2012) Mechanisms of transformation of various forms of carbon and cycling pathways in the Hooghly estuarine system. In: Jordan J (ed) Estuaries: classification, ecology and human impacts. Nova Science Publishers, New York
Mukherjee J, Ray S, Ghosh PB (2013) A system dynamic modeling of carbon cycle from mangrove litter to the adjacent Hooghly estuary, India. Ecol Model 252:185–195
Mukherjee J, Banerjee M, Banerjee A, Roy M, Ghosh PB, Ray S (2014) Impact of environmental factors on the carbon dynamics at Hooghly estuarine region. J Ecosyst 2014:1–10
Mukherjee J, Scharler UM, Fath BD, Ray S (2015) Measuring sensitivity of robustness and network indices for an estuarine food web model under perturbations. Ecol Model 306:160–173
Nee S (2002) Thinking big in ecology. Nature 417:229–230
Nekola JC, Allen CD, Brown JH, Burger JR, Davidson AD, Fristoe TS, Hamilton MJ, Hammond ST, Kodric-Brown A, Mercado-Silva N, Okie JG (2013) The Malthusian-Darwinian dynamic and the trajectory of civilization. Trends Ecol Evol 28:127–130
Nielsen SN, Müller F (2009) Understanding the functional principles of nature-Proposing another type of ecosystem services. Ecol Model 220:1913–1925
Niklas KJ, Hammond ST (2013) Biophysical effects on plant competition and coexistence. Funct Ecol 27:854–864
Odum WE, Tohannes RE (1975) The response of mangroves to man induced environmental stress. In: Ferguson Wood E, Tohannes RE (eds) Tropical marine pollution. Elsevier Oceanography Series, Amsterdam, pp 52–62
Osland MJ, Spivak AC, Nestlerode JA, Lessmann JM, Almario AE, Heitmuller PT, Russell MJ, Krauss KW, Alvarez F, Dantin DD, Harvey JE, From AS, Cormier N, Stagg CL (2012) Ecosystem development after mangrove wetland creation: plant-soil change across a 20-year chronosequence. Ecosystems 15:848–866
Osland MJ, Enwright N, Day RH, Doyle TW (2013) Winter climate change and coastal wetland foundation species: salt marshes vs. mangrove forests in the southeastern United States. Glob Chang Biol 19:1482–1494
Osland MJ, Day RH, Larriviere JC, From AS (2014a) Aboveground allometric models for freeze-affected black mangroves (*Avicennia germinans*): equations for a climate sensitive mangrove-marsh ecotone. PLoS One 9:e99604
Osland MJ, Enwright N, Stagg CL (2014b) Freshwater availability and coastal wetland foundation species: ecological transitions along a rainfall gradient. Ecology 95:2789–2802

Osland MJ, Enwright NM, Day RH, Gabler CA, Stagg CL, Grace JB (2016) Beyond just sea-level rise: considering macroclimatic drivers within coastal wetland vulnerability assessments to climate change. Glob Chang Biol 22:1–11

Osland MJ, Feher LC, Griffith KT, Cavanaugh KC, Enwright NM, Day RH, Stagg CL, Krauss KW, Howard RJ, Grace JB, Rogers K (2017) Climatic controls on the global distribution, abundance, and species richness of mangrove forests. Ecol Monogr 87:341–359

Ouyang X, Lee SY, Connolly RM (2017) Global patterns of root decomposition rates and the fate of root carbon production in mangroves and saltmarsh. Earth-Sci Rev 166:53–63

Peters DPC, Loescher HW, SanClements MD, Havstad KM (2014) Taking the pulse of a continent: expanding site-based research infrastructure for regional- to continental-scale ecology. Ecosphere 5:1–23

Peyronnin N, Green M, Richards CP, Owens A, Reed D, Chamberlain J, Groves DG, Rhinehart WK, Belhadjali K (2013) Louisiana's 2012 coastal master plan: overview of a science-based and publicly informed decision-making process. J Coast Res 67:1–15

Pfeifer M, Disney M, Quaife T, Marchant R (2012) Terrestrial ecosystems from space: a review of earth observation products for macroecology applications. Glob Ecol Biogeogr 21:603–624

Piou C, Berger U, Hildenbrandt H, Feller IC (2008) Testing the intermediate disturbance hypothesis in species-poor systems: a simulation experiment for mangrove forests. J Veg Sci 19:417–424

Polasky S, Segerson K (2009) Integrating ecology and economics in the study of ecosystem services: some lessons learned. Ann Rev Resour Econ 1:409–434

Polidoro BA, Carpenter KE, Collins L, Duke NC, Ellison AM, Ellison JC, Farnsworth EJ, Fernando ES, Kathiresan K, Koedam NE, Livingstone SR, Miyagi T, Moore GE, Vien NN, Ong JE, Primavera JH, Salmo SG III, Sanciangco JC, Sukardjo S, Wang YM, Yong JWH (2010) The loss of species: mangrove extinction risk and geographic areas of global concern. PLoS One 5:e10095

Quisthoudt K, Schmitz N, Randin CF, Dahdouh-Guebas F, Robert EMR, Koedam N (2012) Temperature variation among mangrove latitudinal range limits worldwide. Trees-Struct Func 26:1919–1931

Ray S (2008) Comparative study of virgin and reclaimed islands of Sundarban mangrove ecosystem through network analysis. Ecol Model 215:207–216

Ray S, Straškraba M (2001) The impact of detritivorous fishes on a mangrove estuarine system. Ecol Model 140:207–218

Ray S, Ulanowicz RE, Majee NC, Roy AB (2000) Network analysis of a benthic food web model of a partly reclaimed island in the Sundarban mangrove ecosystem, India. J Biol Syst 8:263–278

Record S, Charney ND, Zakaria RM, Ellison AM (2013) Projecting global mangrove species and community distributions under climate change. Ecosphere 4:1–23

Reef R, Feller IC, Lovelock CE (2010) Nutrition of mangroves. Tree Physiol 30:1148–1160

Rivera-Monroy VH, Torres LA, Nixon B, Newmark F, Twilley RR (1999) The potential use of mangrove forests as nitrogen sinks of shrimp aquaculture pond effluents: the role of denitrification. J World Aquacult Soc 30:12–25

Rivera-Monroy VH, Twilley RR, Bone D, Childers DL, Coronado-Molina C, Feller IC, Herrera-Silveira J, Jaffe R, Mancera E, Rejmankova E, Salisbury JE, Weil E (2004) A conceptual framework to develop long-term ecological research and management objectives in the wider Caribbean region. Bioscience 54:843–856

Rivera-Monroy VH, Twiley RR, Mancera-Pineda JE, Madden CJ, Alcantara-Eguren A, Moser EB, Jonsson BR, Castañeda-Moya E, Casas-Monroy O, Reyes-Forero P, Restrepo J (2011) Salinity and chlorophyll a as performance measures to rehabilitate a mangrove-dominated deltaic coastal region: the Ciénaga Grande de Santa Marta-Pajarales lagoon complex, Colombia. Estuar Coasts 34:1–19

Rivera-Monroy VH, Castañeda-Moya E, Barr JG, Engel V, Fuentes JD, Troxler TG, Twilley RR, Bouillon S, Smith TJ, O'Halloran TL (2013) Current methods to evaluate net primary production and carbon budgets in mangrove forests. In: DeLaune RD, Reddy KR, Megonigal P, Richardson C (eds) Methods in biogeochemistry of wetlands. Soil Science Society of America Book Series, Madison, pp 243–288

Rovai AS, Riul P, Twilley RR, Castañeda-Moya E, Rivera-Monroy VH, Williams AA, Simard M, Cifuentes-Jara M, Lewis RR, Crooks S, Horta PA, Schaeffer-Novelli Y, Cintron G, Pozo-Cajas M, Pagliosa PR (2016) Scaling mangrove aboveground biomass from site-level to continental-scale. Glob Ecol Biogeogr 25:286–298
Roy M, Mandal S, Ray S (2008) Detrital ontogenic model including decomposer diversity. Ecol Model 215:200–206
Ruggiero A, Hawkins BA (2006) Mapping macroecology. Glob Ecol Biogeogr 15:433–437
Saenger P, Snedaker SC (1993) Pantropical trends in mangrove above-ground biomass and annual litterfall. Oecologia 96:293–299
Saintilan N (2009) Biogeography of Australian saltmarsh plants. Austral Ecol 34:929–937
Saintilan N, Wilson NC, Rogers K, Rajkaran A, Krauss KW (2014) Mangrove expansion and salt marsh decline at mangrove poleward limits. Glob Chang Biol 20:147–157
Salmo SG III, Lovelock C, Duke NC (2013) Vegetation and soil characteristics as indicators of restoration trajectories in restored mangroves. Hydrobiologia 720:1–18
Schaeffer-Novelli Y, Cintron G (1990) Status of mangrove research in Latin America and the Caribbean. Boletim do Instituto Oceanográfico 38:93–97
Schramski JR, Dell AI, Grady JM, Silby RM, Brown JH (2015) Metabolic theory predicts whole-ecosystem properties. Proc Natl Acad Sci 112:201423502. https://doi.org/10.1073/pnas.1423502112
Semeniuk V (2013) Predicted response of coastal wetlands to climate changes: a Western Australian model. Hydrobiologia 708:23–43
Shafer SL, Bartlein PJ, Thompson RS (2001) Potential changes in the distributions of western North America tree and shrub taxa under future climate scenarios. Ecosystems 4:200–215
Simard M, Zhang KQ, Rivera-Monroy VH, Ross MS, Ruiz PL, Castañeda-Moya E, Twilley RR, Rodriguez E (2006) Mapping height and biomass of mangrove forests in Everglades National Park with SRTM elevation data. Photogramm Eng Remote Sens 72:299–311
Simberloff DS, Wilson EO (1969) Experimental zoogeography of islands: the colonization of empty islands. Ecology 50:278–296
Smith TJ III, Duke NC (1987) Physical determinants of inter-estuary variation in mangrove species richness around the tropical coastline of Australia. J Biogeogr 14:9–19
Smith FA, Lyons SK, Ernest SKM, Brown JH (2008) Macroecology: more than the division of food and space among species on continents. Prog Phys Geogr 32:115–138
Smith TJ III, Anderson GH, Balentine K, Tiling G, Ward GA, Whelan KRT (2009) Cumulative impacts of hurricanes on Florida mangrove ecosystems: sediment deposition, storm surges and vegetation. Wetlands 29:24–34
Smith FA, Gittleman JL, Brown JH (eds) (2014) Foundations of macroecology: classic papers with commentaries. The University of Chicago Press, Chicago
Staver AC, Archibald S, Levin SA (2011) The global extent and determinants of savanna and forest as alternative biome states. Science 334:230–232
Stuart SA, Choat B, Martin KC, Holbrook NM, Ball MC (2007) The role of freezing in setting the latitudinal limits of mangrove forests. New Phytol 173:576–583
Suárez-Abelenda M, Ferreira TO, Camps-Arbestain M, Rivera-Monroy VH, Macías F, Nóbrega GN, Otero XL (2014) The effect of nutrient-rich effluents from shrimp farming on mangrove soil carbon storage and geochemistry under semi-arid climate conditions in northern Brazil. Geoderma 213:551–559
Thom BG (1967) Mangrove ecology and deltaic geomorphology: Tabasco, Mexico. J Ecol 55:301–343
Thom BG (1982) Mangrove ecology: a geomorphological perspective. In: Clough BF (ed) Mangroves ecosystems in Australia, structure, function, and management. Australian National University Press, Canberra, pp 3–17
Tomlinson PB (1986) The botany of mangroves. Cambridge University Press, Cambridge
Torio DD, Chmura GL (2013) Assessing coastal squeeze of tidal wetlands. J Coast Res 29:1049–1061

Troxler TG, Gaiser E, Barr J, Fuentes JD, Jaffe R, Childers DL, Collado-Vides L, Rivera-Monroy VH, Castañeda-Moya E, Anderson W, Chambers R, Chen ML, Coronado-Molina C, Davis SE, Engel V, Fitz C, Fourqurean J, Frankovich T, Kominoski J, Madden C, Malone SL, Oberbauer SF, Olivas P, Richards J, Saunders C, Schedlbauer J, Scinto LJ, Sklar F, Smith T, Smoak JM, Starr G, Twilley RR, Whelan K (2013) Integrated carbon budget models for the Everglades terrestrial-coastal-oceanic gradient: Current status and needs for inter-site comparisons. Oceanography 26:98–107

Twilley, RR (1995) Propierties of mangroves ecosystems and their relation to the energy signature of coastal environments, Chapter 7 in: Hall, C.A.S. (Ed.), Maximum Power. Unversity Press of Colorado Press, Colorado, pp. 43–62

Twilley RR (1997) Mangrove wetlands. In: Messana G, Connor W (eds) Southern forested wetlands: ecology and management. CRC Press, Boca Raton, pp 445–473

Twilley RR, Chen RH (1998) A water budget and hydrology model of a basin mangrove forest in Rookery Bay, Florida. Mar Freshw Res 49:309–323

Twilley RR, Rivera-Monroy VH, Chen R, Botero L (1999) Adapting an ecological mangrove model to simulate trajectories in restoration ecology. Mar Pollut Bull 37:404–419

Twilley RR, Day JW (2012) Mangrove wetlands. In: Day JW, Crump BC, Kemp MW, Yáñez-Arancibia A (eds) Estuarine ecology, 2nd edn. John Wiley & Sons, Hoboken, pp 165–202

Twilley RR, Rivera-Monroy VH (2005) Developing performance measures of mangrove wetlands using simulation models of hydrology, nutrient biogeochemistry, and community dynamics. J Coast Res 40:79–93

Twilley RR, Rivera-Monroy VH (2009) Ecogeomorphic models of nutrient biogeochemistry for mangrove wetlands. In: Perillo GME, Wolanski E, Cahoon DR, Brinson MM (eds) Coastal wetlands: an integrated ecosystem approach. Elsevier, Amsterdam, pp 641–683

Twilley RR, Chen RH, Hargis T (1992) Carbon sinks in mangroves and their implications to carbon budget of tropical ecosystems. Water Air Soil Pollut 64:265–288

Twilley RR, Rivera-Monroy VH, Chen R, Botero L (1998) Adapting an ecological mangrove model to simulate trajectories in restoration ecology. Mar Pollut Bull 37:404–419

Uchida E, Rivera-Monroy VH, Gold A, Uchida H (2014) U.S-Tanzania: building a research collaboration on the dynamics of mangrove ecosystem services and poverty traps. Final Report to the NSF Catalyzing International Collaborations program, National Science Foundation, University of Rhode Island

Ulanowicz RE, Holt RD, Barfield M (2014) Limits on ecosystem trophic complexity: insights from ecological network analysis. Ecol Lett 17:127–136

Urban DL (2005) Modeling ecological processes across scales. Ecology 86:1996–2006

Valiela I, Bowen JL, York JK (2001) Mangrove forests: one of the world's threatened major tropical environments. Bioscience 51:807–815

Van Groenigen KJ, Qi X, Osenberg CW, Luo Y, Hungate BA (2014) Faster decomposition under increased atmospheric CO_2 limits soil carbon storage. Science 344:508–509

Vega-Cendejas ME, Arreguín-Sánchez F (2001) Energy fluxes in a mangrove ecosystem from a coastal lagoon in Yucatán Peninsula, Mexico. Ecol Model 137:119–133

Vogt J, Kautz M, Herazo MLF, Tran T, Walther D, Saint-Paul U, Diele K, Berger U (2013) Do canopy disturbances drive forest plantations into more natural conditions? – A case study from Can Gio Biosphere Reserve, Viet Nam. Glob Planet Chang 110:249–258

Vogt J, Piou C, Berger U (2014) Comparing the influence of large- and small-scale disturbances on forest heterogeneity: a simulation study for mangroves. Ecol Complex 20:107–115

Vovides AG, Vogt J, Kollert A, Berger U, Grueters U, Peters R, Lara-Domínguez AL, López-Portillo J (2014) Morphological plasticity in mangrove trees: salinity-related changes in the allometry of *Avicennia germinans*. Trees-Struct Func 28:1413–1425

Wang W, Yan Z, You S, Zhang Y, Chen L, Lin G (2011) Mangroves: obligate or facultative halophytes? A review. Trees-Struct Funct 25:953–963

Webb EL, Friess DA, Krauss KW, Cahoon DR, Guntenspergen GR, Phelps J (2013) A global standard for monitoring coastal wetland vulnerability to accelerated sea-level rise. Nat Clim Chang 3:458–465

West RC (1977) Tidal salt-marsh and mangal formations of Middle and South America. In: Chapman VJ (ed) Wet coastal ecosystems. Elsevier, Amsterdam, pp 193–213
Whittaker RH (1970) Communities and ecosystems. The Macmillan Company, New York
Whittaker RJ, Willis KJ, Field R (2001) Scale and species richness: towards a general, hierarchical theory of species diversity. J Biogeogr 28:453–470
Witman JD, Roy K (2009) Experimental marine macroecology: progress and Prospects. In: Witman JD, Roy K (eds) Marine macroecology. The University of Chicago Press, Chicago, pp 341–356
Woodroffe C (1992) Mangrove sediments and geomorphology. In: Robertson AI, Alongi DM (eds) Tropical mangrove ecosystems. American Geophysical Union, Washington, DC, pp 7–42
Woodroffe C (2002) Coasts: form, process and evolution. Cambridge University Press, Cambridge
Xiao X, McGlinn DJ, White EP (2015) A strong test of the maximum entropy theory of ecology. Am Nat 185:E70–E80
Yue TX, Jorgensen SE, Larocque GR (2011) Progress in global ecological modelling. Ecol Model 222:2172–2177
Zedler JB (1982) The ecology of southern California coastal salt marshes: a community profile. U.S. Fish and Wildlife Service, Biological Services Program, FWS/OBS-81/54, Washington, DC

Epilogue

Shing Yip Lee, Victor H. Rivera-Monroy, Erik Kristensen, and Robert R. Twilley

An idea that developed among the editors of this book during the 2011 Coastal and Estuarine Research Federation meeting at Daytona Beach, Florida, has finally come to fruition. Ever since the conception of this book, we never felt it would be an easy task – but we have now completed it! It turned out to be a challenge because mangrove research has grown from its humble beginnings with a restricted geographic and scientific coverage five decades ago, to a diverse global discipline with >1100 publications by researchers from 96 countries in 2016 (Web of Science 2017).

Research on mangrove ecology has come a long way since the days when almost all key theories and paradigms were "borrowed" from the much better-studied Atlantic saltmarshes. As with the development of most scientific disciplines, early paradigms in mangrove ecosystem ecology were guided by bold and insightful

S.Y. Lee
Australian Rivers Institute and School of Environment, Griffith University Gold Coast campus, Southport, QLD, 4222, Australia

School of Life Sciences and Simon F. S. Li Marine Science Laboratory, The Chinese University of Hong Kong, Shatin, Hong Kong SAR

V.H. Rivera-Monroy
College of the Coast and Environment, Department of Oceanography and Coastal Sciences, Louisiana State University, Baton Rouge, LA 70803, USA

E. Kristensen
Department of Biology, University of Southern Denmark, Campusvej 55, 5230, Odense M, Denmark

R.R. Twilley
School of Life Sciences and Simon S.F. Li Marine Science Laboratory, The Chinese University of Hong Kong, Shatin, Hong Kong SAR

Louisiana Sea Grant Program, 39 Sea Grant Building, Baton Rouge, LA 70803-7507, USA

V.H. Rivera-Monroy et al. (eds.), *Mangrove Ecosystems: A Global Biogeographic Perspective*, https://doi.org/10.1007/978-3-319-62206-4

hypotheses, again, mostly inspired by studies from the West Atlantic saltmarshes, but adapted for the tropical Caribbean mangrove environments. This view prompted the portrait of *all* mangrove forests as net exporters of organic matter that is efficiently assimilated by a wide range of estuarine consumers and, therefore, forming the trophic base of nearshore fisheries. This trophic sustenance adds to the complex habitat structure offered by the trees to underpin the nursery value of mangrove wetlands. Mangrove ecologists today owe a great deal to these pioneers in their success of drawing much-needed attention to the Caribbean systems and beyond, and mangrove ecological research started to flourish during the five decades since the emergence of these ground-breaking hypotheses.

Data from the *Web of Science* database suggest that the volume of mangrove research output also trailed that on saltmarsh systems since bibliometric records began, but this pattern was reversed for the first time in 2006. The number of mangrove publications per year has since consistently exceeded that on saltmarshes, and the difference is widening. This dramatic reversal may be attributed to the recent interests on the key role of mangrove wetlands as "Blue Carbon" reservoirs and their amelioration of natural disasters such as tsunamis. As Alongi (2009a) has concluded, mangrove forests may function more akin to tropical humid evergreen forests than their temperate coastal saltmarsh counterparts. While this notion again invites more in-depth analysis of mangrove ecosystem functioning, an indisputable fact is that mangrove ecology is now truly standing on its own feet.

The 11 chapters in this book bear clear evidence of the vast strides that mangrove ecology has taken in the last 50 years – we now have enough information to develop truly global perspectives of mangrove ecosystems with significantly larger spatial scales, as well as greater level of sophistication and diversity in approaches. Sadly, however, parallel to this development of mangrove ecosystem science, the last few decades have experienced unprecedented loss, destruction, and degradation of mangrove forests throughout the world, especially in the Indo-West Pacific region, where the most diverse and expansive mangrove ecosystems occur. Despite facing similar pressures, other debatably more charismatic, but certainly more inviting tropical marine ecosystems such as coral reefs have enjoyed grander research attention and arguably less direct threat compared to mangrove wetlands (just do a quick search on YouTube to see the difference). Despite the late start in mangrove ecosystem science, it is now, more than ever, required that tropical mangrove forests must be scientifically understood before all near-natural habitats of this ecosystem vanish.

The breakthrough of this book for mangrove ecology is not just the updated reviews of topics and issues that dominated early mangrove research 50 years ago, but certainly also the many issues confronting modern mangrove ecosystems and their management for sustainability that have emerged during this period: the impacts of global climate change, rehabilitation, and restoration; the socioeconomics of mangrove loss; and the implications for ecosystem services. Advances in many of these research fronts benefitted immensely from developments of novel tools, such as remote sensing and chemical tracers. For issues that have been long-standing topics of mangrove research, such as the productivity and carbon dynamics, biodiversity, and biogeochemical cycles, popularization of mangrove research to regions beyond the Caribbean in the last 50 years first developed in Australia, followed by Southeast Asia, East Africa, and South America. This development has generated sufficient data to enable a construction of global perspectives of mangrove ecosystem structure and function. Significant paradigm shifts have already been identified in key areas of mangrove biology (Alongi 2009b), but similar revisions if not "revolutions" have occurred in the less-studied aspects of mangrove ecosystems during this period. As a matter of fact, a similar revision process has earlier occurred in saltmarsh research (Weinstein and Kreeger 2000), although still being an overwhelmingly North American perspective. How mangrove ecosystems function and perform their many ecological and environmental roles, as well as their threats and perils, is no longer necessarily treated as a uniform whole – not only can variations due to drivers specific to local, regional, and biogeographic scales be recognized, but also be applied to improve our understanding of their implications for ecosystem responses to threats as well as sustainable management. With contributions from experts covering a comprehensive range of traditional to contemporary research questions from all major continents, as well as socioeconomic and biogeographic settings, we hope this volume has fulfilled the purpose of establishing a milestone in mangrove ecosystem research and will help set the research agenda for the coming decades.

Notwithstanding the effort to include the geographic coverage of both the authorship as well as the science in this book, much is yet to be done to further enrich the global perspective on mangrove ecology and ecosystem science. Despite that publications on mangrove wetlands in 2016 include contributions from 96 countries and territories, >72% of the contributions were from only five countries (USA, China, India, Brazil, and Australia). Among the top 15 countries in global mangrove forest area (Giri et al. 2010), researchers from the 12 countries other than India, Brazil, and Australia only contributed ~17% of the total number of publications in 2016, while their countries support close to 60% of the world's mangrove forests. Notwithstanding, mangrove ecologists today must not forget that some of the pioneering classical work on mangrove ecosystems were conducted in the Indo-West Pacific, for example, Macnae's account of the biotic communities (Macnae 1968) and Watson's hydrological analysis of different mangrove forest types (Watson 1928). While researchers may not necessarily conduct work in their own countries, it is notable also that it is in these under-represented countries where the greatest threats to mangrove forests and their services occur. Therefore, after decades of growth in research effort and increased public attention, a significant mismatch still exists between the geographic distribution of the mangrove forests and the research effort. While this book demonstrates that significant insight has been gained on the diversity in the ecology and services of mangrove ecosystems, there is still much work to do in further developing a truly global perspective of mangrove ecosystems as well as their management and conservation.

References

Alongi DM (2009a) The energetics of mangrove forests. Springer, New York

Alongi DM (2009b) Paradigm shifts in mangrove biology. In: Perillo GME, Wolanski E, Cahoon DR, Brinson MM (eds) Coastal wetlands: an integrated ecosystem approach. Elsevier, Amsterdam, pp 615–640

Giri C, Ochleng E, Tieszen LL, Singh A, Loveland T, Masek J, Duke N (2010) Status and distribution of mangrove forests of the world using earth observation satellite data. Global Ecol Biogeogr 20:154–159

Macnae W (1968) A general account of the flora and fauna of mangrove swamps and forests in the Indo-West Pacific region. Adv Mar Biol 6:73–270

Watson JG (1928) Mangrove forests of the Malay Peninsula. Malayan Forest Records No. 6, Federated Malay States, Kuala Lumpur, Malaysia

Web of Science (2017) Web of science. Institute of Scientific Information Thomson Reuters. http://apps.webofknowledge.com. Accessed 4 Apr 2017

Weinstein MP, Kreeger DA (2000) Concepts and controversies in tidal marsh ecology. Kluwer Academic Press, Amsterdam, 875p

About the Editors

Victor H. Rivera-Monroy is an associate professor in the Department of Oceanography and Coastal Sciences, College of the Coast and Environment at Louisiana State University. He has performed ecological studies in mangrove-dominated ecosystems for the last 20 years in Latin America, particularly research on biogeochemistry, wetland productivity, and restoration issues. Currently, he is a member of both the Florida Coastal Everglades (FCE) Long Term Ecological Research (LTER) program funded by the US National Science Foundation and the Mexican LTER program.

Shing Yip (Joe) Lee is a professor at the School of Life Sciences and director of the Simon F. S. Li Marine Science Laboratory at The Chinese University of Hong Kong. His work on mangroves is focused on the ecosystem processes of mangroves, particularly the role of key fauna in trophodynamics. He has been working on mangroves in Southeast Asia and Australia since 1985. He was appointed founding co-chair of the IUCN Specialist Group on Mangroves in 2013.

Erik Kristensen is a professor at the Institute of Biology, University of Southern Denmark, Denmark. He has during more than three decades traveled to many mangrove areas in Africa, Asia, Australia, and South America to do his research. This has been a simple necessity because his home country lacks mangrove forests. His mangrove research is focused on carbon, sulfur, iron, and nitrogen biogeochemistry in mangrove sediments as well as food web dynamics among mangrove benthos.

Robert R. Twilley is a professor in the Department of Oceanography and Coastal Sciences, College of the Coast and Environment at Louisiana State University. He has performed research in new and old world mangrove ecosystems during the last 25 years with emphasis on system ecology and biogeochemistry, conceptual and simulation models, coastal ecosystem restoration, and nutrient dynamics. He is currently the executive director of the LSU-based Louisiana Sea Grant College Program at LSU.

V.H. Rivera-Monroy et al. (eds.), *Mangrove Ecosystems: A Global Biogeographic Perspective*, https://doi.org/10.1007/978-3-319-62206-4

Contributors

Claudia Agraz-Hernández Instituto de Ecología, Pesquerías y Oceanografía del Golfo de México (EPOMEX). Universidad Autónoma de Campeche-UAC, San Francisco de Campeche, CM, Mexico

Emma Asbridge Centre for Ecosystem Science, School of Biological, Earth and Environmental Science, The University of New South Wales, Kensington, NSW, Australia

Edward Castañeda-Moya Department of Oceanography and Coastal Sciences, School of the Coast and the Environment, Louisiana State University, Baton Rouge, LA, USA

Southeast Environmental Research Center, Florida International University, Miami, FL, USA

Gustavo A. Castellanos-Galindo Departmento de Biología, Grupo de Investigación en Ecología de Estuarios y Manglares, Universidad del Valle, Cali, Colombia

Leibniz Center for Tropical Marine Ecology (ZMT), Fahrenheitstrasse, Bremen, Germany

Luzhen Chen Xiamen University, Xiamen, China

Rod M. Connolly School of Environment & Australian Rivers Institute - Coast & Estuaries, Griffith University, Gold Coast Campus, Gold Coast, QLD, Australia

Farid Dahdouh-Guebas Laboratoire d'Écologie des Systèmes et Gestion des Ressources, Département de Biologie des Organismes, Faculté des Sciences, Université Libre de Bruxelles – ULB, Bruxelles, Belgium

John W. Day Department of Oceanography and Coastal Sciences, College of the Coast and Environment, Louisiana State University, Baton Rouge, LA, USA

Richard H. Day U.S. Geological Survey, Lafayette, LA, USA

V.H. Rivera-Monroy et al. (eds.), *Mangrove Ecosystems: A Global Biogeographic Perspective*, https://doi.org/10.1007/978-3-319-62206-4

Amrit Dencer-Brown Institute for Applied Ecology, Auckland University of Technology, Auckland, New Zealand

Karen Diele School of Life, Sport and Social Sciences, Edinburgh Napier University, Edinburgh, UK

St Abbs Marine Station, St Abbs, UK

Norman C. Duke James Cook University TropWATER Centre, Townsville, QLD, Australia

Tiago O. Ferreira Dept. de Ciência do Solo, ESALQ/USP, São Paulo, Brazil

E. Gilman College of Natural Sciences, Hawaii Pacific University, Honolulu, HI, USA

Mark Huxham School of Applied Sciences, Edinburgh Napier University, Edinburgh, UK

T.C. Jennerjahn Leibniz Centre for Tropical Marine Research (ZMT), Bremen, Germany

E. B. Gareth Jones Department of Botany and Microbiology, College of Science, King Saud University, Riyadh, Kingdom of Saudi Arabia

Muhammad Kamal Remote Sensing Research Centre, School of Geography, Planning and Environmental Management, The University of Queensland, Brisbane, QLD, Australia

Kandasamy Kathiresan Centre of Advanced Study in Marine Biology, Annamalai University, Parangipettai, TN, India

Nico Koedam Plant Biology and Nature Management (APNA), Vrije Universiteit Brussel, Brussels, Belgium

K.W. Krauss U.S. Geological Survey, Wetland and Aquatic Research Center, Lafayette, LA, USA

Erik Kristensen Department of Biology, University of Southern Denmark, Odense M, Denmark

Claudia Kuenzer German Remote Sensing Data Centre, DFD of the German Aerospace Centre, DLR, Wessling, Germany

L.D. Lacerda Labomar, Universidade Federal do Ceara, Fortaleza, Brazil

Shing Yip Lee Simon F. S. Li Marine Science Laboratory, School of Life Sciences, and Earth System Science Programme, The Chinese University of Hong Kong, Shatin, Hong Kong SAR, China

Roy R. "Robin" Lewis III Lewis Environmental Services, Inc., Salt Springs, FL, USA

Jorge López-Portillo Instituto de Ecología, A.C. (INECOL), Red de Ecología Funcional, Veracruz, Mexico

Richard Lucas Centre for Ecosystem Science, School of Biological, Earth and Environmental Science, The University of New South Wales, Kensington, NSW, Australia

Cyril Marchand Institut de Recherche pour le Développement (IRD), Nouméa, New Caledonia, France

Joyita Mukherjee Department of Zoology, System Ecology and Ecological Modelling Laboratory, Visva-Bharati University, Santiniketan, India

Ivan Nagelkerken Southern Seas Ecology Laboratories, School of Biological Sciences, The University of Adelaide, Adelaide, SA, Australia

Inga Nordhaus Leibniz Centre for Tropical Marine Ecology (ZMT), Bremen, Germany

Michael J. Osland U.S. Geological Survey, Lafayette, LA, USA

Victor Osorio Escuela Superior Politécnica y del Litoral, Guayaquil, Ecuador

Universidad de Almeria, Almeria, Spain

Xose L. Otero Dept. de Edafoloxía e Química Agrícola, Facultade de Bioloxía, Universidade de Santiago de Compostela, Santiago de Compostela, Spain

Departamento de Tecnología de Ciencia de los Alimentos y Biotecnología, Universidad Politécnica Nacional, Quito, Ecuador

Santanu Ray Department of Zoology, System Ecology and Ecological Modelling Laboratory, Visva-Bharati University, Santiniketan, India

Victor H. Rivera-Monroy Department of Oceanography and Coastal Sciences, College of the Coast and Environment, Louisiana State University, Baton Rouge, LA, USA

María Teresa Rodríguez Geomatics, National Commission for the Knowledge and Use of Biodiversity (CONABIO), Mexico City, Mexico

André Rovai Departamento de Ecologia e Zoologia Universidade Federal de Santa Catarina, Florianópolis, Brazil

Department of Oceanography and Coastal Sciences, College of the Coast and Environment, Louisiana State University, Baton Rouge, LA, USA

Rinku Roy Chowdhury Clark University, Worcester, MA, USA

Peter Saenger Centre for Coastal Management, Southern Cross University, Lismore, NSW, Australia

Nathan Thomas Geography and Earth Sciences, Aberystwyth University, Aberystwyth, UK

Robert R. Twilley Department of Oceanography and Coastal Sciences, College of the Coast and Environment, Louisiana State University, Baton Rouge, LA, USA

Emi Uchida University of Rhode Island, Kingston, RI, USA

Alma Vázquez Lule Department of Plant and Soil Sciences, University of Delaware, Newark, DE, Mexico

Caroline Wanjiru Kenya Marine and Fisheries Research Institute, Mombasa, Kenya

E. Wolanski James Cook University, Townsville, QLD, Australia

Landon Yoder Indiana University, Bloomington, IN, USA

Index

V.H. Rivera-Monroy et al. (eds.), *Mangrove Ecosystems: A Global Biogeographic Perspective*, https://doi.org/10.1007/978-3-319-62206-4

N

O

P

R

S

Kolibri 360 GmbH
Ferdinand-Jühlke-Straße 7
99095 Erfurt, Deutschland
produktsicherheit@kolibri360.de